Protein Delivery

Physical Systems

Pharmaceutical Biotechnology

Series Editor: Ronald T. Borchardt
The University of Kansas
Lawrence, Kansas

Recent volumes in this series:

Volume 2 STABILITY OF PROTEIN PHARMACEUTICALS, Part A: Chemical and Physical Pathways of Protein Degradation
Edited by Tim J. Ahern and Mark C. Manning

Volume 3 STABILITY OF PROTEIN PHARMACEUTICALS, Part B: *In Vivo* Pathways of Degradation and Strategies for Protein Stabilization
Edited by Tim J. Ahern and Mark C. Manning

Volume 4 BIOLOGICAL BARRIERS TO PROTEIN DELIVERY
Edited by Kenneth L. Audus and Thomas J. Raub

Volume 5 STABILITY AND CHARACTERIZATION OF PROTEIN AND PEPTIDE DRUGS: Case Histories
Edited by Y. John Wang and Rodney Pearlman

Volume 6 VACCINE DESIGN: The Subunit and Adjuvant Approach
Edited by Michael F. Powell and Mark J. Newman

Volume 7 PHYSICAL METHODS TO CHARACTERIZE PHARMACEUTICAL PROTEINS
Edited by James N. Herron, Wim Jiskoot, and Daan J. A. Crommelin

Volume 8 MODELS FOR ASSESSING DRUG ABSORPTION AND METABOLISM
Edited by Ronald T. Borchardt, Philip L. Smith, and Glynn Wilson

Volume 9 FORMULATION, CHARACTERIZATION, AND STABILITY OF PROTEIN DRUGS: Case Histories
Edited by Rodney Pearlman and Y. John Wang

Volume 10 PROTEIN DELIVERY: Physical Systems
Edited by Lynda M. Sanders and R. Wayne Hendren

A Continuation Order Plan is available for this series. A continuation order will bring delivery of each new volume immediately upon publication. Volumes are billed only upon actual shipment. For further information please contact the publisher.

Protein Delivery

Physical Systems

Edited by

Lynda M. Sanders

Consultant in Pharmaceutical Development
Los Altos, California

and

R. Wayne Hendren

Glaxo Wellcome
Research Triangle Park, North Carolina

Plenum Press • New York and London

Library of Congress Cataloging-in-Publication Data

Protein delivery : physical systems / edited by Lynda M. Sanders and
R. Wayne Hendren.
p. cm. -- (Pharmaceutical biotechnology ; v. 10)
Includes bibliographical references and index.
ISBN 0-306-45359-2
1. Protein drugs--Physiological transport. 2. Drug delivery
systems. I. Sanders, Lynda M. II. Hendren, R. Wayne.
III. Series.
RS431.P75P756 1997
615'.7--dc21 97-176
CIP

ISBN 0-306-45359-2

A Division of Plenum Publishing Corporation
233 Spring Street, New York, N. Y. 10013

http://www.plenum.com

10 9 8 7 6 5 4 3 2 1

Printed in the United States of America

Contributors

Mary Tanya am Ende • Pfizer Central Research, Pharmaceutical Research and Development, Groton, Connecticut 06340

D. Bommannan • Cygnus, Inc., Redwood City, California 94063

Jens Brange • Novo Nordisk A/S, DK-2880 Bagsvaerd, Denmark

Ulrike Bremer • Pharmetrix, Inc., Menlo Park, California 94025

Susan M. Cady • Hoechst Roussel Vet, Somerville, New Jersey 08876-1258

Jeffrey L. Cleland • Pharmaceutical Research & Development, Genentech, Inc., 460 Pt. San Bruno Boulevard, South San Francisco, California 94080

S. Mark Cox • Atrix Laboratories, Inc., Fort Collins, Colorado 80525

Mary D. DiBiase • Biogen, Cambridge, Massachusetts 02142

Avi Domb • Department of Pharmaceutical Chemistry, The Hebrew University, Jerusalem, Israel 91120

Richard L. Dunn • Atrix Laboratories, Inc., Fort Collins, Colorado 80525

Ellen G. Duysen • Atrix Laboratories, Inc., Fort Collins, Colorado 80525

Michael L. Francoeur • Pharmetrix, Inc., Menlo Park, California 94025; *current address:* De Novo, Inc., Menlo Park, California 94025

Ruxandra Gref • Laboratoire de Chimie-Physique Macromoléculaire (URA CNRS 494), ENSIC, 54001 Nancy Cedex, France

C. Russell Horres • CyberRx, Inc., San Diego, California 92130

Robert Langer • Department of Chemical Engineering, Massachusetts Institute of Technology, Cambridge, Massachusetts 02139

Lotte Langkjær • Novo Nordisk A/S, DK-2880 Bagsvaerd, Denmark

Richard Maskiewicz • Matrix Pharmaceutical, Inc., Fremont, California 94555

Antonios G. Mikos • Cox Laboratory for Biomedical Engineering, Institute of Biosciences and Bioengineering, Department of Chemical Engineering, Rice University, Houston, Texas 77251

Yoshiharu Minamitake • Suntory Limited, Ohra-Gun, Gunma-Ken 370-05, Japan

Nancy A. Monteiro-Riviere • Cutaneous Pharmacology and Toxicology Center, College of Veterinary Medicine, North Carolina State University, Raleigh, North Carolina 27606

Eric M. Morrel • Ascent Pediatrics, Inc., Wilmington, Massachusetts 01887

Maria Teresa Peracchia • Dipartimento Farmaceutico, Universita degli Studi di Parma, 43100 Parma, Italy

Russell O. Potts • Cygnus, Inc., Redwood City, California 94063

Jim E. Riviere • Cutaneous Pharmacology and Toxicology Center, College of Veterinary Medicine, North Carolina State University, Raleigh, North Carolina 27606

Kathleen V. Roskos • Matrix Pharmaceutical, Inc., Fremont, California 94555

Burton H. Sage, Jr. • Becton Dickinson Research Center, Research Triangle Park, North Carolina 27709

W. Mark Saltzman • Department of Chemical Engineering, The Johns Hopkins University, Baltimore, Maryland 21218; *current address:* School of Chemical Engineering, Cornell University, Ithaca, New York 14853

Merrick L. Shively • NeXagen, Inc., Boulder, Colorado 80301; *current address:* Drug Delivery Solutions, LLC, Louisville, Colorado 80027

William D. Steber • Fort Dodge Animal Health, Cyanamid Agricultural Research Center, Princeton, New Jersey 08543-0400

Janet A. Tamada • Cygnus, Inc., Redwood City, California 94063

Vladimir Torchilin • Department of Radiology, Massachusetts General Hospital-East, Charlestown, Massachusetts 02129

Vladimir Trubetskoy • Department of Radiology, Massachusetts General Hospital-East, Charlestown, Massachusetts 02129

Ooi Wong • Cygnus, Inc., Redwood City, California 94063

Tammy L. Wyatt • Department of Chemical Engineering, The Johns Hopkins University, Baltimore, Maryland 21218

Gerald L. Yewey • Atrix Laboratories, Inc., Fort Collins, Colorado 80525

Preface to the Series

A major challenge confronting pharmaceutical scientists in the future will be to design successful dosage forms for the next generation of drugs. Many of these drugs will be complex polymers of amino acids (e.g., peptides, proteins), nucleosides (e.g., antisense molecules), carbohydrates (e.g., polysaccharides), or complex lipids.

Through rational drug design, synthetic medicinal chemists are preparing very potent and very specific peptides and antisense drug candidates. These molecules are being developed with molecular characteristics that permit optimal interaction with the specific macromolecules (e.g., receptors, enzymes, RNA, DNA) that mediate their therapeutic effects. Rational drug design does not necessarily mean rational drug delivery, however, which strives to incorporate into a molecule the molecular properties necessary for optimal transfer between the point of administration and the pharmacological target site in the body.

Like rational drug design, molecular biology is having a significant impact on the pharmaceutical industry. For the first time, it is possible to produce large quantities of highly pure proteins, polysaccharides, and lipids for possible pharmaceutical applications. Like peptides and antisense molecules, the design of successful dosage forms for these complex biotechnology products represents a major challenge to pharmaceutical scientists.

Development of an acceptable drug dosage form is a complex process requiring strong interactions between scientists from many different divisions in a pharmaceutical company, including discovery, development, and manufacturing. The series editor, the editors of the individual volumes, and the publisher hope that this new series will be particularly helpful to scientists in the development areas of a pharmaceutical company (e.g., drug metabolism, toxicology, pharmacokinetics and pharmacodynamics, drug delivery, preformulation, formulation, and physical and analytical chemistry). In addition, we hope this series will help to build bridges between the development scientists and scientists in discovery (e.g., medicinal chemistry, pharmacology, immunology, cell biology, molecular biology) and in manufacturing (e.g., process chemistry, engineering). The design of successful dosage forms for the next generation of drugs will require not only a high level of expertise by individual scientists, but also a high degree of interaction between scientists in these different divisions of a pharmaceutical company.

Finally, everyone involved with this series hopes that these volumes will also be useful to the educators who are training the next generation of pharmaceutical scientists. In addition to having a high level of expertise in their respective disciplines, these young scientists will need to have the scientific skills necessary to communicate with their peers in other scientific disciplines.

RONALD T. BORCHARDT
Series Editor

Preface

Protein and peptide therapeutics currently represent eight of the top 100 prescription pharmaceuticals in the U.S., and biotechnology products are projected to account for 15% of the total U.S. prescription drug market by 2003. Of the protein and peptide products now on the market, many are administered as daily injections, though several are delivered by noninvasive routes. For example, desmopressin is delivered as a nasal spray, and deoxyribonuclease I is administered by inhalation. Although cyclosporin A is orally active, as yet there are no *general* means to confer oral bioavailability to peptides and proteins. A major advance in delivery of peptides was achieved with the introduction of a monthly injectable, biodegradable microsphere formulation of LHRH.

Despite the enormous success of biotechnology products to date, much effort continues to be focused on the development of more convenient and noninvasive routes of administration for those products that require frequent and prolonged dosing. Here we present an overview of the technologies, both developed and emerging, which are applicable to protein delivery. In addition, chapters 11 through 13 detail case studies on physical methods for delivery of insulin and growth hormone.

Several common themes clearly emerge from the technology-specific chapters and from the case studies. First, the complex, three-dimensional, and somewhat fragile nature of macromolecules imposes certain restrictions on the process parameters that may be employed to package these drugs into sustained-release delivery systems. Second, when multiple doses of proteins are confined within practical injection volumes, the effective protein concentrations are quite high. As a result, intermolecular interactions leading to aggregation, denaturation, and precipitation may pose severe and sometimes unanticipated formulation challenges. Such problems have plagued efforts to develop implantable pumps for insulin delivery and sustained release implants of growth hormone. Third, biocompatibility considerations are critical when selecting materials for sustained release devices and formulations. Safety considerations dictate that sustained release systems for drugs having a narrow therapeutic index (e.g., insulin) must be retrievable or perform in a fail-safe manner. Finally, physical delivery systems are often not available in "one size fits all," nor are all techniques universally applicable, because each protein possesses unique physical chemical properties and unique stability and safety profiles.

The following chapters are a tribute to the innovative approaches to the delivery of proteins developed by the authors and their colleagues over many years. Their efforts have resulted in the development of safe and convenient systems for protein delivery. Continued development of these technologies promises to expand the scope of utility and application of protein therapeutics.

LYNDA M. SANDERS
R. WAYNE HENDREN

Contents

Chapter 1

Protein Delivery from Biodegradable Microspheres
Jeffrey L. Cleland

1. Introduction 1
2. Components for Successful Development of Microsphere Formulations 3
   2.1. Polymer Chemistry 3
   2.2. Engineering of Microsphere Formulations 8
   2.3. Protein Stability 21
3. Case Studies of Drug Delivery from Biodegradable Microspheres 24
   3.1. Lupron Depot® 24
   3.2. MN rgp120 Controlled Release Vaccine 26
4. Immunogenicity and Injection-Site Considerations 30
5. Regulatory Requirements for Development of Protein Delivery from Microspheres 34
   5.1. Toxicology Studies 34
   5.2. Residual Solvent Concerns 35
   5.3. Manufacturing Issues 36
   5.4. Preclinical Animal Models and Clinical Experiments 37
6. Summary 38
   References 39

Chapter 2

Degradable Controlled Release Systems Useful for Protein Delivery
Kathleen V. Roskos and Richard Maskiewicz

1. Introduction 45
2. Definitions 48

3. Synthetic Hydrophobic Degradable Polymers 49
3.1. Poly(lactic acid), Poly(glycolic acid), and Their Copolymers 49
3.2. Polycaprolactone 55
3.3. Poly(hydroxybutyrate), Poly(hydroxyvalerate), and Their Copolymers 56
3.4. Poly(ortho esters) 57
3.5. Polyanhydrides 61
3.6. Polyphosphazenes 65
3.7. Delivery of Vaccines 65
4. Hydrophilic Polymeric Biomaterials and Hydrophobic Nonpolymeric Biomaterials 70
4.1. General Properties 70
4.2. Specific Hydrophilic Polymeric Biomaterials 71
4.3. Specific Hydrophobic Nonpolymeric Biomaterials 79
4.4. Miscellaneous 81
5. Conclusions 82
References 83

Chapter 3

Delivery of Proteins from a Controlled Release Injectable Implant
Gerald L. Yewey, Ellen G. Duysen, S. Mark Cox, and Richard L. Dunn

1. The ATRIGEL™ Drug Delivery System 93
2. Effects of Formulation Variables on Protein Release Kinetics 95
2.1. Polymer Type 96
2.2. Polymer Concentration 97
2.3. Polymer Molecular Weight 98
2.4. Solvent 99
2.5. Protein Load 100
2.6. Additives 101
3. *In Vitro* Characterization 102
3.1. Protein Quantitation in Different Release Media 102
3.2. Protein Structure 105
3.3. Enzyme Activity 107
3.4. Cellular Bioactivity 108
4. *In Vivo* Evaluations 110
4.1. Biocompatibility 110
4.2. Protein Release Kinetics 111
4.3. Bioactivity 113

5. Conclusions 115
References 116

Chapter 4

Protein Delivery from Nondegradable Polymer Matrices
Tammy L. Wyatt and W. Mark Saltzman

1. Introduction 119
1.1. Biocompatible Polymers Used as Hydrophobic Matrices . . 120
1.2. Protein Release from Polymer Matrices 122
2. Mechanisms and Models for Protein Release from Matrices 124
2.1. Macroscopic Models of Diffusion in Porous Polymer Matrices 125
2.2. Microscopic Models of Diffusion in Porous Polymer Matrices 131
3. Applications of Protein/Polymer Matrix Systems 132
3.1. Topical Delivery 133
3.2. Targeted Delivery of Proteins to Specific Tissue Regions 133
3.3. Systemic Delivery for Extended Periods 134
References 134

Chapter 5

Diffusion-Controlled Delivery of Proteins from Hydrogels and Other Hydrophilic Systems
Mary Tanya am Ende and Antonios G. Mikos

1. Introduction 139
1.1. Mechanisms of Protein Diffusion 140
1.2. Structure of Hydrophilic Polymers 142
1.3. Methods for Loading Proteins into Hydrogels 144
2. Diffusion-Controlled Delivery Systems 145
2.1. Reservoir Systems 145
2.2. Matrix Systems 147
2.3. Biodegradable Hydrogels 149
3. Factors Affecting the Diffusion of Proteins 151
3.1. Environmental Conditions 151
3.2. Hydrogel Structure 152

4. Techniques for Measurement of the Diffusion Coefficient 153
4.1. Membrane Permeation Method 155
4.2. Absorption/Desorption Method 157
4.3. Scanning Electron Microscopy (SEM) 160
4.4. Fourier Transform Infrared (FTIR) Spectroscopy 160
4.5. Quasi-Elastic Light Scattering (QELS) Method 161
4.6. Other Techniques 162
References 162

Chapter 6

Poly(ethylene glycol)-Coated Nanospheres: Potential Carriers for Intravenous Drug Administration
Ruxandra Gref, Yoshiharu Minamitake, Maria Teresa Peracchia, Avi Domb, Vladimir Trubetskoy, Vladimir Torchilin, and Robert Langer

1. Introduction 167
1.1. Approaches to Increase Particle Blood Circulation Time 169
1.2. PEG Hydrophilic Coatings: Mechanism of Protein Rejection 170
2. PEG-Coated Long-Circulating Drug Carriers 171
3. PEG-Coated Biodegradable Nanospheres: Potential Long-Circulating Drug Carriers 173
3.1. Biodegradable Polymers Containing PEG Blocks 174
3.2. Preparation of PEG-Coated Nanospheres 176
4. Nanosphere Characterization 177
4.1. Morphology Studies 177
4.2. Size Distribution Measurement 179
4.3. Detection and Stability of the PEG Coating 180
4.4. Surface Hydrophobicity and Charge Determination 181
5. Drug Encapsulation in PEG-Coated Nanospheres 183
5.1. Drug Encapsulation and Release Properties 183
5.2. Parameters Influencing Drug Release 184
6. *Ex Vivo* Studies (Phagocytosis Assay) 187
7. Blood Half-Life and Organ Distribution of PEG-Coated Nanospheres 188
8. Conclusion 192
References 193

Chapter 7

Multiple Emulsions for the Delivery of Proteins

Merrick L. Shively

1. Introduction 199
2. Methods of Preparation 200
3. Stability Issues 201
   3.1. Background 201
   3.2. Surfactant Migration 202
   3.3. Osmotic Gradients 202
   3.4. Process Denaturation of Protein 202
   3.5. Methods to Determine Physical Stability 203
4. Applications 203
   4.1. Parenteral Administration 203
   4.2. Oral Administration 204
5. Solid-State Emulsions 205
   5.1. Method of Preparation 206
   5.2. Physical Properties of Solid-State Emulsions 206
   5.3. Oral Administration of Vancomycin Solid-State Emulsion 207
6. Miscellaneous Applications 208
   6.1. Vaccine Adjuvants 208
   6.2. Enzyme Immobilization 208
7. Summary 209
   References 209

Chapter 8

Transdermal Peptide Delivery Using Electroporation

Russell O. Potts, D. Bommannan, Ooi Wong, Janet A. Tamada, Jim E. Riviere, and Nancy A. Monteiro-Riviere

1. Introduction 213
2. Results and Discussion 217
   2.1. *In Vitro* Transport 217
   2.2. Isolated Perfused Porcine Skin Flap 227
   2.3. Skin Toxicology following Electroporation 232
3. Conclusion 235
   References 235

Chapter 9

Protein Delivery with Infusion Pumps

Ulrike Bremer, C. Russell Horres, and Michael L. Francoeur

1. Introduction 239
   1.1. Rationale for Infusion Therapy 239
   1.2. Limitations of Infusion Therapy 242
2. History of Infusion Therapy 243
3. Stationary and Portable Infusion Pumps 245
   3.1. Stationary Infusion Pumps 246
   3.2. Implantable Infusion Pumps 249
   3.3. External Infusion Pumps 250
4. Summary 252
   References 253

Chapter 10

Oral Delivery of Microencapsulated Proteins

Mary D. DiBiase and Eric M. Morrel

1. Introduction 255
2. Mechanisms of Intestinal Absorption of Proteins and Peptides 257
   2.1. Passive Diffusion 257
   2.2. Carrier-Mediated Transport 259
   2.3. Receptor-Mediated and Non-Receptor-Mediated Endocytosis 261
3. Mechanisms of Intestinal Absorption of Microparticulates 264
   3.1. Transcellular Pathway 265
   3.2. Paracellular Transport 267
   3.3. Liposome Absorption 268
4. Case Studies 269
   4.1. Introduction 269
   4.2. Polyester Microspheres 270
   4.3. Zein Microspheres 271
   4.4. Proteinoid Microspheres 272
   4.5. Polycyanoacrylate Microspheres 273
   4.6. Lipid-Based Systems 275
5. Conclusion 277
   References 277

Chapter 11

Controlled Delivery of Somatotropins
Susan M. Cady and William D. Steber

1. Introduction 289
2. Preformulation Development 291
   2.1. Solution Stability 291
   2.2. Molecular Modification 293
3. Injectables 295
   3.1. Oil-Based Gel Depots 295
   3.2. Microsphere Systems 299
   3.3. Liposomes 301
   3.4. Emulsions 301
   3.5. Aqueous Gels and Complexes 302
4. Implants 303
   4.1. Uncoated Implants 303
   4.2. Coated Implants 305
5. Osmotic Devices 310
6. Miscellaneous Systems 312
   6.1. Wound Healing 312
   6.2. Nasal Delivery Systems 313
7. Conclusions 313
   References 313

Chapter 12

Insulin Iontophoresis
Burton H. Sage, Jr.

1. Introduction 319
2. Specific Drug Delivery Requirements for Insulin 322
   2.1. Duplicating the Function of the Pancreas 322
   2.2. Candidate Systems for Insulin Delivery 323
3. Capabilities of Iontophoresis Related to Insulin Delivery 326
   3.1. Noninvasive Delivery of Insulin 327
   3.2. Control of Delivery Rate of Insulin 327
   3.3. Bolus Administration 327
   3.4. Dose Precision 328
   3.5. Portal Delivery 328
   3.6. Bioavailability 328

3.7. Compliance 329
3.8. Summary of Capabilities Related to Insulin Delivery 330
4. Theoretical Limitations and Published Results 330
4.1. Published Results of Insulin Iontophoresis 330
4.2. Theoretical and Practical Limitations to Insulin Iontophoresis 333
5. Physicochemical Properties of Insulin Related to Iontophoresis 336
5.1. Charge Titration 336
5.2. Solubility 337
5.3. Enzymatic Degradation 338
5.4. Insulin Self-Association 338
6. Future Prospects for Iontophoretic Delivery of Insulin 339
References 340

Chapter 13

Insulin Formulation and Delivery
Jens Brange and Lotte Langkjær

1. Introduction 343
2. Formulation of Insulin 344
2.1. Introduction 344
2.2. Formulation for Parenteral Administration 345
2.3. Formulation for Alternative Routes 351
2.4. Insulin Analogs and Derivatives 352
3. Delivery of Insulin 355
3.1. Introduction 355
3.2. Parenteral Insulin Delivery 357
3.3. Alternative Routes of Insulin Delivery 368
4. Summary and Future Perspectives 385
References 386

Index 411

Protein Delivery

Physical Systems

Chapter 1

Protein Delivery from Biodegradable Microspheres

Jeffrey L. Cleland

1. INTRODUCTION

Most protein pharmaceuticals are delivered by invasive routes such as subcutaneous injections. The protein formulation used in these injections is usually a solution or suspension, as in the case of insulin. These formulations are developed to provide stability of the protein during storage, but they are usually not designed to control the delivery of the protein after injection. In many cases, the formulation can have a significant impact on the pharmacokinetics and pharmacodynamics of the drug. For example, insulin can be formulated with various components such as zinc and protamine to provide different pharmacodynamics, and insulin has a different stability profile in each of these formulations (for a review, see Cleland and Langer, 1994). Unfortunately, the development of conventional formulations that modify pharmacodynamics and pharmacokinetics may not be possible for many proteins, and, even in the case of insulin, the duration of response cannot be extended for more than one day. Thus, protein pharmaceuticals must be repeatedly injected over extended periods (e.g., months or years) for chronic applications. Patients may not comply with repeated injections, and repeated administration in a hospital results in increased medical costs. If frequent administration (e.g., daily or twice a day) is required, patients and physicians may be unwilling to use the drug, depending upon the indication.

Jeffrey L. Cleland • Pharmaceutical Research & Development, Genentech, Inc., South San Francisco, California 94080.

Protein Delivery: Physical Systems, Sanders and Hendren, eds., Plenum Press, New York, 1997.

The frequency of injection for protein parenterals may be reduced by using a matrix to control the release of the drug from the injection site after administration. The rate of release from the site of injection would then be controlled by the properties of the matrix and the drug, as well as the physiology of the injection site. By developing matrices that continuously release the protein over time, the number of injections could be reduced because a sustained blood level of the protein could be achieved by its constant release from the matrix. In addition to less frequent administration, the matrix would also offer the advantage of localized delivery of the protein pharmaceutical. For some cancer therapies and vaccines, delivery of a high dose at the target site would increase the efficacy of the drug and potentially reduce its toxicity since a high systemic concentration of the drug in the circulation may be required to achieve the desired effect. Many vaccines require the use of multiple immunizations to achieve an immune response that is sufficient to provide extended protection; a reduced number of immunizations or a single-immunization vaccine formulation would be more convenient for patients and perhaps provide a better immune response (e.g., longer duration of protective response, higher titers, etc.). The entrapment of proteins in a matrix could also provide protection against degradation *in vivo*. Cells or proteases in the surrounding tissues would not be in contact with the protein until it is released from the matrix. The matrix may also provide stability for labile compounds that are rapidly degraded or cleared *in vivo*. Overall, the encapsulation of proteins in a controlled release matrix could provide reduced administrations, localized delivery, and improved *in vivo* stability.

To maintain these advantages, the controlled release matrix should be easily administered and not require removal. The matrix could be configured into several forms, including implants of various shapes and sizes or small microspheres. However, implants usually require surgical procedures, and this requirement dramatically reduces the commercial value and practical utility of the controlled release formulation unless the drug is released for very long periods (e.g., years). The development of formulations that release proteins for long periods would be difficult because proteins are unstable when maintained in physiological conditions. On the other hand, microspheres of the appropriate size (1–100 μm) could be readily injected subcutaneously or into the target site or could be given orally. Microspheres can also be easily fabricated in an emulsion system consisting of two immiscible phases: polymer dissolved in a solvent and a nonsolvent. The size of the microspheres may be controlled by the droplet size of the polymer phase. By generating a homogeneous and uniform emulsion, a unimodal distribution of microspheres may be produced. These microspheres could then be easily injected or administered orally.

2. COMPONENTS FOR SUCCESSFUL DEVELOPMENT OF MICROSPHERE FORMULATIONS

While controlled release microsphere formulations offer significant advantages over typical parenteral dosage forms, their successful development depends on a thorough understanding of three basic components: polymer chemistry, engineering, and protein stability. Each individual component must be considered in conjunction with the others since they are all intricately related. The basic properties and synthesis of the polymer matrix used to produce the microspheres as well as the interaction of the matrix with various solvents should be well characterized. The properties of the polymer determine the design of the process for manufacturing the microspheres. This design should also include the consideration of the scalability of the process and eventual need to produce clinical-grade material under Good Manufacturing Practices (GMP) conditions. The behavior of the protein in the polymer system should also be well understood, because the solvents used in the process or other process conditions could denature the protein. In addition, the protein could interact with the polymer, thus altering the rate of protein release. All of these issues should be addressed prior to embarking on the development of microsphere formulations for proteins.

2.1. Polymer Chemistry

A polymer must first be chosen for the entrapment of the protein. A biodegradable polymer is usually preferred to avoid the need for surgical removal after the protein is completely released. The polymer also must not alter the pharmacological properties of the drug (e.g., reduced potency, increased immunogenicity, etc.). The protein should not interact with the polymer in an irreversible or uncontrolled manner (e.g., polymer reacts with or denatures protein). The polymer and its degradation products must not produce any adverse side effects and local or systemic toxicity. In addition, the polymer should be produced with consistent properties that are readily quantifiable and reproducible. All of these criteria should be used to select the appropriate polymer for protein encapsulation.

As shown in Table I, there are several natural and synthetic polymers that have been used for delivery of proteins and peptides. Most of the natural polymers are extracted from animal or plant sources and, therefore, may vary in their overall composition. In addition, proteins from these sources and also recombinant products such as albumin must be thoroughly

Table I
Biodegradable Materials for Controlled Delivery of Proteins or Peptides

Material	Degradation mechanism	Reference(s)[a]
Natural		
Starch	Amylase	Arthur *et al.*, 1984; Stjarnkvist *et al.*, 1991
Alginate[b]	pH, enzymes	Wheatley *et al.*, 1991; Downs *et al.*, 1992
Collagen (gelatin)	Collagenase	Takaoka *et al.*, 1991; Lindholm and Gao, 1993; Horisaka *et al.*, 1994
Proteins (cross-linked albumin)	Enzymes	Levy and Andry, 1991; Santiago *et al.*, 1993
Tricalcium phosphate[c] or calcium carbonate (hydroxyapatite)	Dissolves over time	Ripamonti *et al.*, 1992; Herr *et al.*, 1993; Kenley *et al.*, 1993
Synthetic		
Hydrogels	Chemical or enzymatic hydrolysis, solubilization in aqueous media	Kamath and Park, 1993
Polyanhydrides	Hydrolysis	Langer, 1993; Ron *et al.*, 1993; Shieh *et al.*, 1994
Polyesters (polylactides)	Ester hydrolysis, esterases	Heller, 1993; Shah *et al.*, 1993
Poly(ortho esters)	Ester hydrolysis, esterases	Heller, 1993
Polyiminocarbonates	Hydrolysis	Pulapura *et al.*, 1990; Arshady, 1991
Polycaprolactones	Hydrolysis	Marchalheussler *et al.*, 1991; Coffin and McGinity, 1992
Polyamino acids	Enzymes	Li *et al.*, 1993
Polyphosphazenes[b]	Hydrolysis, dissolution	Andrianov *et al.*, 1993; Crommen *et al.*, 1993

[a]Reviews describing the application of these materials for delivery of proteins or peptides. When recent reviews are not available, several examples are listed.
[b]Alginate and polyphosphazene depot systems usually require a cross-linking agent such as calcium or polycations.
[c]These depot systems are often used for mechanical strength in bone formation and are usually used in combination with another carrier such as collagen or starch.

cleaned to remove endotoxins and other contaminants. Although the sugar-based natural sources, such as starch or alginate, should not be immunogenic, the systemic administration of proteins such as collagen or albumin microspheres may cause an unwanted immune response. Further, these protein-based polymers may act as an immune adjuvant to the protein that is intended for delivery. All these natural polymers are water-soluble, and the encapsulation of proteins can be accomplished without the use of organic solvents or elevated temperatures, both of which can denature proteins. Unfortunately, the risks of immunogenicity and contamination as well as the large-scale costs of production make most of the natural polymers poor choices for a drug delivery matrix.

On the other hand, synthetic biodegradable polymers are usually well characterized, reproducibly made in large quantities (e.g., kilograms), and easily purified from contaminants or by-products. Among these polymers, only hydrogels, polyamino acids, and polyphosphazenes do not require organic solvents or elevated temperatures for encapsulation of proteins. To entrap proteins in the biodegradable hydrogels, the protein and a hydrogel polymer are mixed together, and a cross-linking agent is then added (Kamath and Park, 1993). Ideally, the hydrogel cross-linking reaction will not adversely affect the protein (e.g., by forming adducts). The entrapment of proteins in polyamino acids is usually performed by a simple precipitation of the polymer that is caused by the addition of a nonsolvent [e.g., high molecular weight poly(ethylene glycol)] (Li *et al.*, 1993). These polyamino acids, however, may have adjuvant properties since their protein-like composition may elicit an immune response. In addition, the polyphosphazenes may have good adjuvant properties, possibly attributable to their large negative charge (Andrianov *et al.*, 1993). Polyphosphazene microspheres can be made by adding a divalent cation such as calcium to a solution of polyphosphazene and the protein while stirring (Andrianov *et al.*, 1993; Crommen *et al.*, 1993). The release rate from the polyphosphazenes can also be controlled by the addition of poly(L-lysine) to form a coating to stabilize the microspheres (Andrianov *et al.*, 1993). These systems—hydrogels, polyamino acids, and polyphosphazenes—have not been extensively assessed for their compatibility with therapeutic proteins. However, with additional development, these systems may be useful for controlled release microsphere formulations.

The remaining polymers listed in Table I are not soluble in aqueous solutions and require organic solvents or elevated temperatures for fabrication into microspheres and encapsulation of proteins. Two recent reviews describe the use of polylactides, poly(ortho esters), and polyanhydrides, all of which have been used for the controlled release of several proteins and peptides (Langer, 1993; Heller, 1993). Polyiminocarbonates are relatively new biodegradable polymers that, like polycaprolactones, have not yet been extensively characterized as controlled release matrices for therapeutic proteins (Pulapura *et al.*, 1990). While all of these polymers require relatively harsh conditions for entrapment of the protein, their release properties may allow for a prolonged delivery (e.g., up to one year) because

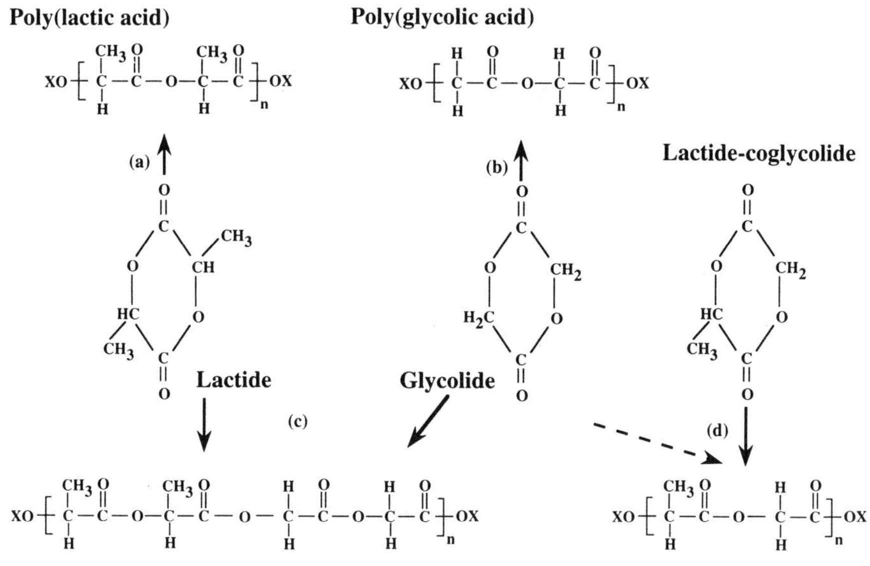

Figure 1. Chemical structure and synthesis of poly(lactic acid), poly(glycolic acid), and poly(lactic-*co*-glycolic acid). All polymers are usually prepared by ring-opening polymerization methods such as heating in the presence of tin chloride ($SnCl_4$) as a catalyst, and the reaction is terminated through the addition of compounds with a free alcohol group (see Nieuwenhuis, 1992, for details). (a) Poly(lactic acid) (PLA) is synthesized from lactide and degrades in water to form lactic acid. (b) Poly(glycolic acid) (PGA) is produced from glycolide and degrades in water to form glycolic acid. (c) Poly(lactic-co-glycolic acid) (PLGA) is often synthesized by mixing lactide and glycolide in different ratios depending upon the desired final ratio in the polymer. (d) Alternatively, PLGA may be prepared by using lactide-*co*-glycolide, and, if the final desired ratio is not 50:50, either lactide or glycolide is added. Preparation of PLGA by either method (c or d) results in the same chemical composition, but each polymer may behave differently due to the ordering of the monomer units. Method c usually results in a block copolymer (e.g., LLLLGGGG) whereas method d yields a more ordered copolymer (e.g., LGLGLG) depending on the final ratio of lactide to glycolides. These structural differences affect the ability of the polymer to form amorphous and crystalline phases.

their degradation is slower in many cases than that of the more hydrophilic polymers. In addition, polylactides [poly(lactic acid), (PLA), poly(glycolic acid), (PGA), and poly(lactic-*co*-glycolic acid), (PLGA)] (Fig. 1) and the polyanhydrides, poly[bis(*p*-carboxyphenoxy)propane] anhydride and sebacic acid (PCPP:SA; Langer, 1991) and poly(fatty acid dimer: sebacic acid) [P(FAD-SA); Tabata *et al.*, 1993], have been administered to humans without any reports of toxicity or safety concerns. Thus, if the difficulties of encapsulation with these polymers (e.g., solvents, process conditions, etc.)

can be overcome, success in the clinic is more likely to be achieved with these formulations than with the hydrophilic polymers. The probability of the successful development of a new polymer, one which is not approved for use in humans or has not been tested in humans, and of a new therapeutic protein simultaneously is minimal, owing to the complexity of both. Therefore, for new protein therapeutics, polylactides and polyanhydrides have the greatest chance of becoming viable commercial products, although many of the biodegradable polymers could be used to add value to an existing protein therapeutic, albeit at the cost of additional development and clinical studies.

Polymers approved for human use may have far fewer clinical hurdles to overcome and will probably not require expensive, long-term toxicity studies in humans. The polylactides are the only biodegradable polymers approved for use in humans by the U.S. Food and Drug Administration (U.S. FDA). Polylactides have been used for over 20 years in resorbable sutures. Recently, several controlled release formulations of luteinizing hormone-releasing hormone (LHRH) agonists were approved for use in humans. These formulations consist of an LHRH agonist encapsulated in PLGA microspheres, providing a continuous release of the agonist for one month (see Table II). The first of these products to reach the market, Lupron Depot®, accumulated approximately $570 million in revenues in 1992 for prostate cancer, precocious puberty, and endometriosis indications (Anonymous, 1993). The use of Lupron Depot® in children for treatment of precocious puberty further strengthens the existing detailed safety and toxicity data on PLGA. Also, polylactides degrade to form innocuous products, lactic and glycolic acid. Besides the polylactides, the polyanhydride PCPP:SA, while not yet approved in a commercial parenteral formulation, has been tested in humans for sustained release of BCNU [*N*,*N*-bis(2-chloroethyl)-*N*-nitrosourea] (Langer, 1991). Overall, these polymers are preferred for controlled drug delivery applications unless the unique

Table II

Commercial Sustained Released Parenteral Formulations of Luteinizing Hormone-Releasing Hormone (LHRH) Agonists in PLGA Microspheres[a]

LHRH agonist	Trade name for depot formulation	Company
Leuprolide acetate	Lupron Depot®	Takeda-Abbott
Goserelin acetate	Zoladex®	I.C.I.
Triptorelin	Decapeptyl®	Debiopharm

[a]Sources: *Pharma Japan*, 1993, **1355**:2; *FDC Reports: The Pink Sheet*, 1991, **53**:T&G-5; *Scrip*, 1992, **1765**:25.

properties of a clinically untested polymer outweigh the advantage of a well-defined safety and toxicity profile as described in Drug Master Files maintained by several polymer companies. In some cases, it may be necessary to choose a clinically untested polymer that is similar to an approved polymer but has unique properties. An excellent example of this type of polymer is poly(ethylene glycol)–PLGA, developed by Langer and co-workers (Gref *et al.*, 1994). The polyethylene glycol prevents uptake of nanospheres by shielding them from the reticuloendothelial system. The desired properties of the polymer must be weighed against the additional requirements for toxicology and development.

2.2. Engineering of Microsphere Formulations

Although most of the polymers listed in Table I can be used to make microsphere formulations, each polymer requires different processing conditions. For the more hydrophobic polymers, such as polylactides and polyanhydrides, similar processes are utilized to make microsphere formulations. Since the polylactides are the only commercially approved biodegradable polymers for human use, the discussion herein of the engineering aspects of producing microsphere formulations will focus on these polymers.

2.2.1. BASIC ENCAPSULATION METHODS

The most basic methods for encapsulation of proteins in polylactides are solvent evaporation and solvent extraction (coacervation). Polylactides are insoluble in water and therefore must be dissolved in an organic solvent such as ethyl acetate or methylene chloride to facilitate protein encapsulation. Several proteins have been encapsulated in polylactides by different methods as shown in Table III. The polymer dissolved in the organic solvent is referred to as the oil phase. A solid or liquid protein formulation is then dispersed in the oil phase by homogenization or sonication, resulting in the formation of a fine suspension or emulsion. The protein-phase droplets or solids must be significantly smaller than the final microspheres to ensure incorporation. For example, if the final microspheres are 30 μm in diameter, the protein phase (solid or liquid) should be 2–5 μm in diameter to ensure a homogeneous dispersion and spherical drug-loaded particles. As the dimensions of the protein-phase droplets or solids approach the size of the final microspheres, the extent of incorporation of the protein will be reduced, a larger initial release of the protein will occur, and an irregular release of the protein may be observed (e.g., multiphase release).

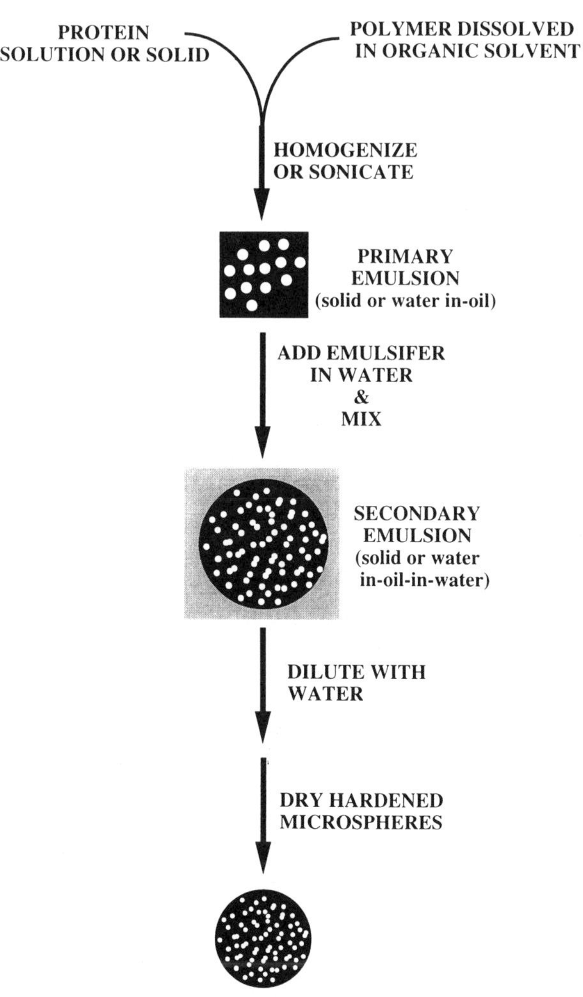

Figure 2. Solvent evaporation method for production of polylactide microspheres. The polymer (PLGA or PLA) is dissolved in an organic solvent (e.g., methylene chloride or ethyl acetate). The aqueous or solid drug is then added, and the solution is mixed by sonication or homogenization to form the primary emulsion (solid or water in-oil). This emulsion is then transferred to water containing an emulsifying agent [e.g., poly(vinyl alcohol)]. Mixing of the primary emulsion in the water phase produces the microspheres, resulting in a secondary emulsion (solid or water in-oil-in-water). The final emulsion is diluted with excess water to facilitate removal of the organic solvent in the oil phase. The microspheres are then dried. (Reproduced from Cleland, 1995, copyright Plenum Press.)

Table III
Some Recent Examples of Polylactide Microsphere Formulations for Controlled Release of Proteins or Peptides

Protein or peptide	Polymer(s)[a]	Summary of results	Application	Reference
Thyrotropin	75:25 PLGA (11 kDa)	*In vitro* correlation to *in vivo* release with 33 mM phosphate as release medium; release dependent on erosion; <20% burst, 30-day continuous release	Central nervous system dysfunction	Heya *et al.*, 1991
Atriopeptin III	50:50 PLGA, PLA	Polymer accelerated degradation rate of peptide; peptide–polymer binding; 7-day continuous release	Natriuretic and vasorelaxant	Johnson *et al.*, 1991
Somatostatin analog	55:45 PLGA (23–76 kDa)	Release decreased with media ionic strength; peptide–polymer interactions; <20% burst, continuous or triphasic release for >30 days	Acromegaly, tumors	Bodmer *et al.*, 1992
Neurotensin analog	PLA (2–6 kDa)	Fatty acid (palmitic) and buffer (borate) improved entrapment; 20% initial burst, 30-day continuous release	Psychothropic	Yamakawa *et al.*, 1992
Cyclosporin A	50:50 PLGA (0.44 and 0.80 dl/g)	No drug–polymer interaction biphasic release; 1–40% burst, >28-day release	Immunosuppression	Sánchez *et al.*, 1993
Colonizing factor antigen (*E. coli*)	PLGA (0.73 dl/g)	25% released in 1 day; triphasic release?; stabilized antigen	Oral vaccine	Reid *et al.*, 1993

Diphtheria toxoid, formilin treated	PLA (49 kDa)	*In vivo* continuous increase in antibody titers to 75 days	Vaccine	Singh *et al.*, 1992
Ovalbumin	50:50 PLGA (22 kDa) 85:15 PLGA (53 kDa)	Small outer aqueous phase yielded 1–2-μm particles with 10% loading	Vaccine	Jeffery *et al.*, 1993
Tetanus toxoid	50:50 PLGA (100 kDa)	Size reduced with lipid; erosion controlled release rate; <30% burst, continuous or triphasic for >30 days	Vaccine	Alonso *et al.*, 1993
Luteinizing hormone-releasing hormone (LHRH) antagonists and agonists	50:50 PLGA, 75:25 PLGA	Maintained steady-state blood levels for 28 days; high initial level for 7 days	Tumor suppression	Stoeckemann and Sandow, 1993
Horseradish peroxidase; bovine serum albumin	75:25 PLGA (10 kDa)	Inner aqueous phase volume and organic phase volume effects on initial release; 20–70% burst, >60-day continuous release	Marker proteins; mechanistic studies	Cohen *et al.*, 1991

[a]X:Y PLGA indicates the mole fraction (%) of lactide (X) and glycolide (Y) in the copolymer. The polymer size is reported as either molecular weight in kilodaltons (kDa) or intrinsic viscosity in decaliters per gram. Some references did not provide complete descriptions of the polymer. All microspheres except those made by Johnson *et al.* (1991) (coacervation) were prepared by the water-in-oil-water process.

The mixture of the protein and oil phases is used to form the final microspheres.

For the solvent evaporation method, the protein and oil mixture (primary emulsion) is usually stirred with an aqueous solution containing an emulsifying agent such as poly(vinyl alcohol) (PVA) to form a secondary emulsion. The addition of the primary emulsion to the aqueous solution results in phase separation and precipitation of the polylactide, entrapping the protein (see Fig. 2). To ensure entrapment of the protein and to remove excess solvent, additional water is then added to the solution, causing a hardening of the microspheres. Finally, excess solvent and emulsifying agent are washed away from microspheres, and any remaining solvent is removed by drying the microspheres (e.g., lyophilization). If the protein phase is aqueous, the preparation is referred to as a water-in-oil-in-water (WOW) double emulsion.

Alternatively, coacervation or solvent extraction is often used to produce microspheres (Fig. 3). The protein and polymer emulsion is stirred with a nonsolvent for the polymer such as silicone oil, resulting in the formation of embryonic microspheres (for a review, see Lewis, 1990). The nonsolvent extracts the methylene chloride or ethyl acetate from the polymer phase, causing precipitation of the polymer and entrapment of the protein in the polymer matrix. To remove the nonsolvent, a volatile second nonsolvent (e.g., heptane) is added, and the microspheres are allowed to harden in the nonsolvent. After repeated extraction with the volatile nonsolvent, the final microspheres are then dried. While this method offers the advantage of avoiding contact between the protein phase and an aqueous phase as in the solvent evaporation method, the additional solvents utilized in this process are often difficult to completely remove and are a safety and toxicity concern.

Both of these basic processes are used to produce microspheres that have different release characteristics. The microspheres may release the protein in either a continuous or a pulsatile manner. Two mechanisms control the release of the protein out of the microspheres. The first mechanism is the simple diffusion of the protein out of the polymer matrix. Typically, the diffusion process occurs in two or more stages comprising an initial release of protein at or near the microsphere surface followed by additional release of protein by diffusion from the microsphere's interior pores. The second mechanism is the erosion of the polymer matrix, which occurs by hydrolysis of the polymer backbone. For continuous release, the diffusion and erosion processes must be complementary to allow the protein to constantly diffuse out of the microspheres (Fig. 4). However, if the initial diffusion phase ends prior to the onset of sufficient polymer erosion to allow pore formation, the protein cannot diffuse out of the microspheres until the

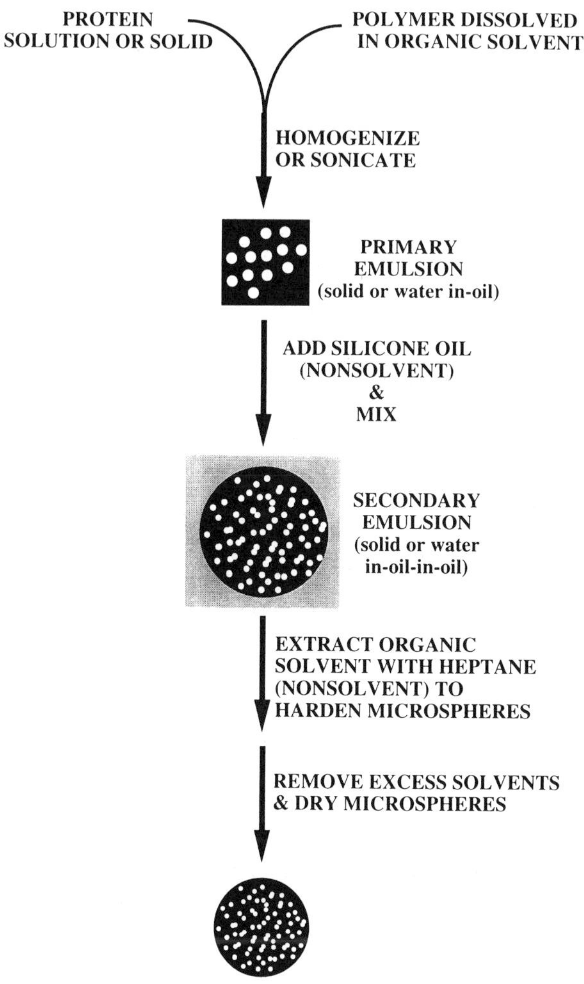

Figure 3. Coacervation method for production of polylactide microspheres. The polymer (PLGA or PLA) is dissolved in an organic solvent (e.g., methylene chloride or ethyl acetate). The aqueous or solid drug is then added, and the solution is mixed by sonication or homogenization to form the primary emulsion (solid or water in-oil). This emulsion is then transferred to a nonsolvent, i.e., a solvent in which the polymer has a low or negligible solubility (e.g., silicone oil). Mixing of the primary emulsion in the nonsolvent produces the microspheres, resulting in a secondary emulsion (solid or water in-oil-in-oil). Another nonsolvent (e.g., heptane) is added to the final emulsion to extract the organic solvent from the first oil phase. The excess solvents in the supernatant are then removed, and the final microspheres are dried. (Reproduced from Cleland, 1995, copyright Plenum Press.)

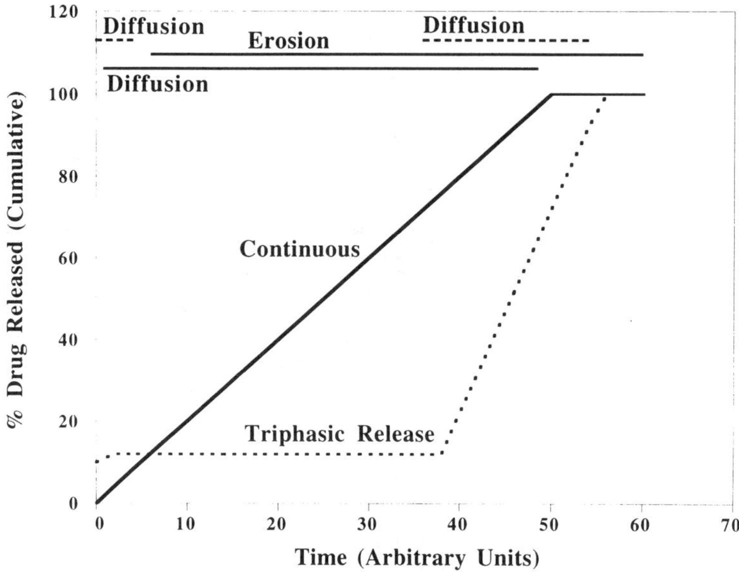

Figure 4. Illustration of the two primary types of release patterns for materials encapsulated in polylactides. Continuous release of the drug may be achieved by the continuous diffusion of the drug out of the polymer matrix. In this case, the rate of drug release is primarily controlled by the diffusivity of the drug. The triphasic release of drug is, however, dependent upon the erosion of the polymer matrix. This type of release (also referred to as the S release pattern) is characterized by an initial diffusion of drug at or near the microsphere surface. After the initial diffusive phase, a lag phase occurs until the polymer achieves bulk erosion, resulting in a significant increase in pores or channels for diffusion of the drug. The remaining drug may then diffuse out of the more porous polymer matrix. The time scale listed for each type of release is directly dependent upon the processing conditions, polymer characteristics, and drug properties as discussed in the text. (Reproduced from Cleland, 1995, copyright Plenum Press.)

erosion phase is nearly complete. This process results in a pulsatile release pattern: an initial release of the protein, little or no protein release during the erosion phase, and release of the remaining protein upon bulk erosion of the microspheres. For most applications, a continuous release of protein is usually desired. However, a pulsatile release of protein may be useful in the delivery of vaccines since they usually require a long lag period (e.g., months) between pulses (immunizations) and a total of only two or three administrations (see Cleland, 1995, for a review of polylactide vaccines). Either of the basic methods may be used to generate these types of release profiles.

2.2.2. NOVEL METHODS OF MICROENCAPSULATION

Recently, several new approaches have been developed to produce an improved process for encapsulation of proteins in polylactide microspheres. Many of these approaches have focused on methods that do not denature the protein. For example, operation of the microencapsulation process at low temperatures and in the absence of a water phase should result in increased stability of the protein during the process. One method employed the spray freeze drying of proteins to obtain solid protein particles of 2–5 μm in diameter (Gombotz *et al.*, 1990). The solid protein was then suspended in the polymer phase, and the suspension was sprayed into a container with solid ethanol and liquid nitrogen. The microspheres formed during the spraying process settled onto the solid ethanol, and the solution was then warmed to -80 °C. The microspheres were suspended in the cold, now liquid, ethanol to extract the organic solvent from the polymer phase. After repeated washing with cold ethanol, the microspheres were dried by lyophilization. This process was claimed to provide a high encapsulation efficiency (i.e., most of the protein is encapsulated) and enhanced stability of the protein (Gombotz *et al.*, 1990). The conventional spray drying approach at high temperatures ($\sim$50 °C) resulted in microspheres that tended to agglomerate and have variable morphology depending upon the polymer (Pavanetto *et al.*, 1993). The high-temperature spray drying process may cause significant protein denaturation.

Another method that utilized low-temperature processing without an aqueous phase involved supercritical fluid extraction techniques (Randolph *et al.*, 1994). The polymer, which was dissolved in an organic solvent, was sprayed into a continuous phase of supercritical carbon dioxide. The carbon dioxide acts as a nonsolvent for the polymer phase and extracts the organic solvent. The microsphere size should correlate to the density of the carbon dioxide phase (e.g., more dense, larger microspheres). This process should also provide a high encapsulation efficiency and improved protein stability.

Other new processes have involved modifications of the solvent evaporation method. To produce a more homogeneous emulsion and stabilize the protein, surfactants have been added to the protein phase. For example, researchers have claimed that the addition of hydrophobic ion pairs to the protein phase allows a more homogeneous mixture of the protein in the polymer phase and provides stabilization of the protein (M. C. Manning, personal communication, 1994). Reversed micelles (sucrose esters of fatty acids) stabilize ultrafine emulsions of the protein in the polymer phase and provide stabilization of the protein (Hayashi *et al.*, 1994). In addition, the use of a multiphase encapsulation system may protect the protein from denaturation. One type of multiphase system involves the suspension of aqueous drug in a heavy oil phase (primary emulsion) which is then emulsified in a light oil phase containing the polymer (secondary emulsion), and the secondary emulsion is then added to an aqueous solution to extract the light oil (e.g., methylene chloride) and form a triple emulsion (Iwata and McGinity, 1993). Alternatively, surfactants have been added to the polymer phase to alter the release properties of the microspheres. The addition of nonionic surfactants such as Pluronics (block copolymers of ethylene oxide and propylene oxide) provides a reduced initial burst of protein and prolongs the duration of release (Park *et al.*, 1992a). The rate of protein release from these microspheres is further reduced by the addition of polyethyleneimine, which coats the microspheres and interacts with the protein to inhibit release (Park *et al.*, 1992b). However, these methods in their present form do not provide high encapsulation efficiencies, and hydrophobic proteins may remain entrapped in the surfactant–polymer microspheres.

2.2.3. DESIGN OF A MANUFACTURING PROCESS

For the design and development of a basic manufacturing process, the type of microencapsulation process must first be chosen. For both the basic microencapsulation methods of solvent evaporation and coacervation, the unit operations used are very similar (see Figs. 2 and 3). The initial process step involves the suspension (solid protein) or emulsion (aqueous protein) of the protein in the polymer phase (polymer dissolved in organic solvent). This process step requires the formation of a homogeneous dispersion of the protein phase. There are two methods commonly used to accomplish this task. The first method, which is used primarily for aqueous protein emulsification in the polymer phase, is sonication and usually employs a probe-type sonicator. One drawback of this method is the poor scalability of the sonication device. For batch processing, the protein/polymer mixture must also be stirred to ensure uniform emulsification of the protein in the polymer phase. In addition, as the polymer phase becomes more viscous, the amount of energy required at the probe tip must also increase to produce a fine emulsion of the aqueous protein phase. A major disadvantage of this technique is that it results in a heating of the emulsion and may result in high temperatures at or near the probe tip, causing thermal denaturation of the protein. An alternative method is to use a rotor stator or homogenization system. These systems are very flexible as there are a number of different configurations available. Both large-scale and small-scale rotor stator assemblies are commercially available, and several types of homogenization tips are used with these systems. The homogenization tip consists of a fixed external cylinder with vertical slits and a rotating inner impeller. The rate of rotation of the inner impeller is controlled by the homogenization device, which is used to modulate the mixing of the fluid as well as the size of the emulsion droplets. Unlike sonicators, rotor stator assemblies are readily scalable and do not generate the same degree of heating. The size of the emulsion droplets generated by these devices is controlled by the tip design, the fluid properties, and the rotation rate of the inner impeller.

The subsequent processing equipment for microencapsulation generally consists of a stirred tank for mixing the polymer/protein solution with a nonsolvent (water or silicone oil), a hardening tank to remove excess solvent (e.g., methylene chloride), and equipment for washing and drying of the hardened microspheres (Fig. 5). Each process step shown in Fig. 5 should be optimized by analyzing several variables (Table IV). While there are a number of variables for each process step and a complex interaction between each of the variables, only a few variables are critical to optimize the overall encapsulation process. These critical variables are evaluated by assessing the final characteristics of the microspheres. These characteristics include the encapsulation efficiency, which is the ratio of the experimental to the theoretical protein loading in the microspheres, the initial burst of protein from the microspheres, and the release rate of the protein from the microspheres. The critical variables are divided into two major categories, defined as the emulsion parameters and the final processing parameters. The emulsion parameters are essential to the first two steps of the process (homogenizer and stirred tank). The viscosity of the polymer phase, the volume ratio of the aqueous or solid and organic phases in the primary emulsion, the temperature of the emulsion steps, and the amount of nonsolvent (volume) in the secondary emulsion step have the greatest impact on the final microspheres. For example, to achieve a high encapsulation efficiency, the polymer viscosity in the primary emulsion should be sufficient to prevent coalescence or agglomeration of the protein phase. In addition, as the volume of the protein phase increases to levels greater than half the polymer phase volume, the encapsulation efficiency decreases and the initial burst increases since the protein phase cannot be homogeneously dispersed under these conditions. For the secondary emulsion, the volume of the nonsolvent relative to the volume of the polymer phase is a critical determinant of the encapsulation efficiency and initial burst. A low volume of nonsolvent (e.g., <10 times the volume of the polymer phase) results in the slow formation of the microspheres because the concentration driving force for solvent extraction is reduced under these conditions and the rate of solvent removal rapidly becomes dependent upon its evaporation into the headspace above the nonsolvent. The release rate is also dependent on these emulsion variables. A homogeneous dispersion of the protein in the polymer

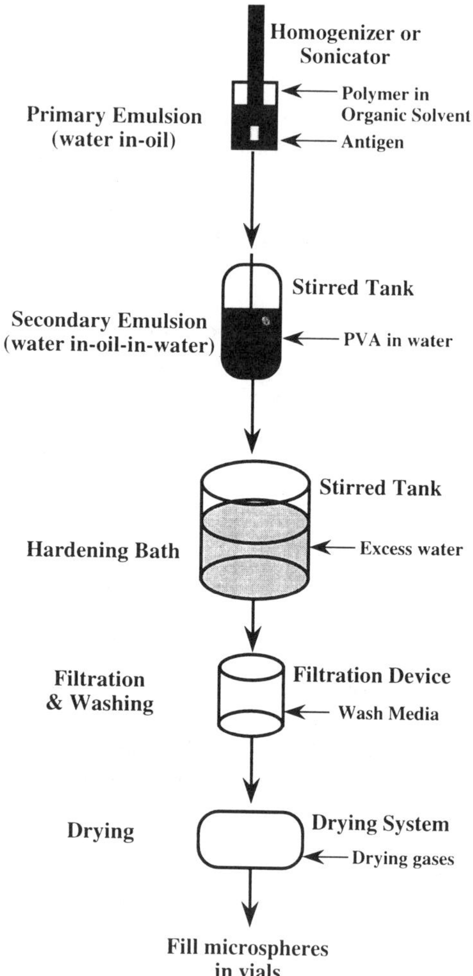

Figure 5. Simplified process diagram for production of polylactide microspheres using the solvent evaporation method. The items listed next to arrows on the right-hand side of the process diagram indicate the materials that must be added by sterile filtration. The formation of the primary emulsion may be performed by homogenization or sonication in a sealed sterile vessel. The emulsion is then transferred to a stirred tank containing the emulsifying agent in water to form the secondary emulsion. After mixing, the secondary emulsion is added to a large excess of water in a stirred tank (hardening bath). The hardening bath step may require incubation and ventilation with nitrogen sparging to facilitate removal of the organic solvent. Filtration and washing of the hardened microspheres is then performed to remove the emulsifying agent, small particulates, and large agglomerates. The final microspheres are then dried (e.g., lyophilization or vacuum drying) to produce a free-flowing powder. Each addition of components as well as the connections between unit operations may compromise the sterility of the system and will require thorough testing for an aseptic process. The process variables for each step are listed in Table IV. (Reproduced from Cleland, 1995, copyright Plenum Press.)

Table IV
Main Process Variables for Double-Emulsion Method in Production of Polylactide Vaccine[a,b]

Process step	Variables
Primary emulsion	Polymer concentration in organic solvent Polymer composition and molecular weight Organic solvent (e.g., methylene chloride or ethyl acetate) Oil-phase volume Concentration of antigen in aqueous solution Antigen solution volume Mixing rate Mixing device (e.g., homogenization or sonication) Rate of antigen addition to oil phase Duration of mixing Temperature and pressure
Secondary emulsion	Water-phase volume Concentration of emulsifying agent in water phase Rate of oil-phase addition to water phase Mixing rate Mixing device (e.g., stirred tank) Duration of mixing Temperature and pressure
Hardening bath	Water volume Presence of additives or stabilizers Incubation time Mixing rate Mixing device (e.g., stirred tank) Temperature and pressure
Filtration and washing	Filtration device (e.g., stirred cell) Rate of filtration Filtration mesh size Wash volume Wash composition Temperature and pressure
Drying	Method of drying (e.g., lyophilization) Facilitated mass transfer (e.g., fluidized bed) Drying time Amount of residual moisture Addition of excipients

[a]Reproduced from Cleland, 1995; copyright Plenum Publishing Corporation.
[b]See Fig. 5 for process diagram.

phase should provide a more uniform and continuous release, whereas a more heterogeneous distribution often results in a large initial burst with little subsequent release. To design microspheres with the desired release properties and protein loading, the key emulsion variables must be optimized for the protein and polymer system.

In addition to the emulsion variables, the final processing variables also affect the characteristics of the final microspheres. The essential parameters in these steps are the hardening bath volume, residence time of the microspheres in the hardening bath, and the final drying steps. If the volume of the hardening bath is not sufficient (e.g., <3 times the volume of the polymer phase) to extract the remaining solvent from the polymer phase, the microspheres will not harden. The rate and extent of solvent removal from the microspheres are dependent upon the solubility of the solvent in the hardening bath solution and the concentration driving force (difference between the solvent concentration in the polymer phase and the hardening bath solution). The residence time of the microspheres in the hardening bath must increase with decreasing volume of the hardening bath. A slow hardening of the microspheres may allow release of the protein or allow the microspheres to agglomerate. After hardening, the microspheres are washed with excess nonsolvent to remove any excess materials remaining from the early processing steps (e.g., PVA or silicone oil). The washed microspheres are then dried for storage. The drying process may alter the characteristics of the microspheres. For example, drying the microspheres above the glass-transition temperature of the polymer can result in agglomeration of the microspheres unless excipients are added to stabilize the microspheres. On the other hand, freezing the microspheres during lyophilization often results in a cracking or fracturing of the microspheres, which is probably the result of ice-crystal formation within the microspheres. This result may only occur with microspheres that contain an inner aqueous protein phase or were produced by the solvent extraction method with water as the nonsolvent. Thus, the final processing parameters must also be optimized to produce microspheres with the desired characteristics.

The other important elements in the engineering of a process to produce microspheres are the reproducibility, scalability, and aseptic operation of the process. The overall process must yield microspheres with the same characteristics from batch to batch. As there are a number of variables in the process (as listed in Table IV), the reproducibility of the overall process may be quite difficult to achieve. However, only a few variables are critical to the final microsphere characteristics, and, thus, if these variables are well regulated, the process should reproducibly manufacture the desired microspheres. One consideration often neglected by developers of novel methods for encapsulation is the scalability of the process. Many processes are difficult to scale up owing to either a failure to completely characterize all the process variables or the lack of availability or cost of large-scale process equipment. The process equipment should not only be amenable to scaling but also be readily adapted to aseptic processing methods. The final microsphere product, unlike most parenterals, cannot be sterile-filtered, and therefore it must either be produced aseptically or terminally sterilized.

Terminal sterilization by gamma irradiation or electron beam often causes irreversible degradation of proteins and polylactides (polymer degradation: Birkinshaw *et al.*, 1992; Hartas *et al.*, 1992; Horacek and Kudlacek, 1993; protein degradation: Haskill and Hunt, 1967a, b). An aseptic process requires that the assembly of process equipment and final dry-powder handling occur under class 100 air handling systems. If possible, process equipment should be steamed in place to reduce the potential for contamination. Typically, bacterial spores are placed in the system to assess sterility. The system is then sterilized and assembled. Next, culture media are passed through the system, and the resulting solution is cultured for several days. If bacteria are not found, the system is assumed to operate aseptically. The assessment of final microspheres for sterility is also very difficult. The surface contamination of the microspheres is usually measured by incubating the microspheres in physiological buffer at 37 °C. The buffer is then removed and checked for endotoxins and bacteria. Unfortunately, it is impossible to check the interior of the microspheres for bacterial contaminants without using harsh solvent conditions or high pH (1 M NaOH), which would degrade the contaminants. Overall, demonstration of aseptic processing and surface sterility of the final product may be sufficient to obtain approval for clinical testing.

2.3. Protein Stability

The majority of studies of protein delivery systems fail to address the issue of protein stability both during encapsulation and during incubation *in vivo*. Many studies utilize either denaturing gel electrophoresis (e.g., SDS-PAGE), enzymatic assays, or antibodies (e.g., ELISA) to demonstrate that the encapsulation process has not altered the protein, but these studies usually do not verify the maintenance of the protein's physical and chemical structure. The protein must maintain its intact three-dimensional structure as well as its chemical integrity during the encapsulation process to allow delivery of the native protein upon administration. Previous studies have shown that several types of delivery systems including PLGA microspheres and gels, can cause physical or chemical degradation of the protein (Jones *et al*, 1995). If a degraded protein is released from the microspheres, it may cause unwanted immunogenicity when administered. In addition, degradation of the protein may affect its rate of release and bioactivity. Thus, a basic understanding of the protein's properties is necessary for the design of microsphere formulations.

The physicochemical properties of the protein released from microspheres have been assessed in a few cases. For example, several proteins have been observed to retain their antibody epitopes and do not aggregate as the result of the encapsulation process (ovalbumin: O'Hagen *et al.*, 1993; MN rgp120: Cleland *et al.*, 1994; *Salmonella enteritidis*, Hazrati *et al.*, 1993; tetanus toxoid, Alonso *et al.*, 1993). Recombinant glycoprotein 120 (MN rgp120), which was originally derived from the surface coat of the human immunodeficiency virus type 1, MN strain, was analyzed in detail after encapsulation. For MN rgp120 encapsulation, the protein was characterized by size-exclusion and reversed-phase chromatography, polyclonal and monoclonal (V3 loop-specific) antibody assays (ELISAs), CD4 binding (receptor for gp120), and circular dichroism (Cleland *et al.*, 1994). These assays indicated that MN rgp120 was encapsulated without any measurable alteration in its physical or chemical properties, and it invoked neutralizing antibodies when administered to guinea pigs. In general, a rigorous analysis of the protein conformation (e.g., circular dichroism, absorbance, ELISAs, activity assays, etc.) and chemical state (e.g., chromatographic assays) should be performed on the protein released from the microspheres.

If the protein is labile to the harsh solvents or processing used in microencapsulation, a stable formulation for encapsulation is required to prevent this degradation. The design of this stable protein formulation is performed in a manner similar to typical protein formulation development for parenterals. However, the protein must be stabilized against denaturation from the microencapsulation process as well as remain essentially unaltered during storage for at least two years, a standard requirement for parenteral protein formulations. The development of this formulation includes screening for the appropriate pH, buffer, and excipients. Each protein is likely to have a unique set of conditions (e.g., pH, excipients, etc.) that provide stability during microencapsulation.

The development of a stable protein formulation for microencapsulation also includes consideration of the potential for protein–polymer interactions. For example, proteins that are very basic (high pI) may interact with the free acid groups generated by the degradation of polylactides. In this case, it may be necessary to add excipients such as polyionic compounds (anionic for protein binding, cationic for polymer binding) that prevent or reduce the interaction between the protein and the polymer. For polylactides, it may be unlikely that the protein will form a covalent adduct with the polymer under normal physiological conditions, but other polymers may react with surface moieties on the protein (e.g., lysines) during degradation. If an adduct of the protein and polymer is formed, the protein may become inactivated or immunogenic. Therefore, it is essential to assess the possible interactions (covalent and noncovalent) between the polymer and protein.

In addition to protein–polymer interactions, the stability of the protein after administration should also be considered. Most proteins degrade upon incubation at 37 °C at physiological conditions (see Cleland and Langer, 1994; Cleland *et al.*, 1993). A few recent studies have been performed to assess the conditions that would occur after administration of microspheres (Hageman *et al.*, 1992; Costantino *et al.*, 1994). These studies have revealed that a significant amount of degradation may occur *in vivo* prior to release of the protein from the microspheres. Degradation may result in a loss of bioactivity or an increased immunogenicity of the protein. The extent of degradation is weighed against the potential side effects to determine the limit for the duration of the protein release from the microspheres. Usually, the protein must remain stable to *in vivo* degradation in the microspheres for 2–4 weeks. Unfortunately, excipients added to stabilize the protein against *in vivo* degradation may diffuse out of the microspheres faster than the protein. Some common degradation routes include deamidation of asparagines, oxidation of methionines, and aggregation of the protein. While it is probably difficult to significantly reduce the chemical degradation processes, the physical degradation and aggregation may be reduced by the addition of surfactants or a reduction in the protein concentration added to the primary emulsion. The amount of protein entrapped within each pore of the microspheres is dependent upon the amount of protein added in the primary emulsion (concentration and volume or mass). A reduction in the protein concentration within the individual pores will likely decrease the extent of aggregation during incubation *in vivo*. The *in vivo* stability of the protein in the microspheres is difficult to control and requires careful consideration in the formulation design and duration of protein release.

Another major issue related to protein stability that has not been addressed by researchers in protein delivery is the stability of the protein in the microspheres during storage. At first, this issue may not seem critical to the delivery system. However, if the protein is not stable in the dry microspheres during storage, it is unlikely that the formulation will become a commercial product. The protein formulation used for encapsulation in the microspheres is required to provide stability of the protein in the dried state for two years of storage. In general, the development of a stable dry protein formulation is an extremely complex task that has only recently been investigated rigorously (see, e.g., Pikal, 1990; Pikal *et al.*, 1991; Carpenter *et al.*, 1991). Again, the development of a stable formulation requires screening of the optimum media conditions, and these conditions are tested by drying the protein under the conditions used for the production of dried microspheres. In general, a protein formulation for microencapsulation should provide stability for the final dried protein and prevent degradation during microencapsulation.

3. CASE STUDIES OF DRUG DELIVERY FROM BIODEGRADABLE MICROSPHERES

A number of proteins have been formulated in biodegradable matrices (see Table III). However, only a few of these proteins are commercially useful. To warrant the development costs of a microsphere formulation, the protein pharmaceutical should have known therapeutic benefit and require chronic or frequent administration. In addition, if the toxicity of the protein can be reduced by maintaining a low steady-state blood level, a microsphere formulation should be considered. There are a number of patents that indicate the potential use of protein microsphere formulations for vaccines (Beck, 1990; Tice *et al.*, 1991) and continuous release of peptides and proteins (Folkman and Langer, 1983; Hutchinson, 1988; Eppstein and Schryver, 1990). However, protein microsphere formulations have not yet been approved for human use. One major concern associated with these formulations is the control of the *in vivo* release of the protein. In particular, proteins with a small therapeutic window (e.g., limited acceptable dose range) could be difficult to deliver with biodegradable microspheres because the initial release (burst) may cause unwanted side effects and the steady-state blood level may vary slightly among individuals. On the other hand, if the therapeutic window is relatively large (e.g., 5- to 100-ng/ml blood level, with the high value corresponding to the initial burst), these formulations should not encounter any dose-related safety concerns. As mentioned previously, several peptide microsphere formulations (Table II) have been approved and are currently sold for use in humans. These formulations are based upon LHRH agonists, which have a large therapeutic window, and provide valuable information on the potential of protein microsphere formulations.

3.1. Lupron Depot®

In the process of developing microsphere formulations for LHRH agonists, several different aspects of the formulation and the process were investigated. Ogawa and colleagues at Takeda Chemical Industries initially analyzed methods to provide a high peptide loading in the microspheres, and they were able to achieve 10–20% loading of leuprolide acetate by increasing the viscosity of the inner aqueous phase in their water-in-oil-in-water process (Ogawa *et al.*, 1988a). The utility of increasing the inner aqueous phase viscosity was also described in their United States patent (Okada *et al.*, 1987). By increasing the aqueous phase viscosity, the protein droplets formed in the polymer phase were stabilized, and the release rate was modulated by the interactions of the viscosity enhancer with the peptide. Unfortunately, a continuous release of the drug was not achieved in this initial study.

After optimizing the method to achieve high peptide loading, Ogawa and co-workers assessed the effect of additives, polymer molecular weight, and copolymer blend ratio (lactide-to-glycolide ratio) on the release of leuprolide acetate from microspheres (Ogawa *et al.*, 1988b). Additives were encapsulated with the peptide in an effort to facilitate release by creating additional pores or channels for diffusion. All of the additives used (glyceryl monooleate, glyceryl monocaprate, methyl *p*-hydroxybenzoate, and D-lactide) decreased the encapsulation efficiency. Ogawa and colleagues attributed this effect to an alteration of the stability of the initial water-in-oil emulsion by the additives. In addition, the additives had little or no effect on the release rate, with the exception of D-lactide, which provided only a slight improvement in the release rate. On the other hand, the molecular weight of the polymer had a significant impact on the release rate. As the PLA molecular weight was decreased from 22.5 kDa to 6 kDa, the rate of release increased such that over 50% of the peptide was released after 28 days as compared to less than 20% released over the same period for the high-molecular-weight PLA. To achieve their goal of a one-month dosage formulation, Ogawa and colleagues also used copolymers (90:10 PLGA, 75:25 PLGA). The copolymer comprised primarily of lactide (90:10 PLGA; 21.2 kDa) did not provide an improvement over the PLA. However, the 75:25 PLGA (14.4 kDa) yielded a formulation that released the peptide at a continuous zero-order rate for 28 days. Blending the two copolymers together resulted in a release rate that was slower than that for the 75:25 PLGA alone.

To verify that the *in vitro* and *in vivo* release rates were comparable, Ogawa and co-workers injected the microspheres into rats (Sprague-Dawley) and measured the amount of peptide remaining at the injection site over time (Ogawa *et al.*, 1988c). The *in vivo* release was then compared to the release *in vitro*, which was performed by incubation of the microspheres in 33 mM phosphate buffer, pH 7.0, 0.05% Tween 80. As shown in Fig. 6, the *in vivo* and *in vitro* release were similar up to 21 days. However, the *in vivo* release decreased after 21 days, and 20% of the peptide remained after 35 days. Since significant bioerosion of the PLGA had occurred by 28 days, it

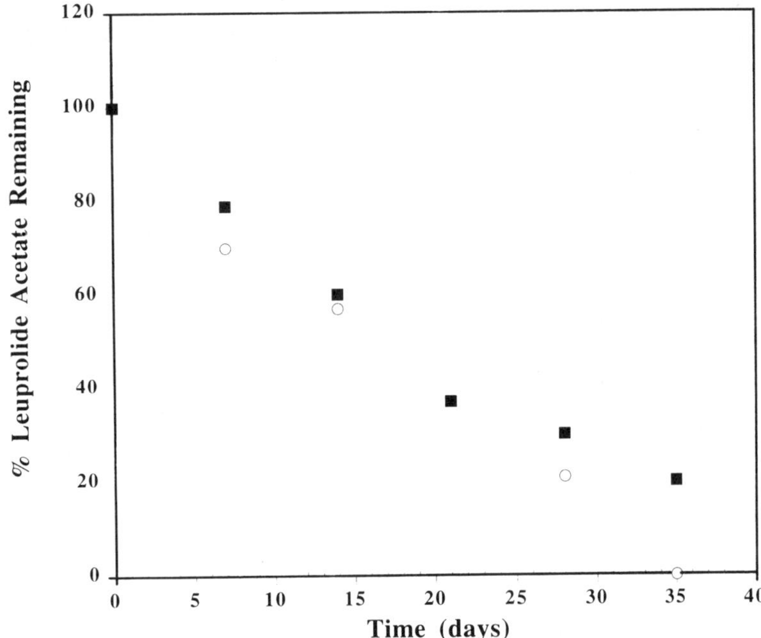

Figure 6. Release of leuprolide acetate from PLGA (75:25; 14.4 kDa) microspheres during *in vitro* incubation (33 mM phosphate buffer, pH 7.0, 0.05% Tween 80) (○) and after administration to rats. (■). The data were obtained from Ogawa *et al.* (1988c).

was assumed that the remaining peptide *in vivo* was associated with the injection-site tissues and not the PLGA microspheres. Additional preclinical studies of the leuprolide acetate–PLGA microspheres in rats and dogs indicated that the serum levels of the peptide decreased to less than 1 ng/ml after 5 weeks (Okada *et al.*, 1991). These studies also demonstrated that the steady-state blood level of peptide was dose-dependent, and repeated administration yielded reproducible blood levels. Overall, this research resulted in clinical testing and eventual approval of Lupron Depot® for use in prostate cancer and endometriosis.

3.2. MN rgp120 Controlled Release Vaccine

The continuous release of proteins and peptides is desirable for chronic diseases or long-term treatments as in the case of cancer. However, vaccine formulations are typically administered three times over the course of a year, with the doses separated in time by several months. Thus, if a biodegradable microsphere formulation is required to mimic this type of dosing regimen, it should provide a pulsatile release, and the pulse of protein (antigen) should occur *in vivo* at the same time as the repeated immunization. While most publications and patents focus on continuous release systems, one study has emphasized the ability to produce a pulsatile release in which the pulses of protein are separated by several weeks or months (Cleland *et al.*, 1994). This type of release is often referred to as S-shaped or triphasic: initial release, little or no release, and a second release phase. The duration of the lag phase is dependent upon the degradation rate of the polymer and the polymer molecular weight because bulk erosion of the polymer must occur to facilitate release of the entrapped protein (Cleland, 1995). Therefore, by varying the polymer composition or molecular weight, the desired immunization schedule can be achieved from a single immunization.

MN rgp120, the surface glycoprotein from the MN strain of the human immunodeficiency virus type 1 (HIV-1_{MN}), was encapsulated in PLGA microspheres to address the potential for producing a single-immunization

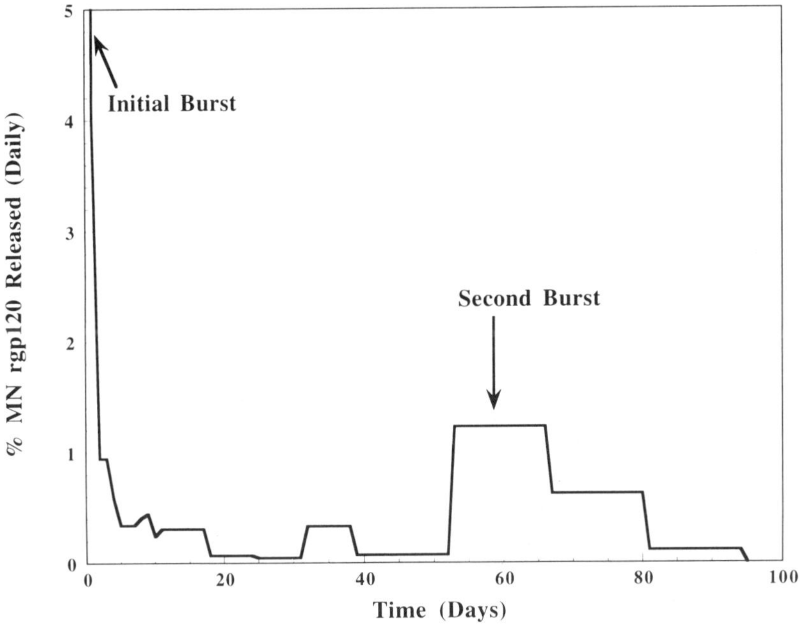

Figure 7. *In vitro* release of MN rgp120 from PLGA microspheres. Microspheres were incubated at 37 °C in 10 mM HEPES buffer, 120 mM NaCl, pH 7.4, and the amount of protein released was quantitated by using a dye binding assay (BCA, Pierce Chemicals) with MN rgp120 as the standard.

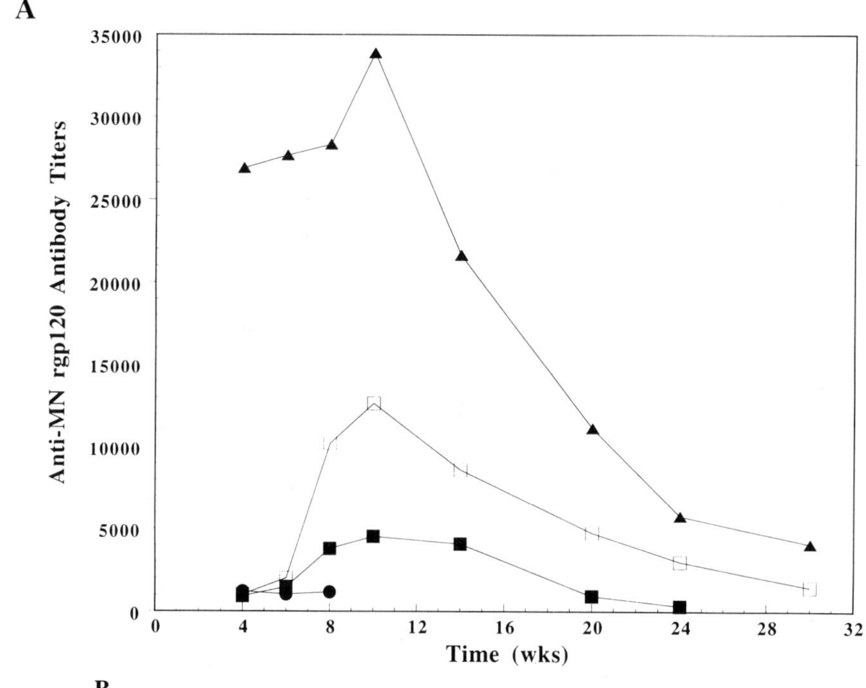
A
Anti-MN rgp120 Antibody Titers
35000
30000
25000
20000
15000
10000
5000
0
0
4
8
12
16
20
24
28
32
Time (wks)

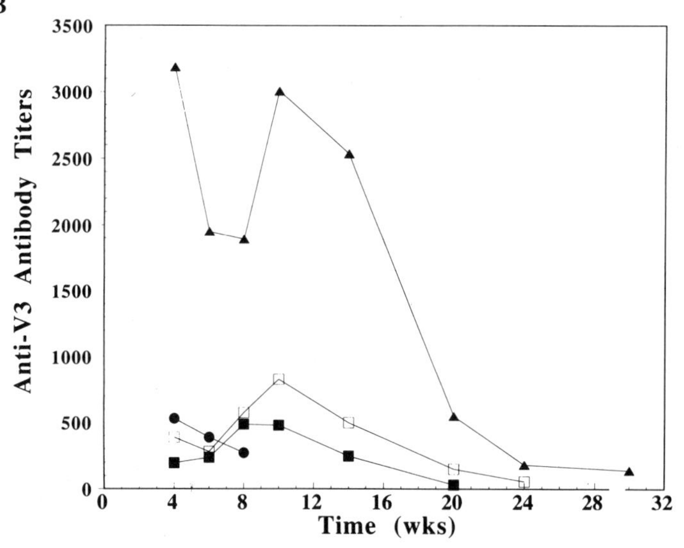
B
Anti-V3 Antibody Titers
3500
3000
2500
2000
1500
1000
500
0
0
4
8
12
16
20
24
28
32
Time (wks)

vaccine for HIV-1 as well as to assess the possible increased immune response to the protein or any coencapsulated adjuvants (Cleland *et al.*, 1994). A solvent evaporation process was used to produce PLGA microspheres that released MN rgp120 in a pulsatile manner as shown in Fig. 7. These microspheres had a large initial release (56%) that was followed by insignificant release of protein until 8 weeks of incubation. After 8 weeks, bulk erosion of the polymer caused new channels to form, allowing diffusion of the protein from the microspheres. This second diffusive phase is often referred to as the "second burst" because the protein is released over a few weeks (e.g., 2–4 weeks). The *in vivo* release of these microspheres was assessed by administering them to guinea pigs and measuring the antibody response to both the whole protein and one region of the MN rgp120 referred to as the V3 loop, which is the principal neutralizing determinant for HIV-1. As shown in Fig. 8, the initial antibody response was dose-dependent (e.g., more antigen, greater response). The antibody titers then decreased over time until the microspheres released additional antigen. The MN rgp120 was released from the microspheres prior to the rise in antibody levels after 8 weeks. The initial antibody response followed by a decay and subsequent rise in titers can only be achieved by a pulsatile release of the protein, and the time of release correlates well to the observed *in vitro* second burst phase. Thus, the pulsatile release of a subunit antigen, MN rgp120, was achieved with these microspheres.

The ultimate goal of this vaccine preparation was the generation of a neutralizing antibody response to HIV-1. To invoke neutralizing antibodies upon administration, gp120 must maintain its native conformation (Haigwood *et al.*, 1992). It was therefore a requirement that MN rgp120 be delivered in its native state. Analysis of the initial protein released from the microspheres indicated that it had maintained its native conformation after encapsulation (Cleland *et al.*, 1994). To provide an improved humoral and cell-mediated response to MN rgp120, the adjuvant, QS-21 from *Quillaja saponaria* was added to the primary emulsion and encapsulated in the microspheres with MN rgp120. These microspheres were administered to guinea pigs, and the amount of antibodies that neutralized the MN strain

Figure 8. Dose response of *in vivo* autoboost from PLGA formulations as measured by the antibody titers to the MN rgp120 (A) and the V3 loop of MN rgp120 (B). Guinea pigs were dosed with varying amounts of a gp120–PLGA formulation. The total antigen dose delivered from the PLGA formulations was 14 (■), 42 (□), or 112 (▲) μg of MN rgp120. A control group dosed with 30 μg of MN rgp120 formulated with 60 μg of alum (Rehydrogel HPA®) was also included (●). All animals were given a single injection initially, and antibody titers in the sera were measured over time. (Reproduced from Cleland *et al.*, 1994, copyright Mary Ann Liebert, Inc. Publishers.)

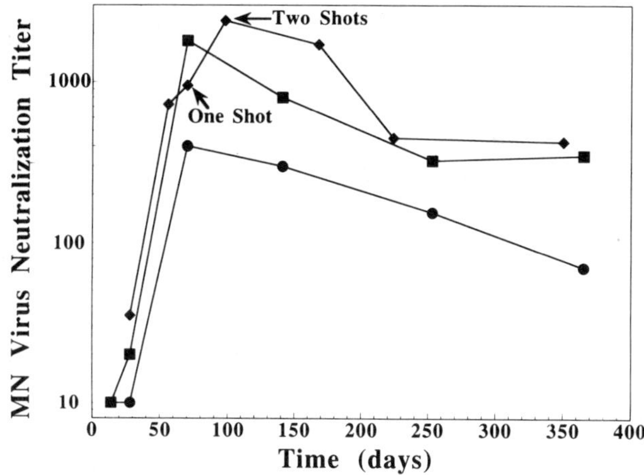

Figure 9. Comparison of virus neutralization titers for MN rgp120 in different formulations. Guinea pigs were immunized at 0, 1, and 2 months with 30 μg of MN rgp120 (90 μg total) and two different formulations, 60 μg of alum (Rehydrogel HPA®) per dose without QS-21 (180 μg total; Alum; ●) and with 50 μg of QS-21 per dose (150 μg total; QS-21 + Alum; ■). In a separate experiment, guinea pigs were immunized with coencapsulated MN rgp120/QS-21 at a dose of 25 μg of MN rgp120 and 19 μg of QS-21 at 0 and 2 months (50 μg MN rgp120 and 38 μg QS-21 total dose; ♦). (Reproduced from Cleland *et al.*, 1994, copyright Mary Ann Liebert, Inc. Publishers.)

of the human immunodeficiency virus type 1 (HIV-1_{MN}) was measured. A single administration of microspheres containing both MN rgp120 and QS-21 provided equivalent neutralizing titers to three conventional immunizations with MN rgp120 formulated with both QS-21 and alum (Fig. 9; assay standard error is twice the titer value). Thus, the MN rgp120 released from the microspheres after *in vivo* incubation for several weeks must also have a native conformation since neutralizing antibody titers increased prior to a second immunization with the microspheres. A similar approach to encapsulating subunit antigens should have great utility in reducing the number of immunizations and, perhaps, may lead to single-shot vaccines (Cleland, 1995).

4. IMMUNOGENICITY AND INJECTION-SITE CONSIDERATIONS

While the generation of antibodies is essential for a subunit vaccine, antibodies to a therapeutic protein may result in a loss of efficacy, autoimmune responses, and other adverse side effects. If a heterologous protein (e.g., murine) is administered to humans, it is likely that this protein, if different in chemical composition from the human form, will invoke an antibody response. A microsphere formulation containing a heterologous protein may further enhance this response by providing a constant stimulus as described previously for vaccines (Cleland, 1995). However, it is unclear whether a homologous protein would become immunogenic if administered in a microsphere formulation. Previous studies have indicated that PLGA or PLA microspheres may act as an adjuvant (Eldridge *et al.*, 1991a,b; O'Hagen *et al.*, 1993; Hazrati *et al.*, 1993). The primary adjuvant effect of these polymers has not been clearly demonstrated, but it is most likely attributable to their depot properties. In other words, if the antigen is formulated as a depot, which is often achieved with alum, the antigen is presented to the immune system for a longer period of time, perhaps providing a greater opportunity for the antibody formation. A homologous protein in PLGA or PLA microspheres has not been administered to humans, and therefore the issue of immunogenicity may only be resolved with clinical trials of microsphere formulations containing therapeutic proteins. Alternatively, the administration of a protein that is either species-independent (e.g., same composition in most or all species) or homologous (e.g., murine delivered to mouse) in microspheres could be performed to assess the potential immunogenicity of the preparation.

The tissue response to the microspheres at the site of injection may also contribute to the greater immune response observed for heterologous protein microsphere formulations. Upon administration of the microspheres, a foreign-body response occurs, resulting in an acute initial inflammation (Visscher *et al.*, 1987). This initial inflammation is followed by the infiltration of small foreign-body giant cells and neutrophils (Visscher *et al.*, 1985, 1987). These immune cells could consume the released protein and produce an immune response. However, if the protein is recognized as a self protein (e.g., homologous), the probability of an immune response by these cells is reduced. It is therefore always essential to release the protein in its native conformation. The release of aggregated or denatured protein from the microspheres may result in an unwanted immune response (see Cleland *et al.*, 1993, for a discussion of immunogenicity of degraded proteins). After the immune cells invade the injection site, fibroblasts begin to line the periphery of the site and produce collagen (Visscher *et al.*, 1985). Previous studies with PLGA and PLA microspheres have shown that the fibrous tissue formation around the site of injection is not significant (Visscher *et al.*, 1985). The presence of the collagen matrix around the injection site may affect the rate of protein release into the circulation if the injection site is not vascularized by capillaries. Thus, if microspheres do not release an equivalent bioavailable dose to a subcutaneous pump (e.g., Alzet minipump), it may be the result of protein–fibrous tissue interactions. This

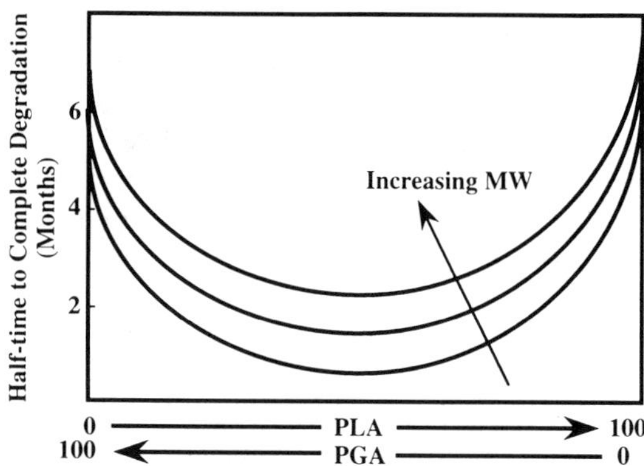

Figure 10. A modified version of the *in vivo* degradation profile for polylactides in rats as described by Miller *et al.* (1977). As the relative amount of either monomer, glycolide or lactide, increases, the degradation time increases due to the differences in the hydrophobicity and crystallinity of the polymer (see Table V). (Reproduced from Cleland, 1995, copyright Plenum Press.)

possibility should be investigated in the early stage of development of a microsphere formulation.

With polylactide or polyglycolide microspheres, the resolution of the injection site is then dependent upon the polymer molecular weight and composition, both of which dictate the rate of polymer degradation (Miller *et al.*, 1977). The injection site eventually resolves and returns to normal tissue (no scarring) after degradation of the microspheres is at or near completion (Visscher *et al.*, 1987). When designing a microsphere formulation, one must consider the amount of time required for degradation of the polymer. Clearly, one would not want a one-month release from a polymer that required one year to degrade, since the result would be the accumulation of the polymer in the body. As shown in Fig. 10 and Table V, the appropriate molecular weight and composition of a polylactide may be chosen based on its degradation rate *in vivo*. As mentioned above, the duration of protein release from the microspheres is also dependent upon these parameters. A microsphere formulation that is administered once a month should utilize a polymer that degrades within two months. The leuprolide acetate–PLGA microspheres discussed in Section 3.1 are degraded completely *in vivo* after approximately 60 days (Okada *et al.*, 1991; Miller *et al.*, 1977). For vaccines, the degradation coincides with the *in vivo* release of the antigen (second burst), providing an autoboost effect, and therefore the microsphere formulation is designed to utilize the differences in the degradation rates of the polymers (Cleland, 1995). In this case, the

Table V
In Vivo Degradation Times of Polylactides in Rats[a]

Polymer composition (lactide:glycolide)	MW (kDa)	Half-time to complete degradation (months)	Crystallinity
PLA	85	6.1	Moderate
75:25 PLGA	50	0.6	Low/moderate
50:50 PLGA	46	0.24	Low
25:75 PLGA	N.A.	0.55	Moderate
PGA (slow cured)	N.A.	5.0	N.A.

[a]Data from Miller *et al.*, 1977.
[b]N.A., Not available.

polymer may not degrade for several months, but the microspheres are not administered more than once or twice. The polylactides have been used in humans for over 20 years in resorbable sutures and for the past few years in depot formulations with LHRH agonists. There have not been any adverse events reported regarding the polymer. The administration of polylactide microspheres is generally well tolerated with minimal irritation or inflammation.

The dose administered also has an effect on the local events at the site of injection. In particular, the mass of microspheres is often limited because high doses of microspheres result in the formation of palpable or visible bumps under the skin. For subcutaneous administration, the choice of injection site affects the mass of microspheres that can be delivered. The fascial plane of subcutaneous tissue space determines the amount of material administered. Thus, injections into the thigh, buttocks, or back of the arm provide sufficient space for a reasonable dose of microspheres in the subcutaneous region. Large doses in a small subcutaneous space (e.g., >100 mg in a rat) result in the formation of small nodules because the microspheres are confined to a small volume. The dose is also limited by the amount of microspheres that can be readily suspended and injected. The use of a 27- or 25-gauge needle is usually preferred for subcutaneous injections of small-volume ($\leqslant 1$ ml) protein pharmaceuticals. The microspheres are then required to flow through the needle and into the subcutaneous space. Often, dilatancy effects occur, causing the microspheres to clog the needle, and the microsphere dose is incompletely administered. The use of excipients in the suspending vehicle such as carboxymethylcellulose (CMC), dextran, or sorbitol is useful in preventing the agglomeration of the microspheres. The delivery vehicle for injection of the microspheres may also contain surfactants and salts to alter the microsphere's fluid properties.

A careful screening of several excipients is required to optimize delivery of the microspheres. The Lupron Depot® (leuprolide acetate–PLGA) package insert indicates the use of CMC, Tween, and phosphate-buffered saline, but the overall doses are relatively small (e.g., 30–60 mg of microspheres). In general, it is difficult to achieve greater than 200 mg/ml of microspheres for injection through a 21- or 23-gauge needle, and lower concentrations are required for the smaller gauge needles. If one assumes a maximum subcutaneous dose of 1 ml, then the maximum microsphere dose is 200 mg. This maximum dose should be used to calculate the maximum amount of protein that can be administered, and the protein dose is dependent upon the protein loading (protein loading mass fraction times 200 mg). This value, in turn, should be divided by the duration of release to provide an estimate of the maximum achievable daily dose of protein for a continuous-release microsphere formulation. These calculations are critical in determining the feasibility of a microsphere formulation for a therapeutic protein.

5. REGULATORY REQUIREMENTS FOR DEVELOPMENT OF PROTEIN DELIVERY FROM MICROSPHERES

In developing a microsphere formulation for proteins, it is important to consider the regulatory requirements for the approval of an Investigational New Drug (IND) application by the U.S. FDA. These requirements include toxicology testing, manufacturing reviews, quality assurance and quality control issues, and preclinical studies.

5.1. Toxicology Studies

The repeated administration and long-term exposure to the protein and the microspheres are assessed in preclinical animal toxicology studies. Typically, the microsphere preparation, which is representative of the clinical form, is administered to two different species (e.g., mouse and monkey). The repeated administration of the formulation is also performed to demonstrate both safety and a reproducible physiological response (e.g., same steady-state blood level achieved for same dose). The toxicology studies are performed under Good Laboratory Practice (GLP) guidelines as set forth by the U.S. FDA. These studies include a thorough histological examination of the injection site and complete clinical pathology. In addition, any adverse side effects are recorded during the study. It is essential to perform preliminary toxicology studies on a potential microsphere formulation well in advance of the final development because this early screening may eliminate formulations that are too toxic for human use. Further, the continuous infusion of the protein from an implantable device (e.g., Alzet minipumps) should be performed very early in the development of microsphere formulations. If a continuous release of the protein is toxic, it would preclude development of a sustained release microsphere formulation. Generally, the continuous administration of a protein should provide reduced toxicity because the high peak concentrations observed for bolus administration are not obtained and lower blood levels of the protein when maintained for long times (weeks) may achieve the same efficacy as higher doses given as a daily injection.

5.2. Residual Solvent Concerns

A major toxicity-related issue for polylactide microsphere preparations is the presence of any residual solvents used in the manufacturing process. In particular, organic solvents such as methylene chloride and ethyl acetate used to dissolve the polymer may pose significant health risks for long-term exposure. The regulatory agency may set the acceptable level of these solvents based on its previous experience or published literature on the toxicity. As shown in Table VI, the previous toxicology studies for administration of methylene chloride and ethyl acetate indicate that ethyl acetate is less toxic and may be better tolerated if given at the same dose. Low doses of ethyl acetate are often used to enhance food flavor (apple flavor) and are well tolerated when delivered orally. However, some polylactides, notably 50:50 lactide:glycolide PLGA, are less soluble in ethyl acetate, and therefore methylene chloride is required in the manufacturing process. The level of methylene chloride in the final product should be reduced as much as possible without compromising the quality or properties of the microspheres. The regulatory agency often requires that the manufacturer attempt to reduce the residual solvent levels as much as possible. For Lupron Depot®, the acceptable levels of methylene chloride per dose are less than 50 ppm, and usually levels of less than 10 ppm are achieved. These levels of residual solvent are often difficult to obtain because the solvents tend to remain bound to the polymer even after lyophilization. These solvents are more tightly bound to the more hydrophobic polymers (higher lactide content). High-temperature ($>20\,^{\circ}C$) vacuum drying can remove the bound solvent, but temperatures above the glass transition of the polymer alter the microsphere properties and may result in agglomeration of the microspheres and denaturation of the protein (e.g., above 25–40 °C). Ultimately, the

Table VI
Toxicity Data for Methylene Chloride and Ethyl Acetate[a,b]

Component	Toxicity[c]
Methylene chloride	Rat: LD_{50} 1.6 g/kg (oral) Human: LDLo 357 mg/kg (oral); narcotic effect Suspected carcinogen and mutagen Damage to liver and kidneys Metabolized to carbon monoxide Nervous system disorders Skin irritation Therapeutic category: Pharmaceutical aid (solvent)[d]
Ethyl acetate	Mouse: LD_{50} 709 mg/kg (i.p.); 4.1 g/kg (oral) Rat: LD_{50} 5.62 g/kg (oral); 5.0 g/kg (s.q.) Rabbit: LD_{50} 4.9 g/kg (oral) Cat: LD_{50} 3.0 g/kg (s.q.) Skin and eye irritation Target organs: Liver, kidneys, central nervous system, and blood Therapeutic category: Pharmaceutical aid (flavor)[d]

[a]Reproduced from Cleland, 1995, copyright Plenum Press. Data are taken from Sigma-Aldrich material safety data sheets.
[b]Material safety data sheets.
[c]Abbreviations: LD_{50}, Lethal dose that causes 50% of the treated animals to die; LDLo, lethal dose resulting in a few deaths; i.p., intraperitoneal; e.g., subcutaneous.
[d]*Merck Index* (M. Windholz, S. Budavari, R. F. Blumetti, and E. S. Otterbein, eds.), Tenth edition, Merck & Co., Rahway, New Jersey, 1983, pp. 545, 869.

removal of excess solvent from the microspheres must be balanced with the quality and safety requirements of the final product. Regardless of the final solvent level, the final microsphere formulation should be subjected to rigorous toxicological assessment.

5.3. Manufacturing Issues

In addition to a review of toxicological results, the regulatory agency assesses the manufacturing process and the final product quality. As discussed in Section 2.2.3, the microspheres must be manufactured in a reproducible manner and meet specified quality standards agreed upon by the regulatory agency and the manufacturer. Generally, the microspheres are required to have the same characteristics from batch to batch (e.g., protein loading, size, release rate, initial burst, residual solvents, etc.) within the specified limits for the product. Assays for protein loading, protein release, protein quality (e.g., bioactivity, etc.), microsphere size, residual solvents, and sterility must be developed and used under GLP guidelines. The final product in the vials should also be inspected for fill consistency and content uniformity. Both the assays and the manufacturing process should have standard operating procedures (SOPs) for documentation of each batch. In addition, each preparation of final microspheres should have a batch record. The process validation also needs to be well documented for regulatory review. With the appropriate quality assurance and quality control assessment of the final product, the manufacturing process and final product should meet the regulatory requirements for an IND application.

5.4. Preclinical Animal Models and Clinical Experiments

Beyond toxicology studies and manufacturing concerns, the regulatory agency is interested in a demonstration of the potential efficacy of the final product. In particular, the biological activity or potency of the protein is usually shown in an appropriate animal model prior to submission of an IND application. If the protein is already in clinical trials or an approved pharmaceutical, a bioequivalence study is required to demonstrate that the protein delivered from the microsphere formulation provides equivalent efficacy to the current formulation. This study can be performed in the same animal model used to assess the original protein formulation. In addition, early experiments utilizing continuous-infusion pumps (e.g., Alzet minipumps) provide valuable information on the dose response for a continuous administration as well as the expected pharmacodynamics of the protein. These studies are then used to determine the required dose administered from the microsphere preparation. If the pump rate is set to the same level as the predicted release rate from the microsphere preparation, a direct comparison is possible and provides an indication of the relative bioavailability of the microsphere formulation.

Following preclinical animal experiments, a Phase I clinical experiment should be performed to assess the dose and effects of continuous administration of the protein. These clinical experiments are particularly useful for analysis of proteins that are already approved for human use (e.g., insulin or growth hormone) or that have been previously observed to be very toxic (e.g., cytokines). Patients are administered a continuous infusion of the protein from either implanted or external pumps. This infusion should mimic the expected rate of delivery from the microsphere formulation. The overall biological effects of maintaining a steady-state blood level of the protein are also assessed in this type of clinical experiment. Of course, for proteins that have not been previously tested in humans, the development of a continuous-release formulation becomes more difficult since the physiological effects of the protein are unknown and a safe and efficacious dose when delivered conventionally has not been established. In this case, it is essential to perform a Phase I clinical experiment to test the concept of continuous infusion and to define a clinical dose. The combination of preclinical studies in appropriate animal models and clinical experiments should provide sufficient data to indicate whether a continuous delivery system such as biodegradable microspheres is necessary and feasible.

6. SUMMARY

The key components to the successful development of a biodegradable microsphere formulation for the delivery of proteins are polymer chemistry, engineering, and protein stability. These areas are intricately related and require a thorough investigation prior to embarking on the encapsulation of proteins. While each of these components is important for the development of a biodegradable microsphere formulation for protein delivery, other critical issues should also be considered. In particular, preclinical studies in the appropriate animal model are usually necessary to assess the potential feasibility of a continuous-release dosage form. These studies should be performed at the earliest possible stage of development to validate the feasibility of a controlled release formulation. After the utility of a controlled release formulation has been demonstrated, the polymer matrix should be chosen and bench-scale production of microspheres initiated. The only polymers presently approved for human use for controlled delivery are the polylactides [poly(lactic acid), poly(glycolic acid), and poly(lactic-*co*-glycolic) acid]. These polymers require multiphase processes involving several steps to produce microspheres containing the desired protein. A thorough review of previous work on encapsulation with these polymers should provide some insight into conditions to be assessed in developing a process. Once a process is chosen, it must be optimized to provide the highest possible yield of microspheres with the desired characteristics (e.g., loading, release, size, etc.). Finally, the final aseptic process should be validated and methods generated to assess the final product. The clinical studies should then start upon approval of the IND application.

In the future, the biotechnology industry, and the pharmaceutical industry in general, will be seeking new methods to improve the delivery of therapeutic agents such as proteins and peptides. Formulations like biodegradable microspheres significantly reduce health-care costs since fewer administrations are needed, and they provide a competitive advantage in markets with several competing products (e.g., LHRH agonist market). Further, many new indications such as neurological diseases may require a long-term delivery system. The future success of biodegradable microsphere formulations will primarily depend on the commitment of the pharmaceutical and biotechnology industries to the development of this technology.

ACKNOWLEDGMENTS

The author appreciates the comments, support, and guidance of Drs. Tue Nguyen, Rodney Pearlman, and Andy Jones in the preparation of this manuscript. In addition, the timely assistance of Juliana Monroe is greatly appreciated. The work of Janet Yang, Eileen Duenas, Anne Mac, Melissa Roussakis, Dennis Brooks, Brooks Boyd, Yu-Fun Maa, Doug Yeung, and Chung Hsu has been essential to the development of the knowledge incorporated in this manuscript. The author also greatly appreciates the editorial comments and assistance of Jessica Burdman.

REFERENCES

Alonso, M. J., Cohen, S., Park, T. G., Gupta, R. K., Siber, G. R., and Langer, R., 1993, Determinants of release rate of tetanus vaccine from polyester microspheres, *Pharm. Res.* **10:**945–953.

Andrianov, A. K., Cohen, S., Visscher, K. B., Payne, L. G., Allcock, H. R., and Langer, R., 1993, Controlled release using ionotropic polyphosphazene hydrogels, *J. Controlled Release* **27:**69–77.

Anonymous, 1993, *Pharma Japan* **1355:**2.

Arshady, R., 1991, Preparation of biodegradable microspheres and microcapsules, 2. Polylactides and related polyesters, *J. Controlled Release* **17:**1–21.

Arthursson, P., Edman, P., Laakso, T., and Sjölm, I., 1984, Characterization of polyacryl starch microparticles as carriers for proteins and drugs, *J. Pharm. Sci.* **73:**1507–1513.

Beck, L. R., 1990, Mammal immunization, U.S. Patent 4,919, 929.

Birkinshaw, C., Buggy, M., Henn, G. G., and Jones, E., 1992, Irradiation of poly(D,L-lactide), *Polym. Degrad. Stab.* **38:**249–253.

Bodmer, D., Kissel, T., and Traechslin, E., 1992, Factors influencing the release of peptides and proteins from biodegradable parenteral depot systems, *J. Controlled Release* **21:**129–138.

Carpenter, J. F., Arakawa, T., and Crowe, J. H., 1991, Interactions of stabilizing additives with proteins during freeze-thawing and freeze-drying, *Dev. Biol. Stand.* **74:**225–231.

Cleland, J. L., 1995, Design and production of single immunization vaccines using polylactide polyglycolide microsphere systems, in: *Vaccine Design* (M. F. Powell and M. J. Newman, eds.), Plenum, New York, pp. 439–472.

Cleland, J. L., and Langer, R., 1994, Formulation and delivery of proteins and peptides: Design and development strategies, in: *Formulation and Delivery of Proteins and Peptides* (J. L. Cleland and R. Langer, eds.), ACS Symposium Series, American Chemical Society, Washington, D.C., pp. 1–21.

Cleland, J. L., Powell, M. F., and Shire, S. J., 1993, Development of stable protein formulations: A close look at protein aggregation, deamidation, and oxidation, *CRC Crit. Rev. Ther. Drug Carrier Syst.* **10:**307–377.

Cleland, J. L., Powell, M. F., Lim, A., Barrón, L., Berman, P. W., Eastman, D. J., Nunberg, J. H., Wrin, T., and Vennari, J. C., 1994, Development of a single shot subunit vaccine for HIV-1, *AIDS Res. Human Retrovirus,* **10:**521–525.

Coffin, M. D., and McGinity, J. W., 1992, Biodegradable pseudolatexes—the chemical stability of poly(D,L-lactide) and poly(epsilon-caprolactone) nanoparticles in aqueous media, *Pharm. Res.* **9:**200–205.

Cohen, S., Yoshioka, T., Lucarelli, M., Hwang, L. H., and Langer, R., 1991, Controlled delivery systems for proteins based on poly(lactic/glycolic acid) microspheres, *Pharm. Res.* **8:**713–720.

Costantino, W. R., Langer, R., and Klibanov, A., 1994, Moisture-induced aggregation of lyophilized insulin, *Pharm. Res.* **11:**21–29.

Crommen, J., Vandorpe, J., and Schacht, E., 1993, Degradable polyphosphazenes for biomedical applications, *J. Controlled Release* **24:**167–180.

Downs, E. C., Robertson, N. E., Riss, T. L., and Plunkett, M. L., 1992, Calcium alginate beads as a slow-release system for delivering angiogenic molecules *in vivo* and *in vitro, J. Cell. Physiol.* **152:**422–429.

Eldridge, J. H., Staas, J. K., Meulbroek, J. A., Tice, T. R., and Gilley, R. M., 1991a, Biodegradable and biocompatible poly(DL-lactide-*co*-glycolide) microspheres as an adjuvant for staphylococcal enterotoxin B toxoid which enhances the level of toxin-neutralizing antibodies, *Infect. Immun.* **59:**2978–2986.

Eldridge, J. H., Staas, J. K., Meulbroek, J. A., McGhee, J. R., Tice, T. R., and Gilley, R. M., 1991b, Biodegradable microspheres as a vaccine delivery, *Mol. Immunol.* **28:**287–294.

Eppstein, D. A., and Schryver, B. B., 1990, Controlled release of macromolecular polypeptides, U.S. Patent 4,962,091.

Folkman, M. J., and Langer, R. S., 1983, Systems for the controlled release of macromolecules, U.S. Patent 4,391,797.

Gombotz, W. R., Healy, M. S., Brown, L. R., and Auer, H. E., 1990, Process for producing small particles of biologically active molecules, Patent Application WO 90/13285.

Gref, R., Minamitake, Y., Peracchia, M. T., Trubetskoy, V., Torchilin, V., and Langer, R., 1994, Biodegradable long-circulating polymeric nanospheres, *Science* **263:**1600–1603.

Hageman, M. J., Bauer, J. M., Possert, P. L., and Darrington, R. T., 1992, Preformulation studies oriented toward sustained delivery of recombinant somatotropins, *J. Agric. Food Chem.* **40:**348–355.

Haigwood, N. L., Nara, P. L., Brooks, E., Van Nest, G. A., Ott, G., Higgins, K. W., Dunlop, N., Scandella, C. J., Eichberg, J. W., and Steimer, K. S., 1992, Native but not denatured recombinant human immunodeficiency virus type 1 gp120 generates broad-spectrum neutralizing antibodies in baboons, *J. Virol.* **66:**172–182.

Hartas, S. R., Collett, J. H., and Booth, C., 1992, The influence of gamma-irradiation on the release of melatonin from poly(lactide-coglycolide) microspheres, *Proc. Int. Symp. Control. Rel. Bioact. Mater.* **19:**321–322.

Haskill, J. S., and Hunt, J. W., 1967a, Radiation damage to crystalline ribonuclease: Identification of polypeptide chain breakage in the denatured and aggregated products, *Radiat. Res.* **32:**827–848.

Haskill, J. S., and Hunt, J. W., 1967b, Radiation damage to crystalline ribonuclease: Importance of free radicals in the formation of denatured and aggregated products, *Radiat. Res.* **32:**606–624.

Hayashi, Y., Yoshioka, S., Aso, Y., Po, A. L. W., and Terao, T., 1994, Entrapment of proteins in poly(L-lactide) microspheres using reversed micelle solvent extraction, *Pharm. Res.* **11:**337–340.

Hazrati, A. M., Lewis, D. H., Atkins, T. J., Stohrer, R. C., McPhillips, C. A., and Little, J. E., 1993, *Salmonella enteritidis* vaccine utilizing biodegradable microspheres, *Proc. Int. Symp. Control. Rel. Bioact. Mater.* **20:**101–102.

Heller, J., 1993, Polymers for controlled parenteral delivery of peptides and proteins, *Adv. Drug Delivery Rev.* **10:**163–204.

Herr, G., Wahl, D., and Kusswetter, W., 1993, Osteogenic activity of bone morphogenic protein and hydroxyapatite composite implants, *Ann. Chir. Gynaecol.* **82**(S207)**:**99–107.

Heya, T., Okada, H., Ogawa, Y., and Toguchi, H., 1991, *in vitro* and *in vivo* evaluation of thyrotropin releasing hormone release from copoly(dl- lactic/glycolic acid) microspheres, *Int. J. Pharm.* **72:**199–205.

Horacek, I., and Kudlacek, L., 1993, Influence of molecular weight on the resistance of polylactide fibers by radiation sterilization, *J. Appl. Polym. Sci.* **50:**1–5.

Horisaka, Y., Okamoto, Y., Matsumoto, N., Yoshimura, Y., Hirano, A., Nishida, M., Kawada, J., Yamashita, K., and Takagi, T., 1994, Histological changes of implanted collagen material during bone induction, *J. Biomed. Mater. Res.* **28:**97–103.

Hutchinson, F. G., 1988, Continuous release pharmaceutical compositions, U.S. Patent 4,767,628.

Iwata, M., and McGinity, J. W., 1993, Dissolution, stability, and morphological properties of conventional and multiphase poly(DL-lactic-*co*-glycolic acid) microspheres containing water-soluble compounds, *Pharm. Res.* **10:**1219–1227.

Jeffery, H., Davis, S. S., and O'Hagan, D. T., 1993, The preparation and characterization of poly(lactide-*co*-glycolide) microparticles. II. The entrapment of a model protein using a (water-in-oil)-in-water emulsion solvent evaporation technique, *Pharm. Res.* **10:**362–368.

Johnson, R. E., Lanaski, L. A., Gupta, V., Griffin, M. J., Gaud, H. T., Needham, T. E., and Hossein, Z., 1991, Stability of atriopeptin III in poly(D,L-lactide-*co*-glycolide) microspheres, *J. Controlled Release* **17:**61–68.

Jones, A. J. S., Nguyen, T. H., Cleland, J. L., and Pearlman, R., 1995, New delivery systems for recombinant proteins—practical issues from proof of concept to clinic, in: *Trends and Future Perspectives in Peptide and Protein Drug Delivery* (V. H. L. Lee, M. Hashida, and Y. Mizushima, eds.), Drug Targeting & Delivery Series, Harwood Academic Publishers, Amsterdam, pp. 57–67.

Kamath, K. R., and Park, K., 1993, Biodegradable hydrogels in drug delivery, *Adv. Drug Delivery Rev.* **11:**59–84.

Kenley, R. A., Yim, K., Abrams, J., Ron, E., Turek, T., Marden, L. J., and Hollinger, J. O., 1993, Biotechnology and bone graft substitutes, *Pharm. Res.* **10:**1393–1401.

Langer, R., 1991, Polymer implants for drug delivery in the brain, *J. Controlled Release* **16:**53–60.

Langer, R., 1993, Polymer-controlled drug delivery systems, *Acc. Chem. Res.* **26:**537–542.

Levy, M. C., and Andry, M. C., 1991, Mixed-walled microcapsules made of cross-linked proteins and polysaccharides—Preparation and properties, *J. Microencapsulation* **8:**335–347.

Lewis, D. H., 1990, Controlled release of bioactive agents from lactide glycolide polymers, in: *Drugs and the Pharmaceutical Sciences* (M. Chasin and R. Langer, eds.), Vol. 45, *Biodegradable Polymers as Drug Delivery Systems*, Marcel Dekker, New York, pp. 1–42.

Li, C., Yang, D. J., Kuang, L. R., and Wallace, S., 1993, Polyamino acid microspheres—preparation, characterization and distribution after intravenous injection in rats, *Int. J. Pharm.* **94:**143–152.

Lindholm, T. S., and Gao, T. J., 1993, Functional carriers for bone morphogenetic proteins, *Ann. Chir. Gynaecol* **82**(S207): 3–12.

Marchalheussler, L., Sirbat, D., Hoffman, M., and Maincent, P., 1991, Nanocapsules containing beta-blocking agents—a new drug delivery system for the eye—application to the medical treatment of glaucoma in rabbits, *J. Fr. Ophthal.* **14:**371–375.

Miller, R. A., Brady, J. M., and Cutright, D. E., 1977, Degradation rates of oral resorbable implants (polylactates and polyglycolates): Rate modification with changes in PLA/PGA copolymer ratios, *J. Biomed. Mater. Res.* **11:**711–719.

Nieuwenhuis, J., 1992, Synthesis of polylactides, polyglycolides and their copolymers, *Clin. Mater.* **10:**59–67.

Ogawa, Y., Yamamoto, M., Okada, H., Yashiki, T., and Shimamoto, T., 1988a, A new technique to efficiently entrap leuprolide acetate into microcapsules of polylactic acid or copoly(lactic/glycolic) acid, *Chem. Pharm. Bull.* **36:**1095–1103.

Ogawa, Y., Yamamoto, M., Okada, H., Yashiki, T., and Shimamoto, T., 1988b, Controlled-release of leuprolide acetate from polylactic acid or copoly(lactic/glycolic) acid microcapsules: Influence of molecular weight and copolymer ratio of polymer, *Chem. Pharm. Bull.* **36:**1502–1507.

Ogawa, Y., Yamamoto, M., Okada, H., Yashiki, T., and Shimamoto, T., 1988c, *In vivo* release profiles of leuprolide acetate from microcapsules prepared with polylactic acids or copoly(lactic/glycolic) acids and *in vivo* degradation of these polymers, *Chem. Pharm. Bull. (Japan)* **36:**2576–2581.

O'Hagen, D. T., Jeffery, H., and Davis, S. S., 1993, Poly(lactide-*co*-glycolide) microparticles as controlled release vaccines, *Proc. Int. Symp. Control. Rel. Bioact. Mater.* **20:**59–60.

Okada, H., Ogawa, Y., and Yashiki, T., 1987, Prolonged release microcapsule and its production, U.S. Patent 4,652,441.

Okada, H., Inoue, Y., Heya, T., Ueno, H., Ogawa, Y., and Toguchi, H., 1991, Pharmacokinetics of once-a-month injectable microspheres of leuprolide acetate, *Pharm. Res.* **8:**787–791.

Park, T. G., Cohen, S., and Langer, R., 1992a, Poly(L-lactic acid)/pluronic blends: Characterization of phase separation behavior, degradation, and morphology and use as protein-releasing matrices, *Macromolecules* **25:**116–122.

Park, T. G., Cohen, S., and Langer, R., 1992b, Controlled protein release from polyethyleneimine-coated poly(L-lactic acid)/pluronic blend matrices, *Pharm. Res.* **9:**37–39.

Pavanetto, F., Genta, I., Giunchedi, P., and Conti, B., 1993, Evaluation of spray drying as a method for polylactide and polylactide-coglycolide microsphere preparation, *J. Microencapsulation* **10:**487–497.

Pikal, M. J., 1990, Freeze-drying of proteins, Part I: Process design, *Biopharm* **3:**18–22.

Pikal, M. J., Dellerman, K., and Roy, M. L., 1991, Formulation and stability of freeze-dried proteins: Effects of moisture and oxygen on the stability of freeze-dried formulations of human growth hormone, *Dev. Biol. Stand.* **74:**21–26.

Pulapura, S., Li, C., and Kohn, J., 1990, Structure–property relationships for the design of polyiminocarbonates, *Biomaterials* **11:**666–678.

Randolph, T. W., Randolph, A. D., Mebes, M., and Yeung, S., 1994, Sub-micron sized particles of poly(L-lactic acid) via the gas antisolvent spray precipitation process, *Biotech. Prog.* **9:**429–435.

Reid, R. H., Boedeker, E. C., McQueen, C. E., Davis, D., Tseng, L. Y., Kodak, J., and Sau, K., 1993, Preclinical evaluation of microencapsulated CFA/II oral vaccine against enterotoxigenic *E. coli, Vaccine* **11:**159–167.

Ripamonti, U., Ma, S., and Reddi, A. H., 1992, The critical role of geometry of porous hydroxyapatite delivery system in induction of bone by osteogenin, a bone morphogenetic protein, *Matrix* **12**:202–212.

Ron, E., Turek, T., Mathiowitz, E., Chasin, M., Hageman, M., and Langer, R., 1993, Controlled release of polypeptides from polyanhydrides, *Proc. Natl. Acad. Sci. USA* **90**:4176–4180.

Sánchez, A., Vila-Jato, J. L., and Alonso, M. J., 1993, Development of biodegradable microspheres and nanospheres for the controlled release of cyclosporin A, *Int. J. Pharm.* **99**:263–273.

Santiago, N., Milstein, S., Rivera, T., Garcia, E., Zaidi, T., Hong, H., and Bucher, D., 1993, Oral immunization of rats with proteinoid microspheres encapsulating influenza antigens, *Pharm. Res.* **10**:1243–1247.

Shah, N. H., Railkar, A. S., Chen, A. S., Chen, F. C., Tarantino, R., Kumar, S., Murjani, M., Plamer, D., Infeld, M. H., and Malick, A. W., 1993, A biodegradable injectable implant for delivering micromolecules and macromolecules using poly(lactic-*co*-glycolic) acid (PLGA) copolymers, *J. Controlled Release* **27**:139–147.

Shieh, L., Tamada, J., Tabata, Y., Domb, A., and Langer, R., 1994, Drug release from a new family of biodegradable polyanhydrides, *J. Controlled Release* **29**:73–82.

Singh, M., Singh, O., Singh, A., and Talwar, G. P., 1992, Immunogenicity studies on diphtheria toxoid loaded biodegradable microspheres, *Int. J. Pharm.* **85**:R5–R8.

Stjarnkvist, P., Degling, L., and Sjholm, I., 1991, Biodegradable microspheres. 13. Immune response to the DNP hapten conjugated to polyacryl starch microparticles, *J. Pharm. Sci.* **80**:436–440.

Stoeckemann, K., and Sandow, J., 1993, Effects of the luteinizing-hormone-releasing hormone (LHRH) antagonist ramolrelix (hoe013) and the LHRH agonist buserelin on dimethylbenz[*a*]anthracene-induced mammary carcinoma: Studies with slow-release formulations, *J. Cancer Res. Clin. Oncol.* **119**:457–462.

Tabata, Y., Gutta, S., and Langer, R., 1993, Controlled delivery systems for proteins using polyanhydride microspheres, *Pharm. Res.* **10**:487–496.

Takaoka, K., Koezuka, M., and Nakahara, H., 1991, Telopeptide-depleted bovine skin collagen as a carrier for bone morphogenetic protein, *J. Orthop. Res.* **9**:902–907.

Tice, T. R., Gilley, R. M., Eldridge, J. H., Staac, J. K., Hollingshead, M. G., and Shannon, W. M., 1991, Method of potentiating an immune response, U.S. Patent 5,075,109.

Visscher, G. E., Robinson, R. L., Maulding, H. V., Fong, J. W., Pearson, J. E., and Argentieri, G. J., 1985, Biodegradation of and tissue reaction to 50:50 poly(DL-lactide-*co*-glycolide) microcapsules, *J. Biomed. Mater. Res.* **19**:349–365.

Visscher, G. E., Robinson, R. L., and Argentieri, G. J., 1987, Tissue response to biodegradable injectable microcapsules, *J. Biomater. Appl.* **2**:118–131.

Wheatley, M. A., Chang, M., Park, E., and Langer, R., 1991, Coated alginate microspheres—factors influencing the controlled delivery of macromolecules, *J. Appl. Polym. Sci.* **43**:2123–2135.

Yamakawa, I., Tsushima, Y., Machida, R., and Watanabe, S., 1992, Preparation of neurotensin analogue-containing poly(dl-lactic acid) microspheres formed by oil-in-water solvent evaporation, *J. Pharm Sci.* **81**:899–903.

Chapter 2

Degradable Controlled Release Systems Useful for Protein Delivery

Kathleen V. Roskos and Richard Maskiewicz

1. INTRODUCTION

In the very short time since their emergence, peptide and protein pharmaceuticals have created a renaissance in controlled drug delivery. The dosing of protein-based therapeutic agents by various routes has focused on four areas for potential efficacy enhancement (Lee, 1991): Sustained release—Release of the active agent, although slower than with conventional formulations, is still substantially affected by the external environment into which it is released; Controlled release—The release profile is predominantly controlled by the design of the system itself and may provide a near zero-order profile; Pulsatile release—The release profile occurs in multiple, discrete, and controlled "pulses" of time following a single injection; and Release of drug in a temporal manner consistent with the biochemistry and pathology of the disease state.

For polypeptides having short *in vivo* half-lives, controlled release systems can also offer protection from proteolytic degradation.

There are several major types of controlled release device designs. One of the distinct benefits of using a degradable matrix in the preparation of a controlled release system is that these polymeric vehicles require no subse-

---

Kathleen V. Roskos and Richard Maskiewicz • Matrix Pharmaceutical, Inc., Fremont, California 94555.

Protein Delivery: Physical Systems, Sanders and Hendren, eds., Plenum Press, New York, 1997.

quent surgical removal of the device once the therapeutic agent is depleted. The selection of the matrix of choice must take into account the particular application and factors such as the cost, the potency, and the properties of the active agent, environment and site of use, and the requirements for degradation time and release rate. The high molecular weight and secondary, tertiary, and quaternary structure of proteins pose unique, basic formulation challenges. The physical restrictions of potential low membrane and/or tissue diffusivity are compounded by the instability of many proteins during and after incorporation into hydrophobic matrices and in concentrated aqueous solutions often present in delivery systems (Pitt, 1990a), reinforcing the need to mitigate their tendency to bind to hydrophobic surfaces. In addition to the potential loss of biologic activity due to conformational changes and partial denaturation of proteins, even subtle changes, for example, in vaccines, which do not *per se* destroy biological activity can result in altered immunogenicity due to exposure of new epitopes.

Because of these restrictions, novel strategies for the predictable, controlled delivery of proteins are required. The high molecular weight and hydrophobicity of proteins cause the release rates obtained by predominantly diffusional processes to frequently be too slow in synthetic hydrophobic polymers to allow practical application. Significant current research activity therefore centers around the achievement of enhanced release by erosional breakdown of the polymer matrix.

Several excellent review articles have recently been published outlining the various strategies for controlled delivery of proteins (and peptides)—discussed with an emphasis on the difficulties associated with protein delivery, namely, the low permeability and rapid proteolysis of proteins and their denaturation due to interaction with the delivery vehicle itself (Holland *et al.*, 1986; Leong and Langer, 1987; Eppstein and Longenecker, 1988; Sanders, 1990; Pitt, 1990a; Heller, 1993a).

During the past several years, researchers have synthesized and developed many biodegradable polymers and employed them in several device designs useful for both small molecules and macromolecular therapeutic agents. A number of natural biopolymer and lipid-based matrices have also been examined for delivery of polypeptides. Our brief discussion in this chapter provides a qualitative description of release patterns derived from the more common delivery systems.

Initial work in this area sought to produce monolithic systems with release being degradation-, dissolution-, or diffusion-controlled. Monolithic systems are characterized by having a more or less uniform dispersion of the therapeutic agent within the polymer or lipid matrix. In this device design, the active agent is essentially immobilized in the matrix until released by the degradation or disruption of the surrounding diffusion-limiting material. At the present time, it is clear that the number of polymers that degrade by a surface mechanism, as is desirable for this type of device, is relatively small. Monolithic lipid matrices can, on the other hand, be dispersed/disrupted by either surface or bulk processes. In more recent work, the emphasis has been placed on diffusion-controlled monolithic systems in which the active agent is released by diffusion concurrent with or prior to the degradation of the polymer matrix. The polymeric matrix, in this case, erodes only after its delivery role is completed. Further, descriptions of diffusion-controlled reservoir systems, in which the active agent is encapsulated by a rate-controlling membrane through which the agent escapes by diffusion, can be found in the literature, and examples are provided in Section 4.

The dissolution or solubilization of a polymeric matrix and subsequent drug release is effected by hydrolytic or enzymatic cleavage of the backbone of the polymer. The cleavage can also occur in bridging bonds, rendering soluble an initially cross-linked polymer. Alternatively, dissolution of a drug-containing matrix may also originate from hydrolysis, ionization, or protonation of the side chains of the polymer. Each type of polymer erosion has certain advantages and limitations, and the choice of a particular mechanism is dictated by the specific application. Several excellent general articles and books have been published discussing various polymers useful in controlled delivery (Chasin, and Langer, 1990; Heller, 1987). For lipid matrices, dissolution is typically a physical rather than a chemical process.

The transformation of a solid implant into water-soluble (or dispersable) material is best described by the term "erosion." This process is associated with changes in the appearance of the device (i.e., deformation, structural disintegration, or swelling), as well as changes in the physical properties of the matrix itself (decrease in molecular weight, increase in porosity, and increase in bulk water content). Two mechanisms of erosion have been described, "bulk erosion" and "surface erosion" (Baker, 1987). In "bulk erosion," water penetrates into the matrix at a rate that exceeds the rate at which the polymer hydrolyzes or lipid dissolves, and consequent erosion occurs uniformly throughout the matrix (also sometimes termed homogeneous erosion; Baker, 1987). During "surface erosion," water penetrates into the matrix at a rate slower than the rate of polymer hydrolysis or lipid dispersion, and erosion occurs preferentially at the surface of the matrix (or heterogeneously; Baker, 1987). However, the rate of hydrolysis of most polymers is not constant and is usually not confined to the surface layers of a polymeric device. Rather, the erosion of a device usually occurs by some combination of these mechanisms.

Finally, implantable degradable matrices must incorporate the following desirable characteristics: the systems must be well tolerated by the body; they must provide well-defined (reproducible) *in vivo* degradation, dissolution, and/or drug diffusion rates; and they must degrade *in vivo* to well-defined nontoxic and readily metabolized or excreted products. In addition, these types of vehicles must maintain these characteristics after purification to remove toxic impurities or residual chemicals used in their preparation.

The objective of this chapter is to review degradable materials, including polymers, and the resulting delivery systems fabricated from them that are useful for the delivery of proteins and peptides. Owing to the diverse nature of the subject area, we have chosen to divide the chapter into sections on hydrophobic synthetic polymers, hydrophilic polymeric biomaterials, and hydrophobic nonpolymeric biomaterials. Each section seeks to briefly highlight the chemistry and characteristics of the polymer or matrix and provide recent examples of their use in the delivery of proteins.

2. DEFINITIONS

In the field of biomaterials research, as in other multidisciplinary research areas, much of the terminology employed has not been precisely defined. Thus, when a careful review of the literature is conducted, a clear consensus on the meaning of "degradation," "biodegradation," "bioabsorbable," "bioresorption," or "bioerosion" cannot be readily established. Recently, efforts have been made to establish generally accepted definitions for all aspects of biomaterials research. In conformance with the usage suggested by the Consensus Conference of the European Society for Biomaterials (Williams, 1987), polymer *degradation* is defined here as a deleterious change in the properties of a polymer due to a change in the chemical structure. *Biodegradation* will be referred to when we wish to emphasize that a biological agent (e.g., enzyme or microbe) is a dominant component in the degradation process. *Bioabsorption* and *bioresorption* will be used interchangeably and are often used to imply that the polymer or its degradation products are removed by cellular activity (e.g., phagocytosis) in a biological environment.

However, in concurrence with Heller's suggestion (Heller, 1987), *bioerosion* is herein defined as changes in polymer or matrix structure that occur under *physiologic* conditions as a consequence of a chemical reaction, dissolution of a water-soluble polymer, dissolution of a water-insoluble lipid, or dissolution of a polymer promoted by ionization or protonation of functional groups. Bioerosion therefore includes both physical processes (i.e., dissolution) and chemical processes (i.e., polymer backbone cleavage).

3. SYNTHETIC HYDROPHOBIC DEGRADABLE POLYMERS

3.1. Poly(lactic acid), Poly(glycolic acid), and Their Copolymers

The linear polyesters, particularly copolymers of glycolic and lactic acids (Scheme I), are currently the most widely investigated degradable

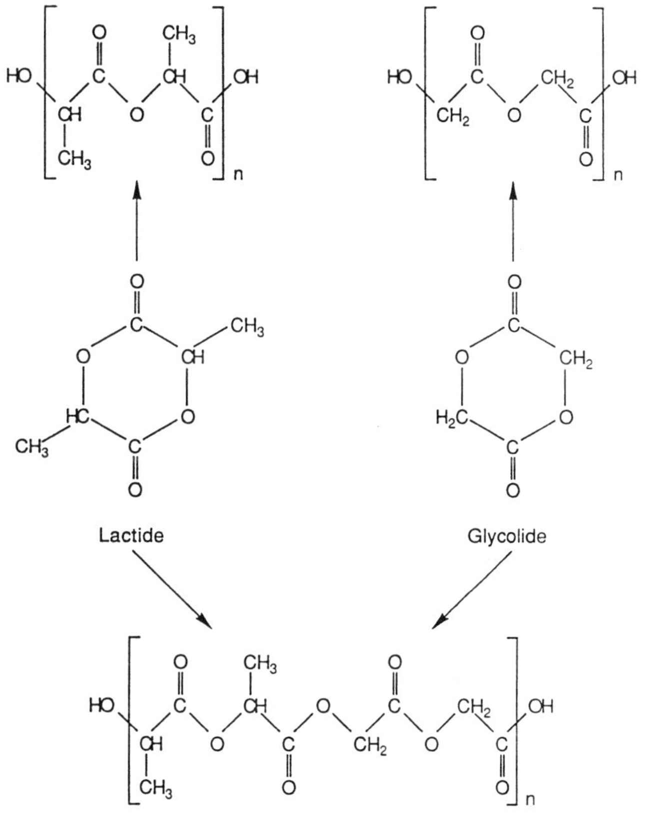

Lactide

Glycolide

Scheme I

polymers. The family of homo- and copolymers derived from these monomers has received considerable attention in recent years, primarily due to their favorable biocompatibility and toxicological characteristics, ease of fabrication, predictability of *in vivo* degradation kinetics, and regulatory approval in commercial suture applications (Lewis, 1990). While the initial data obtained for these polymers resulted from studies aimed at suture applications (Frazza and Schmitt, 1971; Brady *et al.*, 1973), interest in the use of these polymers in drug delivery applications has increased dramatically in the last 10 years. This is particularly true at the present time owing to relatively wide commercial availability of the lactide/glycolide polymers as off-the-shelf items.

Lactic acid is optically active and can be produced as poly(L-lactide), poly(D-lactide), and the racemic poly(D,L-lactide). Polylactides (PLA) are soluble in common organic solvents (e.g., halogenated hydrocarbons, ethyl acetate, tetrahydrofuran; Medisorb Technologies International, 1990). The presence of the methyl group of lactic acid produces a more hydrophobic polymer than that produced upon polymerization of glycolic acid. Polyglycolides (PGA) are highly crystalline solids (Lewis, 1990) with a high melting point and are almost insoluble in common solvents. When randomly copolymerized (30–50%) with polylactide, the resulting copolymer (PLGA) retains physical properties more readily amenable to processing (those of a low-melting thermoplastic with good solubility in common solvents) (Lewis, 1990). The preferred method of producing high-molecular-weight homopolymers or copolymers is via the ring-opening melt condensation polymerization of the cyclic diester—lactide or glycolide—using a suitable catalyst (Kulkarni *et al.*, 1971; Dittrich and Schulz, 1971).

An advantage of the lactide/glycolide copolymers is the well-documented versatility in polymer properties (via manipulation of comonomer ratio and polymer molecular weight) and corresponding performance characteristics (predictable *in vivo* degradation rates). These have been extensively reviewed elsewhere (Lewis, 1990). The 50:50 DL-lactide/glycolide copolymer is the vehicle of choice for many drug delivery systems designed for a 30-day duration of action using this polymer system.

Degradation of the aliphatic polyesters occurs predominantly by bulk erosion (Baker, 1987). The lactide/glycolide polymer chains are cleaved by hydrolysis to their monomeric acids, which in the body are eliminated via the Krebs cycle (Brady *et al.*, 1973). The role of enzymatic involvement in the biodegradation of these materials remains controversial—with early literature concluding that spontaneous hydrolysis was the sole mechanism of degradation. Further work supported the conclusion that little or no enzymatic involvement is expected in the early stages of degradation, while the polymers are in the glassy state. However, enzymes could potentially play a role in degradation for polymers in the rubbery state (Holland *et al.*, 1986).

A distinct advantage of lactide/glycolide materials for use in drug delivery is their relative flexibility of fabrication. In general, delivery systems based on implants and microparticles predominate. Microparticles are generally produced by encapsulation methodologies incorporating solvent evaporation or phase separation (coacervation) techniques (Lewis, 1990; Arshady, 1991). Both techniques, however, require solubility of the polymer in an organic solvent. Implants are generally fabricated by compression molding and melt processing methods (injection molding or screw extrusion). As with all hydrolytically unstable polymeric materials, it is extremely important to thoroughly dry the bulk polymer and the bioactive agent and to provide a carefully maintained ultradry processing environment throughout device fabrication. However, with regard to melt processing methods, the thermal stability of the active agent is of critical importance, with many macromolecules of interest being unstable under the conditions employed (Lewis, 1990).

Sterilization of these delivery systems, via aseptic processing and terminal sterilization, can be used to produce acceptably sterile products. This subject has been adequately reviewed by Lewis (1990). Briefly, both of these methods appear suitable for products based on PLGA copolymers—if proper care and early design considerations are exercised in processing. Aseptic processing is feasible primarily because of the adequate solubility of PLGA copolymers in organic solvents, allowing filter sterilization of drug/polymer solutions in a clean-room environment. This method has proved particularly useful with microencapsulated products, which almost always involve solutions of the polymer in organic solvents. Further, some macromolecules may prove sensitive to terminal sterilization—rendering aseptic processing as the only alternative. Ethylene oxide (EO) has been found to plasticize some lactide polymer compositions, in addition to the potential problem of EO residue in the delivery system (Lewis, 1990). Gamma irradiation has proven to be useful for PLGA formulations; however, for each new drug delivery product, the appropriate dose of gamma irradiation must be determined, and many drug compounds are themselves not stable to irradiation. It has been reported that gamma irradiation decreases the inherent viscosity of PLGA copolymers and increases the degradation rate, with drug release kinetics often changed (increased) (Lewis, 1990). This only serves to highlight the importance of adequate consideration of sterilization procedures early in the development of these drug delivery systems.

Macromolecules and peptides including bovine serum albumin (BSA; Bodmer *et al.*, 1992; Cohen *et al.*, 1991; Wang *et al.*, 1991; Sah and Chien, 1993), human serum albumin (Hora *et al.*, 1990), calcitonin (Lee *et al.*, 1991), bovine growth hormone (Lewis, 1990), horseradish peroxidase (Cohen *et al.*, 1991), interleukin-2 (Hora *et al.*, 1989), interferon-β (Eppstein, 1986), insulin (Kwong *et al.*, 1986), lypressin (Maulding, 1987), lysozyme (Tabata *et al.*, 1993a), luteinizing hormone-releasing hormone (LHRH) analogs (Anik *et al.*, 1984; Sanders *et al.*, 1984, 1985, 1986; Hutchinson and Furr, 1985, 1987, 1990, 1991; Furr and Hutchinson, 1992; Okada, 1989; Okada *et al.*, 1988, 1991; Ogawa *et al.*, 1988; Asano *et al.*, 1989a, b, 1991), and nerve growth factor (Camarata *et al.*, 1992) have been formulated in PLA or PGA homopolymers or PLGA copolymers as controlled release formulations.

Early work in this area was published in several articles by Sanders and co-workers, who described the use of poly(D,L-lactide-*co*-glycolide) (PLGA) microcapsules for the delivery of the peptide nafarelin acetate, an LHRH analog [(D-Nal(2)^6, Aza-Gly^{10})-LHRH; Syntex] (Anik *et al.*, 1984; Sanders *et al.*, 1984, 1985, 1986). Microcapsules were produced by a coacervation microencapsulation technique, and the release of the peptide was shown to be predominately triphasic (Fig. 1) — high initial release lasting for several hours, followed by a period of low or limited release, and finally several days

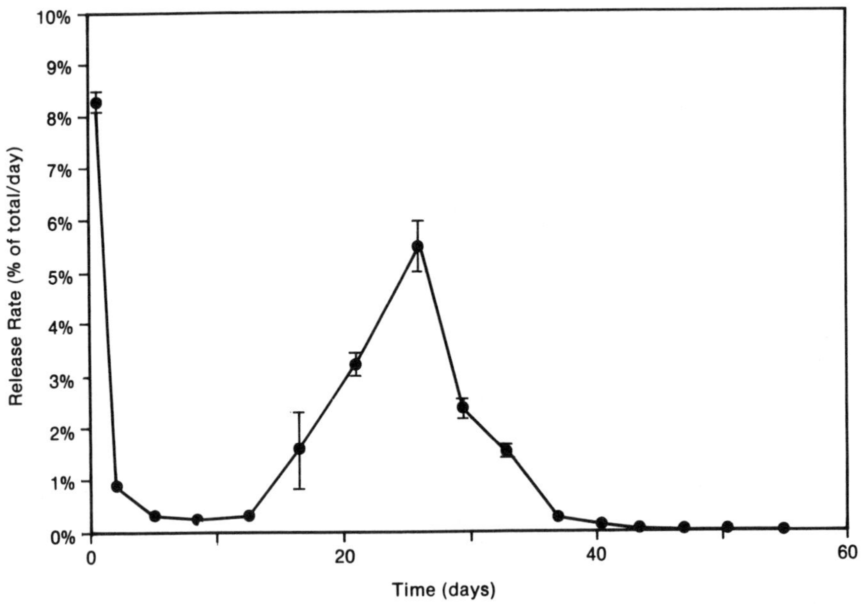

Figure 1. *In vitro* release of nafarelin from 45:56 poly(DL-lactide-*co*-glycolide) microspheres. Conditions: Agitated in ethanolic phosphate buffer (pH 7.4) at 37 °C. (After Sanders *et al.*, 1985.)

of measurable release. The initial phase was attributed to rapid diffusional release of the peptide from the surface of the spheres; the second phase was presumably due to depletion of the surface drug with ongoing bulk hydrolysis of the polymer, which progressed to the point of erosion; the final breakup of the spheres allowed adequate porosity of the matrix for final diffusional release of the peptide in the third phase. As with other ester copolymers, the rates of polymer chain scission and onset of matrix erosion were determined by initial polymer weight and copolymer ratios (Pitt, 1990a). The biologic activity, as measured by estrus suppression of regularly cycling mature female rats, is also typical of a triphasic release profile. Estrus is completely suppressed during initial LHRH release, followed by normal cycling during low or limited LHRH release, and finally estrus is again suppressed when release of the peptide resumes. Careful experimentation and choice of polymer properties eventually resulted in a delivery system that provided a minimum initial and secondary phase, allowing continuous efficacy of estrus suppression in the rat for greater than 8 months and minimal efficacy beyond 15 months (Sanders *et al.*, 1986).

Similar work has been published for the LHRH analog goserelin [(D-Ser(Bu^t)^6, Azgly^{10})-LHRH; Zoladex®, Zeneca (Hutchinson and Furr, 1985, 1987, 1990, 1991; Furr and Hutchinson, 1992). In addition to estrus suppression, these authors demonstrated that a 50:50 PLGA 28-day implant (1 mm × 3–6 mm rod) containing goserelin was effective in causing regression of hormone-responsive dimethylbenzanthracene-induced rat mammary carcinoma and in inhibiting growth of hormone-responsive transplantable Dunning R3327 rat prostate adenocarcinoma. Previous studies had demonstrated that efficacy of Zoladex®, in these two hormone-dependent rat tumor models was equivalent to that of the peptide administered by daily injection. Zoladex® depot is approved in the United States for the palliative treatment of advanced carcinoma of the prostate [Physicians' Desk Reference (PDR), 1994]. In addition, Zoladex® depot is approved for the management of endometriosis, including pain relief and reduction of endometric lesions for the duration of therapy (PDR, 1994).

Okada and co-workers (1988, 1989, 1991) and Ogawa *et al.* (1988) have also described a 75:25 PLGA microencapsulated 30-day release system for the delivery of another LHRH analog—leuprolide acetate [(D-Leu^6, Pro^{9}NEt)-LHRH; Lupron Depot®, Takeda-Abbott]. Their results indicate that a single injection of microcapsules maintains constant and effective serum levels of the analog for one month. Sufficient therapeutic efficacy in the treatment of advanced prostatic cancer has also been shown for leuprolide acetate, and the product has recently been approved for use by the U.S. Food and Drug Administration (Lupron Depot®, 7.5 mg; PDR, 1994). This treatment has also shown to cause a dramatic regression of growth of experimental rat endometriosis, leading to approval of the one-month release system for management of endometriosis in humans (Lupron Depot®, 3.75 mg; PDR, 1994).

A potential problem encountered with the use of high-molecular-weight lactide/glycolide copolymers as drug delivery matrices is the presence of residual catalyst used in the polymerization procedure. Further, the processing and fabrication of protein delivery systems often requires the use of solvents and high temperatures. Asano *et al.* (1989a, b, 1991) have prepared interesting low-molecular-weight homo- and copolymers of L-lactic acid with D-lactic acid, L-lactic acid with glycolic acid, and DL-lactic acid in the absence of catalysts. Other waxy-type copolyesters have also been prepared (Fukuzaki *et al.*, 1989; Imasaka *et al.*, 1991). The polymers were prepared by heating the appropriate monomers, under nitrogen in a sealed tube at 200 °C, in the absence of catalysts. Implants containing leuprolide acetate [des-Gly^{10}-(D-Leu^{6})-LHRH ethylamide acetate] were easily prepared in melt-pressed cylindrical copolymer implants at relatively low temperatures ($\sim$ 70 °C); however slightly higher temperatures are needed to process some poly(DL-lactic acid) blends. *In vivo* release of the LHRH analog was demonstrated over a period of several weeks.

Recently, degradable systems designated "injectable implant" delivery systems have been described (Dunn *et al.*, 1991; Duysen *et al.*, 1992; Radomsky *et al.*, 1993; Duysen *et al.*, 1993; Yewey *et al.*, 1993). These low-viscosity systems are readily injectable solutions or dispersions containing polymer and drug dissolved (or dispersed) in a biocompatible organic solvent. Upon injection through a standard needle into an aqueous environment (i.e. the subcutaneous space), the solvent is diluted by the surrounding water and the water-insoluble degradable polymer precipitates to trap the bioactive agent in a solid implant. This solution-based system has been trademarked ATRIGEL™ (Atrix Laboratories) and is described in more detail later in this text. This system has been studied using poly(DL-lactide), poly(DL-lactide-*co*-glycolide), and poly(DL-lactide-*co*-caprolactone) as the polymer and *N*-methylpyrrolidone as the biocompatible organic solvent to incorporate and deliver ganirelix acetate (another LHRH analog) and various cytokines.

Similarly, Shah *et al.* (1993) have used 50:50 poly(DL-lactide-*co*-glycolide) as the polymer but have expanded the organic solvents useful in this type of system to include triacetin and triethyl citrate in order to study the implant and release characteristics of dispersed myoglobin and cytochrome *c*. While this type of system has inherent advantages over traditional implants (i.e., obviating the need for microsurgery for implantation) and injectable microsphere formulations (i.e., since, once injected, microsphere retrieval and removal has limited success), the successful use of this system is restricted to appropriate drug delivery applications—injection into a uniformly restricted body compartment—allowing reproducible surface area development and release characteristics, upon precipitation of the implant. A thorough summary describing this type of system is found in Chapter 3 of this volume.

Another area of investigation is the use of PLA homopolymers (Lovell *et al.*, 1989; Miyamoto *et al.*, 1992) or PLGA copolymers (Schmitz and Hollinger, 1988; Hollinger and Battistone, 1986; Hollinger *et al.*, 1990; Kenley *et al.*, 1993) in the delivery of osteoinductive bone morphogenetic proteins (BMPs) for promotion of bone regeneration in orthopedic applications. Various BMPs are all present in bone matrix and may function in a complex synergy to stimulate osseous regeneration (Kenley *et al.*, 1993). It is known that, in the absence of other BMPs, recombinant human bone morphogenetic protein-2 (rhBMP-2) effectively regenerates osseous tissue in a bony site (Toriumi *et al.*, 1991; Yasko *et al.*, 1992). Recently, Hollinger and colleagues have examined PLGA copolymers for the delivery of rhBMP-2. Delivery of this protein from PLGA/rhBMP-2 implantable semisolids formulated with allogeneic blood clot or other undisclosed thickening agents resulted in osseous regeneration in (critical-size) rat calvarial defects, indicating that sustained release of active rhBMP-2 is possible from lactide/glycolide copolymeric delivery systems.

3.2. Polycaprolactone

Polycaprolactone (**1**), a semicrystalline polymer with high solubility and low melting point (59–64 °C), is reported to have surface eroding properties and has shown utility for controlled release of steroids (Pitt, 1990b). The toxicology of polycaprolactone has been extensively studied, and it is currently regarded as a nontoxic and tissue-compatible material, degrading to the monomeric hydroxyacid (Pitt, 1992). Whereas this polymer is very permeable to low-molecular-weight drugs ($M_w < 500$), diffusion of peptides and proteins through polycaprolactone is too slow for practical

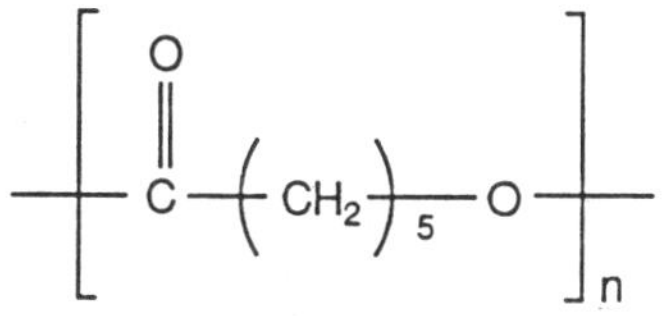

$$\left[\overset{O}{\overset{\|}{C}} - (CH_2)_5 - O \right]_n$$

(**1**)

application. This limitation was circumvented by the preparation of polycaprolactone capsules with controlled porosity [via a leaching procedure (Pitt, 1990b)]. Using this preparation technique, a constant rate of 40–50 μg/day per centimeter of length was maintained for the peptide [D-Trp^6, des-Gly^{10}]-LHRH diethylamide for >60 days. (Pitt *et al.*, 1987; Schindler, 1987).

3.3. Poly(hydroxybutyrate), Poly(hydroxyvalerate), and Their Copolymers

Poly(hydroxybutyrate) (PHB), poly(hydroxyvalerate) (PHV), and their copolymers (**2**) are other examples of degradable polyesters. These materials, derived from the bacterium *Alcaligenes eutrophus*, are intracellular storage polymers providing reserve energy for the organisms. PHB homopolymer is a highly crystalline material, usually producing highly brittle polymers. Copolymers of PHB and PHV are more flexible, less crystalline, and easier to fabricate (Holmes, 1985), with the rate of degradation being controlled by the copolymer composition (Holland *et al.*, 1987; Doi *et al.*, 1990). Degradation of these materials yields hydroxybutyric acid and hydroxyvaleric acid. These polymers are commercially available under the trade name Biopol®, (ICI Biological Products, U.K). PHB has been formulated with buserelin [(D-Ser(Bu^t)^6, Pro^9 NEt)-LHRH; Hoechst], another LHRH analog, in a degradable implant. The fabrication procedure uses an organic solvent (i.e., methanol) to mix the compound and the polymer. Following drying, this coarse combination is compression-molded to produce "tablets." This system has shown utility in rats (Konig *et al.*, 1985) and clinically (Waxman *et al.*, 1985) for the controlled release of buserelin. A significant advantage of the PHB/PHV system is the relatively low cost of very pure material, owing to its being made by fermentation. However, the slow rate of degradation and consequent long life *in vivo* have limited its practical utility to date.

$$\left[-C(=O)-CH_2-CH(CH_3)-O- \right]\left[-C(=O)-CH_2-CH(CH_2CH_3)-O- \right]$$

(**2**)

3.4. Poly(ortho esters)

Major activity in the area of protein delivery from synthetic degradable matrices has focused on the use of aliphatic polyesters, prinicipally owing to their readily manipulated polymer characteristics and the favorable toxicology of their degradation products. However, the development of new polymers and polymeric delivery systems where drug release is predominantly controlled by surface polymer hydrolysis is also desirable. Poly(ortho esters) (POE), of the general structure **3**, are an example of this class of synthetic erodible polymers and have been under extensive development for approximately 20 years.

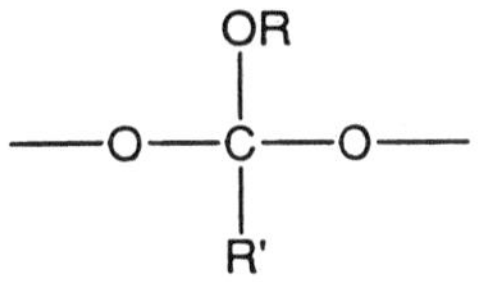

OR

—O—C—O—

R'

(3)

Initial work in this area described polymers prepared with an ortho ester linkage in their backbone (Choi and Heller, 1978a, b, 1979). These proprietary materials, designated by the trade name Alzamer (Alza Corporation), have been recently reviewed (Heller *et al.*, 1990a; Heller, 1993b).

Further effort resulted in the development of a second major family of POE polymers based on the addition of polyols to diketene acetals (Heller *et al.*, 1980), prepared by the procedure shown in Scheme II.

Cross-linked polymeric materials can also be easily produced from these polymers. The therapeutic agent and cross-linking triol(s) are mixed into the ketene acetal-terminated prepolymer, most commonly a viscous

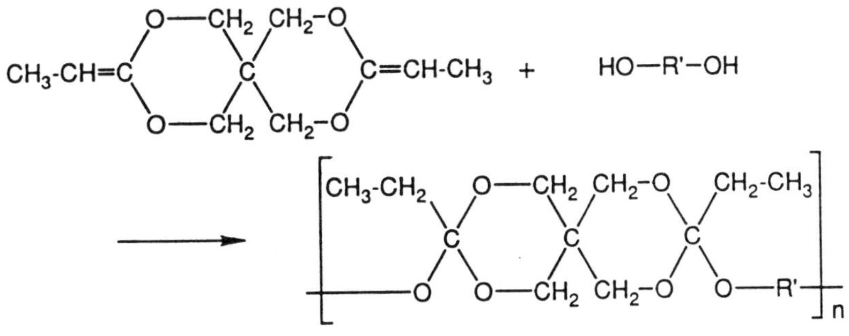

CH_3-CH=C(O—CH_2)(O—CH_2)C(CH_2-O)(CH_2-O)C=CH-CH_3 + HO—R'—OH

⟶ [CH_3-CH_2 C (O—CH_2 CH_2-O) C (O—CH_2 CH_2-O) C CH_2-CH_3 —O O—R']$_n$

Scheme II

liquid at room temperature, and the mixture cross-linked at low temperatures (as low as 40 °C) (Heller *et al.*, 1990a). This relatively mild processing allows incorporation of thermally sensitive therapeutic agents into a solid polymer. However, owing to the presence of reactive ketene acetal groups on the prepolymer, therapeutic agents containing hydroxyl groups will be covalently attached to the matrix via ortho ester bonds (Heller, 1987b).

While the pure polymers degrade by bulk hydrolysis, the degradation pattern can be substantially altered through the inclusion of additives (basic or acidic salts) to suppress bulk hydrolysis and promote surface erosion. The rate of surface erosion of POEs is controlled in part by the hydrophobicity of the polymer composition and, in some cases, the crosslink density—increasing the cross-link density also serves to eliminate diffusional loss of drug. The hydrolysis of the ortho ester linkage is acid-catalyzed; therefore, inclusion of basic excipients [e.g., $Mg(OH)_2$] physically incorporated into the polymeric matrix has been used to stabilize the interior of the matrix, causing degradation to occur preferentially at the surface of the polymeric matrix. Conversely, addition of acidic excipients or the use of copolymerized acidic monomers causes the matrix to degrade faster. Finally, addition of neutral salts can also alter the erosion rate of the polymer; inclusion of salts such as sodium chloride will significantly accelerate the degradation rate, perhaps owing to increased osmotic imbibition of water into the matrix (Baker, 1987). Hydrolysis of these materials to constituent products has been extensively reviewed (Heller *et al.*, 1990a; Heller, 1993b). Briefly, initial hydrolysis produces neutral products—the diol (or mixture of diols used in the reaction with the diketene acetal) and a pentaerythritol diester. These pentaerythritol esters slowly undergo further hydrolysis to pentaerythritol and the corresponding aliphatic acid.

A third class of POEs has been recently described (Heller *et al.*, 1990b). Preparation of the polymer is conducted by reacting a triol containing two vicinal hydroxyl groups and one hydroxyl group at least four methylene groups away (e.g., 1,2,6-hexanetriol) with a triethylorthoalkyl ester (**4**). The intermediate does not have to be isolated, and continuous reaction produces

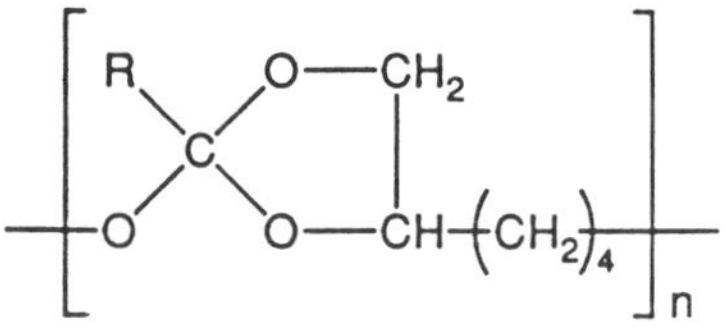

(**4**)

a polymer. These materials have an ointmentlike consistency at room temperature, and properties such as viscosity and hydrophobicity can be varied by controlling the molecular weight of the polymer and the size of the alkyl R– group. The potential use(s) of these POEs is described in Section 3.7. These materials also undergo initial hydrolysis at the labile ortho ester linkage to generate neutral products, i.e., one or more esters of the triol used in synthesis. These esters slowly undergo further hydrolysis to produce a carboxylic acid and the original triol, as previously described. It is unlikely that the carboxylic acid produced will catalyze hydrolysis of the acid-sensitive POE, since the low-molecular-weight triol monoesters produced in the initial hydrolysis will presumably diffuse away from an implant site before further hydrolysis can occur.

As with other hydrolytically unstable polymers, these highly hydrophobic materials should be stored at ambient temperature with the exclusion of moisture. The materials must be processed and packaged with rigorous control of the environmental conditions, preferably with a relative humidity of less than 20% at 70 °C (Heller *et al.*, 1990a)—with lower relative humidity leading to improved device properties. Prior to fabrication, polymer, drug, and any excipient employed must be dried, and the final device packaged in a dry environment using packaging material exhibiting high resistance to moisture diffusion (Heller *et al.*, 1990a).

Processing of linear POE materials is conducted by standard techniques used for nonerodible thermoplastic polymers and has been reviewed elsewhere (Tadmor and Gogos, 1979). When drug, catalyst, and other additives are compounded in the melt, the possibility of reactions between catalyst, polymer, and drug must be considered. Since these procedures are usually carried out at relatively high temperatures, protein stability issues predominate when these methods are used. Alternatives to compounding in the melt include solution mixing or powder blending of dry solids; however, protein stability and protein/polymer interaction must be carefully evaluated for each individual case.

The release of the potent LHRH analog nafarelin has also been studied using a cross-linked POE (Heller *et al.*, 1987). For this approach, a semisolid ketene acetal-terminated prepolymer was prepared as previously described (Heller *et al.*, 1980). Following this, the LHRH analog and the cross-linking agent (1,2,6-hexanetriol) were mixed into the prepolymer at room temperature and in the absence of solvents. The viscous paste was then extruded into a mold, and the mixture cross-linked (40 °C) for 24 hours. As previously stated, a disadvantage of this incorporation procedure is that the prepolymer contains ketene acetal end groups which can react with hydroxyl (and amine) groups of proteins or peptides in addition to the hydroxyl groups of the triol cross-linker. Some proteins can therefore

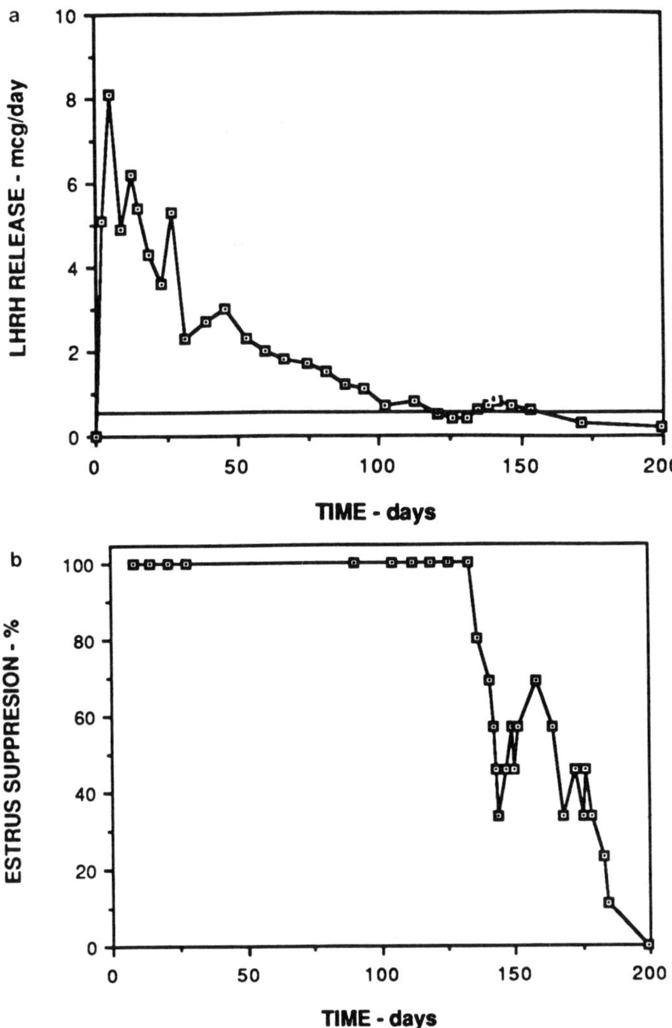

Figure 2. (a) Daily release of nafarelin from a cross-linked poly(ortho ester) (POE). Crossbar denotes minimum daily requirement of 0.4 μg/day for estrus suppression. Conditions: 7.4 mm × 20 mm rods prepared from a 3-methyl-1,5-pentanediol prepolymer cross-linked with 1,2,6-hexanetriol and placed into pH 7.4 buffer at 37 °C. (b) Percent of female rats showing estrus suppression as a function of treatment time with nafarelin-containing POE rods. Devices were prepared from a 3-methyl-1,5-pentanediol prepolymer cross-linked with 1,2,6-hexanetriol and contained 3 wt.% LHRH analog. (After Heller *et al.*, 1987.)

become chemically bound to the matrix. However, ultimate hydrolysis of the POE matrix will also liberate the unchanged protein.

Figure 2a shows the daily *in vitro* release rate of nafarelin from a cross-linked POE polymer (Heller *et al.*, 1987). The kinetics of release indicate a burst of drug released in the first days of testing, with the minimum level of 0.4 μg/day necessary for estrus suppression in rats maintained for approximately 130 days. These results are in good agreement with results in Fig. 2b, which shows data obtained in *in vivo* studies of estrus suppression, indicating release of active LHRH analog.

3.5. Polyanhydrides

The synthesis of polyanhydrides, another class of surface eroding polymers, was first published by Buchner and Slade (1909) and further explored by Hill and Carothers (1932). These polymers were initially probed as possible substitutes for polyesters in textile applications but ultimately were found unsuitable due to inherent hydrolytic instability (Conix, 1958; Yoda, 1963). Langer and co-workers began to explore polyanhydrides as early as 1983, seeking to exploit the inherent hydrolytic instability for drug delivery applications (Chasin *et al.*, 1990; Rosen *et al.*, 1983; Leong *et al.*, 1985, 1986a, b, 1987). The versatility of the polyanhydride system is based on the large differences in hydrolysis rates between aliphatic and aromatic polyanhydrides; aliphatic polyanhyrides degrade within days, whereas selected aromatic polyanhydrides degrade slowly over several years, allowing synthesis of copolymers with a wide range of erosion rates. Langer and co-workers have synthesized aliphatic–aromatic copolymers which exhibit intermediate rates of degradation depending on starting monomer composition. Figure 3 illustrates this concept for copolymers of sebacic acid (SA) and bis(*p*-carboxyphenoxy)propane (PCPP) (**5**).

These polymers are capable of undergoing a hydrolysis process primarily confined to the surface of devices. Domb and Langer (1987) have indicated that the hydrolysis of anhydride linkages is inhibited by the presence of acids; bulk erosion of these materials is therefore suppressed by

$$-\left[\overset{O}{\overset{\|}{C}}-(CH_2)_8-\overset{O}{\overset{\|}{C}}-O\right]_m-\left[\overset{O}{\overset{\|}{C}}-C_6H_4-O-(CH_2)_3-O-C_6H_4-\overset{O}{\overset{\|}{C}}-O\right]_n-$$

(**5**)

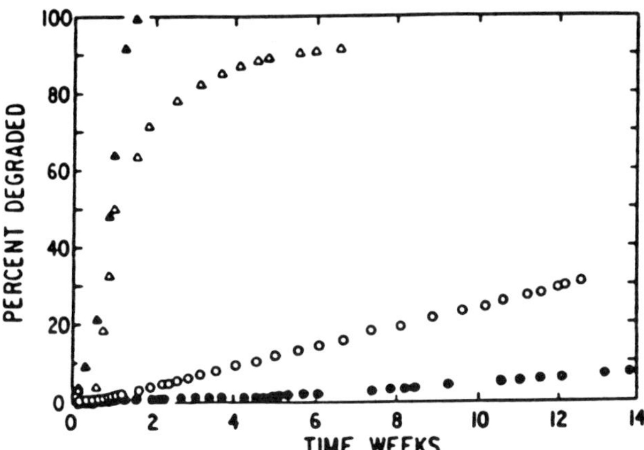

Figure 3. Degradation profiles of compression-molded poly[(bis-*p*-carboxyphenoxy)propane (PCPP) anhydride] and its copolymers with sebacic acid (SA). Conditions: 1.4-cm-diameter disks, approximately 1 mm thick, placed into 0.1 M phosphate buffer, pH 7.4 at 37 °C. Cumulative percentage of the polymer that degraded was measured by absorbance at 250 nm (Leong *et al*., 1985.)

the acidity of the carboxylic products produced upon hydrolysis, and the erosion process occurs preferentially toward the outer polymeric surface.

These materials are somewhat reactive and hydrolytically unstable—resulting in polymer properties having both advantages and disadvantages. Owing to their high rate of degradation, many polyanhydrides degrade by surface erosion without the need for incorporated excipients. However, under some processing conditions (high temperature processing at $>150\,^{\circ}C$), these materials will react with drugs containing free amino groups or other nucleophilic moieties.

Recently, a new family of polyanhydrides has been explored (Domb *et al*., 1987; Tabata and Langer, 1993). These copolymers are based on fatty acid dimers derived from oleic and sebacic acids **(6)** and are termed

$$\left[\overset{O}{\overset{\|}{C}}-(CH_2)_7-\underset{\underset{CH_3}{|}}{\overset{\overset{CH_3}{|}}{\underset{}{CH}}}\!\;-CH-(CH_2)_8-\overset{O}{\overset{\|}{C}}-O\right]_m\left[\overset{O}{\overset{\|}{C}}-(CH_2)_8-\overset{O}{\overset{\|}{C}}-O\right]_n$$

(CH with $(CH_2)_7CH_3$ branch; CH with $(CH_2)_8CH_3$ branch)

(6)

poly(fatty acid dimer) (PFAD)–sebacic acid (SA) copolymers [P(FAD-SA)]. These materials seek to obviate the brittleness and possible fragmentation encountered with the aromatic diacid polyanhydride copolymers upon exposure to water, greatly reducing the release rate of incorporated water-soluble bioactive molecules at high drug loadings.

Drug-loaded polyanhydride polymeric devices are easily prepared by injection and compression molding (Leong *et al.*, 1986a) or "hot melt" (Mathiowitz and Langer, 1987), solvent removal (Mathiowitz *et al.*, 1988), or double-emulsion (Tabata and Langer, 1993; Tabata *et al.*, 1993b) microencapsulation techniques. Briefly, in "hot melt" encapsulation, microspheres are prepared by mixing the drug and melted polymer—suspending the mixture in a hot nonmiscible solvent—and cooling the spheres until solid. In the solvent removal method, the protein is dispersed in a polymer solution prepared in a volatile organic solvent. This mixture is then suspended in an organic oil, and the organic solvent extracted into the oil, with the microspheres collected by filtration. The double-emulsion techique also uses solvent evaporation. In this technique, an aqueous solution of the protein is added to a methylene chloride solution of the polymer. This mixture is emulsified and poured into a vigorously stirring aqueous solution of 1% poly(vinyl alcohol) saturated with methylene chloride; the resulting double emulsion is then stirred extensively, enabling complete evaporation of the methylene chloride.

The incorporation of a variety of proteins into polyanhydride microspheres has been examined, including insulin (Mathiowitz *et al.*, 1985; Mathiowitz and Langer, 1987), bovine somatotropin (Ron *et al.*, 1989), chondrogenic stimulating proteins (Lucas *et al.*, 1990), and several enzymes (Chasin *et al.*, 1990). These matrices have been extensively characterized *in vitro* and efforts for *in vivo* characterization continue (Chasin *et al.*, 1990).

Toxicological evaluation of these materials, including (but not limited to) mutation and teratogenic assays, rabbit corneal implants, and rat implants (in brain and subcutaneous tissue), has indicated that, in general, the polyanhydrides exhibit excellent *in vivo* biocompatibility for several copolymers (Leong *et al.*, 1986a; Chasin *et al.*, 1990). Recently, polyanhydrides prepared from bis(*p*-carboxyphenoxy)propane and sebacic acid have been accepted by the U.S. FDA for human clinical trials (Langer and Moses, 1991).

The polymers represented by **6** have been used to incorporate proteins of different molecular sizes into microspheres by a double-emulsion technique (Tabata *et al.*, 1993b). The proteins—lysozyme, trypsin, heparinase, ovalbumin, BSA, and immunoglobulin—were primarily incorporated into a 25:75 fatty acid/sebacic acid copolymer at a loading of 2 wt %. The microspheres produced by this method were spherical, irrespective of the

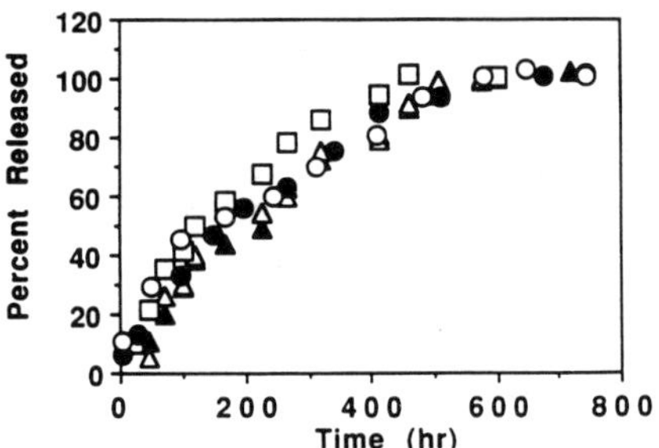

Figure 4. Release of proteins from [P(FAD-SA)] 25:75 microspheres: ○, lysozyme; ●, trypsin; △, ovalbumin; ▲, BSA; □, immunoglobulin, $M_w = 42{,}900$. Conditions: Release of proteins from microspheres into 0.1 M phosphate buffer, pH 7.4 at 37 °C; 2% protein loading. (After Tabata *et al.*, 1993b.)

type of proteins encapsulated. All proteins were released at a near-constant rate without any large burst for up to three weeks (Fig. 4). While relatively gentle, this method of protein encapsulation did ultimately result in loss of enzymatic activity for one protein: For trypsin-loaded microspheres, trypsin lost 40% of its activity during the microsphere preparation, primarily as a result of the sonication process. Activity studies demonstrated that reducing the period of ultrasound exposure reduced the loss of protein activity to approximately 20%.

Use of a hydrophobic polyanhydride to control the release of an unstable protein has also been examined (Ron *et al.*, 1993). Poly[1,3-bis(*p*-carboxyphenoxy)hexane] has been used as a delivery matrix for bovine somatotropin (BST). The instability of BST presents an intriguing problem for controlled release technology. Perturbation of the tertiary structure of BST can result in partially unfolded intermediates and potential formation of insoluble aggregates (Brems, 1988). Additionally, chemical instability of BST in aqueous environment can lead to deamidated, chain-cleaved, and covalently bonded oligomeric products (Hageman *et al.*, 1992). To attempt to overcome the instability problems encountered with this protein, particles of polymer (130–150 μm), lyophilized protein (150 μm), and sucrose as a stabilizer were mixed thoroughly in dry powder form at a loading of 10 wt.%. Compression-molded disks of polyanhydride with coarsely dispersed drug were fabricated at 25 °C using a Carver press. The authors indicated that this processing method was chosen over solvent casting for two reasons: (1) to avoid possible organic solvent denaturation of the proteins, and (2) to limit the interaction between protein amine groups and the carbonyl groups of the polyanhydride, thus allowing release of the BST over approximately 2 weeks. The released protein appeared to maintain its tertiary integrity as assessed by acidic reversed phase high-pressure liquid chromatography (HPLC), size-exclusion HPLC, radioimmunoassay, and conformation-sensitive immunoassay.

3.6. Polyphosphazenes

Polyphosphazenes are a group of inorganic polymers, having the general molecular structure represented by **7**. The backbone consists of

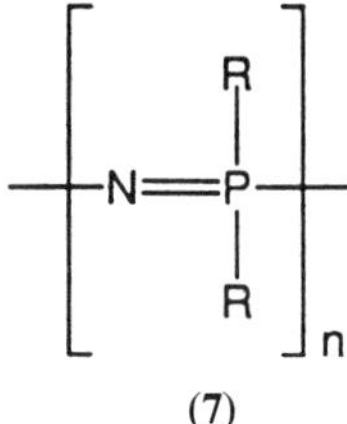

(7)

nitrogen–phosphorus bonds, and the hydrolytic instability or stability of these polymers is determined by changes in the side groups attached to the backbone.

Synthetic approaches to produce polymers with desirable biomedical characteristics for this class of materials have been extensively reviewed (Allcock, 1990; Crommen *et al.*, 1993). Poly[(amino acid ester) phosphazenes] are known to be susceptible toward hydrolytic degradation and hold promise as degradable materials. Recently, Laurencin *et al.* (1987) used a poly(imidazole methylphenoxy) phosphazene to study the release characteristics of BSA. Protein release was demonstrated using ^{14}C-labeled BSA in a 20% imidazole-substituted polyphosphazene. Release from this matrix consisted of an initial burst of almost 25% of the protein, followed by release over several hundred hours in which a total of 55% of the protein was released. Polymer degradation for the 20% imidazole-substituted polyphosphazene was also studied and found to be quite slow, with 4% of the polymer degraded in 600 hr.

3.7. Delivery of Vaccines

The relatively recent explosion of the recombinant DNA field has led to the identification, cloning, expression, and large-scale production of many previously unavailable proteins, including antigens and vaccines. The newest approaches to vaccine development yield vaccines with several advantages over more traditional immunogens; they are chemically well defined, can be prepared reproducibly and assayed readily, and are usually relatively inexpensive to manufacture. However, a general drawback is poor immunogenicity, which results in the need for repeated immunizations. Increased production of antigen and vaccine molecules has created a need for potent immunological adjuvant/delivery systems to boost the immune response of the host to these "clean," isolated antigens (Wise *et al.*, 1987).

The vaccine delivery systems under development should deliver the immunogen in such a manner as to enhance the response of the immune system and to achieve a long-lasting effect in a single administration, obviating the need for a return visit for multiple or "booster" injections. Incorporation of the immunogen into the delivery system should be compatible with the stability considerations of the particular immunogen. As in conventional vaccines, it is obligatory that the immunogen be properly stabilized prior to release and that the epitope sites that evoke protective immunity be presented to the immune apparatus in an optimal manner. One of the most attractive features of polymeric-based drug-delivery systems is that the carrier materials can be engineered to release the vaccines in a sustained fashion or in a predetermined sequence such that the vaccine is released within precise and well-controlled periods, thus enabling "pulses" of drug release to be created, in a manner sufficient to stimulate lasting immunity. If successful, this approach would enable single-dose mass vaccination of those at risk of exposure to unusual pathogens (usually military personnel) and enhance the effectiveness of worldwide vaccination programs against conventional diseases, such as tetanus, malaria, and other diseases endemic in the Third World, where logistical considerations and follow-up visits all too often present a major obstacle to successful vaccination programs (Aguado, 1993).

Antigen delivery systems (such as microcapsules) are widely regarded as carriers with immunoadjuvant properties. Most of these dosage forms exhibit adjuvant properties based on their ability to release the antigen over a protracted period of time relative to when the antigen is delivered in a "free" form. Whereas other systems are capable of modulating or stimulating the immune system—in addition to possessing sustained or controlled release properties. Khan *et al.* (1994) have recently published an excellent review article outlining current approaches for single-step immunization—the controlled release of antigens to induce prolonged immunity following a single dose.

Polymers shown to possess immunostimulatory properties in the form of their degraded products are the polyiminocarbonates (Kohn *et al.*, 1986), based on a variation of classical polycarbonates (**8**). The only structural

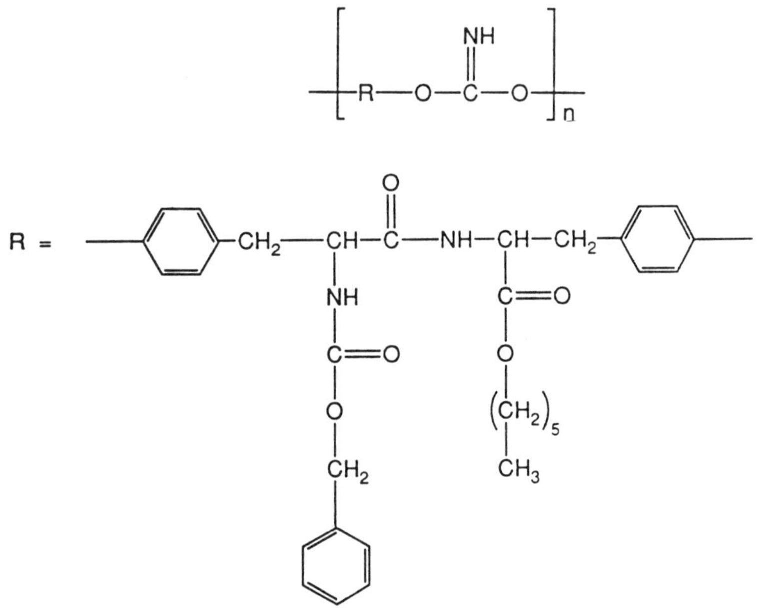

poly(CTTH-iminocarbonate)

(8)

difference between carbonates and iminocarbonates is the replacement of the carbonyl oxygen by an imino group.

Implantable antigen delivery devices were prepared (via solvent casting) using poly(CTTH-iminocarbonate) polymer (**8**) and BSA. *In vitro* release profiles showed sustained release of BSA (10% loading) from this polymer device for greater than 600 hr. Devices were implanted subcutaneously into mice, and anti-BSA antibody titers were followed over the course of 56 weeks. Mean anti-BSA antibody titers in the animals treated with poly(CTTH-iminocarbonate) were significantly higher than those in the control group (iminocarbonate polymeric implant with no adjuvant properties). Histologic examination confirmed that the higher antibody titers were not caused by the immunostimulatory effect of a strong local inflammatory response to the implant material itself. An erosion product of the polymer, CTTH, was found to be as potent as Freund's complete adjuvant and muramyl dipeptide in enhancing the immune response.

Further, certain polyiminocarbonate polymers have also shown the potential for delayed release of a small molecule, *p*-nitroaniline (pNA) (Pulapura *et al.*, 1990). The release profile of pNA from poly(bisphenol

A-iminocarbonate) had a lag period of about 70 days, during which time no release of pNA occurred. The release of pNA started abruptly and continued in a linear fashion for about 80 days. The onset of pNA release coincided with the onset of significant water uptake into the device and corresponding degradation to low-molecular-weight residues. Although no protein release *per se* has been studied, this unusual release profile may have the potential for exploitation in the design of delayed-release vaccine delivery systems.

Lactide/glycolide copolymers are another class of polymers under development as an injectable vaccine delivery system. Again, this class of polymers has the proven advantage of biocompatibility and provides the ability to control and predict the rate at which the material is released. Microencapsulated formulations based on these synthetic polymers are under active development with immunogens entrapped within the polymer matrix, providing a system that can be readily administered by injection. Several model antigens have been encapsulated using PLA or PLGA polymer-based systems [such as diphtheria toxoid (Singh *et al.*, 1991a, 1992), formalinized staphylococcal enterotoxin B (fSEB) (Eldridge *et al.*, 1991), ovalbumin (O'Hagan *et al.*, 1991), ricin toxoid (Yan *et al.*, 1993), tetanus toxoid (Raghuvanshi *et al.*, 1993; Hazrati *et al.*, 1992; Alonso *et al.*, 1993), synthetic human chorionic gonadotropin subunit (Stevens *et al.*, 1992), and malarial antigen (Bathurst *et al.*, 1992)]. Of the antigens studied, only fSEB and the toxoids are altered protein preparations and thus present minimal stability problems following polymer incorporation.

One system receiving considerable attention is the pulsatile release of fSEB (Eldridge *et al.*, 1991) from microspheres made from PLGA. This system was designed to provide distinct "pulses" of antigen release after injection of a mixture of vaccine-containing microspheres of various sizes and degradation times, providing discrete primary and booster doses following a single injection. A mixture of two microsphere size distributions, 1–10 μm and 20–125 μm, containing fSEB encapsulated in a (50:50) copolymer of PLGA induced both a primary and an anamnestic anti-SEB response following single-dose administration. Microspheres of <10 μm were reportedly phagocytosed and released the antigen at an accelerated rate, providing the primary response, while the larger microspheres released the antigen at a slower rate, resulting in a stimulatory secondary response in the mice.

The technique for fabrication of the previously described vaccine delivery systems requires the use of organic solvents for incorporation of the antigen into the polymeric matrix. Whereas some antigens may prove relatively robust and may not be adversely affected by this type of treatment, some immunogens may be more labile and require more delicate processing.

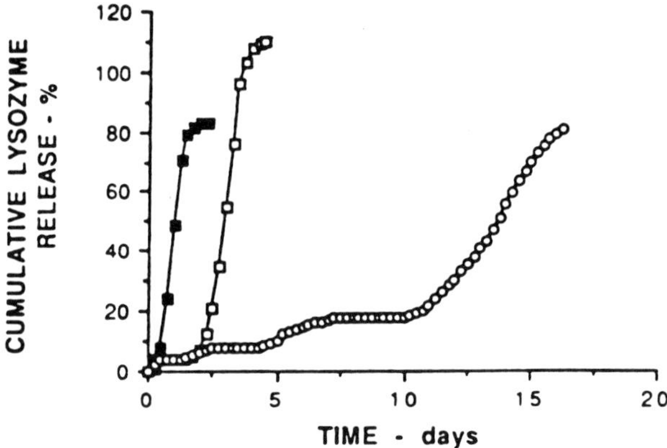

Figure 5. Effect of alkyl group on release of lysozyme from ointmentlike poly(ortho esters) (POE). R-groups and molecular weights are as follows: ■, methyl, 5350; □, propyl, 4600; ○, pentyl, 5500. Conditions: 0.1 M phosphate buffer, pH 7.4, at room temperature; 5 wt.% lysozyme loading. (After Wuthrich *et al.*, 1992.)

The effect of exposure to organic solvents is likely to be both antigen- and solvent-dependent and cannot be predicted *a priori*. For this reason, Heller and co-workers (Wuthrich *et al.*, 1992) investigated an ointmentlike hydrophobic polymer into which antigens could be incorporated at room temperature and without the use of solvents. The polymer synthesis and characterization have been previously described (Heller *et al.*, 1990b). Solid therapeutic agents can be readily incorporated into the material by a simple mixing procedure. These workers showed a delayed release of a model protein, lysozyme, mixed into an acetate derivative of the polymer. The delay time could be reproducibly controlled by varying the molecular weight or by varying the nature of the substituent R– group, as seen in Fig. 5. However, it should be noted the release experiments were performed at room temperature. When polymer molecular weight is followed concurrently with lysozyme release, the data are consistent with a bulk polymer hydrolysis mechanism of release. Since lysozyme is a fairly large molecule, diffusional release of the protein is restricted from the intact polymer. However, at the first appearance of polymer degradation product, the monoester of 1,2,6-hexanetriol, lysozyme is released due to polymer solubilization (Heller, 1993a). Current studies are aimed at achieving a "pulsed" release by encapsulating these polymers—having different delay times—within a macroporous, erodible membrane to obtain the desired pulsatile release profile at the physiologic temperature of 37 °C.

4. HYDROPHILIC POLYMERIC BIOMATERIALS AND HYDROPHOBIC NONPOLYMERIC BIOMATERIALS

4.1. General Properties

For the purposes of this review, biodegradable polymeric biomaterials are defined as nonsynthetic hydrophilic macromolecules, primarily polypeptides and polysaccharides, which are natural substrates for specific enzymes, and where in most cases the enzymes are endogenous in mammals. Biodegradable lipids are defined as nonsynthetic, nonpolymeric hydrophobic substances which can slowly dissolve or disperse in an aqueous intercellular environment and which can be metabolized by pathways typically employed in lipid processing. Another defining and advantageous aspect of polymeric biomaterial- or lipid-based matrices for delivery and sustained release of proteins is the very high biocompatibility of the natural or slightly modified molecules employed, compared to that of the synthetic polymers. Immunogenicity is, however, a potential issue for several of the polymeric biomaterials that have been employed for protein delivery.

Degradation of proteins during incorporation into and fabrication of a delivery system, instability during storage and during *in vivo* release, and incomplete release due to binding to matrix components can be significant problems for delivery systems using synthetic polymers. The mild conditions, in terms of temperature and solvents (aqueous buffers), employed during preparation of polymeric biomaterial-based delivery systems can minimize the degradation or inactivation of labile proteins. The fact that most preparation procedures employ physical rather than chemical processes during protein incorporation also minimizes drug loss. An environment providing reasonably acceptable stability during storage and release is often achieved due to either the high water contents of the matrices employed or the hydrophilic nature of polymeric biomaterials utilized as matrix components.

Several of the polymeric biomaterials that can potentially be employed for delivery and sustained release of protein drugs have the ability to form ordered linear aggregates, yielding microscopic and/or submicroscopic fibers. Dispersion of either individual biomaterial molecules or agglomerated systems in aqueous solution can yield a gel matrix, with a mesh size dependent on polymer or fiber loading. Mesh size can be defined as the average distance between neighboring chains of biomaterial or rods of fiber in a matrix confining a diffusing protein molecule. Depending on the mesh size and dimensions of the polypeptide drug being delivered, sustained release via hindered diffusion can be achieved. Mesh size can also be reduced by cross-linking, employing either natural chemistry, as in the case of fibrin, or synthetic cross-linking agents such as glutaraldehyde.

Lipid-based systems can achieve sustained release by providing a barrier to diffusion of water, necessary for dissolution of solid protein dispersed in a hydrophobic matrix, and by a subsequent barrier to diffusion or convection of protein solution entrapped within the matrix. The rate-limiting step for aqueous solute release is often dissolution or disintegration of the lipid mass or particle. Factors affecting release of proteins from various matrices have been reviewed in detail by Pitt (1990a) and Park *et al.* (1993).

The gels that can be formed using polymeric biomaterials (such as collagen) can achieve high viscosities and exhibit useful rheological behavior such as shear thinning, allowing parenteral injection in addition to implantation. Sustained release of proteins (in terms of time required for redistribution of a bolus dose into the systemic circulation) in high-viscosity systems can be achieved by the resistance to displacement or dispersal of the drug solution within the matrix, due to the mechanical strength of the high-viscosity gel. High-viscosity gels can also be formed in lipid systems by incorporating either high-melting-point ($>37\,^{\circ}C$) species or water to promote lipid structuring.

4.2. Specific Hydrophilic Polymeric Biomaterials

4.2.1. ALBUMIN

Human serum albumin has been extensively used for the preparation of microspheres providing sustained drug release. One of the earlier examples of the utility of albumin microspheres for the delivery of protein drugs is the work of Goosen *et al.* (1982). *In vivo* release of insulin from microbeads (prepared via incorporation of insulin crystals into an albumin matrix prior to cross-linking with 1% glutaraldehyde) was examined in rats. Sustained levels of insulin exceeding 20 μU/ml were measured for 3 months after subcutaneous administration of 20 mg of microbeads containing 200 mg of insulin per gram of matrix. Peak levels did not exceed 60 μU/ml, and complete biodegradation was achieved within 5 months. A study conducted with a comparable system by Royer *et al.* (1983) showed a prolonged release phase for 30 days following intramuscular injection in rabbits, after a burst release of about 70% of the insulin dose within 7 min. No inflammation or pyrogenic response was observed after repeated injection of cross-linked microspheres consisting of bovine serum albumin into calves. More recently, Bhargava and Ando (1992) incorporated urokinase in glutaraldehyde-cross-linked albumin microspheres and showed that enzymatic activity was maintained but decreased with increased cross-link density.

The use of albumin as a component of heparin-containing microspheres cross-linked with carbodiimide has been described by Feijen (1990). Release of model protein drugs such as lysozyme and ovalbumin was shown to occur *in vitro* for up to 100 h. Investigation of the mechanism of protein retention within and release from albumin–heparin microspheres was conducted by Kwon *et al.* (1992). Incorporation of lysozyme into albumin and albumin–heparin matrices occurred via ion-exchange-mediated adsorption, with protein loading into albumin–heparin microspheres being threefold greater than that for albumin particles. Measured "apparent" diffusion coefficients for lysozyme of $2.1 \times 10^{-12}\,cm^2/sec$ in albumin–heparin microspheres and $3.9 \times 10^{-11}\,cm^2/sec$ for albumin microspheres indicate that release is adsorption/desorption-limited and independent of true diffusion. Fifty percent release of incorporated lysozyme (7% loading) occurred within 70 hr for albumin microspheres, whereas less than 25% release occurred from the microspheres containing heparin plus albumin over the same time period.

4.2.2. COLLAGEN

Collagen is a three-subunit polypeptide forming a triple helix, which can associate into fibers of various sizes and degrees of structural order. Such fibers are a component of the connective tissue matrix, and, as such, protein delivery systems based on collagen can be expected to be highly biocompatible and biodegradable, with hydrolysis occurring over several months due to endogenous collagenases. The collagen typically employed for drug delivery is bovine derived and has been enzymatically processed to minimize immunogenicity. The hydrophilic nature of collagen and the high water content of gels consisting of collagen fibers may also lead to delivery systems in which problems with drug binding and stability would be comparable to the behavior of a given protein in simple solution.

Collagen gels consisting of dispersions of fibrils possess viscoelastic properties characterized by shear thinning. This allows the preparation of high-viscosity matrices containing protein drugs in aqueous solution which can be readily extruded via fine-gauge needles during parenteral administration. General molecular and physicochemical properties of collagen have been reviewed by Piez (1985), Miller and Gay, (1987), and Wallace *et al.* (1988).

Collagen gels containing proteins and other solutes can also be dried to yield implantable disks or rods. Release profiles in such systems can be modified by the rate and extent of matrix hydration prior to protein diffusion out of the resultant gel. Since collagen gels typically possess water contents exceeding 80%, achieving sustained release of proteins via hindrance of diffusion can be problematic. Diffusional release rates of macromolecules from collagen gels as a function of mesh size, i.e., collagen structure and fibril size distribution, have been studied by Rosenblatt *et al.* (1989). Diffusivities of chymotrypsinogen (MW 25,000) and fibrinogen (MW 400,000) were measured in fibrillar (large fiber) collagen containing 9.5×10^{14} fibers/ml and randomly structured succinylated collagen containing 7.6×10^{16} fibers/ml. Hindered diffusion through fibrillar collagen was observed for fibrinogen [4.6×10^{-8} cm^2/sec in the matrix versus 2.0×10^{-7} cm^2/sec in water (23 °C)] but not for chymotrypsinogen. Diffusivity of the smaller protein was, however, reduced—2.0×10^{-7} cm^2/sec in the matrix versus 9.5×10^{-7} cm^2/sec in water (37 °C)—when incorporated in the collagen having a higher fiber density.

Diffusion coefficients of a series of proteins of increasing molecular weight were studied by Gilbert *et al.* (1988) in collagens having different structures, hydration states, and glutaraldehyde cross-link densities. Diffusion coefficients of lysozyme (MW 14,500) through a matrix of nonfibrillar (i.e., small fiber size) collagen were found to decrease from 1.4×10^{-7} cm^2/Usec (23 °C) at a 2% cross-link density to 5.6×10^{-10} cm^2/sec at 45% cross-linking. For fibrillar collagen at an equivalent (45%) cross-link density, the diffusion coefficient for lysozyme was found to increase to 8.2×10^{-7} cm^2/sec, and remained high (3.40×10^{-7} cm^2/sec) even for proteins as large as BSA (MW 66,000). Other studies by Gilbert and Kim (1990) have examined collagenase-mediated degradation of collagen as a function of both unmodified collagen structure and cross-link density of glutaraldehyde-modified matrix. Native (fibrillar) collagen was found to hydrolyze more rapidly than randomly oriented associations of small fibers. As expected, rates of degradation decreased with increasing percentage of cross-links. *In vivo* biocompatibility of cross-linked collagen was also examined and found to be comparable to that of surgical Dacron.

Utility of collagen for sustained release of protein drugs was demonstrated by Weiner *et al.* (1985), who reported that enhanced retention of intramuscularly administered human growth hormone was measured in rats, upon incorporation of liposomal hormone in a dilute collagen gel. Increased duration of serum insulin levels in diabetic rats occurred with insulin-containing liposomes entrapped in 0.9% collagen, relative to liposomal insulin alone. Hori *et al.* (1989) showed that subcutaneous injection of insulin in a collagen gel increased plasma levels in rats (over a period of 60 min), relative to injection of free insulin solution. Gradual release of bone morphogenic protein from fibrillar collagen resulting in a stimulation of osteogenesis has been described for lyophilized collagen disks by Horisaka *et al.* (1994). Similar results have been described by Takaoka *et al.* (1991), Lucas *et al.* (1989), and Deatherage and Miller (1987).

Several other examples of sustained delivery or retention of pharmaceutically active proteins have been described in the patent literature. Enhanced wound healing via delivery of polypeptide growth factors from a lyophilized wafer consisting of soluble (nonfibrillar) collagen soaked into a cross-linked gelatin sponge has been described by Song and Morawiecki (1993). Fifty percent of biologically active platelet-derived growth factor was released within 3 hr from a gelatin sponge. The time for equivalent release increased to 14 hr for wafers impregnated with 4% collagen solution and to 20 hr for wafers that had been lyophilized.

Yamahira *et al.* (1984, 1988, 1989) have shown that sustained release of both interferon-α and granulocyte-macrophage colony-stimulating factor can be achieved for lyophilized collagen compression-molded into rods/pellets or pulverized and suspended in oil. Steady-state (>30 IU/ml) serum levels of interferon were achieved in rabbits for >48 hr after intramuscular injection of 10^6 U/kg. Steady-state interferon levels exceeding 500 U/ml were measured for 72 hr in mice after subcutaneous injection. Follow-up patents by Fujioka *et al.* (1989, 1993) describe a 25% release of growth hormone-releasing factor after 14 days (*in vitro* into phosphate-buffered saline) from compression-molded rods of lyophilized collagen containing release-modifying amino acids.

Collagen, either as a lyophilized matrix or an aqueous gel, appears to hold significant promise as a delivery system for protein drugs. For lower-molecular-weight proteins, the large mesh size of gels consisting of fibrillar collagen (which typically exists in a physiological solute environment) results in diffusivities comparable to those in bulk water. Sustained release of very high molecular weight proteins via hindered diffusion is, however, possible. Use of nonphysiological collagens, where mesh size is reduced owing to higher densities of smaller fibers or to cross-linking, may affect hindered diffusion of smaller proteins.

4.2.3. FIBRINOGEN/FIBRIN

Fibrinogen is a highly water soluble protein present in blood that is responsible for (and the primary component of) trauma-induced blood clots. Clot formation occurs via a thrombin-catalyzed conversion of fibrinogen to fibrin, with subsequent polymerization and cross-linking of fibrin "monomers" (catalyzed by factor XIII) to yield a mechanically strong fibrillar gel. Final gel structure and properties can be influenced by other serum proteins such as fibronectin. Drug delivery systems based on a bolus injection or implant of human fibrinogen/fibrin would be expected to be fully biocompatible. Biodegradation should also be rapid owing to the ubiquitous presence of plasminogen/plasmin. General molecular and physicochemical properties of fibrinogen and fibrin have been reviewed (Doolittle, 1973, 1984).

An early study demonstrating the use of fibrin as a delivery system for macromolecules was the work of Kawamura and Urist (1988). The study showed a fivefold increase in bone yield in mice upon administration of bone morphogenic protein via lyophilized fibrin pellets, relative to the effect produced by the implantation of an equivalent dose (5 mg) of the lyophilized growth factor alone. The release of several model protein drugs from dried fibrin sheets has also been investigated (Senderoff *et al.*, 1991). Levels of α-lactalbumin, ovalbumin, and lysozyme large enough to be detectable by HPLC were shown to permeate out of rehydrated sheets, after 24 hr at 37 °C.

The role of solute molecular weight on the permeability of proteins through fibrin membranes was recently documented by Ho and Chen (1993). Diffusion coefficients of 1.1×10^{-7} and 1.7×10^{-8} cm^2/sec were measured for lysozyme (MW 14,500) and BSA (MW 66,000) at 25 °C in a 6% fibrin gel. These values are smaller than those obtained for the same solutes in cross-linked fibrillar collagen (5% fiber content Gilbert, 1988), suggesting that the average mesh size in natural fibrin gels is smaller than that for native fibrillar collagen even when highly cross-linked. Subsequent cross-linking of fibrin films with glutaraldehyde retarded permeation further, with a larger effect being measured for higher-molecular-weight proteins. More extensive use of fibrin for protein drug delivery can be anticipated in the future, based on the high fiber loading achievable with an aqueous dispersion, with concomitant reduction in mesh size to a level at which hindrance of macromolecule diffusion can be realized.

4.2.4. GELATIN

Gelatin consists of denatured collagen in which the subunits of the triple helix have dissociated and have been reduced to random-coil conformations. Gelatin therefore shares the biocompatibility and biodegradability attributes of the parent molecule. The high viscosities and beneficial viscoelastic properties achievable with collagen are not, however, possible. On the other hand, high loading levels of relatively nonassociated (nonfibrillar) gelatin molecules are possible under physiological conditions, potentially resulting in gels with a mesh size smaller than those achievable with collagen. A study by Brooks *et al.* (1988) showed that intramuscular administration of porcine calcitonin in gelatin solution into humans altered the pharmacokinetics relative to those for a simple solution in saline. While both dosage forms exhibited equivalent bioavailability, calcitonin in gelatin yielded peak serum levels of 4.8 ng/ml at 120 min, compared to 8.3 ng/ml at 13 min for the saline formulation.

Gelatin has also been used as a matrix for microsphere delivery systems. Tabata and Ikada (1989) have shown that interferon-α can be efficiently ($>50\%$) entrapped in glutaraldehyde-cross-linked microspheres, with subsequent protein release being dependent upon the presence of collagenase. The rates of gelatin degradation and interferon release were inversely related to the extent of glutaraldehyde cross-linking. Fifty percent release of interferon from microspheres phagocytosed by macrophages occurred over a period of 6 hr to 4 days, depending on cross-link density. Recent work which provides an additional example of gelatin as a sustained release system for protein drugs is that of Golumbek *et al.* (1993). Microspheres containing either interferon-γ or granulocyte-macrophage colony-stimulating factor were prepared by coacervation, upon addition of gelatin solution to a chondroitin sulfate plus protein drug solution. The microspheres were then cross-linked with glutaraldehyde. Both proteins delivered from the microspheres were found to stimulate a systemic antitumor response in mice, whereas subcutaneous administration of a much larger dose (1 mg) of free drug had virtually no effect.

4.2.5. HYALURONIC ACID

Hyaluronic acid is a polysaccharide composed of glucuronic acid and *N*-acetylglucosamine. This linear polymer is present in the extracellular matrix and various body fluids and possesses high solubility in water. *In vitro* and *in vivo* release of insulin-like growth factor (MW 7500) from partially deprotonated hyaluronic acid has been studied (Prisell *et al.*, 1992). Fifty percent release of radiolabeled protein dissolved in 2% hyaluronic acid (into phosphate-buffered saline, separated from the protein solution by a 10-μm nylon membrane) occurred within 5 hr. Reduction of the polymer concentration to 0.5% resulted in a decrease in retention time, with 50% release occurring within 1.5 hr. Subcutaneous injection of radioiodinated insulin-like growth factor dissolved in 2% hyaluronic acid into rats showed that 50% of the dose was retained at the injection site after 4 hr. In comparison, 50% loss of free drug occurred within 1 hr of injection.

Use of biocompatible and biodegradable esters of hyaluronic acid for sustained delivery of nerve growth factor has also been examined (Ghezzo, *et al.*, 1992). Nerve growth factor (0.02%) was incorporated into microspheres consisting of either a partially water-soluble ethyl ester or water-insoluble benzyl ester. Fifty percent release of protein occurred *in vitro* (phosphate-buffered saline, 37 °C) after 4 hr from benzyl ester microspheres containing 20% monosialoganglioside surfactant. Approximately 60% of the incorporated nerve growth factor could be recovered from microspheres when the surfactant was coincorporated during microsphere manufacture, as opposed to $<10\%$ recovery in its absence.

More recent work has studied polypeptide diffusion through membranes consisting of ethyl and benzyl esters of hyaluronic acid. Papini *et al.* (1993) have calculated apparent diffusion coefficients (at 37 °C) for a series of proteins, including lysozyme (MW 14,500; $D = 0.2 \times 10^{-8}\,\mathrm{cm}^2/\mathrm{sec}$) and BSA (MW 66,000; $D < 0.006 \times 10^{-8}\,\mathrm{cm}^2/\mathrm{sec}$), through hydrated membranes of the ethyl ester. In all cases, the measured values were considerably lower than estimated aqueous diffusion coefficients, indicating that hindered diffusion could be achieved (perhaps coupled with various levels of binding to the polymer). Membranes prepared from the benzyl ester of hyaluronic acid showed a significant reduction in apparent diffusion coefficient ($D < 0.2 \times 10^{-11}\,\mathrm{cm}^2/\mathrm{sec}$ for ribonuclease) by comparison with those prepared from the ethyl ester ($D = 0.2 \times 10^{-8}\,\mathrm{cm}^2/\mathrm{sec}$), due in part to reduced hydration and smaller polymer matrix mesh size.

Polysaccharides may therefore be very useful as components of sustained release systems for protein delivery. High biocompatibility and general biodegradability, coupled with mild aqueous fabrication conditions and high water content, should help resolve safety and stability issues. The ability to prepare systems with high polymer loadings, thereby achieving release matrices or barriers with small mesh size, may allow sustained release of proteins that are too small or labile to be effectively delivered by other technologies.

4.2.6. OTHER POLYSACCHARIDES

A number of additional examples of protein delivery via polysaccharides are available in the scientific and patent literature. Edman *et al.* (1980) have demonstrated release of carbonic anhydrase, catalase, human serum albumin, and immunoglobulin G from cross-linked biodegradable polyacryl dextrans having molecular weights ranging from 10,000 to 2,000,000. Thermal stability of carbonic anhydrase was found to be enhanced via entrapment in microspheres of the above polymer. Schroder (1984) has shown that sustained release of biologically active insulin and interferon-α can be achieved from microspheres of crystallized dextran. Six days were required for 50% release of insulin, and 12 days for 50% release of ovalbumin.

Artursson *et al.* (1984) have prepared maltodextrin microparticles cross-linked with acrylic acid glycidyl ester which provided sustained release of bioactive proteins. *In vitro* release (phosphate-buffered saline, 37 °C) showed that approximately three weeks were required for 50% release of proteins as diverse in weights and properties as carbonic anhydrase, human serum albumin, and immunoglobulin G. Complete *in vivo* biodegradation occurs after lysosomal sequestration. Kost and Shefer (1990) have developed procedures for cross-linking starch granules with calcium chloride to form a matrix for protein release. Sustained release of myoglobin and BSA was demonstrated, with 2% and 1% of each protein being released (into phosphate-buffered saline, 37 °C) after 6 hr. Fifty percent release occurred after only 5 hr when 0.5 U/ml of α-amylase was present in the release medium.

Porous chitosan has recently been evaluated as a protein delivery system by Cardinal *et al.* (1990). Sustained release was demonstrated for bovine growth hormone, alkaline phosphatase, and BSA. Treatment of dried particles of chitosan with methanol for brief periods extended release duration, with a 10-min treatment increasing the time required for 50% release of BSA from <1 day to 2 weeks. Jamas *et al.* (1991) have documented preparations of β-glucan microparticles, where time to 50% release ranged from 30 min for cytochrome *c* (MW 14,000) to 200 min for alcohol dehydrogenase (MW 150,000).

Recently, Cady *et al.* (1993) have described water-dispersible and soluble polymeric carbohydrate compositions for parenteral administration of growth hormones, somatomedins, and growth factors. One-week release of bovine growth hormone from a 25% aqueous dispersion of dextrin was demonstrated, owing to complexation between protein and carbohydrate. Release rates from both dextrin and dextran could be increased by the addition of surfactants.

While issues of *in vivo* biodegradation have not been resolved, use of water-soluble cellulose derivatives for topical and/or postsurgical delivery of proteins has been described. Aspenberg and Lohmander (1989) have shown that fibroblast growth factor stimulates bone formation by sustained release from a methyl cellulose matrix. Delivery and prolonged local retention of transforming growth factor-β and relaxin from a lyophilizable hydroxyethyl cellulose gel have been described (Hsu *et al.*, 1993). An additional application of modified cellulose for topical protein delivery, discussed by Yamazaki *et al.* (1992), is the adsorption and release of trypsin and plasmin from cellulose fiber sponges that have been surface-modified with diethylaminoethyl and sulfate groups.

4.2.7. ZEIN

Recent patent publications and disclosures discuss the use of the protein zein (obtained from corn) in the delivery of protein drugs. Mathiowitz *et al.* (1993) have described the release of insulin (5% loading) from zein microspheres intended for oral delivery; linear release for over 50 hr was achieved after a 5% burst release. Bernstein *et al.* (1993) have described the sustained delivery of erythropoietin from zein microspheres in dogs. Subcutaneous injection of 30 mg of microspheres with 1% drug loading resulted in peak erythropoietin concentrations at 10 hr and plasma levels greater than 250 mU/ml for up to 80 hr.

4.3. Specific Hydrophobic Nonpolymeric Biomaterials

4.3.1. CHOLESTEROL

Use of lipids and other natural hydrophobic substances for sustained delivery of protein drugs has been the subject of both investigative and developmental activities for several years. Biodegradation of matrices of such materials can be viewed as a process of *in situ* dissolution by endogenous solvents and surfactants with subsequent normal metabolism of the dispersed material.

An early description of a cholesterol-based system for protein delivery was provided by Kent (1984). Sustained release of human growth hormone and insulin from compression-molded disks prepared from powdered cholesterol and lyophilized drug substance was described. Surfactants could be added during powder blending, to modulate release behavior. Wang (1987) also showed the utility of cholesterol disks for insulin delivery. *In vitro* release (into phosphate-buffered saline, 21 °C) exhibited zero-order kinetics, with 50% release from a disk containing 10% insulin occurring over 20 days. A release duration of 25 days was measured *in vivo*. Increase in insulin content decreased release duration *in vivo*, with 20% loading reducing duration to 15 days, and 50% loading reducing measurable release to 10 days.

Use of cholesterol pellets for sustained release of protein antigens was recently examined by Opdebeeck and Tucker (1993). Utilizing matrices prepared by dry blending and compression molding, sustained delivery of BSA could be achieved for 77 days upon subcutaneous administration of a 30-mg pellet in mice. Release rate and degree of burst release (during the first 6 days) increased with albumin loading (0.15 to 3.0 mg per 30-mg pellet). Greater than 80% of the total protein was released in all cases. The effect of lecithin incorporation into similar pellets was examined by Khan *et al.* (1993). *In vivo* release rates for BSA (subcutaneously in mice) were found to increase with increase in lecithin content.

4.3.2. GLYCERIDES

Several examples of the use of natural oils and waxes for delivery of proteins have been described in the patent literature. In some cases, a lipid matrix was employed merely to enhance retention of an injected dose. In other cases, the lipid was employed as a barrier to diffusion of water into particles of suspended protein drug and to subsequent release of dissolved protein. Mitchell (1985) described suspensions (10–40% loading) of zinc somatotropin in triglyceride oil gelled with aluminum monostearate. The system provided effective (>12 ng/ml) levels in cows for >15 days. A similar system for somatotropin employing a wax plus oil mixture has been described (Ferguson *et al.*, 1986). A wax plus oil system for delivery of bovine growth hormone was described by Steber *et al.* (1987). Use of wax shell beadlets for delivery of porcine somatotropin has been detailed by Sivaramakrishnan and Miller (1990, 1993). Average time to beadlet rupture could be varied between 1 and 14 days.

Composite systems containing lipids for the delivery of proteins have also been developed. Yamahira *et al.* (1989) have demonstrated sustained release of interferon-α from suspensions of lyophilized collagen plus interferon in triglyceride oils. Engstrom *et al.* (1992) discussed delivery of insulin from a cubic liquid-crystalline phase consisting of monolinolein and aqueous solutes. Other examples of protein delivery via structured lipid/water systems include the delivery of nerve growth factor from colloidal gangliosides (Marshall *et al.*, 1991), where altered biodistribution was achieved. Finally, the work of Janoff *et al.* (1993) shows that enhanced intramuscular retention of human growth hormone can be achieved in rabbits via colloidal mixtures of cholesterol and structured lipids. Injection of free solution resulted in only 5% retention of protein after 3 hr, whereas 56% was retained when growth hormone was delivered via the described system. Twenty-two percent of the dose was still retained after 168 hr.

4.4. Miscellaneous

4.4.1. SYNTHETIC HYDROGELS

Erodible synthetic hydrogels have been studied as delivery systems for proteins that are stable, or can be stabilized, in an aqueous environment. A number of systems have been described (Torchilin *et al.*, 1977; Edman *et al.*, 1980; Heller *et al.*, 1983; Singh, 1991b). The reader is referred to an excellent and comprehensive review of degradable hydrogels for drug delivery by Park *et al.* (1993).

Recently, synthesis of a novel polyphosphazene (**9**) has been reported

$$\left[-N{=}P(-O-C_6H_4-COOH)_2- \right]_n$$

(9)

(Allcock and Kwon, 1989). This polymer, poly[bis(carboxylatophenoxy)-phosphazene] (PCPP) undergoes a liquid–gel phase transition upon contact with a bivalent cation (i.e., Ca^{2+}) at or below room temperature (Allcock and Kwon, 1989; Cohen *et al.*, 1990; Bano *et al.*, 1991; Andrianov *et al.*, 1993). Owing to its polyelectrolyte nature, this water-soluble polymer can be converted to a cross-linked hydrogel by treatment with dissolved cations (presumably via salt bridges between carboxylic groups of adjacent polymers), creating an ionically cross-linked hydrogel matrix. This unusual polymeric property has allowed development of mild encapsulation conditions for macromolecules. Microspheres with controlled size (0.5–1.5 mm) and spherical shape are prepared by spraying the polymer solution through a droplet-forming apparatus into the gelation solution (aqueous calcium chloride). A delayed release of up to 90% of incorporated BSA from PCPP-gel microspheres occurred over 24 hr. Coating of the microspheres with poly(L-lysine) (PLL), a positively charged polyelectrolyte, resulted in the formation of a stable polyelectrolyte complex and effectively reduced the release of BSA over the first half-hour. However, approximately 80% of the protein was still released from the PLL-coated PCPP-gel microspheres at 24 hr. Slightly better results were obtained when β-galactosidase was encapsulated—PLL-coated spheres still released almost 20% of the protein as a burst in the first 5 hr, however, these coated microspheres were capable of extending protein release for up to 150 hr, indicating the potential for a permselective membrane for specific macromolecules.

5. CONCLUSIONS

Despite the number of advances recently made in the area of controlled release technology for the delivery of macromolecules, much remains to be done to successfully develop delivery systems where desireable release kinetics are maintained with preservation of biological activity. However, the field is growing more rapidly now than ever before. One major stimulus for this expansive growth has been the advent of genetic engineering—which has made possible large-scale production of complex biomolecules. Due to the extreme susceptibility of proteins to proteolysis and rapid clearance from the bloodstream, the potential of these polypeptide drugs will be realized only with suitable delivery systems. Bioerodible and biodegradable materials offer the great advantage of enabling either site-specific or systemic administration of macromolecules without the need for subsequent retrieval of the delivery system. The progress of controlled release systems should develop simultaneously with new drug discovery and evaluation. This may allow drugs that are considered unsuitable by conventional routes of administration to be delivered sucessfully by controlled release formulations.

The development of peptides and proteins as commercially viable therapeutic agents has presented some unique challenges to the drug delivery scientist. Areas where expanded research efforts are important include development of delivery systems with precisely controlled release kinetics—with a particular emphasis on a fundamental understanding of the mechanism of drug release (Shah *et al.*, 1992)—and further development of pulsatile (e.g., for vaccine delivery), and temporal release systems where the release of drug is consistent with the biochemistry of the disease state.

REFERENCES

Aguado, M. T., 1993, Future approaches to vaccine development: Single-dose vaccines using controlled-release delivery systems, *Vaccine* **11**:596–597.

Allcock, H. R., 1990, Polyphosphazenes as new biomedical and bioactive materials, in: *Biodegradable Polymers as Drug Delivery Systems* (M. Chasin and R. Langer, eds.), Marcel Dekker, New York, pp. 163–193.

Allcock, H. R., and Kwon, S., 1989, An ionically cross-linkable polyphosphazene: Poly(bis(carboxylatophenoxy)phosphazene) and its hydrogel and membranes, *Macromolecules* **22**:75–79.

Alonso, M. J., Cohen, S., Park, T. G., Gupta, R. K., Siber, G. R., and Langer, R., 1993, Determinants of release rate of tetanus vaccine from polyester microspheres, *Pharm. Res.* **10**:945–953.

Andrianov, A. K., Cohen, S., Visscher, K. B., Payne, L. G., Allcock, H. R., and Langer, R., 1993, Controlled release using ionotropic polyphosphazene hydrogels, *J. Controlled Release* **27**:69–77.

Anik, S. T., Sanders, L. M., Chaplin, M. D., Kushinsky, S., and Nurenberg, C., 1984, Delivery systems for LHRH and analogs, in: *LHRH and Its Analogs. Contraceptive and Therapeutic Applications* (B. H. Vickery, J. J. Nestor, Jr., and E. S. E. Hafez, eds.), MTP Press, Boston, pp. 421–438.

Arshady, R., 1991, Preparation of biodegradable microspheres and microcapsules: 2. Polylactides and related polyesters, *J. Controlled Release* **17**:1–22.

Artursson, P., Edman, P., Laakso, T., and Sjoholm, I., 1984, Characterization of polyacryl starch microparticles as carriers for proteins and drugs, *J. Pharm. Sci.* **73**:1507–1513.

Asano, M., Fukuzaki, H., Yoshida, M., Kumakura, M., Mashimo, T., Yuasa, H., Imai, K., and Yamanaka, H., 1989a, *In vivo* characteristics of low molecular weight copoly(D,L-lactic acid) formulations with controlled release of LH-RH agonist, *Biomaterials* **10**:569–573.

Asano, M., Fukuzaki, H., Yoshida, M., Kumakura, M., Mashimo, T., Yuasa, H., Imai, K., Yamanaka, H., and Suzuki, K., 1989b, *In vivo* characteristics of low molecular weight copoly(L-lactic acid/glycolic acid) formulations with controlled release of luteinizing hormone-releasing hormone agonist, *J. Controlled Release* **9**:111–122.

Asano, M., Fukuzaki, H., Yoshida, M., Kumakura, M., Mashimo, T., Yuasa, H., Imai, K., Yamanaka, H., Kawaharada, U., and Suzuki, K., 1991, *In vivo* controlled release of a luteinizing hormone-releasing hormone agonist from poly(DL-lactic acid) formulations of varying degradation pattern, *Int. J. Pharm.* **67**:67–77.

Aspenberg, P., and Lohmander, S., 1989, Fibroblast growth factor stimulates bone formation, *Acta Orthoped. Scand.* **60**:473–476.

Baker, R. W., 1987, *Controlled Release of Biologically Active Agents*, John Wiley & Sons, New York, pp. 1–275.

Bano, M. C., Cohen, S., Allcock, H. R., and Langer, R., 1990, Novel polyphosphazene system for drug delivery and cell microencapsulation, Proc. Int. Symp. Control. Rel. Bioact. Mater. **17**:206–207.

Bathurst, I. C., Barr, P. J., Kaslow, D. C., Lewis, D. H., Atkins, T. J., and Rickey, M. E., 1992, Development of a single injection transmission-blocking malaria vaccine using biodegradable microspheres, *Proc. Int. Symp. Control. Rel. Bioact. Mater.* **19**:120–121.

Bernstein, H., Mathiowitz, E., Morrel, E., and Brickner, A., 1993, Erythropoietin drug delivery system, International Patent Application WO 93/25221, date of publication: December 23, 1993.

Bhargava, K., and Ando H., 1992, Immobilization of active urokinase on albumin microspheres: Use of chemical dehydrant and process monitoring, *Pharm. Res.* **9:**776–781.

Bodmer, D., Kissel, T., and Traechslin, E., 1992, Factors influencing the release of peptides and proteins from biodegradable parenteral depot systems, *J. Controlled Release* **21:**129–138.

Brady, J. M., Cutright, D. E., Miller, R. A., and Battistone, G.C., 1973, Resorption rate, route of elimination, and ultrastructure of the implant site of polylactic acid in the abdominal wall of the rat, *J. Biomed. Mater. Res.* **7:**155–166.

Brems, D. N., 1988, Solubility of different folding conformers of bovine growth hormone, *Biochemistry* **27:**4541–4546.

Brooks, S., Butterworth, K., Christie, R., and Fox, J., 1988, Comparison of the pharmacokinetics of porcine calcitonin in saline and in gelatin diluents in healthy volunteers, *Eur. J. Drug Metab. Pharmacokinet.* **13:**91–97.

Buchner, J. E., and Slade, W. C., 1909, The anhydrides of isophthalic and terephthalic acids, *J. Am. Chem. Soc.* **31:**1319–1321.

Cady, S., Fishbein, R., Schroder, U., Eriksson, H., and Probasco, B., 1993, Water dispersible and water soluble carbohydrate polymer compositions for parenteral administration of growth hormone, U.S. Patent 5,266,333, November 30, 1993.

Camarata, P. J., Suryanarayanan, R., Turner, D. A., Parker, R. G., and Ebner, T. J., 1992, Sustained release of nerve growth factor from biodegradable polymer microspheres, *Neurosurgery* **30:**313–319.

Cardinal, J., Curatolo, W., and Ebert, C., 1990, Chitosan compositions for controlled and prolonged release of macromolecules, U.S. Patent 4,895,724, January 23, 1990.

Chasin, M and Langer, R. (eds.), 1990, *Biodegradable Polymers as Drug Delivery Systems*, Marcel Dekker, New York.

Chasin, M., Domb, A., Ron, E., Mathiowitz, E., Langer, R., Leong, K. W., Laurencin, C., Brem, H., and Grossman, S., 1990, Polyanhydrides as drug delivery systems, in: *Biodegradable Polymers as Drug Delivery Systems* (M. Chasin and R. Langer, eds.), Marcel Dekker, New York, pp. 43–70.

Choi, N. S., and Heller, J., 1978a, Drug delivery devices manufactured from polyorthoesters and polyorthocarbonates, U.S. Patent 4,093,709, June 6, 1978.

Choi, N. S., and Heller, J., 1978b, Structured orthoesters and orthocarbonate drug delivery devices, U.S. Patent 4,131,648, December 26, 1978.

Choi, N. S., and Heller, J., 1979, Erodible agent releasing device comprising poly(orthoesters) and poly(orthocarbonates), U.S. Patent 4,138,344, December 26, 1979.

Cohen, S., Bano, M. C., Visscher, K. B., Chow, M., Allcock, H. R., and Langer, R., 1990, Ionically cross-linkable polyphosphazene: A novel polymer for microencapsulation, *J. Am. Chem. Soc.* **112:**7832–7833.

Cohen, S., Yoshioka, T., Lucarelli, M., Hwang, L. H., and Langer, R., 1991, Controlled delivery systems for proteins based on poly(lactic/glycolic acid) microspheres, *Pharm. Res.* **8:**713–720.

Conix, A., 1958, Aromatic polyanhydrides, a new class of high melting fibre–forming polymers, *J. Polym. Sci.* **29:**343–353.

Crommen, J., Vandorpe, J., and Schacht, E., 1993, Degradable polyphosphazenes for biomedical applications, *J. Controlled Release* **24:**167–180.

Deatherage, J., and Miller, E., 1987, Packaging and delivery of bone induction factors in a collagenous implant, *Collagen Relat. Res.* **7:**225–231.

Dittrich, V. W., and Schulz, R. C., 1971, Kinetics and mechanism of the ring-opening polymerization of L-lactide, *Angew. Makromol. Chem.* **15:**109–126.

Doi, Y., Kanesawa, Y., Kunioka, M., and Saito, T., 1990, Biodegradation of microbial copolyesters: Poly (3-hydroxybutyrate-*co*-3-hydroxyvalerate) and poly(3-hydroxybutyrate-*co*-4-hydroxyvalerate), *Macromolecules* **23:**26–31.

Domb, A. J., and Langer, R., 1987, Polyanhydrides. I. Preparation of high molecular weight polyanhydrides, *J. Polym. Sci.* **25:**3373–3386.

Doolittle, R., 1973, Structural aspects of the fibrinogen to fibrin conversion, *Adv. Protein Chem.* **27:**1–109.

Doolittle, R., 1984, Fibrinogen and fibrin, *Annu. Rev. Biochem.* **53:**195–229.

Dunn, R. L., Tipton, A. J., and Menardi, E. M., 1991, A biodegradable *in-situ* forming drug delivery system, *Proc. Int. Symp. Control. Rel. Bioact. Mater.* **18:**465–466.

Duysen, E. G., Yewey, G. L., and Dunn, R. L., 1992, Bioactivity of polypeptide growth factors released from the ATRIGEL™ drug delivery system, *Pharm. Res.* **9:**S–73.

Duysen, E. G., Yewey, G. L., Southard, J. L., Dunn, R. L., and Huffer, W., 1993, Release of bioactive growth factors from the ATRIGEL™ delivery system in tibial defect and dermal wound models, *Pharm. Res.* **10:**S–83.

Edman, P., Ekman, B., and Sjoholm, I., 1980, Immobilization of proteins in microspheres of biodegradable polyacryldextran, *J. Pharm. Sci.* **69:**838–842.

Eldridge, J. H., Staas, J. K., Meulbroek, J. A., McGhee, J. R., Tice, T. R., and Gilley, R. M., 1991, Biodegradable microspheres as a vaccine delivery system, *Mol. Immunol.* **28:**287–294.

Engstrom, S., Lindman, B., and Larsson, K., 1992, Method of preparing controlled release preparations for biologically active materials and resulting compositions, U.S. Patent 5,151,272, September 29, 1992.

Eppstein, D. A., 1986, Alternative delivery of interferons, in: *Targeting of Drugs with Synthetic Systems* (G. Gregoriadis, J. Senior, and G. Poste, eds.), Plenum Press, New York, p. 207.

Eppstein, D. A., and Longenecker, J. P., 1988, Alternative delivery systems for peptides and proteins as drugs, *Crit. Rev. Ther. Drug Carrier Syst.* **5:**99–139.

Feijen, J., 1990, Biodegradable hydrogel matrices for the controlled release of pharmacologically active agents, U.S. Patent 4,925,677, May 15, 1990.

Ferguson, T., Harrison, R., and Moore, D., 1986, Injectable sustained release formulation, European Patent Application 0 211 691 A2, date of filing: August 20, 1986.

Frazza, E. J., and Schmitt, E. E., 1971, A new absorbable suture, *J. Biomed. Mater. Res. Symp.* **1:**43–58.

Fujioka, K., Sato, S., Sasaki, Y., Miyata, T., Furuse, M., and Naito, H., 1989, Method for producing sustained release formulation, U.S. Patent 4,849,141, July 18, 1989.

Fujioka, K., Sato, S., Tamura, N., Takada, Y., Sasaki, Y., and Maeda, M., 1993, Controlled release formulation, U.S. Patent 5,236,704, August 17, 1993.

Fukuzaki, H., Yoshida, M., Asano, M., Kumakura, M., Yamanaka, H., Mashimo, T., Yuasa, H., Imai, K., Kawaharada, U., and Suzuki, K., 1989, A new biodegradable pasty-type of L-lactic acid and D-valerolactone with relatively low molecular weight for application in drug delivery systems, *J. Controlled Release* **10:**293–303.

Furr, B. J. A., and Hutchinson, F. G, 1992, A biodegradable delivery system for peptides: Preclinical experience with the gonadotrophin–releasing hormone agonist, Zoladex®, *J. Controlled Release* **21:**117–128.

Ghezzo, E., Benedetti, L., Rochira, M., Biviano, F., and Callegaro, L., 1992, Hyaluronane derivative microspheres as NGF delivery devices: Preparation methods and *in vitro* release characterization, *Int. J. Pharm.* **87:**21–29.

Gilbert, D., and Kim S., 1990, Macromolecular release from collagen monolithic devices, *J. Biomed. Mater. Res.* **24:**1221–1239.

Gilbert, D., Okano, T., Miyata, T., and Kim S., 1988, Macromolecular diffusion through collagen membranes, *Int. J. Pharm.* **47:**79–88.

Golumbek, P., Azhari, R., Jaffee. E., Levitsky, H., Lazenby, A., Leong, K., and Pardoll, D., 1993, Controlled release, biodegradable cytokine depots: A new approach in cancer vaccine design, *Cancer Res.* **53**:5841–5844.

Goosen, M., Leung, Y., Chou, S., and Sun, A., 1982, Insulin–albumin microbeads: An implantable, biodegradable system, *Biomater. Med. Devices Artif. Organs* **10**:205–218.

Hageman, M. J., Bauer, J. M., Possert, P. L., and Darrington, R. T., 1992, Preformulation studies oriented toward sustained delivery of recombinant somatotropins, *J. Agric, Food Chem.* **40**:348–355.

Hazrati, A. M., Lewis, D. H., Atkins, T. J., Stohrer, R. C., Little, J. E., and Meyer, L., 1992, Studies of controlled delivery tetanus vaccine in mice, *Proc. Int. Symp. Control. Rel. Bioact. Mater.* **20**:67–68.

Heller, J., 1987, Use of polymers in controlled release of active agents, in: *Controlled Drug Delivery. Fundamentals and Applications* (J. R. Robinson and V. H. L. Lee, eds.), 2nd ed., Marcel Dekker, New York, pp. 180–210.

Heller, J., 1993a, Polymers for controlled parenteral delivery of peptides and proteins, *Adv. Drug Delivery Rev.* **10**:163–204.

Heller, J., 1993b, Poly(ortho esters), *Adv. Polym. Sci.* **107**:43–92

Heller, J., Penhale, D. W. H., and Helwing, R. F., 1980, Preparation of poly(ortho esters) by the reaction of ketene acetals and polyols, *J. Polym. Sci., Polym. Lett. Ed.* **18**:82–83.

Heller, J., Helwing, R. F., Baker, R. W., and Tuttle, M. E., 1983, Controlled release of water-soluble macromolecules from bioerodible hydrogels, *Biomaterials* **4**:262–266.

Heller, J., Ng, S. Y., Penhale, D. W., Fritzinger, B. K., Sanders, L. M., Burns, R. A., Gaynon, M. G., and Bhosale, S. S., 1987, Use of poly(ortho esters) for the controlled release of 5-fluorouracil and a LHRH analogue, *J. Controlled Release,* **6**:217–224.

Heller, J., Sparer, R. V., and Zentner, G. M., 1990a, Poly(ortho esters) in: *Biodegradable Polymers as Drug Delivery Systems* (M. Chasin and R. Langer, eds.), Marcel Dekker, New York, pp. 121–161.

Heller, J., Ng, S. Y., Fritzinger, B. K., and Roskos, K. V., 1990b, Controlled drug release from bioerodible hydrophobic ointments, *Biomaterials* **11**:235–237.

Hill, J., and Carothers, W. H., 1932, Studies on polymerization and ring formation. XIV. A linear superpolyanhydride and cyclic dimeric anhydride from sebacic acid, *J. Am. Chem. Soc.* **54**:1569–1579.

Ho, H., and Chen C., 1993, Diffusion characteristics of fibrin film, *Int. J. Pharm.* **90**:95–104.

Holland, S. J., Tighe, B. J., and Gould, P. L., 1986, Polymers for biodegradable medical devices. 1. The potential of polyesters as controlled macromolecular release systems, *J. Controlled Release* **4**:155–180.

Holland, S. J., Jolly, A. M., Yasin, M., and Tighe, B.J., 1987, Polymers for biodegradable medical devices II. Hydroxybutyrate–hydroxyvalerate colpolymers: Hydrolytic degradation studies, *Biomaterials* **8**:289–295.

Hollinger, J. O., and Battistone, G. C., 1986, Biodegradable bone repair materials, *Clin. Orthop. Relat. Res.* **207**:290–305.

Hollinger, J. O., Schmitz, J. P., Mark, D. E., and Seyfer, A. E., 1990, Osseous wound healing with xenogeneic bone implants with a biodegradable carrier, *Surgery* **107**:50–54.

Holmes, P. A., 1985, Applications of PHB—A microbially produced biodegradable thermoplastic, *Phys. Technol.* **16**:32–36.

Hora, M. S., Rana, R. K., Taforo, T. A., Nunberg, J. H., Tice, T. R., Gilley, R. M., and Hudson, M. E., 1989, Development of a controlled release microsphere formulation of interleukin-2, *Proc. Int. Symp. Control. Rel. Bioact. Mater.* **16**:509–510.

Hora, M. S., Rana, R. K., Nunberg, J. H., Tice, T. R., Gilley, R. M., and Hudson, M. E., 1990, Release of human serum albumin from poly(lactide-*co*-glycolide) microspheres, *Pharm. Res.* **7:**1190–1194.

Hori, R., Komada, F., Iwakawa, S., Seino, Y., and Okumura, K., 1989, Enhanced bioavailability of subcutaneously injected insulin coadministered with collagen in rats and humans, *Pharm. Res.* **6:**813–816.

Horisaka, Y., Okamoto, Y., Matsumoto, N., Yoshimura, Y., Hirano, A., Nishida, M., Kawada, J., Yamashita, K., and Takagi, T., 1994, Histological changes of implanted collagen material during bone induction, *J. Biomed. Mater. Res.* **28:**97–103.

Hsu, C., Nguyen, H., and Wu, S., 1993, Reconstitutable lyophilized protein formulation, U.S. Patent 5,192,743, March 9, 1993.

Hutchinson, F. G., and Furr, B. J. A., 1985, Biodegradable polymers for the sustained release of peptides, *Biochem. Soc. Trans.* **12:**520–523.

Hutchinson, F. G., and Furr, B. J. A., 1987, Biodegradable carriers for the sustained release of polypeptides, *TibTech.* **5:**102–106.

Hutchinson, F. G., and Burr, B. J. A., 1990, Biodegradable polymer systems for the sustained release of polypeptides, *J. Controlled Release* **13:**279–294.

Hutchinson, F. G., and Furr, B. J. A., 1991, Biodegradable polymer systems for the sustained release of polypeptides, in: *High Value Polymers* (A. H. Fawcett, ed.), *Proc. Symp. R. Soc. Chem.* **87:**58–78.

Imasaka, K., Yoshida, M., Fukuzaki, H., Asano, M., Kumakura, M., Mashimo, T., Yamanaka, H., and Nagai, T., 1991, A new biodegradable implant consisting of waxy-type poly(ε-caprolactone-*co*-D-valerolactone) and estramustine, *Int. J. Pharm.* **68:**87–95.

Jamas, S., Ostroff, G., and Easson, D., 1991, Glucan drug delivery system and adjuvant, U.S. Patent 5,032,401, July 16, 1991.

Janoff, A., Popescu, M., Weiner, A., Bolcsak, L., Tremblay, P., and Swenson, C., 1993, Compositions containing tris salt of cholesterol hemisuccinate and antifungal, U.S. Patent 5,231,112, July 27, 1993.

Kawamura, M., and Urist, M., 1988, Human fibrin is a physiologic delivery system for bone morphogenetic protein, *Clin. Orthop. Relat. Res.* **235:**302–310.

Kenley, R. A., Yim, K., Abrams, J., Ron, E., Turek, T., Marden, L. J., and Hollinger, J. O., 1993, Biotechnology and bone graft substitutes, *Pharm. Res.* **10:**1393–1401.

Kent, J., Cholesterol matrix delivery system for sustained release of macromolecules, U.S. Patent 4,452,775, June 5, 1984.

Khan, M., Tucker, I., and Opdebeeck, J., 1993, Evaluation of cholesterol–lecithin implants for sustained delivery of antigen: Release *in vivo* and single step immunization of mice, *Int. J. Pharm.* **90:**255–262.

Khan, M. Z. I., Opdebeeck, J. P., and Tucker, I. G., 1994, Immunopotentiation and delivery systems for antigens for single-step immunization: Recent trends and progress, *Pharm. Res.* **11:**2–11.

Kohn, J., Niemi, S. M., Albert, E. C., Murphy, J. C., Langer, R., and Fox, J. G., 1986, Single-step immunization using a controlled release biodegradable polymer with sustained adjuvant activity, *J. Immunol. Methods* **95:**31–38.

Konig, W., Seidel, J. R., and Sandow, J. K., 1985, European Patent Application 0133,988 (C1A61K37/02), March 13, 1985.

Kost, J., and Shefer, S., 1990, Chemically modified polysaccharides for enzymatically controlled oral drug delivery, *Biomaterials* **11:**695–698.

Kulkarni, R. K., Moore, E. G., Hegyelli, A. F., and Leonard, F., 1971, Biodegradable polylactic acid polymers, *J. Biomed. Mater. Res.* **5:**169–181.

Kwon, G., Bae, Y., Cremers, H., Feijen, J., and Kim, S., 1992, Release of proteins via ion exchange from albumin–heparin microspheres, *J. Controlled Release* **22:**83–94.

Kwong, A. K., Chou, S., Sun, A. M., Sefton, M. V., and Goosen, M. F. A., 1986, *In vitro* and *in vivo* release of insulin from poly(lactic acid) microbeads and pellets, *J. Controlled Release* **4:**47–62.

Langer, R., and Moses, M., 1991, Biocompatible controlled release polymers for delivery of polypeptides and growth factors, *J. Cell. Biochem.* **45:**340–345.

Laurencin, C. T., Koh, H. J., Neenan, T. X., Allcock, H. R., and Langer, R., 1987, Controlled release using a new bioerodible polyphosphazene matrix system, *J. Biomed. Mater. Res.* **21:**1231–1246.

Lee, K. C., Soltis, E. E., Newman, P. S., Burton, K. W., Mehta, R. C., and DeLuca, P. P., 1991, *In vivo* assessment of salmon calcitonin sustained release from biodegradable microspheres, *J. Controlled Release* **17:**199–206.

Lee, V. H. L., 1991, Trends in peptide and protein drug delivery, *Biopharm* **1991(March):**22–25.

Leong, K. W., and Langer, R., 1987, Polymeric controlled drug delivery, *Adv. Drug Delivery Rev.* **1:**199–233.

Leong, K. W., Brott, B. C., and Langer, R., 1985, Bioerodible polyanhydrides as drug-carrier matrices. I. Characterization, degradation and release characteristics, *J. Biomed. Mater. Res.* **19:**945–955.

Leong, K. W., D'Amore, P., Marletta, M., and Langer, R., 1986a, Bioerodible polyanhydrides as drug-carrier matrices. II. Biocompatibility and chemical reactivity, *J. Biomed. Mater. Res.* **20:**51–64.

Leong, K. W., Kost, J., Mathiowitz, E., and Langer, R., 1986b, Polyanhydrides for controlled release of bioactive agents, *Biomaterials* **7:**364–371.

Leong, K. W., Simonte, V., and Langer, R., 1987, Synthesis of polyanhydrides: Melt-polycondensation, dehydrochlorination, and derivative coupling, *Macromolecules* **20:**705–712.

Lewis, D. H., 1990, Controlled release of bioactive agents from lactide/glycolide polymers, in: *Biodegradable Polymers as Drug Delivery Systems* (M. Chasin and R. Langer, eds.), Marcel Dekker, New York, pp. 1–41.

Lovell, T. P., Dawson, E. G., Nilsson, O. S., and Urist, M. R., 1989, Augmentation of spinal fusion with bone morphogenetic protein in dogs, *Clin. Orthop. Relat. Res.* **243:**266–274.

Lucas, P., Syftestad, G., Goldberg, V., and Caplan, A., 1989, Ectopic induction of cartilage and bone by water soluble proteins from bovine bone using a collagenous delivery vehicle, *J. Biomed. Mater. Res.: Appl. Biomater.* **23:**(A1):23–39.

Lucas, P. A., Laurencin, C., Syftestad, G. T., Domb, A., Goldberg, V. M., Caplan, A. I., and R. Langer, 1990, Ectopic induction of cartilage and bone by water–soluble proteins from bovine bone using a polyanhydride delivery vehicle, *J. Biomed. Mater. Res.* **24:**901–911.

Marshall, L., Patel, K., and Roufa, D., 1991, Controlled release formulations of trophic factors in ganglioside liposome vehicle, U.S. Patent 5073,543, December 17, 1991.

Mathiowitz, E., and Langer, R., 1987, Polyanhydride microspheres as drug carriers I. Hot-melt microencapsulation, *J. Controlled Release* **5:**13–22.

Mathiowitz, E., Leong, K., and Langer, R., 1985, Macromolecular drug release from biodegradable polyanhydride microspheres, *Proc. Int. Symp. Control. Rel. Bioact. Mater.* **12:**183–184.

Mathiowitz, E., Saltzman, W. M., Domb, A., Dor, P., and Langer, R., 1988, Polyanhydride microspheres as drug carriers II. Microencapsulation by solvent removal, *J. Appl. Polym. Sci.* **35:**755–774.

Mathiowitz, E., Bernstein, H., Morrel, E., and Schwaller, K., 1993, Method for producing protein microspheres, U.S. Patent 5,271,961, December 21, 1993.

Maulding, H. V., 1987, Prolonged delivery of peptides by microcapsules, *J. Controlled Release* **6:**167–176.

Medisorb Technologies International, 1990, Bioabsorbable Polymers, Properties, Uses, Storage and Handling, Medisorb Technical Bulletin.

Miller, E., and Gay, S., 1987, The collagens: An overview and update, *Methods Enzymol.* **144:**3–41.

Mitchell, J., 1985, Prolonged release by biologically active polypeptides, European Patent Application 0 177 478 A2, date of filing: October 3, 1985.

Miyamoto, S., Takaoka, K., Okada, T., Yoshikawa, H., Hashimoto, J., Suzuki, S., and Ono, K., 1992, Evaluation of polylactic acid homopolymers as carriers for bone morphogenetic protein, *Clin. Orthop. Relat. Res.* **278:**274–285.

O'Hagan, D. T., Rahman, D., McGee, J. P., Jeffery, H., Davies, M. C., Williams, P., Davis, S. S., and Challacombe, S. J., 1991, Biodegradable microparticles as controlled release antigen delivery systems, *Immunology* **73:**239–242.

Okada, H., 1989, One-month release injectable microspheres of leuprolide acetate, a superactive agonist of LHRH, *Proc. Int. Symp. Control. Rel. Bioact. Mater.* **16:**12–13.

Okada, H., Heya, T., Ogawa, Y., and Shimamoto, T., 1988, One-month release injectable microcapsules of a luteinizing hormone-releasing hormone agonist (leuprolide acetate) for treating experimental endometriosis in rats, *J. Pharm. Exp. Ther.* **244:**744–750.

Okada, H., Inoue, Y., Heya, T., Ueno, H., Ogawa, Y., and Toguchi, H., 1991, Pharmacokinetics of once-a-month injectable microspheres of leuprolide acetate, *Pharm. Res.* **8:**787–791.

Ogawa, Y., Okada, H., Yamamoto, M., and Shimamoto, T., 1988, *In vivo* release profiles of leuprolide acetate from microcapsules prepared with polylactic acids or copoly(lactic/glycolic) acids and *in vivo* degradation of these polymers, *Chem. Pharm. Bull.* **36:**2576–2581.

Opdebeeck, J., and Tucker I., 1993, A cholesterol implant used as a delivery system to immunize mice with bovine serum albumin, *J. Controlled Release* **23:**271–279.

Papini, D., Stella, V., and Topp, E., 1993, Diffusion of macromolecules in membranes of hyaluronic acid esters, *J. Controlled Release* **27:**47–57.

Park, K., Shalaby, W., and Park, H., 1993, *Biodegradable Hydrogels for Drug Delivery*, Technomic Publishing, Basle.

Physicians' Desk Reference, 1994, 48th ed., Medical Economics Data Production Company, Montvale, New Jersey.

Piez, K., 1985, *Encyclopedia of Polymer Science and Engineering*, Vol. 3, 2nd ed., John Wiley & Sons, New York, pp. 699–727.

Pitt, C., 1990a, The controlled parenteral delivery of polypeptides and proteins, *Int. J. Pharm.* **59:**173–196.

Pitt, C. G., 1990b, Poly-ε-caprolactone and its copolymers, in: *Biodegradable Polymers as Drug Delivery Systems* (M. Chasin and R. Langer, eds.), Marcel Dekker, New York, pp. 71–120.

Pitt, C. G., 1992, Non–microbial degradation of polyesters: Mechanisms and modifications, in: *Biodegradable Polymers and Plastics* (M. Vert, J. Feijen, A. Albertsson, G. Scott, and E. Chiellini, eds.), Royal Society of Chemistry, Cambridge, pp. 7–19.

Pitt, C. G., Cha, Y., Hendren, R. W., Holloman, M., and Schindler, A., 1987, Manipulation of the permeability and degradability of polymers, *Proc. Int. Symp. Control. Rel. Bioact. Mater.* **14:**75–76.

Prisell, P., Camber, O., Hiselius, J., and Norstedt, G., 1992, Evaluation of hyaluronan as a vehicle for peptide growth factors, *Int. J. Pharm.* **85:**51–56.

Pulapura, S., Li, C., and Kohn, J., 1990, Structure–property relationships for the design of polyiminocarbonates, *Biomaterials* **11:**666–678.

Radomsky, M. L., Brouwer, G., Floy, B. J., Loury, D. J., Chu, F., Tipton, A. J., and Sanders, L. M., 1993, The controlled release of Ganirelix from the Atrigel™ injectable implant system, *Proc. Int. Symp. Control. Rel. Bioact. Mater.* **20:**458–459.

Raghuvanshi, R. S., Singh, M., and Talwar, G. P., 1993, Biodegradable delivery system for single step immunization with tetanus toxoid, *Int. J. Pharm.* **93:**R1–R5.

Ron, E., Turek, T., Mathiowitz, E., Chasin, M., and Langer, R., 1989, Release of polypeptides from poly(anhydride) implants, *Proc. Int. Symp. Control. Rel. Bioact. Mater.* **16:**338–339.

Ron, E., Turek, T., Mathiowitz, E., Chasin, M., Hageman, M., and Langer, R., 1993, Controlled release of polypeptides from polyanhydrides, *Proc. Natl. Acad. Sci. USA* **90:**4176–4180.

Rosen, H. G., Chang, J., Wnek, G. E., Linhardt, R. J., and Langer, R., 1983, Bioerodible polyanhydrides for controlled drug delivery, *Biomaterials* **4:**131–133.

Rosenblatt, J., Rhee, W., and Wallace, D., 1989, The effect of collagen fiber size distribution on the release rate of proteins from collagen matrices by diffusion, *J. Controlled Release* **9:**195–203.

Royer, G., Lee, T., and Sokoloski, T., 1983, Entrapment of bioactive compounds within native albumin beads, *J. Pharm. Sci. Technol.* **37:**34–37.

Sah, H. K., and Chien, Y. W., 1993, Evaluation of a microreservoir-type biodegradable microcapsule for controlled release of proteins, *Drug Dev. Ind. Pharm.* **19:**1243–1263.

Sanders, L. M., 1990, Controlled delivery systems for peptides, in: *Peptide and Protein Drug Delivery* (V. H. L. Lee, ed.), Marcel Dekker, New York, pp. 785–806.

Sanders, L. M., Kent, J. S., McRae, G. I., Vickery, B. H., Tice, T. R., and Lewis, D. H., 1984, Controlled release of luteinizing hormone-releasing hormone analogue from poly(*d,l*-lactide-*co*-glycolide), *J. Pharm. Sci.* **73:**1294–1297.

Sanders, L. M., McRae, G. I., Vitale, K. M., and Kell, B. A., 1985, Controlled delivery of an LHRH analogue from biodegradable injectable microspheres, *J. Controlled Release* **2:**187–195.

Sanders, L. M., Kell, B. A., McRae, G. I., and Whitehead, G. W., 1986, Prolonged controlled-release of Nafarelin, a luteinizing hormone-releasing hormone analogue, from biodegradable polymeric implants: Influence of composition and molecular weight of polymer, *J. Pharm. Sci.* **75:**356–360.

Schindler, A., Porous bioabsorbable polyesters as controlled-release reservoirs for high molecular weight drugs, European Patent Application EP 223708 A2, May 27, 1987.

Schmitz, J. P., and Hollinger, J. O., 1988, A preliminary study of the osteogenic potential of a biodegradable alloplastic-osteoinductive alloimplant, *Clin. Orthop. Relat. Res.* **237:**245–255.

Schroder, U., 1984, A crystallised carbohydrate matrix for biologically active substances, a process of preparing said matrix, and the use thereof, International Patent Application WO 84/00294, date of publication: February 2, 1984.

Senderoff, R., Sheu, M., and Sokoloski, T., 1991, Fibrin based drug delivery systems, *J. Parenter. Sci. Technol.* **45:**2–6.

Shah, N. H., Railkar, A. S., Chen, F. C., Tarantino, R., Kumar, S., Murjani, M., Palmer, D., Infeld, M. H., and Malick, A. W., 1993, A biodegradable injectable implant for delivering micro and macromolecules using poly(lactic-*co*-glycolic) acid (PLGA) copolymers, *J. Controlled Release* **27:**139–147.

Shah, S. S., Cha, Y., and Pitt, C. G., 1992, Poly(glycolic acid-*co*-lactic acid): Diffusion or degradation controlled drug delivery?, *J. Controlled Release* **18:**261–270.

Singh, M., Singh, A., and Talwar, G. P., 1991a, Controlled delivery of diphtheria toxoid using biodegradable poly(D,L-lactide) microcapsules, *Pharm. Res.* **8:**958–961.

Singh, M., Rathi, R., Singh, A., Heller, J., Talwar, G. P., and Kopecek, J., 1991b, Controlled release of LHRH–DT from bioerodible hydrogel microspheres, *Int. J. Pharm.* **76:**R5–R8.

Singh, M., Singh, O., Singh, A., and Talwar, G. P., 1992, Immunogenicity studies on diphtheria toxoid loaded biodegradable microspheres, *Int. J. Pharm.* **85:**R5–R8.

Sivaramakrishnan, K., and Miller, L., 1990, Controlled release delivery device for macromolecular proteins, International Patent Application WO 90/11070, date of publication: October 4, 1990.

Sivaramakrishnan, K., and Miller, L., 1993, Controlled release delivery device for macromolecular proteins, United States Patent 5,219,572, June 15,1993.

Song, S., and Morawiecki, A., 1993, Collagen containing sponges as drug delivery for proteins, European Patent Application 0 568 334 A1, date of filing: April 28, 1993.

Steber, W., Fishbein, R.M. and Cady, S., 1987, Compositions for parenteral administration and their use, European Patent Application 0 257 368 A1, date of filing: August 4, 1987.

Stevens, V. C., Powell, J. E., Lee, A. E., Kaumaya, P. T. P., Lewis, D. H., Rickey, M., and Atkins, T. J., 1992, Development of a delivery system for a birth control vaccine using biodegradable microspheres, *Proc. Int. Symp. Control. Rel. Bioact. Mater.* **19:**112–113.

Tabata, Y., and Ikada, Y., 1989, Synthesis of gelatin microspheres containing interferon, *Pharm. Res.* **6:**422–427.

Tabata, Y., and Langer, R., 1993, Polyanhydride microspheres that display near-constant release of water-soluble model drug compounds, *Pharm Res.* **10:**391–399.

Tabata, Y., Takebayashi, Y., Ueda, T., and Ikada, Y., 1993a, A formulation method using D,L-lactic acid oligomer for protein release with reduced initial burst, *J. Controlled Release* **23:**55–64.

Tabata, Y., Gutta, S., and Langer, R., 1993b, Controlled delivery systems for proteins using polyanhydride microspheres, *Pharm Res.* **10:**487–496.

Tadmor, Z., and Gogos, C. G., 1979, *Principles of Polymer Processing*, John Wiley & Sons, New York.

Takaoka, K., Koezuka, M., and Makahara, H., 1991, Telopeptide-depleted bovine skin collagen as a carrier for bone morphogenetic protein, *J. Orthoped. Res.* **9:**902–907.

Torchilin, V. P., Tischenko, E. G., Smirnov, V. N., and Chazov, E. I., 1977, Immobilization of enzymes on slowly soluble carriers, *J. Biomed. Mater. Res.* **11:**223–235.

Toriumi, D. M., Kotler, H. S., Luxenberg, D. P., Holtrop, M. E., and Wang, E. A., 1991, Mandibular reconstruction with a recombinant bone-inducing factor, *Arch. Otolaryngol. Neck Surg.* **177:**1101–1121.

Wallace, D., McPherson, J., Ellingsworth, L., Cooperman L., Armstrong R., and Piez K., 1988, Injectable collagen for tissue augmentation, in: *Collagen*, Vol. 3 (M. E. Nimni, ed.), CRC Press, Boca Raton, Florida, pp. 117–141.

Wang, H. T., Schmitt, E., Flanagan, D. R., and Linhardt, R. J., 1991, Influence of formulation methods on the *in vitro* controlled release of protein from poly(ester) microspheres, *J. Controlled Release* **17:**23–32.

Wang, P., 1987, Prolonged release of insulin by cholesterol-matrix implant, *Diabetes* **36:**1068–1072.

Waxman, J. H., Sandow, J., Magill, P. J. and Oliver, R. T. D., 1985, Symposium on Treatment Advances: Prostatic Cancer Role LHRH–Superagonists, Baden, June 1985.

Weiner, A., Carpenter-Green, S., Soehngen, E., Lenk, R., and Popescu, M., 1985, Liposome-collagen gel matrix: A novel sustained drug delivery system, *J. Pharm. Sci.* **74:**922–925.

Williams, D. F., 1987, *Definitions in Biomaterials*, Proceedings of a Consensus Conference of the European Society for Biomaterials, Elsevier, Amsterdam.

Wise, D. L., Trantolo, D. J., Marino, R. T., and Kitchell, J. P., 1987, Opportunities and challenges in the design of implantable biodegradable polymeric systems for the delivery of antimicrobial agents and vaccines, *Adv. Drug Delivery Rev.* **1:**19–39.

Wuthrich, P., Ng, S. Y., Fritzinger, B. K., Roskos, K. V., and Heller, J., 1992, Pulsatile and delayed release of lysozyme from ointment-like poly(ortho esters), *J. Controlled Release* **21:**191–200.

Yamahira, Y., Fujioka,K., Sato, S., and Yoshida, N., 1984, Sustained release preparation, European Patent Application 0 138 216 A2, date of filing: October 12, 1984.

Yamahira, Y., Fujioka, K., Sato, S., and Takada, Y., 1988, Long term sustained release preparation, U.S. Patent 4,774,091, September 27, 1988.

Yamahira, Y., Fujioka, K., and Sato, S., 1989, Sustained release preparation, U.S. Patent 4,855,134, August 3, 1989.

Yamazaki, H., Miyazaki, M., and Matsumoto, K., 1992, Cellulosic wound dressing with an active agent ionically absorbed thereon, U.S. Patent 5,098,417, March 24, 1992.

Yan, C., Hewetson, J., Creasia, D., Nelson, E., Rill, W., Tammariello, R., Mereish, K., and Kende, M., 1993, Enhancement of ricin toxoid efficacy by controlled rate-release from microcapsules, *Proc. Int. Symp. Control. Rel. Bioact. Mater.* **20:**71–72.

Yasko, A. W., Lane, J. M., Fellinger, E. J., Rosen, V., Wozney, J. M., and Wang, E. A., 1992, The healing of segmental defects induced by recombinant human bone morphogenetic protein-2, *J. Bone Joint Surg.* **74-A:**659–671.

Yewey, G. L., Duysen, E. G., Southard, J. L., and Dunn, R. L., 1993, Controlled release of growth factors from a biodegradable delivery system, in: *Portland Bone Symposium 1993* (J. Hollinger and A. E. Seyfer, eds.), Oregon Health Sciences University, Portland, pp. 453–454.

Yoda, N., 1963, Syntheses of polyanhydrides. XII. Crystalline and high melting polyamide-polyanhydride of methylenebis(*p*-carboxyphenyl) amide, *J. Polym. Sci. Part A* 1323–1338.

Chapter 3

Delivery of Proteins from a Controlled Release Injectable Implant

Gerald L. Yewey, Ellen G. Duysen, S. Mark Cox, and Richard L. Dunn

1. THE ATRIGEL™ DRUG DELIVERY SYSTEM

Development of controlled release systems for the delivery of recombinant proteins remains a critical research challenge for the biotechnology industry. Current therapies with these biopharmaceutical agents require frequent injections or infusion owing to the short half-lives of the proteins (Bodmer *et al.*, 1992). Biodegradable implants and microspheres for parenteral administration could extend the half-life of serum-labile proteins and provide an effective mechanism for localized as well as systemic delivery. Although such sustained release therapies may result in higher formulation costs, they have the potential to reduce overall medical costs by decreasing the frequency of administration. They are also more convenient for the patient to use, with a resulting improvement in compliance. Biodegradable systems that allow repetitive courses of therapy to be administered without the need for a subsequent medical procedure to remove the device contribute even more to lower costs.

Recently, a liquid polymer system (ATRIGEL™) has been developed which has both the simplicity and control of solid biodegradable implants

---

Gerald L. Yewey, Ellen G. Duysen, S. Mark Cox, and Richard L. Dunn • Atrix Laboratories, Inc., Fort Collins, Colorado 80525.

Protein Delivery: Physical Systems, Sanders and Hendren, eds., Plenum Press, New York, 1997.

and the injectability of microspheres for delivering drugs (Dunn *et al.*, 1992). This drug delivery system combines a biodegradable polymer with a biocompatible solvent, resulting in a solution that can be injected using standard syringes and needles. When the system contacts physiologic fluid, the polymer precipitates as the solvent diffuses into the surrounding tissues. As a result, a biodegradable polymeric implant is formed. For controlled release applications, a drug can be incorporated into the delivery system. The incorporated drug is physically entrapped within the precipitated polymer matrix and is then slowly released. The polymer type, concentration, and molecular weight as well as the carrier solvent, drug load and formulation additives each influence the release kinetics. Manipulation of these formulation variables provides diverse drug delivery profiles as well as polymer biodegradation rates for specific applications.

Candidate biodegradable polymers for use in the drug delivery system include homopolymers of poly(DL-lactide) (PLA) and copolymers of poly(DL-lactide-*co*-glycolide) (PLG) and poly(DL-lactide-*co*-caprolactone) (PLC). These polymers are similar in chemical composition to biodegradable sutures and have been well characterized in the literature (Kulkarni *et al.*, 1971; Cutright *et al.*, 1971; Gourlay *et al.*, 1978; Rice *et al.*, 1978; Nakamura *et al.*, 1989). They are well tolerated in the body and generally accepted as safe by the medical/pharmaceutical community. Biodegradation of the polymers is effected by their hydrolysis to lactic, glycolic, and hydroxycaproic acids, respectively. These are either metabolized by the Krebs (or tricarboxylic acid) cycle to CO_2 and H_2O (Brady *et al.*, 1973; Gilding, 1981; Woodward *et al.*, 1985; Hollinger and Battistone, 1986) or, in the case of D-lactic acid, are excreted unchanged by the kidney. Biocompatible solvents utilized with the system include *N*-methyl-2-pyrrolidone (NMP) and dimethyl sulfoxide (DMSO). Safety studies conducted with pharmaceutical-grade solvents provide extensive toxicological profiles that support substantial margins of safety for both the neat solvents and ATRIGEL™ formulations prepared with these solvents (Wilson *et al.*, 1965; Jacob and Wood, 1971; David, 1972; Bartsch *et al.*, 1976; Wells and Digenis, 1988; Shirley *et al.*, 1988; Wells *et al.*, 1992; International Specialty Products, unpublished results).

In the following sections, we describe research regarding the delivery of proteins, including model proteins, enzymes, hormones, growth factors, and cytokines, from the ATRIGEL™ system. (A listing of the proteins studied in this system is presented in Table I). The methods of protein analysis and the factors used to control release kinetics are summarized. In addition, *in vitro* and *in vivo* systems employed to determine bioactivity of protein formulations are presented. The implications regarding the controlled delivery of proteins are discussed.

Table I
Proteins Studied in the ATRIGEL™ Drug Delivery System

Model proteins
Ovalbumin
Bovine serum albumin (BSA)
Myoglobin
Cytochrome *c*
Fibronectin (FN)
Enzymes
Trypsin
Lysozyme
Horseradish peroxidase (HRP)
Hormones
Follicle-stimulating hormone (FSH)
Somatotropin (ST)
Growth hormone-releasing factor (GHRF)
Insulin
Cytokines
Fibroblast growth factor (bFGF)
Transforming growth factor-β (TGF-β)
Tumor necrosis factor-β (TNF-β)
Epidermal growth factor (EGF)
Platelet-derived growth factor (PDGF-BB)
Insulin-like growth factor-I (IGF-I)
Interleukin-2 (IL-2)
Interferon-β (IFN-β)

2. EFFECTS OF FORMULATION VARIABLES ON PROTEIN RELEASE KINETICS

The release kinetics of peptides and proteins from any polymeric drug delivery system are a function of many factors including the type of peptide or protein used, its water solubility, crystallinity, water content, etc., and the characteristics of the polymer system. We have found that the release rates of peptides and proteins from the ATRIGEL™ system may be affected by varying a number of parameters of the polymer formulations. These parameters include the type of polymer, the concentration of polymer, the molecular weight of the polymer, the type of solvent, the amount of protein loaded in the formulation, and the addition of additives.

2.1. Polymer Type

The type of polymer used in the formulation can significantly affect release rates of certain proteins and peptides. Two polymers used extensively in ATRIGEL™ formulations are PLA and PLG. Copolymers of PLG are available in different lactide:glycolide ratios, and each ratio gives polymers with different physical characteristics. PLA polymers are more hydrophobic than PLG polymers, and as the lactide:glycolide ratio becomes smaller, the copolymer becomes more hydrophilic. As a result, the release kinetics of certain proteins and peptides may be altered simply by using different copolymer ratios. Figure 1 shows the *in vitro* release profiles of a zona pellucida protein antigen released from PLA and PLG formulations. The protein preparation was mixed in the formulation and drawn into a syringe. A 30–50-mg drop of the formulation was then dispensed into vials containing 3 ml of phosphate-buffered saline (PBS), pH 7.4. The samples were next incubated at 37 °C in a shaking environmental chamber. At specific time points, the PBS was decanted and replaced with fresh buffer, and the vials

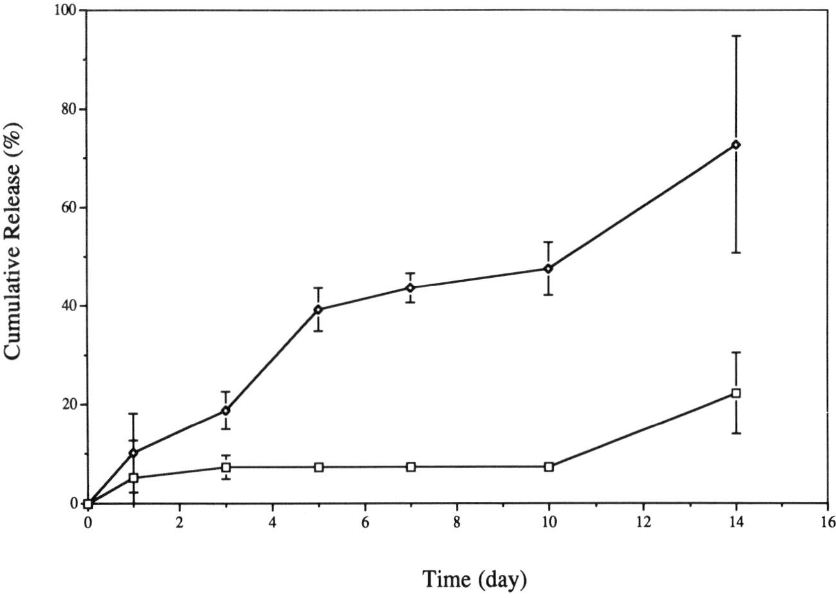

Figure 1. Effect of polymer type on the release of a zona pellucida protein antigen from ATRIGEL™ formulations: □, 45% PLA (inherent viscosity, 0.2); ◇, 45% PLG (inherent viscosity, 0.2). Formulation solvent was NMP, and the protein antigen load was 0.2%. Cumulative release profiles were generated in PBS, pH 7.4.

were replaced in the incubator until the next time point. The decanted buffer was then analyzed for protein content using the bicinchoninic acid assay (BCA; Pierce Chemical Co.). As illustrated in Fig. 1, the PLG formulation displayed a more sustained release of the protein antigen than did a comparable PLA formulation.

2.2. Polymer Concentration

The release kinetics of proteins may also be increased or decreased by varying the concentration of polymer in the formulation. In general, as the concentration of polymer increases in a formulation, the release of any protein is retarded as the density of the polymeric matrix increases. This effect is more prominent in the initial release of drug and is exemplified in Fig. 2 which shows the cumulative release profiles for formulations of increasing PLA concentration that were incorporated with follicle-stimulating hormone (FSH). *In vitro* analysis was done as discussed in Section 2.1,

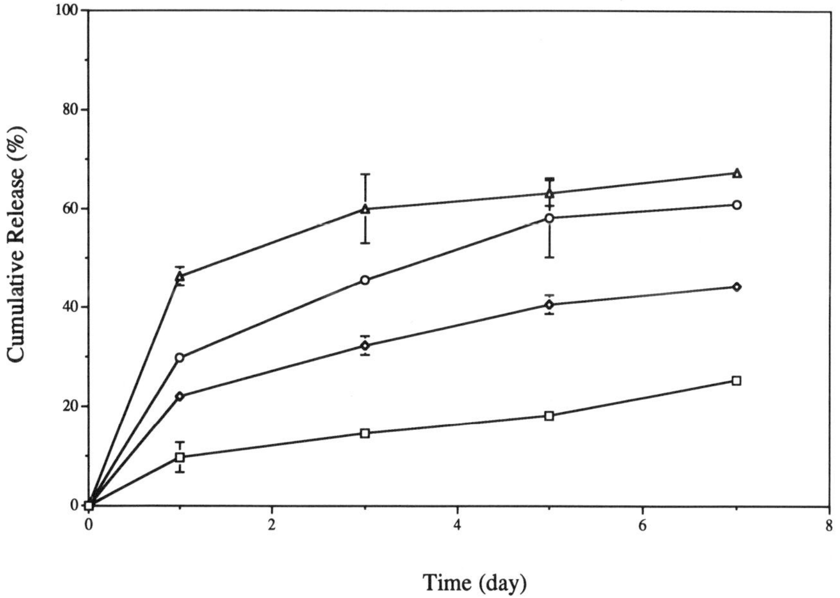

Figure 2. Effect of polymer concentration on the release of follicle-stimulating hormone from ATRIGEL™ formulations: △, 5% PLA; ○, 10% PLA: ◇, 20% PLA; □, 30% PLA. Inherent viscosity of the PLA was 0.75, and the solvent used was NMP. Protein load was 1%. Cumulative release profiles were generated in PBS, pH 7.4.

and total protein was measured by using the BCA assay. As shown in Fig. 2, as the concentration of PLA in the formulation increased from 5% to 30%, the cumulative release of FSH decreased accordingly. This same effect was observed with the ATRIGEL™ system in the release of ganirelix, a GnRH antagonist peptide (Radomsky *et al.*, 1993). Varying the polymer concentration is often a predictable method of controlling the release of peptides and proteins from ATRIGEL™ formulations.

2.3. Polymer Molecular Weight

The molecular weight of the different polymers used in the formulation also plays a role in the release kinetics of proteins. Polymer molecular weights are proportional to their inherent viscosity (iv), which is their flow rate in a solvent expressed in units of deciliters per gram. As polymer chain lengths become longer and more entangled, a characteristic of high-molecular-weight polymers, proteins are hindered in their ability to be released

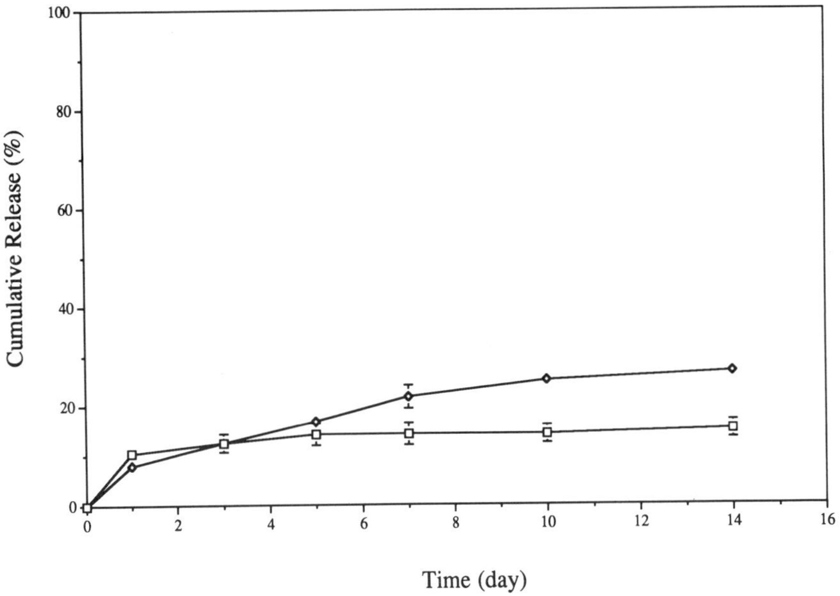

Figure 3. Effect of polymer molecular weight on the release of myoglobin from ATRIGEL™ formulations: ◇, inherent viscosity 0.05; □, inherent viscosity 0.33. The polymer used was 45% PLA. Formulation solvent was NMP, and the protein load was 10%. Cumulative release profiles were generated in PBS, pH 7.4.

from the matrix. To illustrate this, two PLA formulations of equal polymer concentration but different molecular weights were mixed with equivalent loads of myoglobin. *In vitro* release of the protein was carried out as mentioned previously, and protein concentration was determined by BCA analysis. Figure 3 depicts the cumulative release profiles of the two PLA formulations, one of low molecular weight and the other of medium molecular weight. Initially, the two formulations released the protein at similar rates, but with time the lower-molecular-weight polymer released roughly 10% more myoglobin than did its high-molecular-weight counterpart. It is also possible that some degradation of the lower-molecular-weight polymer may have occurred and affected the release of the protein.

2.4. Solvent

The biocompatible solvents used in the system often have different effects on the release of certain peptides and proteins. The two principal

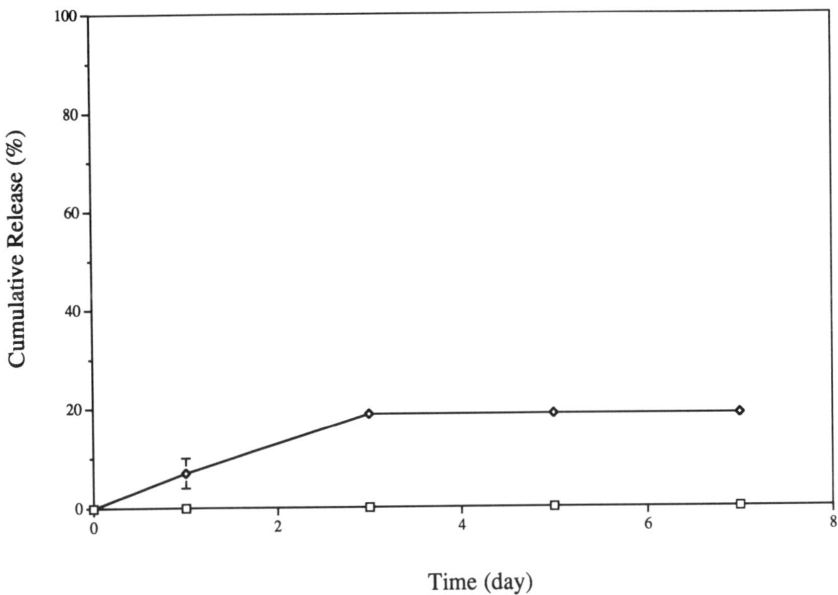

Figure 4. Effects of formulation solvent on the release of a peptide hormone from ATRIGEL™ formulations: ◇, 59% DMSO; □, 59% NMP. The polymer used was 40% PLA (inherent viscosity, 0.22), and the peptide load was 1%. Cumulative release profiles were generated in PBS pH 7.4.

solvents used are NMP and DMSO. Polymers dissolved in NMP often have different coagulation rates than polymers dissolved in DMSO. Also, proteins dissolved in the two solvents behave differently owing to solution or aggregation effects. As a result, the release rates of certain proteins are often affected by using different solvents. Figure 4 shows the release kinetics of bovine growth hormone-releasing factor (Sigma) from two PLA formulations prepared with NMP or DMSO as solvent. The DMSO formulation showed an initial release of approximately 20%, while the NMP formulation displayed no release of the hormone. We suspect that the growth hormone-releasing factor may have aggregated in the NMP solvent and thus became insoluble in water.

2.5. Protein Load

It is also possible to control the release of proteins by varying the protein load within the formulation. The effect of changing the protein load of a formulation was examined by incorporating bovine serum albumin

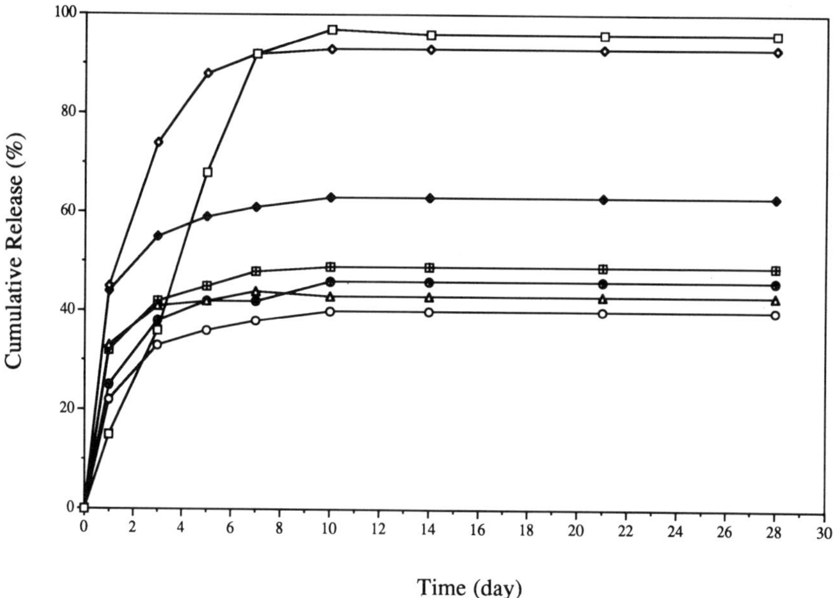

Figure 5. Effect of protein load on the release of bovine serum albumin from ATRIGEL™ formulations: □, 0.01% BSA; ◇, 0.1% BSA; ○, 1.0% BSA; △, 2.5% BSA; ⊞, 5% BSA; ♦, 10% BSA; ⊕, 20% BSA. Formulations consisted of PLA (inherent viscosity, 0.05), with polyvinylpyrrolidone and calcium phosphate as additives. NMP was used as the solvent. Cumulative release profiles were generated in PBS, pH 7.4.

(BSA) in increasing amounts into a PLA formulation. The samples were set up in the *in vitro* model described in Section 2.1, and the amount of protein released was determined using the BCA assay. Figure 5 shows that, in general, as the protein load is increased, a smaller percentage of the total protein in the formulation is released. This effect may be due to increasing protein–protein interactions within the polymer as the load is increased. While this observation does not hold true for the entire range of formulations shown in Fig. 5, it is a valid assessment of protein behavior in the system within certain limits.

2.6. Additives

Sometimes it is necessary to incorporate an additive such as a surfactant into the formulation in order to prevent aggregation of the protein. Because proteins used at relatively high loads in the polymer formulation are prone to giving a poor overall release, surfactants such as sodium dodecyl sulfate (SDS) were added to the formulations. Figure 6 shows the

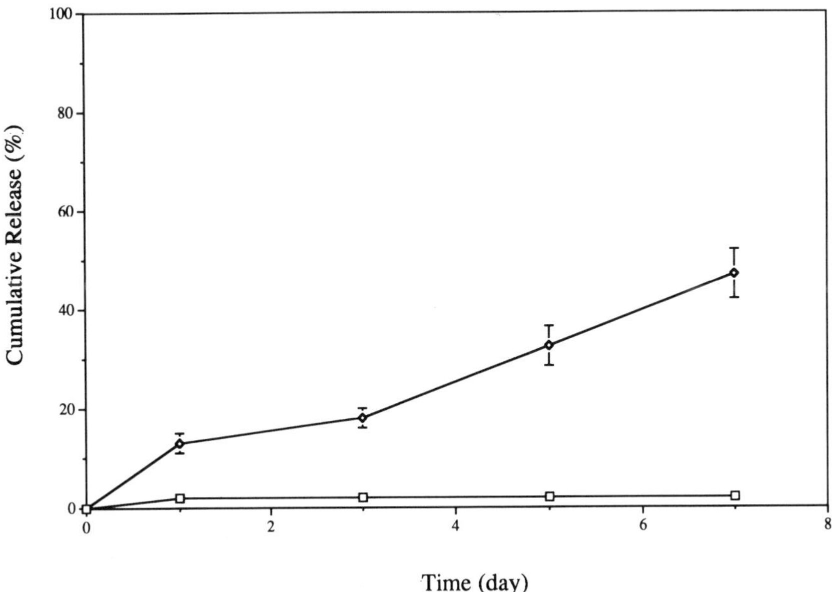

Figure 6. Effect of an additive on the release of ovalbumin from ATRIGEL™ formulations: □, 40% PLA (inherent viscosity, 0.36); ◇, 40% PLA (inherent viscosity, 0.36) with 5% sodium dodecyl sulfate. Formulation solvent was NMP, and the protein load was 2.5%. Cumulative release profiles were generated in PBS, pH 7.4.

release profiles of two PLA formulations containing ovalbumin, one with 5% SDS incorporated and one without. The release profiles were generated using the same *in vitro* model described in Section 2.1. Total protein released was then quantitated using the BCA assay. Approximately 47% of the protein load was released from the formulation with SDS, whereas only about 2% of the protein was released from the formulation without SDS.

3. *IN VITRO* CHARACTERIZATION

3.1. Protein Quantitation in Different Release Media

The *in vitro* release kinetics of numerous growth factors and cytokines were initially determined in phosphate-buffered saline containing azide (PBSA, pH 7.4). Figure 7 depicts the release kinetics of recombinant human tumor necrosis factor-β (rhTNF-β), recombinant human transforming

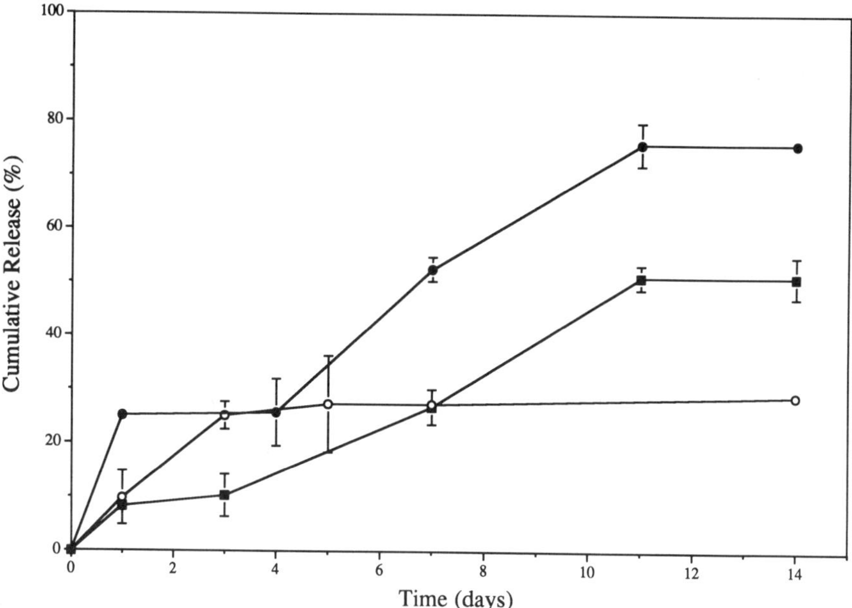

Figure 7. *In vitro* cumulative release profiles of rhTNF-β (●), rhTGF-β (○), and rhbFGF(■) delivered into PBSA receiving fluid, pH 7.4, from ATRIGEL™ formulations. Formulations consisted of 65:35 PLG (inherent viscosity 0.69). NMP was the carrier solvent. Protein loads for the rhTNF-β and the rhTGF-β formulations were 0.002%; the load for the rhbFGF formulation was 0.004%.

growth factor-β (rhTGF-β), and recombinant human fibroblast growth factor (rhbFGF) released from the polymer formulations into PBSA. The procedure for release of proteins into PBSA is the same as that outlined in Section 2.1 for PBS. All of the vials and transfer pipets were siliconized to reduce loss of these hydrophobic proteins during transfer and analysis. In the case of rhTNF-β and rhbFGF, sandwich ELISA methods developed at Atrix Laboratories to facilitate detection of the proteins in PBSA and in the presence of any carrier solvent residue were employed for quantitation. In PBSA, the standard curve for rhTNF-β ranged from 500 to 2000 pg/ml; the range for rhbFGF standards in PBSA was 500–900 pg/ml. Detection and quantitation of rhTGF-β was by densitometric analysis of SDS-polyacrylamide gel electrophoresis (PAGE) gels. Specified amounts of release solutions and rhTGF-β standards were run on 15.0% SDS-PAGE gels (Laemmli, 1990). The gels were then silver-stained (Ohsawa and Ebata, 1983) to allow visualization of the protein bands. Following the silver stain procedure and before drying, the bands were scanned on a Bio-Rad 1-D analyst densitometer. The area of each peak (OD · mm) was integrated, and

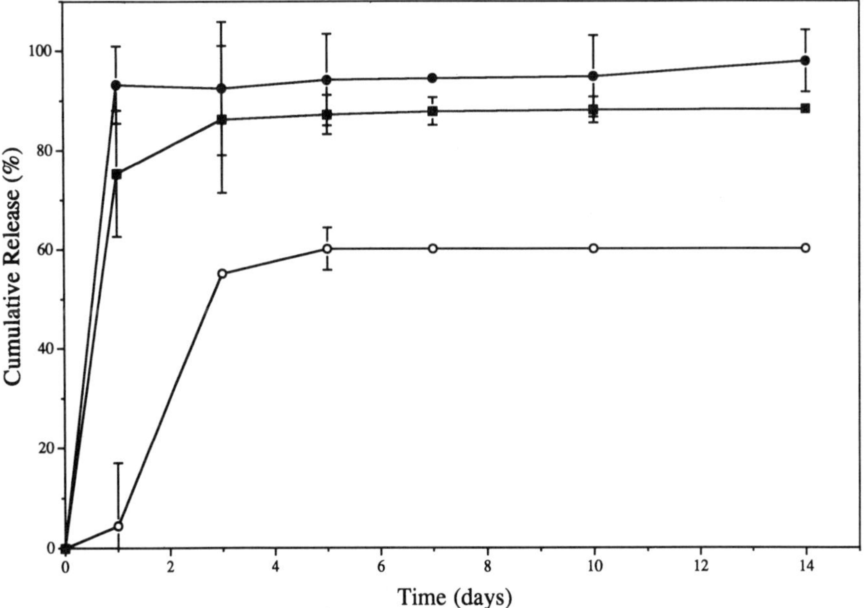

Figure 8. Cumulative release profiles of [^{125}I]-EGF delivered subcutaneously in rats (●), into PBSA receiving fluid, pH 7.4 (○), and into horse serum (■) from a 50:50 PLG (inherent viscosity, 0.19) ATRIGEL™ formulation with DMSO as the carrier solvent. Protein loads in the formulations were 6.7×10^{-6}% [^{125}I]-EGF.

a standard curve was generated from the rhTGF-β standards (range 62.5–500 ng). The release samples were analyzed, and the protein was quantitated from the standard curve.

As *in vivo* release kinetics were determined using radiolabeled proteins, it became apparent that *in vitro* release of proteins into PBSA (pH 7.4) did not properly simulate *in vivo* release. A low initial release of protein by day 1 is characteristically seen in release studies performed in PBSA, whereas in living systems the initial release was much larger. Release into serum provided a more appropriate model for simulating *in vivo* release. Figure 8 shows the cumulative release of epidermal growth factor (EGF) in a serum model and in a PBSA model and an *in vivo* release from a subcutaneous injection of a formulation with radiolabeled EGF. Quantification of the EGF in horse serum was accomplished by use of [^{125}I]-EGF. Detection of [^{125}I]-EGF was by gamma counter. Decay values were considered in the release calculations. Figure 8 depicts the correlation which is seen between *in vivo* release and serum release versus PBSA release of growth factors from the system. The serum release model allows formulations to be tested *in vitro* with somewhat better correlation to *in vivo* results.

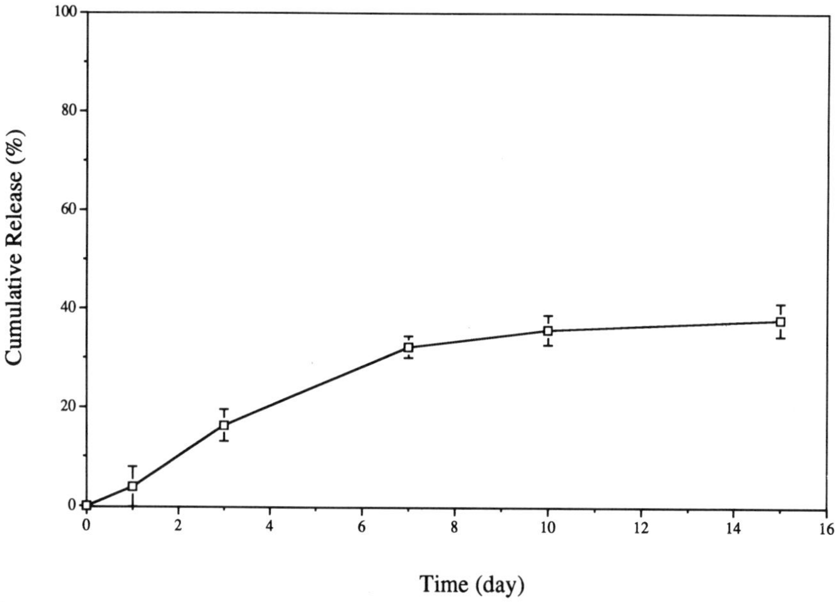

Figure 9. Release kinetics of bovine insulin released from an ATRIGEL™ formulation (60% PLG, inherent viscosity 0.13). Formulation solvent was NMP, and the peptide load was 1%. Cumulative release profiles were generated in PBS, pH 7.4.

Other analytical methods such as reversed-phase high-pressure liquid chromatography (RPHPLC) have also been used to determine the quantity of protein released from the drug delivery system. Figure 9 shows a release profile of insulin from an ATRIGEL™ formulation. Insulin was detected in its native form by RPHPLC through day 15.

3.2. Protein Structure

Maintaining the overall structure of the protein or peptide is of critical importance after its incorporation into and subsequent release from the delivery system. Chromatography and electrophoresis are commonly used in the determination of protein structure. RPHPLC is a very powerful method of detecting changes in protein or peptide structure. Figure 10 shows a pair of chromatograms of bovine insulin. Chromatogram A is that of an 80-μg/ml standard in phosphate-buffered saline, pH 7.4. Chromatogram B shows an insulin sample released from a formulation. The released insulin has the same retention time as that of the standard

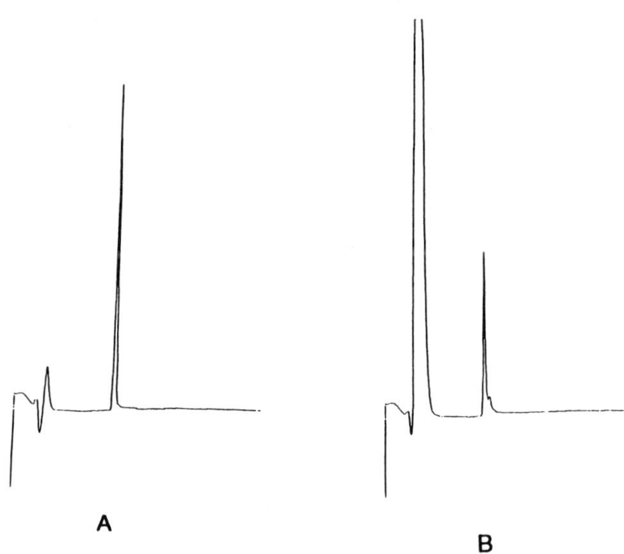

Figure 10. FPLC™ chromatograms of bovine insulin eluted from a C_{18} column: (A) 80-μg/ml standard of bovine insulin; (b) bovine insulin released from a 60% PLG ATRIGEL™ formulation having NMP as the solvent. Cumulative release studies were performed in PBS, pH 7.4.

and appears as a single peak. In general, analysis of protein structure by native gel electrophoresis (Hames and Richwood, 1990) showed that proteins released from the delivery system maintained their native conformation.

A trial was designed to determine the protective effect for proteins incorporated into the system in a proteolytic environment. The samples in this trial were: (1) BSA incorporated at a 5.0% (w/w) level into ATRIGEL™ formulations which were added dropwise to vials containing trypsin (1000 NFU) in 3.0 ml of PBSA; (2) BSA equivalent in concentration to that in the formulations, added to vials containing trypsin (1000 NFU) in 3.0 ml of PBSA; (3) trypsin (1000 NFU) in 3.0 ml of PBSA; and (4) ATRIGEL™ formulation alone added dropwise to vials containing trypsin (1000 NFU) in 3.0 ml of PBSA. The vials containing the different samples were incubated at 37 °C for 24 hr in an environmental shaker. Following the 24-hr incubation, trichloroacetic acid was added to stop the proteolytic activity of the trypsin. Samples of the release solution were removed from each vial and stored at −20 °C until analyzed. The polymer formulations (samples 1 and 4) were washed with water, frozen at 70 °C, and lyophilized overnight. The lyophilized polymer was then added to 1.0 ml of NMP. After the polymer was dissolved, the solution was centrifuged to form a protein pellet. The solution was pipetted off, and the pellet was dissolved in 10 μl of SDS-PAGE gel sample buffer and 10 μl of deionized water. The dissolved protein pellet and the release solutions were then run on a 15% SDS-PAGE gel system (Laemmli, 1970). Protein visualization was by silver stain (Ohsawa and Ebata, 1983).

Analysis of the release solutions by the SDS-PAGE gel method revealed that the trypsin degraded any BSA released from the ATRIGEL™ formulations (sample 1). The free BSA in sample 2 was also degraded, and some self-degradation of the trypsin when used alone (sample 3) was also evident. The protein samples extracted from the precipitated formulations showed only slight protein degradation, which could have resulted from the extraction procedure itself. The extracted fraction from the polymer control without added BSA (sample 4) showed no interference on the gel.

The results demonstrate that the polymer allowed the incorporated BSA to remain intact while the trypsin that surrounded the polymer degraded free BSA readily. The polymer apparently did not allow the influx of the trypsin molecule which was present in the release solution into the polymer implant. The data demonstrate that ATRIGEL™ polymer formulations can protect an incorporated protein or polypeptide from proteolytic degradation. As such, they have the potential to increase its apparent half-life.

3.3. Enzyme Activity

The release of enzymes from the drug delivery system was investigated in order to examine the effects of the delivery system on the tertiary structure and activity of the enzymes. Three enzymes—horseradish peroxidase, lysozyme, and trypsin—were released from ATRIGEL™ formulations, and their activities after release were calculated. The release kinetics of the three enzymes were evaluated in the *in vitro* model described in Section 2.1. Figure 11 shows the release profiles of these three enzymes from PLA formulations. By day 10, the formulation containing horseradish peroxidase had released about 70% of its initial load, the trypsin formulation had released 63%, and the formulation containing lysozyme had released approximately 46% of its original protein load. The activity of these enzymes was also determined following their release from the system. The activity of horseradish peroxidase was measured using the pyrogallol activity assay (Sigma Chemical Co.), the activity of lysozyme was determined by using the *Micrococcus*

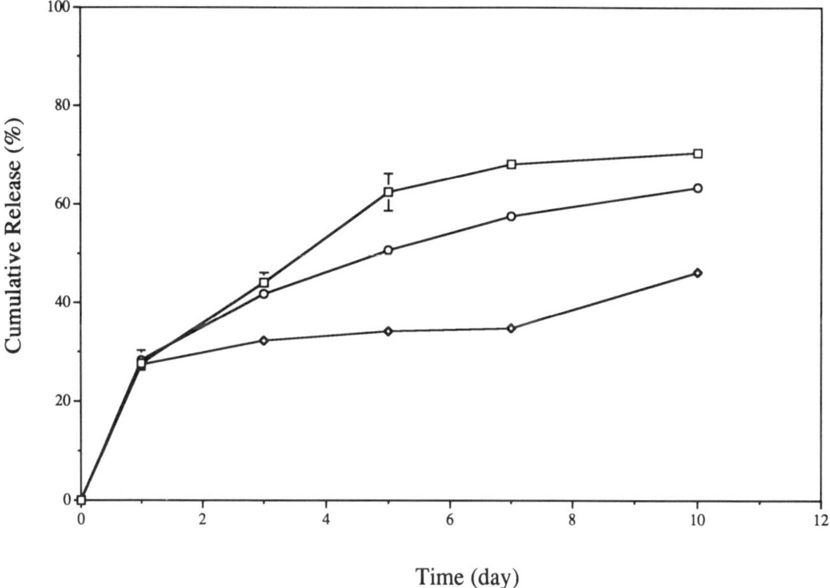

Figure 11. Release kinetics of enzymes released from ATRIGEL™ formulations: □, horseradish peroxidase; ○, trypsin; ◇, lysozyme. Formulations consisted of PLA (inherent viscosity, 0.05), using polyvinylpyrrolidone and calcium phosphate as additives. Formulation solvent was NMP, and the protein load was 5%. Cumulative release profiles were generated in PBS, pH 7.4.

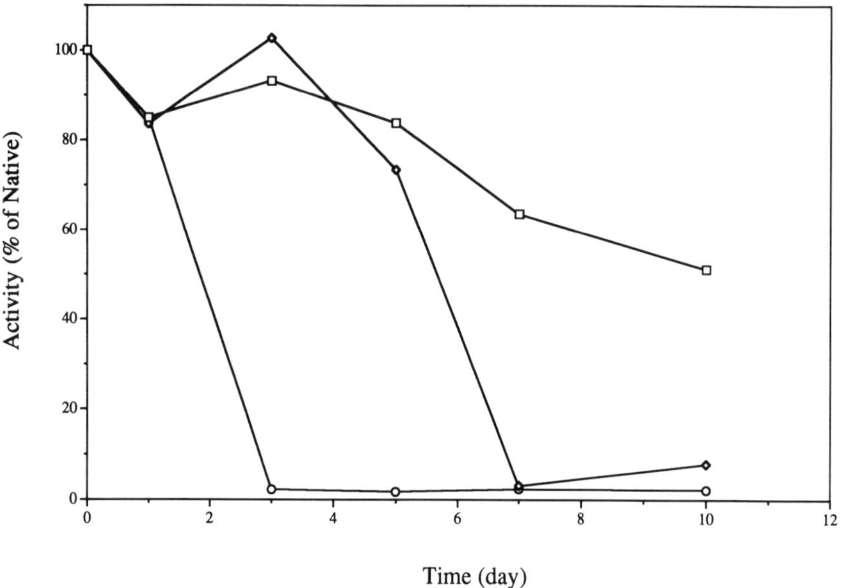

Figure 12. Relative activities of enzymes released from ATRIGEL™ formulations: □, horseradish peroxidase; ○, trypsin; ◇, lysozyme. Formulations consisted of PLA (inherent viscosity, 0.05), using polyvinylpyrrolidone and calcium phosphate as additives. The solvent used was NMP, and the protein load was 5%. Cumulative release profiles were generated in PBS, pH 7.4.

lysodeikticus method, and the activity of trypsin was monitored using the azocoll assay (Chavira *et al.*, 1984). The specific activities of each enzyme solution at indicated time points were determined as a percent of that of the original native enzyme. As shown in Fig. 12, the three enzymes retained their activities to different degrees. Trypsin displayed the shortest retention of activity after release, although most of the activity loss was due to the self-proteolytic nature of trypsin. Lysozyme retained approximately 73% of its activity after 5 days of incorporation in a formulation before losing most of its activity by day 7. Horseradish peroxidase displayed the best retention of activity of the three enzymes examined, retaining about 51% of its native activity after 10 days.

3.4. Cellular Bioactivity

One of the critical factors to be initially determined was the effect that the polymer formulation matrix had on the bioactivity of an incorporated protein. A number of different proteins were tested for retention of activity after release from the liquid polymer system. Following release into PBS and quantitation of the protein, test samples and controls were added to the appropriate cell culture system. The bioactivities of platelet-derived growth factor-BB (PDGF-BB) and insulin-like growth factor-I (IGF-I) were measured by the ability of these growth factors to stimulate [^{3}H]-thymidine incorporation in quiescent BALB/c 3T3 cells (Raines and Ross, 1985). Bioactivity determinations for fibroblast growth factor (FGF) were obtained by [^{3}H]-thymidine incorporation into periodontal ligament (PDL) cells (Somerman, *et al*, 1988). Interleukin-2 (IL-2) bioactivity determinations were made by measuring proliferation of the HT-2 cell line following exposure to samples released from the IL-2 formulation (Gearing and Bird, 1987). Proliferation was measured by MTT dye assay (Freshney, 1994). Cytotoxicity assays were used in determining the bioactivity of TNF-β following release from the system. Cytotoxicity was determined with TNF-susceptible murine L929 cells in the presence of the metabolic inhibitor actinomycin D (Matthews and Neale, 1987). The bioactivity of fibronectin (FN) was determined by stimulation of PDL cell attachment quantified by a Coulter Counter (Somerman *et al.*, 1988).

Table II

Cellular Bioactivity of Proteins Released from the ATRIGEL™ Delivery System

Protein tested[a]	Test system[b]	Days tested/days of retained bioactivity
PDGF-BB	BALB/c 3T3 cells, ^{3}H incorporation	9/9
IGF-I	BALB/c 3T3 cells, ^{3}H incorporation	9/9
FGF	PDL cells,[b] ^{3}H incorporation	3/3
IL-2	HT-2 cells, MTT dye assay	5/5
TNF-β	L929 cells/actinomycin D, MTT dye assay	7/7
FN	PDL cells (cell attachment), Coulter Counter	10/10

[a]For definitions of abbreviations, see Table I.
[b]PDL, Periodontal ligament.

The bioactivity results are summarized in Table II. The length of the bioactivity test periods varied from 3 to 10 days. For all of the samples tested in each group, the protein retained some degree of bioactivity for the duration of the trial although specific bioactivities were not determined. The apparent half-life of each protein was significantly increased following incorporation and release from the system. The results support a protective mechanism of the polymer in the case of these proteins.

4. *IN VIVO* EVALUATIONS

4.1. Biocompatibility

The safety and biocompatibility of the drug delivery system and its components have been extensively tested according to Tripartite Biocompatibility Testing Guidelines (Center for Devices and Radiological Health, 1993). Specifically, these studies have shown that PLA is nontoxic and the hazard potential of NMP is insignificant. Additional preclinical tests to evaluate tissue irritation potential, implantation effects, and biodegradation have been completed for formulations prepared with PLA, PLG, and PLC polymers dissolved in NMP or DMSO. The pharmacokinetics of these formulations have also been tested for specific drug delivery applications.

The subchronic irritation potentials of different formulations of PLA, PLG, and PLC dissolved in NMP or DMSO have been evaluated following subcutaneous and intramuscular injection in mice, rats, and rabbits (Atrix Laboratories, Inc., unpublished results). Based on the Draize scoring method (Draize, 1959) of macroscopic observations taken at necropsy and histopathological analysis, irritation was not remarkable and was characterized by mild to moderate inflammation relative to USP negative control plastic implanted surgically. There were no significant differences between formulations of the different polymer types prepared with NMP or DMSO. The irritation potential of the formulations was not affected following sterilization by gamma irradiation. Further studies have shown that the polymer formulations can decrease the local tissue response to irritating drugs (Atrix Laboratories, Inc., unpublished results).

With respect to biodegradation, a wide spectrum of rates can be achieved by varying the polymer type, molecular weight, and concentration in the formulation. Degradation rates ranging from one week to greater than one year have been observed following subcutaneous surgical implantation or intramuscular and subcutaneous injection of formulations in rats and rabbits (Atrix Laboratories, Inc., unpublished results). In general, for formulations prepared with matched polymer molecular weights and concentrations, the rates of degradation decrease in the order PLG > PLC > PLA. Tissue compatibility of the different formulations was observed throughout the period of biodegradation. However, drug release kinetics are also affected by these formulation variables.

4.2. Protein Release Kinetics

The *in vivo* release kinetics were determined for [^{125}I]-PDGF-BB, [^{125}I]-IGF-I (Institute of Molecular Biology), and [^{125}I]-EGF (Dupont\ NEN) delivered from different formulations in albino rats. Two 50-mg implants were injected subcutaneously per animal in each of the six groups. Formulations were composed of 75:25 PLG with DMSO or NMP as the solvent containing 0.25 μCi of [^{125}I]-PDGF, 0.25 μCi of [^{125}I]-IGF-I, or

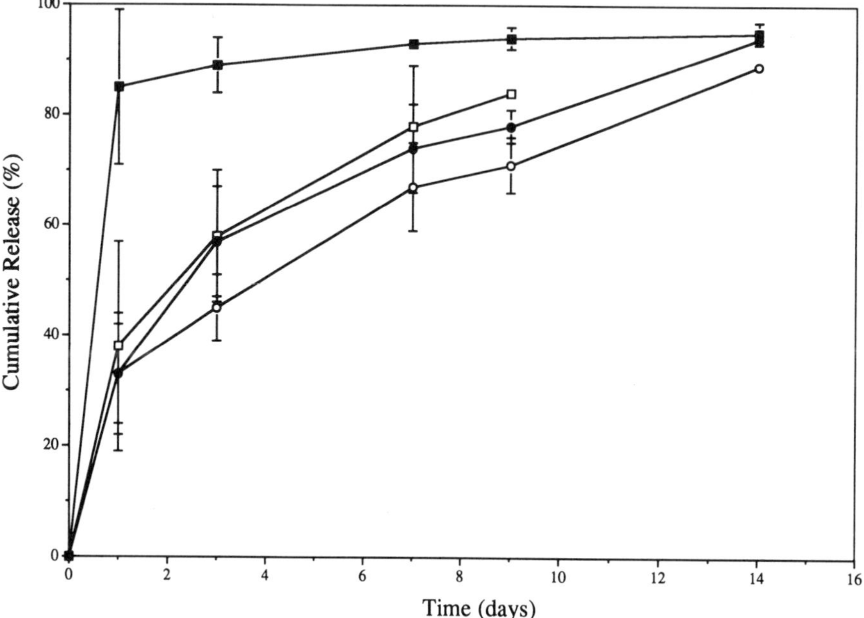

Figure 13. *In vivo* cumulative release profiles of [^{125}I]-PDGF-BB or [^{125}I]-IGF-I from a 75:25 PLG (inherent viscosity, 0.11) ATRIGEL™ formulation prepared with DMSO or NMP as a carrier solvent: ●, PLG/NMP/PDGF-BB; ○, PLG/DMSO/PDGF-BB; ■, PLG/NMP/IGF-I; □, PLG/DMSO/IGF-I. Protein loads in the formulations were 6.7×10^{-6}%.

0.25 μCi of [^{125}I]-EGF. One male and one female from each of six groups were sacrificed on days 1, 3, 7, 9, and 14. The implants were removed from the injection sites, and the remaining radioactivity was determined by gamma-counter detection. The results were then used to calculate the cumulative release kinetics of the six formulations as shown in Figs. 13 and 14.

Figure 13 shows that the [^{125}I]-PDGF-BB formulations in both DMSO and NMP as well as the [^{125}I]-IGF-1 formulation in DMSO have similar release profiles. These formulations exhibited sustained release over a 10- to 14-day period. The [^{125}I]-IGF formulation dissolved in NMP had a large amount of the labeled protein released by day 1, with the remainder being released slowly over the next 13 days.

Figure 14 demonstrates that the formulation dissolved in DMSO released the [^{125}I]-EGF in a somewhat more sustained manner over 7 days than did the formulation in NMP, which exhibited a high initial release *in vivo*.

This trial demonstrated that sustained *in vivo* delivery of protein can be achieved from formulations over a 7-day to 2-week period. Also, the choice

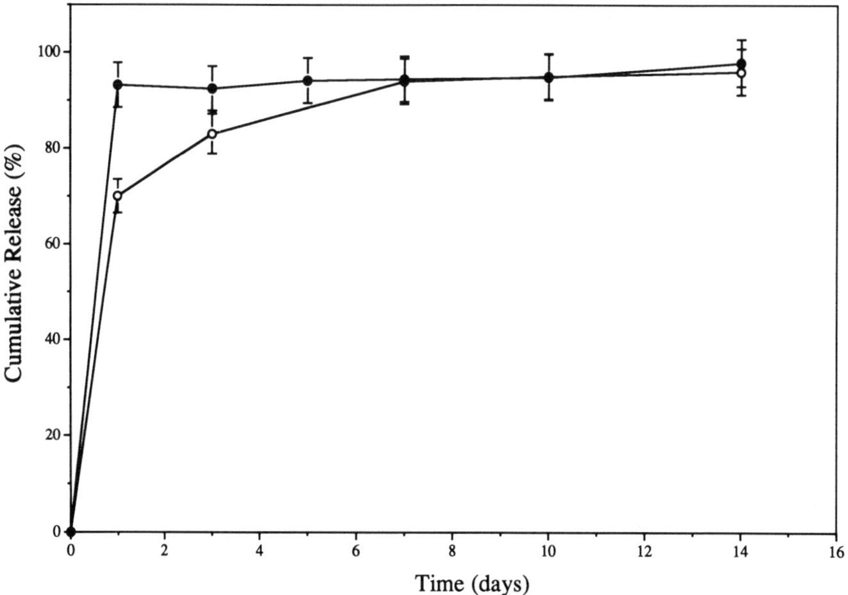

Figure 14. *In vivo* cumulative release profiles of [^{125}I]-EGF from a 75:25 PLG (inherent viscosity, 0.11) ATRIGEL™ formulation prepared with NMP (●) or DMSO (○) as a carrier solvent. Protein loads in the formulations were 6.7×10^{-6}%.

of carrier solvent and the type of protein can have significant effects on the *in vivo* release kinetics.

4.3. Bioactivity

The *in vivo* bioactivity of selected growth factors delivered from formulations was determined utilizing an osteogenic model in rabbits and a dermal wound model in guinea pigs.

The bioactivity of PDGF-BB/IGF-I in saline was compared with that of PDGF-BB/IGF-I in polymer formulation utilizing a rabbit tibia–fibula osteogenic model. Formulation alone and saline alone were used as controls. Tibia–fibula defects (4–5 mm) were induced in healthy New Zealand albino rabbits using conical dental burrs. The marrow cavity was barely exposed in each of the defects, and bleeding was minimal. The fascia and dermis were then sutured over the defect. Following the suturing procedure, 100 μl of test article was injected via a 25-gauge needle into the defect through the sutured site. At given time points, the tibia–fibula region was removed and placed in ice-cold methanol. The samples were then analyzed by histomorphometric analysis to determine new bone growth in the tibial gaps and the internal and external callus. Figure 15 shows the results of the samples retrieved on day 14.

The PDGF-BB/IGF-1 polymer formulation increased new bone formation in the tibial gaps and the internal and external callus compared to the other groups tested. The group containing polymer alone exhibited a greater percentage of bone formation in the tibial gap compared to the growth factor in saline and the saline alone. In the internal and external callus, the growth factor in saline showed increased osteogenesis compared to the formulation alone and saline.

PDGF-BB/IGF-1 delivered from the polymer formulation was found to be more efficacious in the stimulation of new bone growth by day 14 in all regions of the defect analyzed than was a one-dose application of PDGF-BB/IGF-1.

The *in vivo* bioactivity of EGF was determined utilizing a dermal wound model in guinea pigs. A segment of skin 5 mm in diameter was surgically excised from two adjacent sites on opposite sides of the midline in noncompromised albino guinea pigs. EGF in saline, EGF in formulation, or saline alone was injected directly below the wound site into the fascia. The sites were covered with telfa pads, which are nonadherent dressings with a cotton backing that does not stick to wounds and allows ventilation and wicking to keep the wounds dry. Animals were sacrificed on days 1, 3, 5, 7,

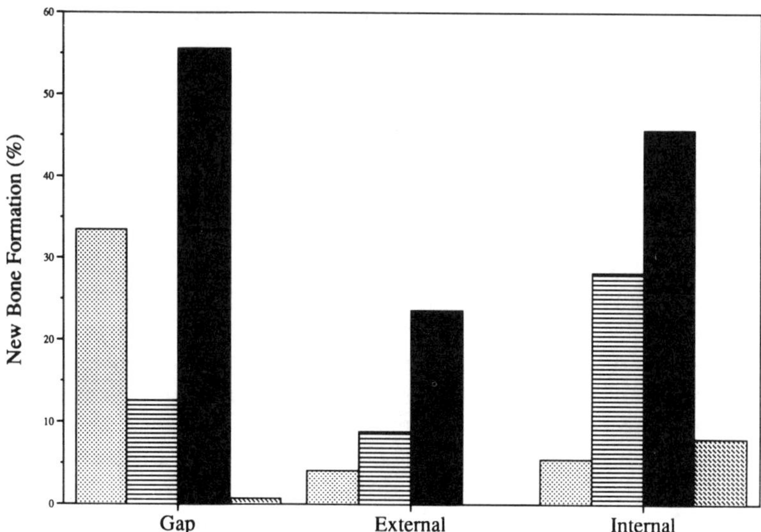

Figure 15. New bone growth in the tibial gap and internal and external callus of a rabbit tibia–fibula model on day 14 post treatment with either a polymer control (PLG/NMP) (▒), PDGF-BB/IGF-I growth factor control (▤), ATRIGEL™ formulation containing PDGF-BB/IGF-I (■), or saline (▧). ATRIGEL™ formulations were composed of 50:50 PLG (inherent viscosity, 0.19) with NMP as the carrier solvent. Protein loads were 6.0 μg of PDGF-BB and 6.0 μg of IGF-I delivered in either saline or ATRIGEL™ formulation per test site.

10, and 14. Measurements of the wound margins (dorsal to ventral and anterior to posterior) were taken immediately following surgery and at the time of necropsy using electronic calipers. In addition, the sites were excised for histological examination.

Figure 16 shows the percent of wound closure for the three groups on each day of analysis. The group containing the EGF released from the polymer formulation exhibited a greater percentage wound closure on days 3, 5, 7, and 10 compared to groups treated with the EGF in saline and the saline alone. As anticipated, by day 14 the wounds in each group showed total closure in this normal, noncompromised wound model. The histopathological summary of each test site stated that the samples with EGF that had been incorporated in the polymer formation showed more advanced healing by day 14, as characterized by increased epithelialization and minimal fibrotic tissue relative to the saline and polymer control groups. In addition, the polymer formulation elicited no irritation response and did not adversely affect the healing process.

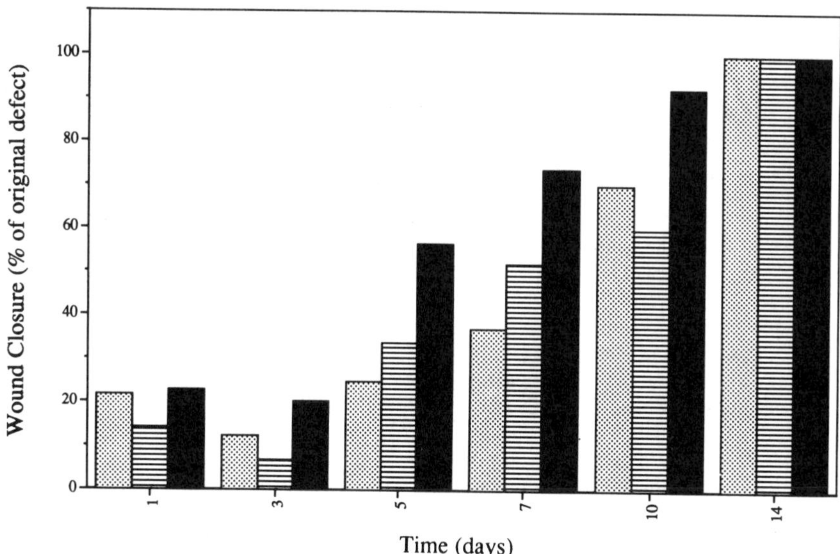

Figure 16. Wound closure in a guinea pig dermal wound model for groups treated with saline (⊡), a control ATRIGEL™ formulation (⊟), or EGF in an ATRIGEL™ formulation (■). ATRIGEL™ formulations were composed of 50:50 PLG (inherent viscosity 0.19) with NMP as the carrier solvent. Protein loads were 1.0 μg of EGF delivered in the ATRIGEL™ formulation per test site.

5. CONCLUSIONS

We have developed a drug delivery system for the controlled release of recombinant proteins. The formulations prepared based on the ATRIGEL™ polymer technology are injectable solutions that form biodegradable implants upon contact with physiological fluids. The system can be formulated to control the release kinetics of the protein and the biodegradation rate of the implant. Proteins are released from the implant structurally intact and in a bioactive form. In addition, the system can protect an incorporated protein from proteolysis and can significantly extend the apparent half-life of the protein through sustained release. The polymeric delivery system is biocompatible following subcutaneous or intramuscular administration and does not interfere with noncompromised dermal or orthopedic wound healing. The results suggest that the system has significant clinical potential for the delivery of bioactive proteins.

REFERENCES

Bartsch, W., Sponer, G., Dietmann, D., and Fuchs, G., 1976, Acute toxicity of various solvents in the mouse and rat. Use of ethanol, dimethylacetamide, dimethylformamide, dimethylsulfoxide, glycerine, *N*-methylpyrrolidone, polyethylene glycol 400, 1,2-propanediol and Tween 20, *Arzneim-Forsch.* **26:**1581–1583.

Bodmer, D., Kissell, T., and Traechslin, E., 1992, Factors influencing the release of peptides and proteins from biodegradable parenteral depot systems, *J. Controlled Release* **21:**129–138.

Brady, J. M., Cutright, D. E., Miller, R. A., Battistone, G. C., and Hunsuck, E. E., 1973, Resorption rate, route of elimination and ultrastructure of the implant site of polylactic acid in the abdominal wall of the rat, *J. Biomed. Mater. Res.* **7:**155–166.

Center for Devices and Radiological Health, 1993, Tripartite biocompatibility testing guidelines, Division of Small Manufacturers Assistance, Food and Drug Administration, Washington, D.C.

Chavira, R., Jr., Burnett, T. J. and Hageman, J. H., 1984, Assaying proteinases with azocoll, *Anal. Biochem.* **136:**446–450.

Cutright, D. E., Beasley, J. D., III, and Perez, B., 1971, Histologic comparison of polylactic and polyglycolic acid sutures, *Oral Surg.* **32:**165–173.

David, N. A., 1972, The pharmacology of dimethyl sulfoxide, *Annu. Rev. Pharm.* **12:**353–374.

Draize, J. H., 1959, *The Appraisal of Chemicals in Foods, Drugs and Cosmetics,* Association of Food and Drug Officials of the United States, Austin, Texas, pp. 46–59.

Dunn, R. L., Yewey, G. L., and Tipton, A. J., 1992, An injectable implant delivery system for tissue growth factors, American Association of Pharmaceutical Scientists Western Regional Meeting, Las Vegas.

Freshney, R. I., 1994, *Culture of Animal Cells,* Wiley-Liss, New York, pp. 296–299.

Gearing, A. J. H., and Bird, C. B., 1987, Production and assay of interleukine 2, in: *Lymphokines and Interferons, A Practical Approach,* (M. J. Clements, A. G. Morris, and A. J. H. Gearing, eds.) IRL Press, Oxford, England, pp. 281–301.

Gilding, D. K., 1981, Degradation of polymers: Mechanisms and implications for biomedical applications, in: *Fundamental Aspects of Biocompatibility,* Vol. 1 (D. F. Williams, ed.), CRC Press, Boca Raton, Florida, pp. 44–65.

Gourlay, S. J., Rice, R. M., Hegyeli, A. F., Wade, C. W. R., Dillon J. C., Jaffe, H., and Kulkarni, R. K., 1978, Biocompatibility testing of polymers: *In vivo* implantation studies. *J. Biomed. Mater. Res.* **12:**219–232.

Hames, B. D. and Richwood, D., 1990, *Gel Electrophoresis of Proteins, A Practical Approach,* IRL Press, Oxford, England, p. 16.

Hollinger, J. O., and Battistone, G. C., 1986, Biodegradable bone repair materials. *Clin. Orthop. Relat. Res.* **207:**290–305.

Jacob, S. W., and Wood, D. C., 1971, Dimethyl sulfoxide—A status report, *Clin. Med.* **78:**21–31.

Kulkarni, R. K., Moore, E. G., Hegyeli, A. F., and Leonard, F., 1971, Biodegradable poly(lactic acid) polymers, *J. Biomed. Mater. Res.* **5:**169–181.

Laemmli, U. K., 1970, Cleavage of structural proteins during the assembly of the head of bacteriophage T4, *Nature* **227:**680–685.

Matthews, N., and Neale, M. L., 1987, Cytotoxicity assays for tumour necrosis factor and lymphotoxin, in: *Lymphokines and Interferons, a Practical Approach,* (M. J. Clemens, A. G. Morris, and A. J. H. Gearing, eds.) IRL Press, Oxford, pp. 221–225.

Nakamura, T., Hitomi, S., Watanabe, S., Shimizu, Y., Jamshidi, K., Hyon, S. H., and Ikada, Y., 1989, Bioabsorption of polylactides with different molecular properties, *J. Biomed. Mater. Res.* **23**:1115–1130.

Ohsawa, K., and Ebata, N., 1983, Silver stain for detecting 10 femtogram quantities of protein after polyacrylamide gel electrophoresis, *Anal. Biochem.*, **135**:409–415.

Radomsky, M. L., Brouwer, G., Floy, B. J., Loury, D. J., Chu, F., Tipton, A. J., and Sanders, L. M., 1993, The controlled release of ganirelix from the ATRIGEL™ injectable implant system, *Proc. Int. Symp. Control. Rel. Bioact. Mater.* **20**:458–459.

Raines, E., and Ross, R., 1985, Purification of human plateletderived growth factor, *Methods Enzymol.* **109**:749–772.

Rice, R. M., Hegyeli, A. F., Gourlay, S. J., Wade, C. W. R., Dillon, J. C., Jaffe, H., and Kulkarni, R.K., 1978, Biocompatibility testing of polymers: *In vitro* studies with *in vivo* correlation, *J. Biomed. Mater. Res.* **12**:43–54.

Shirley, H. H., Lundergan, M. K., Williams, H. J., and Spruance, S. L., 1989, Lack of ocular changes with dimethylsulfoxide therapy of scleroderma, *Pharmacotherapy* **9**:165–168.

Somerman, M. J., Foster, R. A., Vorsteg, G., Progebin, K., and Wynn, R. L., 1988, Effects of minocycline on fibroblast attachment and spreading, *J. Periodont. Res.* **23**:154–159.

Wells, D. A., and Digenis, G. A., 1988, Disposition and metabolism of double-labeled (^{3}H and ^{14}C) *N*-methyl-2-pyrrolidone in the rat, *Drug Metab. Dispos.* **16**:243–249.

Wells, D. A., Hawi, A. A., and Digenis, G. A., 1992, Isolation and identification of the major urinary metabolite of *N*-methylpyrrolidone in the rat, *Drug Metab. Dispos.* **20**:124–126.

Wilson, J. E., Brown, D. E., and Timmens, E. K., 1965, A toxicologic study of dimethyl sulfoxide, *Toxicol. Appl. Pharmacol.* **7**:104–112.

Woodward, S. C., Brewer, P. S., Moatamed, F., Schindler, A., and Pitt. C. C., 1985, The intracellular degradation of poly(ε-caprolactone), *J. Biomed. Mater. Res.* **19**:437–444.

Chapter 4

Protein Delivery from Nondegradable Polymer Matrices

Tammy L. Wyatt and W. Mark Saltzman

1. INTRODUCTION

Among the many technologies for controlled release of bioactive compounds, controlled release systems based on nondegradable,* hydrophobic polymers are the most successful. Silicone elastomer tubes were first used to provide controlled release of small molecules 30 years ago (Folkman and Long, 1964). This discovery led to the Norplant® (Wyeth-Ayerst Laboratories) contraceptive delivery system, which provides reliable delivery of levonorgestrel for five years following subcutaneous implantation. Norplant® has been available to women in the United States since 1990, where it has been well received (Frank *et al.*, 1992). Similar polymers, most notably poly(ethylene-*co*-vinyl acetate) (EVAc), have been used to control the delivery of contraceptive hormones to the female reproductive tract (Progestasert®, Alza Corporation) and lipophilic drugs to the eye or the skin (Ocusert®, Alza Corporation; Estraderm® and Transderm Nitro®, Ciba-Geigy Corporation). Because macromolecules like proteins do not diffuse through films of silicone and EVAc, many investigators believed that it was

*Nondegrable, in the context of this chapter, refers to polymers that are not broken down by the body during the period of implantation, which may be several years.

Tammy L. Wyatt and W. Mark Saltzman • Department of Chemical Engineering, The Johns Hopkins University, Baltimore, Maryland 21218. *Current address of W.M.S.:* School of Chemical Engineering, Cornell University, Ithaca, New York, 14853.

Protein Delivery: Physical Systems, Sanders and Hendren, eds., Plenum Press, New York, 1997.

impossible to develop similar systems capable of releasing proteins (Stannet *et al.*, 1979). In the late 1970s, however, a method for achieving controlled release of proteins from nondegradable polymers was described (Langer and Folkman, 1976). These polymers provide sustained release of biologically active molecules for extended periods of time, up to several years in some cases (Saltzman and Langer, 1989).

1.1. Biocompatible Polymers Used as Hydrophobic Matrices

One particular hydrophobic polymer, EVAc, has been investigated extensively as a matrix system for protein delivery. This polymer is biocompatible, a major consideration because of the interest in developing systems for human health. Other classes of hydrophobic polymers, like silicone elastomers and polyurethanes, may also be useful for controlled protein delivery, although there are fewer examples available in the literature. Nondegradable, hydrophilic polymers, such as poly(2-hydroxyethyl methacrylate) [p(HEMA)], are also biocompatible but usually release proteins over a relatively short period. However, a few examples of long-term release of peptides and proteins from hydrophilic polymers are available. Long-term release of peptides from devices that employ cross-linked p(HEMA) as rate-limiting barriers has been reported (Davidson *et al.*, 1988). The use of hydrophilic polymers for protein release is discussed in more detail elsewhere in this volume.

1.1.1. POLY(ETHYLENE-*co*-VINYL ACETATE)

EVAc is a random copolymer with the structure

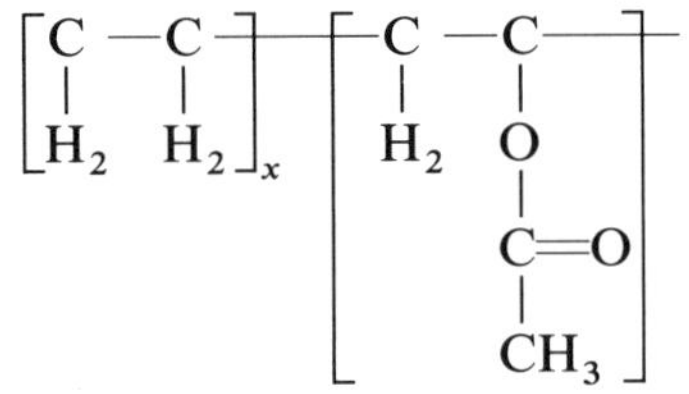

$$\left[\begin{array}{cc} \mathrm{C} & -\mathrm{C} \\ | & | \\ \mathrm{H_2} & \mathrm{H_2} \end{array}\right]_x \left[\begin{array}{cc} \mathrm{C} & -\mathrm{C} \\ | & | \\ \mathrm{H_2} & \mathrm{O} \\ & | \\ & \mathrm{C{=}O} \\ & | \\ & \mathrm{CH_3} \end{array}\right]$$

The most commonly used EVAc (ELVAX-40, Du Pont) consists of approximately 40% vinyl acetate, with a low degree of crystallinity (5–20%). EVAc is hydrophobic; it swells less than 0.8% in water (Hsu and Langer, 1985).

Matrices composed of EVAc and protein can be fabricated by solvent evaporation or compression molding (Siegal and Langer, 1984). In solvent evaporation, EVAc, which has been extensively washed to remove low-molecular-weight oligomers and impurities, is dissolved in methylene chloride. The protein of interest is typically lyophilized, ground and sieved to a desired particle size range, and suspended in the polymer solution. The suspension is poured into a chilled mold and allowed to solidify. The matrix is then removed from the mold and dried at atmospheric pressure and −20 °C for 48 hr and then dried under vacuum at 20 °C for 48 hr.

The biocompatibility of EVAc matrices has been studied quite extensively. When implanted in the cornea of rabbits, which is sensitive to the edema, white-cell infiltration, and neovascularization associated with inflammation, washed EVAc caused no inflammation; unwashed EVAc caused mild inflammation (Langer and Folkman, 1977). After seven months of subcutaneous implantation, only a thin capsule of connective tissue surrounded EVAc implants; no inflammation was present and the adjacent loose connective tissue was normal (Brown *et al.*, 1983). When implanted in the brains of rats, EVAc matrices produce only mild gliosis (During *et al.*, 1989). EVAc has shown good biocompatibility in humans over the years and has been approved by the U.S. Food and Drug Administration (FDA) for use in a variety of implanted and topically applied devices.

1.1.2. SILICONE

Silicone elastomers, also called polysiloxanes, have the general structure

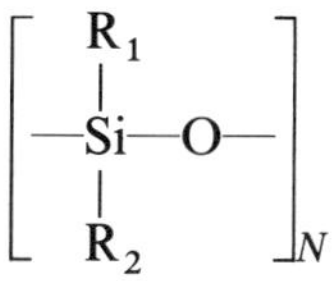

$$\left[-\underset{R_2}{\overset{R_1}{\underset{|}{\overset{|}{Si}}}}-O- \right]_N$$

where R_1 and R_2 represent possible substituent groups. The properties of silicone are determined by these substituents. The most commonly used silicone in biomedical applications is polydimethylsiloxane, in which both R_1 and R_2 are methyl groups (Silastic®, Dow Chemical Corporation). For the Norplant® delivery system, a copolymer of dimethylsiloxane and methylvinylsiloxane was used. These polymers are characterized by high chain flexibility and unusually high oxygen permeability. Polysiloxanes are very stable toward hydrolysis, probably as a result of their hydrophobicity.

Silicone matrix systems have good biocompatibility when implanted subcutaneously (Folkman and Long, 1964) or intercranially (Becker *et al.*, 1990). Until recently, it was assumed that silicone polymers were almost completely inert in biological systems. In fact, silicone has received approval from the FDA for many biomedical applications including breast prostheses, heart valve prostheses, and drug delivery systems such as Norplant®. Based on several laboratory studies and many clinical reports, the FDA is reevaluating the use of silicone gel breast implants (Kessler, 1992). Most of the reported problems appear to be caused by leakage of silicone gel from within the breast prostheses or degradation of polyurethane coatings. Therefore, silicone elastomers are still considered safe for use in drug delivery devices.

1.1.3. POLYURETHANES

Polyurethanes were first suggested for use as biomaterials in 1967 (Boretos and Pierce, 1967). Currently, a variety of polyurethanes are used in biomedical devices such as coatings for catheters and pacemaker leads. Because of the long experience with implanted polyurethane devices, there is some interest in using polyurethane matrices to release peptide and protein drugs (Horbett *et al.*, 1993).

The biocompatibility of biomedical polyurethanes appears to be determined by their purity (i.e., effective removal from the polymer of catalyst residues and low-molecular-weight oligomers) (Gogolewski, 1989). The surface properties of commercially available polyurethanes, which are critically important in determining biocompatibility, can vary considerably, even among lots of the same commercially available preparation (Tyler *et al.*, 1992). In some cases, most famously the polyurethane-foam-coated breast prostheses, implanted polyurethanes have been reported to degrade into carcinogenic compounds (Batich *et al.*, 1989).

1.2. Protein Release from Polymer Matrices

When protein-loaded EVAc matrices are immersed in water, the proteins are slowly released (Fig. 1). The initial rate of release from the matrix is higher for matrices with higher loading (higher initial mass fraction of protein particles within the matrix) (Fig. 1a). This release is frequently linear with respect to the square root of time (Fig. 1b), suggesting a diffusive release mechanism. The kinetics of release can be characterized by an effective diffusion coefficient for the protein in the polymer matrix, as described in Section 2.1. The diffusion coefficient, which is typically much less than the diffusion coefficient of the protein in water, provides a

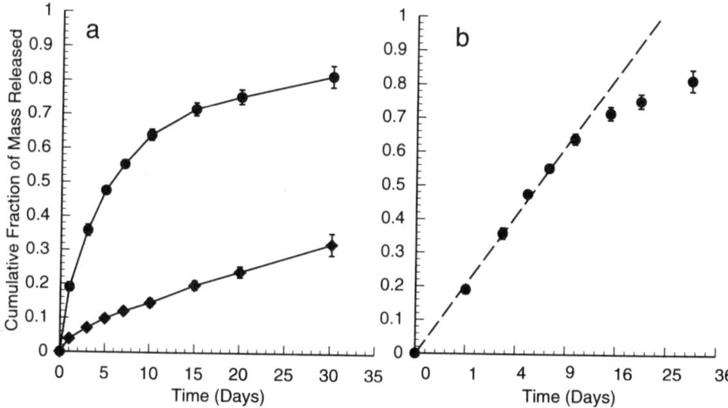

Figure 1. Release of ferritin (500-kDa protein) from a matrix of EVAc. (a) The cumulative fraction of mass released from matrices containing 35% (●) or 50% (♦) ferritin by mass is plotted versus time. (b) The same cumulative mass fraction released from the 50% loaded matrices is plotted versus the square root of time. The dashed line represents the fit to the linear model of desorption from a slab, Eq. (5). Data points represent the mean cumulative fraction of mass of ferritin released from four EVAc matrices incubated in buffered saline at 37 °C. The error bars represent ±1 SD of the mean. Some error bars are smaller than the symbols.

quantitative measure of the rate of protein release, decreasing as the rate of protein release from the matrix decreases.

EVAc matrix systems have been used to release a variety of macromolecules, such as polypeptide and protein hormones (Fischel-Ghodsian *et al.*, 1988; Brown *et al.*, 1986), heparin (Edelman *et al.*, 1990), growth factors (Hoffman *et al.*, 1990; Edelman *et al.*, 1991; Murray *et al.*, 1983; Powell *et al.*, 1990; Beaty and Saltzman, 1992), inhibitors of tumor angiogenesis (Lee and Langer, 1983), polyclonal antibodies (Saltzman and Langer, 1989), monoclonal antibodies (mAb; Radomsky *et al.*, 1992; Sherwood *et al.*, 1992, 1996; Saltzman *et al.*, 1993), and antigens (Preis and Langer, 1979; Wyatt, 1994). Macromolecules have been shown to retain their biological activity after release from EVAc. For example, a mAb against a human chorionic gonadatropin (hCG) retained its ability to bind to hCG after release from an EVAc matrix (Randomsky *et al.*, 1992; Sherwood *et al.*, 1992). In addition, when released from EVAc matrices, nerve growth factor stimulated neurite outgrowth in cultured cells (Powell *et al.*, 1990), insulin altered the blood glucose levels in diabetic rats (Langer and Folkman, 1977; Brown *et al.*, 1986), and angiogenesis inhibitors blocked new blood vessel growth (Lee and Langer, 1983).

2. MECHANISMS AND MODELS FOR PROTEIN RELEASE FROM MATRICES

EVAc matrices prepared by solvent evaporation consist of protein particles dispersed throughout a continuous polymer phase (Fig. 2a). When matrices are placed in an aqueous environment, particles at the surface of the matrix can dissolve. Since macromolecules cannot diffuse through the continuous polymer phase, and since the polymer is hydrophobic and does not swell in water, protein release must occur through pores in the polymer, which form as the dispersed protein particles dissolve. In fact, microscopic observations of the matrix structure reveal a network of interconnected pores in which large pores (diameters of 100–400 μm) are connected by smaller pores or channels (1–10 μm in diameter) (Fig. 2b; Saltzman *et al.*, 1987; Hsu and Langer, 1985; Siegel and Langer, 1984). Connected clusters of pores that contact the matrix boundary can release protein to the surrounding environment. At loadings higher than 35%, most protein particles are found in clusters that reach the matrix surface, while at lower loadings most particles are disconnected (Fig. 2a; Saltzman and Langer, 1989).

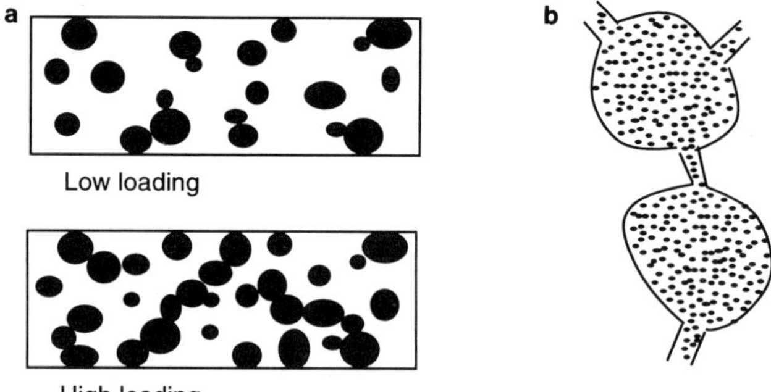

Figure 2. Schematic view of a protein/hydrophobic polymer matrix system. (a) Cross-sectional views of matrices at two different loadings. At higher loadings, connected clusters of pores are formed, providing continuous paths for diffusion to the surface. (b) Typical pores within the matrix are enlarged. Large pores (diameters of 100–400 μm) are connected by smaller channels (diameters of 1–10 μm). The constricted diffusional pathway contributes significantly to the slowness of release.

2.1. Macroscopic Models of Diffusion in Porous Polymer Matrices

Consider a protein-loaded matrix constructed as a thin slab. Release of the protein occurs essentially through the top and bottom faces of this slab because the thickness of the slab is small compared to the other dimensions. The desorption of protein from this slab can be described by Fick's second law of diffusion:

$$\frac{\partial C}{\partial t} = D_{\text{eff}} \frac{\partial^2 C}{\partial x^2} \tag{1}$$

where t is the time since the start of release, x is the direction normal to the top and bottom faces of the slab, C is the concentration of protein at position x and time t, and D_{eff} is the effective diffusion coefficient for the protein in the polymer matrix. As described above, definition of an effective diffusion coefficient is necessary because protein molecules do not diffuse through the pure polymer phase but must find a path out of the slab by diffusing through a tortuous, water-filled network of pores. Note that D_{eff} is assumed to be independent of position in the slab; Eq. (1) is obtained by local averaging over a volume that is large compared to a single pore.

Since protein is initially distributed uniformly throughout the slab, and since we assume that protein is rapidly removed from the surface of the matrix, the following initial and boundary conditions apply for this geometry:

$$\begin{array}{lll} C = C_0 & 0 \leqslant x \leqslant L & t = 0 \\ C = 0 & x = 0 & t > 0 \\ C = 0 & x = L & t > 0 \end{array} \tag{2}$$

where L is the thickness of the slab. The solution to Eq. (1), using the above initial and boundary conditions, is (Vergnaud, 1991)

$$\frac{C}{C_0} = \frac{4}{\pi} \sum_{n=0}^{\infty} \frac{1}{2n+1} \sin\left(\frac{(2n+1)\pi x}{L}\right) \exp\left(-\frac{(2n+1)^2 \pi^2 D_{\text{eff}} t}{L^2}\right) \tag{3}$$

The cumulative fraction of protein released (i.e., the amount released at time t divided by the amount originally dispersed in the matrix) is found by integrating Eq. (3) over the thickness of the matrix:

$$\frac{M_t}{M_\infty} = 1 - \frac{8}{\pi^2} \sum_{n=0}^{\infty} \frac{1}{(2n+1)^2} \exp\left(-\frac{(2n+1)^2\pi^2}{L^2} D_{\text{eff}} t\right) \tag{4}$$

For cumulative fractions released of <0.6, an approximate solution can be used:

$$\frac{M_t}{M_\infty} = \frac{4}{L}\left(\frac{D_{\text{eff}} t}{\pi}\right)^{1/2} \tag{5}$$

Characteristic desorption curves, plots of Eq. (4), are shown in Fig. 3 for D_{eff} between 1×10^{-7} and $1 \times 10^{-9}\,\text{cm}^2/\text{sec}$. The similarity between these curves and the experimental release profile in Fig. 1b is obvious. By fitting Eq. (5) to the experimental data for fractional release with the square root of time, D_{eff} can be estimated (Fig. 1b).

The macroscopic geometry of a matrix can influence the rate and pattern of protein release. Increasing the surface-area-to-volume ratio of the matrix increases the release rate by allowing more particles direct access to the matrix exterior. The common matrix form of a slab has a release rate proportional to the square root of time (Fig. 1b). However, if the matrix is

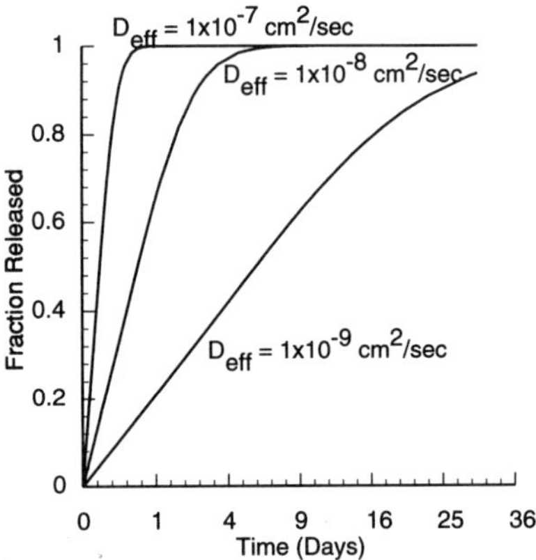

Figure 3. Retardation of release from a matrix. The cumulative fraction released from a 1-mm-thick matrix is plotted versus the square root of time for different assumed values of the diffusion coefficient. As the diffusion coefficient decreases, the release rate decreases. The diffusion coefficient is an appropriate parameter for characterizing the rate of release from these systems.

formed as a hemisphere with an impermeable surface coating, which has a small defect in its planar surface, essentially zero-order kinetics can be obtained (Rhine *et al.*, 1980b).

The effective diffusion coefficient is related to the molecular diffusion coefficient of the protein in water, D_0, by

$$D_{\text{eff}} = \frac{D_0}{F\tau} \tag{6}$$

where τ is the tortuosity and F is the shape factor. The effective diffusion coefficient is less than the molecular diffusion coefficient because of the increased diffusional path length due to windiness of the pore structure, τ, and constrictions in the pore structure, F. The effects of F and τ on the effective diffusion coefficient are difficult to separate experimentally. Because of this, empirical values of the product $F\tau$ are often reported as "tortuosity," a quantitative measure of the retardation in protein diffusion through the polymer matrix. Using overall tortuosity ($F\tau$) as an adjustable parameter—and assuming that the diffusion coefficient in water, D_0, is known—Eq. (4) or (5) can be fit to experimental release data for different proteins. The fits to individual release profiles, which represent one protein (albumin, β-lactoglobulin, or lysozyme) at fixed fabrication conditions (loading of matrix, particle size range of protein, and polymer molecular weight), are reasonable (Bawa *et al.*, 1985).

For a protein incorporated in the usual slab matrix, the rate of release, and therefore the effective diffusion coefficient and $F\tau$, depend on the molecular weight of the protein, the size of the dispersed particles, the loading, and the molecular weight of the polymer. Tables I and II list the tortuosity values calculated from fits of the desorption equation [Eq. (4) or (5)] to available literature data of protein release from EVAc slabs.

Increasing protein particle size increases the rate of protein release (Table I and Fig. 4b). The size of the protein particles in the matrix affects the size of the water-filled channels formed as the particles dissolve. Larger particles occupy more volume in a matrix, increasing the connectivity with other pores. Increased connectivity provides simpler pathways (i.e., less tortuous and less constricted pathways) for diffusion of protein molecules (Rhine *et al.*, 1980a; Siegel and Langer, 1984; Saltzman and Langer, 1989). This can be seen in Table I for ferritin, bovine serum albumin (BSA), γ-globulin, and β-lactoglobulin. As the average particle size increases, the overall tortuosity decreases.

Increasing the loading of protein in the matrix also increases the release rate of proteins (Table I and Fig. 4c). For ferritin, BSA, γ-globulin, and

Table I
Tortuosity Values of Macromolecules Released from EVAc Matrices at Various Particle Sizes and Loadings

Macromolecule	Particle size (μm)	Tortuosity value at a loading (%) of						
		10	20	25	30	35	40	50
Ferritin (500 K)	< 178[a]					2,800		200
γ-Globulin (150 K)	150–250[b]	30,000	56,000	47,000	9,300	12,000	5,900	580
BSA (68 K)	< 75[c]	37,500		28,000				220
	45–75[b]	20,000	46,000	19,000	29,000	22,000	4,500	600
	106–150[d]		11,000	2,300	690			
	150–180[d]		11,000	2,300	860			
	150–250[b]	26,000	7,400	5,700	9,400	1,500	870	160
	75–250[c]	5,700		1,300				59
	250–425[c]	1,700		660				47
	250–425[d]			560	180			
	300–425[e]		95		55		19	
β-Lactoglobulin (18 K)	75–150[d]			10,000			1,600	1,100
	106–250[c]			1,500				
Lysozyme (14 K)	75–250[c]			19,000				
	106–150[d]						2,100	
Heparin (14 K)	< 104[f]						30,000	18,000
Insulin (6 K)	< 75[g]		69			4,800		3,700
	75–250[g]		560			760		790
	250–450[g]		270			180		190

[a]Wyatt, 1994.
[b]Saltzman, 1987.
[c]Rhine *et al.*, 1980a.
[d]Bawa *et al.*, 1985.
[e]Miller and Peppas, 1983.
[f]Unpublished data.
[g]Siegel and Langer, 1984.

Table II
Tortuosity Values of Proteins Released from Different Molecular Weights EVAc Matrices

EVAc molecular weight (M_n/M_w)	Tortuosity value: γ-globulin[a] @ loading, 117–180 μm	Tortuosity value: BSA[b] @ 25% loading, 106–150 μm
167K/253K	44,000	
76K/106K	490	
71K/169K	490	
56K/82K	74	
33K/72K		43
32K/54K	44	
23K/33K	22	
14K/42K		6.2
13K/29K		1.5

[a]Saltzman *et al.*, 1993.
[b]Hsu and Langer, 1985.

β-lactoglobulin with a fixed particle size, the tortuosity value generally decreases as the loading increases. Increasing the loading also provides protein molecules with less tortuous routes to the exterior of the matrix, thus increasing the diffusion coefficient and, therefore, the release rate (Rhine *et al.*, 1980a; Saltzman and Langer, 1989; Siegel and Langer, 1984; Miller and Peppas, 1983).

When particles of a fixed size are dispersed at a fixed loading, the molecular weight of EVAc can be changed to influence the rate of protein release (Table II and Fig. 4a). As the average molecular weight of the polymer in the matrix increases, the rate of albumin or immunoglobulin G (IgG) release decreases (Hsu and Langer, 1985; Saltzman *et al.*, 1993). The rate of release correlates with mechanical properties of the polymer (Saltzman *et al.*, 1993), suggesting that the mechanical properties of the polymer influence the structure of the pore network within the polymer. High-molecular-weight polymers have a low modulus, forming a relatively non-deformable matrix, limiting the rate of formation of small connecting channels. In low-molecular-weight polymers, the modulus is higher, the matrix is more deformable, and osmotic pressure within the pore space can cause the connecting pores to expand. This idea is supported by previous demonstrations that the ratio of channel size to pore size can influence the

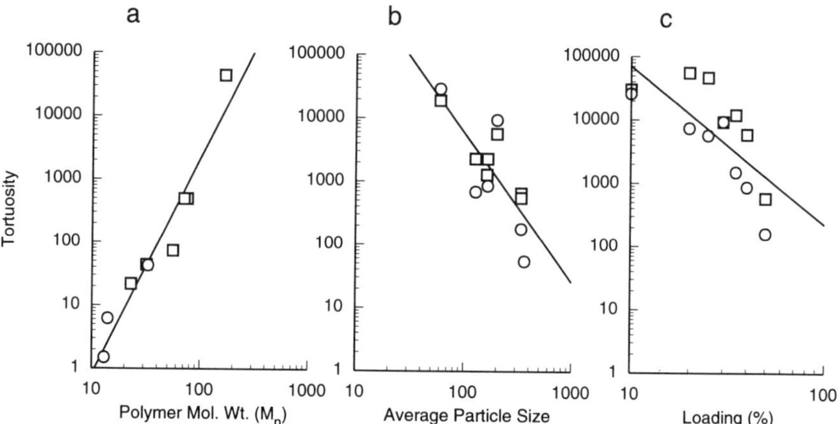

Figure 4. The dependence of tortuosity values on various matrix fabrication parameters. (a) The tortuosity dependence on the molecular weight of the EVAc polymer used in the matrix system. The tortuosity values ($F\tau$) for γ-globulin at 40% loading and a particle size of 117–180 μm (□) and BSA at 25% loading and a particle size of 106–150 μm (○) are plotted versus the number-averaged molecular weight of the EVAc. (b) The tortuosity dependence on the particle size of the macromolecule. Tortuosity values of BSA at 25% loading (□) and BSA at 30% loading (○) are plotted versus the average of the particle size. (c) The effect of loading on the tortuosity values. The $F\tau$ values for γ-globulin at 150–250 μm (□) and BSA at 150–250 μm (○) are plotted versus the percent loading. The lines are the power curve fits to the data.

pore-to-pore transport rate (Siegel and Langer, 1985; Ballal and Zygourakis, 1985).

To quantitate the influence of polymer molecular weight, protein loading, and protein particle size on the rate of release, calculated overall tortuosities ($F\tau$) were compared for release from matrices where only one of these parameters was varied (Fig. 4). The tortuosity was greatly influenced by the molecular weight of the polymer ($F\tau \propto M_n^{3.4}$; Fig. 4a). Particle size and the loading also strongly influence the overall tortuosity ($F\tau \propto \text{size}^{-2.4}$ and $F\tau \propto \text{loading}^{-2.5}$; Fig. 4b, c).

Effective diffusion coefficients or overall tortuosities are useful for evaluating trends in the observed protein release rates. The relationship between $F\tau$ and simple fabrication parameters (as shown most succinctly in Fig. 4) can be used to design a protein-releasing EVAc matrix. To understand how microscopic properties of the material influence these phenomena, it is necessary to develop more complex models of protein release, as described in the next section.

2.2. Microscopic Models of Diffusion in Porous Polymer Matrices

When macroscopic diffusion models are compared to experimental data, high values of tortuosity (corresponding to low values of D_{eff}) are obtained (Tables I and II). For a random porous medium, tortuosity values due to windiness of the diffusional path should be between 1 and 3 (Pismen, 1974; Bhatia, 1986). Because the overall tortuosities predicted by the macroscopic models are much larger, other physical properties of the matrix must influence the rate of protein diffusion within the matrix. Microgeometric models and percolation theory have been used to study the factors that might control protein diffusion within these polymer matrices.

2.2.1. GEOMETRIC DESCRIPTIONS

When EVAc matrices are visualized by light or electron microscopy, a characteristic pore morphology is observed (Fig. 2b). Large pores are connected by much smaller channels. Lattice-walk simulations (Balazs *et al.*, 1985; Saltzman, 1990), random-walk simulations (Siegel and Langer, 1984), and analytical diffusion models (Ballal and Zygourakis, 1985) have been used to estimate the influence of highly constricted pore geometries on local rates of diffusion. Results from all of these techniques suggest that the constricted microgeometry of the pore network is a critical determinant of overall tortuosity in the matrix; reasonable values for the extent of constriction yield high shape factors: $1 < F < 100$ (for a review, see Saltzman, 1990).

2.2.2. PERCOLATION DESCRIPTIONS

Percolation theory can be used to describe diffusion in a porous polymer matrix (Saltzman and Langer, 1989; Siegel *et al.*, 1989; Saltzman, 1990). As a simple example of a percolation process, consider a porous material represented as a two-dimensional square lattice (Fig. 5). In this simple model, the squares are called sites and the edges of the squares are bonds. A certain number of the sites are assumed to represent pores in the polymer (black squares in Fig. 5), and the remaining sites represent the backbone material or polymer (white squares). In site percolation, the porosity of the material is equal to the probability, ρ, that a given site will be a pore. A percolation lattice for simulating a material of porosity ρ is created by randomly assigning each site in the lattice as pore or polymer, based on this probability. For each lattice there exists a critical probability,

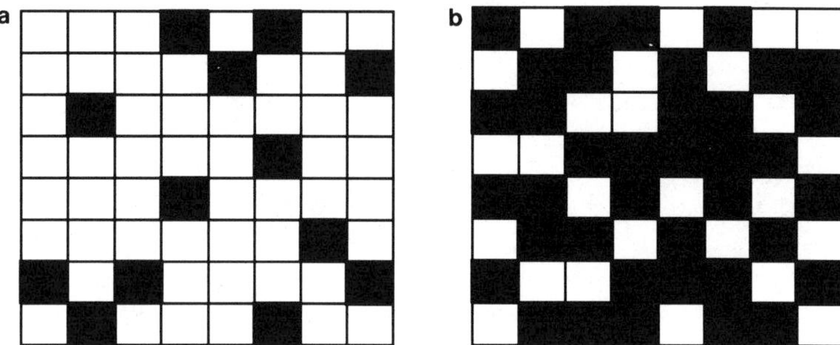

Figure 5. Square lattice representation of two-dimensional space for percolation theory. Black squares represent protein pores; white squares represent hydrophobic polymer backbone. Lattices with two different porosities (percentage of sites designated as pores) are shown: (a) 20% porosity; (b) 60% porosity.

or critical porosity, ρ_c. At this porosity, an infinite cluster (i.e., a of set of connected pores that spans the material) is formed. The critical porosity for a two-dimensional square lattice is 0.5. Figure 5 shows two square lattices with different porosities: one below ρ_c and one above. The difference in the connectivity for these two situations is obvious.

The percolation description of porous systems provides two important insights into protein release from porous polymers. First, the rapid change in connectivity observed for percolation lattices near the critical porosity is remarkably similar to the rapid increase in the total fraction of protein released from an EVAc matrix at ~35% loading (Siegel *et al.*, 1989; Saltzman and Langer, 1989). Second, rates of diffusion on percolation lattices vary predictably with porosity. Therefore, percolation theory provides a means for predicting changes in protein release rate with protein loading. In one case, pore size distributions were determined from several protein-loaded polymers (Saltzman and Langer, 1989). By combining measured pore sizes with models for diffusion in constricted microgeometries and simple percolation lattice models, the kinetics of albumin or IgG release were predicted (Saltzman and Langer, 1989).

3. APPLICATIONS OF PROTEIN/POLYMER MATRIX SYSTEMS

Protein-releasing matrix systems offer the possibility of long-term and localized delivery of proteins. Hydrophobic matrices protect the proteins from degradation following administration. Since modern biotechnology has made it possible to produce sizable quantities of many important therapeutic proteins, these delivery systems have a wide range of possible applications. Here, we briefly summarize the use of EVAc matrices for topical, localized, and systemic delivery of proteins.

It is important to note that the relationship between protein release rates measured *in vitro* (as in Fig. 1) and rate of protein delivery to tissues following implantation is not yet known. In fact, local rates of protein delivery are likely to vary with both properties of the polymer matrix and physiology of the implantation site as well as with properties of the protein. Several new techniques, however, permit the analysis of protein release under conditions that closely simulate different tissues in the body (Radomsky *et al.*, 1990; Beaty and Saltzman, 1992). In addition, experimental techniques and pharmacokinetic models appropriate for determining rates of delivery to tissue are being developed (Saltzman and Radomsky, 1991; Sherwood, 1993; Krewson and Saltzman, 1994; Salehi-Had and Saltzman, 1994).

3.1. Topical Delivery

The mucosal surfaces of the body are constantly exposed to the external environment. As a result, these surfaces are a major entry site for pathogens, such as those causing sexually transmitted diseases (STD). Oral administration of antibodies can prevent intestinal bacterial infections (Tackett *et al.*, 1988); topical application of mAb can prevent genital viral infections (Whaley *et al.*, 1994). To provide long-term passive immunization, nondegradable protein-releasing matrices can be used to deliver antibodies directly to a mucosal surface in need of protection (Sherwood et al., 1996). For example, vaginal rings composed of EVAc delivered biologically active mAb into the vaginal secretions for 30 days (Radomsky *et al.*, 1992). When fluorescently labeled mAbs were released from the matrices, the mAbs were found to be uniformly distributed through the vagina (Radomsky *et al.*, 1992). The same delivery systems may permit stimulation of mucosal immunity, by permitting localized delivery of protein antigens directly to mucosal surfaces (Wyatt, 1994).

3.2. Targeted Delivery of Proteins to Specific Tissue Regions

To achieve effective therapeutic levels of drugs at a local site, large, often toxic, doses must be administered systematically. This is particularly true for treatment of diseases of the brain. The brain capillaries have limited permeability to most drugs, making it difficult to treat many diseases of the brain. An alternate approach, which is capable of achieving high local concentrations of macromolecules in the interstitial space of the brain tissue, is implantation of protein-loaded polymers directly into the intracranial space (Hoffman *et al.*, 1990; Saltzman and Radomsky, 1991; Salehi-Had and Saltzman, 1994).

3.3. Systemic Delivery for Extended Periods

Many proteins have short half-lives and must be administered frequently to produce continuous therapeutic concentrations in the blood. Nondegradable polymer matrices can be used for extended delivery of proteins to the systemic circulation. For example, insulin has been delivered systemically using EVAc matrix systems (Fischel-Ghodsian *et al.*, 1988; Brown *et al.*, 1986; Langer and Folkman, 1977, Creque *et al.*, 1980).

REFERENCES

Balazs, A., Calef, D., Deutch, J., Siegel, R., and Langer, R., 1985, The role of polymer matrix structure and interparticle interactions in diffusion-limited drug release, *Biophys. J.* **47:**97–104.

Ballal, G., and Zygourakis, K., 1985, Diffusion in particles with varying cross-section, *Chem. Eng. Sci.* **40:**1477–1483.

Batich, C., Williams, J., and King, R., 1989, Toxic hydrolysis product from a biodegradable foam implant, *J. Biomed. Mater. Res.* **23**(A3)**:**311–319.

Bawa, R., Siegel, R. A., Marasca, B., Karel, M., and Langer, R., 1985, An explosion for the controlled release of macromolecules from polymers. *J. Controlled Release* **1:**259–267.

Beaty, C., and Saltzman, W. M., 1992, Controlled growth factor delivery induces differential neurite outgrowth in three-dimensional cell cultures, *J. Controlled Release* **24:**15–23.

Becker, J. B., Robinson, T. E., Barton, P., Sintov, A., Siden, R., and Levy, R. J., 1990, Sustained behavioral recovery from unilateral nigrostriatal damage produced by the controlled release of dopamine from a silicone polymer pellet placed into the denervated striatum, *Brain Res.* **508:**60–64.

Bhatia, S. K., 1986, Stochastic theory of transport in inhomogeneous media, *Chem. Eng. Sci.* **411:**1311–1324.

Boretos, J. W., and Pierce, W. S., 1967, Segmented polyurethane: A new elastomer for biomedical applications, *Science* **158:**1481–1482.

Brown, L. R., Wei, C. L., and Langer, R., 1983, *In vivo* and *in vitro* release of macromolecules from polymeric drug delivery systems, *J. Pharm. Sci.* **72:**1181–1185.

Brown, L., Munoz, C., Siemer, L., Edelman, E., and Langer, R., 1986, Controlled release of insulin from polymer matrices: Control of diabetes in rats, *Diabetes* **35:**692–697.

Creque, H. M., Langer, R., and Folkman, J., 1980, One month of sustained release of insulin from a polymer implant, *Diabetes* **29:**37–40.

Davidson, G. W., Domb, A., Sanders, L. M., and McRae, G., 1988, Hydrogels for controlled release of proteins, *Proc. Int. Symp. Control. Rel. Bioact. Mater.* **15:**66–67.

During, M. J., Sabel, B. A., Freese, A., Saltzman, W. M., Deutz, A. Roth, R. H., and Langer, R., 1989, Controlled release of dopamine from a polymeric brain implant: *In vivo* characterization, *Ann. Neurol.* **25:**351–356.

Edelman, E. R., Adams, D., and Karnovsky, M., 1990, Effect of controlled adventitial heparin delivery on smooth muscle cell proliferation following endothelial injury, *Proc. Natl. Acad. Sci. USA* **87:**3773–3777.

Edelman, E. R., Mathiowitz, E., Langer, R., and Klagsbrun, M., 1991, Controlled and modulated release of basic fibroblast growth factor, *Biomaterials* **12:**619–626.

Fischel-Ghodsian, F., Brown, L., Mathiowitz, E., Brandenburg, D., and Langer, R., 1988, Enzymatically controlled drug delivery, *Proc. Natl. Acad. Sci. USA* **85:**2403–2406.

Folkman, J., and Long, D. M., 1964, The use of silicone rubber as a carrier for prolonged drug therapy, *J. Surg. Res.* **4:**139–142.

Frank, M. L., Poindexter, A. N., Johnson, M. L., and Bateman, L., 1992, Characteristics and attitudes of early contraceptive implant acceptors, *Family Planning Perspect.* **24:**208.

Gogolewski, S., 1989, Selected topics in biomedical polyurethanes: A review, *Colloid Polym. Sci.* **267:**757–785.

Hoffman, D., Wahlberg, L., and Aebischer, P., 1990, NGF released from a polymer matrix prevents loss of ChAT expression in basal forebrain neurons following a fimbria-fornix lesion, *Exp. Neurol.* **110:**39–44.

Horbett, T. A., Takeno, M., and Ratner, B., 1993, Anti-thrombotic peptide release from new biomaterials, *AIChE Extended Abstracts*, Abstract 118f.

Hsu, T. T.-P., and Langer, R., 1985, Polymers for the controlled release of macromolecules: Effect of molecular weight of ethylene–vinyl acetate copolymer, *J. Biomed. Mater. Res.* **19:**445–460.

Kessler, D. A., 1992, The basis of the FDA's decision on breast implants, *N. Eng. J. Med.* **326:**1713–1715.

Krewson, C., and Saltzman, W. M., 1994, Targetting of peptides and proteins in the brain following release from implantable polymers, in: *Trends and Future Perspectives in Peptide and Protein Drug Delivery* (V. H. L. Lee, M. Hashida, and Y. Mizushima, eds.), Harlans Academic Publishers, London, pp. 273–294.

Langer, R., and Folkman, J., 1976, Polymers for the sustained release of proteins and other macromolecules, *Nature* **263:**797–800.

Langer, R., and Folkman, J., 1977, The use of polymers for the sustained release of proteins and other large molecules, *Polym. Prepr. (Am. Chem. Soc., Div. Polym. Chem.)* **18**(2)**:**379–384.

Lee, A., and Langer, R., 1983, Shark cartilage contains inhibitors of tumor angiogenesis, *Science* **221:**1185–1187.

Miller, E. S., and Peppas, N. A., 1983, Diffusional release of water-soluble bioactive agents from ethylene–vinyl acetate copolymers, *Chem. Eng. Commun.* **22:**303–315.

Murray, J. B., Brown, L., Langer, R., and Klagsburn, M., 1983, A micro sustained release system for epidermal growth factor, *In Vitro* **19:**743–748.

Pismen, L. M., 1974, Diffusion in porous media of a random structure, *Chem. Eng. Sci.* **29:**1227–1236.

Powell, E. M., Sobarzo, M. R., and Saltzman, W. M., 1990, Controlled release of nerve growth factor from a polymeric implant, *Brain Res.* **515:**309–311.

Preis, I., and Langer, R., 1979, A single-step immunization by sustained antigen release, *J. Immunol. Methods* **28:**193–197.

Radomsky, M. L., Whaley, K. J., Cone, R. A., and Saltzman, W. M., 1992, Controlled vaginal delivery of antibodies in the mouse, *Biol. Reprod.* **47:**133–140.

Rhine, W. D., Hsieh, D. S. T., and Langer, R., 1980a, Polymers for sustained macromolecule release: Procedures to fabricate reproducible delivery systems and control release kinetics, *J. Pharm. Sci.* **69:**265–270.

Rhine, W. D., Sukhatme, V., Hsieh, D. S. T., and Langer, R., 1980b, A new approach to achieve zero-order release kinetics from diffusion-controlled polymer matrix systems, in: *Controlled Release of Bioactive Materials* (R. W. Baker, ed.), Academic Press, New York, pp. 177–187.

Salehi-Had, S., and Saltzman, W. M., 1994, Controlled intracranial delivery of antibodies in the rat, in: *Protein Formulations and Delivery* (J. Cleland and R. Langer, eds.), ACS Symposium Series No. 567, American Clemical Society, Washington, D.C., pp. 278–291.

Saltzman, W. M., 1987, A Microstructural Approach for Modelling Diffusion of Bioactive Macromolecules in Porous Polymers, Ph.D. thesis, Massachusetts Institute of Technology.

Saltzman, W. M., 1990, Transport in porous polymers, in: *Absorbent Polymer Technology* (L. Brannon-Peppas and R. S. Harland, eds.), Elsevier, Amsterdam, pp. 171–199.

Saltzman, W. M., and Langer, R., 1989, Transport rates of proteins in porous materials with known microgeometry, *Biophys. J.* **55:**163–171.

Saltzman, W. M., and Radomsky, M. L., 1991, Drugs released from polymers: Diffusion and elimination in brain tissue, *Chem. Eng. Sci.* **46:**2429–2444.

Saltzman, W. M., Pasternak, S. H., and Langer, R., 1987, Microstructural models for diffusive transport in porous polymers, in: *Controlled-Release Technology: Pharmaceutical Applications* (P. I. Lee and W. R. Good, eds.), American Chemical Society, Washington, D.C., pp. 16–33.

Saltzman, W. M., Sheppard, N. F., McHugh, M. A., Dause, R. B., Pratt, J. A., and Dodrill, A. M., 1993, Controlled antibody release from a matrix of poly(ethylene-*co*-vinyl-acetate) fractionated with a supercritical fluid, *J. Appl. Polym. Sci.* **48:**1439–1500.

Sherwood, J. K., 1993, Controlled Release of Antibodies and Antibody Fragments for Immunoprotection of the Mucus Epithelia, Master of Science essay, The Johns Hopkins University.

Sherwood, J. K., Dause, R. B., and Saltzman, W. M., 1992, Controlled antibody delivery systems, *Bio/Technology* **10:**1446–1449.

Sherwood, J. K., Zeitlin, L., Whaley, K. J., Cone, R. A., and Saltzman, M., 1996, Controlled release of antibodies for long-term topical passive immunoprotection of female mice against genital herpes, *Nature Biotech.* **14:**468–471.

Siegel, R. A., and Langer, R., 1984, Controlled release of polypeptides and other macromolecules, *Pharm. Res.* **1:**2–10.

Siegel, R. A., and Langer, R., 1985, A new Monte Carlo approach to diffusion in constricted porous geometries, *J. Colloid Interface Sci.* **109:**426–440.

Siegel, R. A., Kost, J., and Langer, R., 1989, Mechanistic studies of macromolecular drug release from macroporous polymers. I. Experiments and preliminary theory concerning completeness of drug release, *J. Controlled Release* **8:**223–236.

Stannet, V. T., Koros, W. J., Paul, D. R., Lonsdale, H. K., and Baker, R. W., 1979, Recent advances in membrane science and technology, in: *Advances in Polymer Science*, Vol. 32 (H. J. Cantow, G. Dall' asta, K. Dusek, J. D. Ferry, H. Fujita, M. Gordon, W. Kern, S. Okamura, C. G. Overberger, T. Saegusa, G. V. Shulz, W. P. Slichter, and J. K. Stille, eds.), Springer-Verlag, Berlin, pp. 69–121.

Tackett, C. O., Losonsky, G., Link, H, Hoang, Y., Guesry, P., Hilpert, H., and Levine, M. M., 1988, Protection by milk immunoglobulin concentrate against oral challenge with enterotoxigenic *E. coli, N. Engl. J. Med.* **318:**1240.

Tyler, B. J., Ratner, B. D., Castner, D. G., and Briggs, D., 1992, Variations between Biomer™ lots. I. Significant differences in the surface chemistry of two lots of a commercial poly(ether urethane), *J. Biomed. Mater. Res.* **26:**273–289.

Vergnaud, J. M., 1991, *Liquid Transport Processes in Polymeric Materials*, Prentice-Hall, Englewood Cliffs, New Jersey.

Whaley, K. J., Zeitlin, L., Barrat, R. A., Hoen, T. E., and Cone, R. A., 1994, Passive immunization of the vagina protects mice against vaginal transmission of genital herpes infections, *J. Infect. Dis.* **169:**647–649.

Wyatt, T. L., 1994, Controlled Vaginal Delivery of Antigens to Stimulate Mucosal Immunity, Master of Science essay, The John Hopkins University.

Chapter 5

Diffusion-Controlled Delivery of Proteins from Hydrogels and Other Hydrophilic Systems

Mary Tanya am Ende and Antonios G. Mikos

1. INTRODUCTION

The short *in vivo* half-life of many pharmaceutically active proteins necessitates the need for multiple administrations to produce a therapeutic response, emphasizing the applicability of controlled release formulations (Lee, 1992). The key variables that affect protein transport through hydrophilic polymers depend on the delivery mechanism and device properties, in a similar manner as in the case of lower-molecular-weight drug substances. However, in the case of proteins, the role of solute molecular size is much more dramatic in hindering the diffusion and release from hydrophilic polymers (am Ende, 1993).

Another critical consideration in protein delivery from hydrogel systems is the potential for protein denaturation in the device. For diffusion-controlled delivery systems, where water is the main transporting medium, the protein solution stability governs the type of device. Extended releasing times can be achieved with reservoir systems (Fig. 1) for highly stable proteins (Langer, 1990). Alternatively, dehydrated delivery systems

---

Mary Tanya am Ende • Pfizer Central Research, Pharmaceutical Research and Development, Groton, Connecticut 06340. *Antonios G. Mikos* • Cox Laboratory for Biomedical Engineering, Institute of Biosciences and Bioengineering, Department of Chemical Engineering, Rice University, Houston, Texas 77251.

Protein Delivery: Physical Systems, Sanders and Hendren, eds., Plenum Press, New York, 1997.

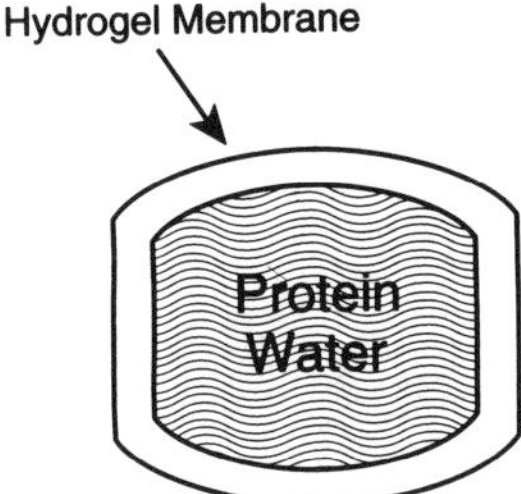

Figure 1. The hydrated state of a protein reservoir system. Key parameters controlling release are membrane composition, area, thickness, and concentration gradient.

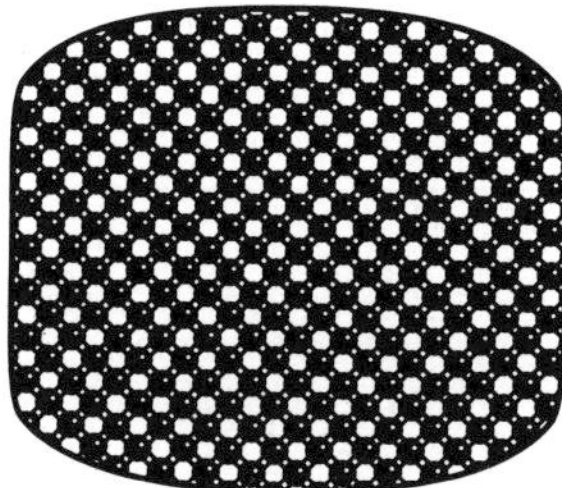

Figure 2. The dehydrated delivery device produced by compression of protein and hydrophilic polymer powder.

can be created by direct compression of protein and hydrophilic polymer powder blends to form a matrix tablet (Fig. 2). To delay or prevent contact of biological fluid with the protein, additional measures must be taken such as laminating hydrophobic layers onto the hydrophilic polymer/protein core (Langer and Folkman, 1976). Such a device prevents contact of the protein with the biological fluids until diffusion and release occurs at the appropriate time, pH, or location and hence prolongs storage and shelf life. The reservoir, matrix, and biodegradable (initial release) systems are examples of diffusion-controlled delivery. The discussion of diffusion-controlled systems in this chapter encompasses hormones, polypeptides, and proteins.

1.1. Mechanisms of Protein Diffusion

Controlled release delivery systems were categorized by Langer and Peppas (1983) according to the releasing mechanism as bioerodible or biodegradable, swelling controlled, osmotically controlled, and diffusion controlled. Surface erodible systems function by erosion at the polymer surface, either by chemical degradation or by dissolution, that directly releases the protein attached to a polymer side chain or entrapped within the device. The released protein then dissolves in the biological fluids. Heller (1988) has developed well-characterized bioerodible, hydrophobic polymers composed of poly(ortho esters) and acid-labile linkages.

Biodegradable systems provide means for chemical reactions to occur within the bulk material, as well as at the surface. Poly(glycolic acid-*co*-DL-lactic acid), (PGLA) is a biodegradable, hydrophilic polymer that is widely used in controlled release applications (Jeffery *et al.*, 1993). These systems function by hydrolytic degradation of the ester bonds. Biodegradable PGLA devices exhibit diffusion-controlled delivery during the induction time of protein release (Shah *et al.*, 1992).

Swelling-controlled delivery systems involve three processes: absorption of water into the polymer to form the hydrogel, dissolution of the protein, and subsequent release of the protein from the device. Hydrogels exhibit this type of release behavior when placed in an aqueous medium initially in the dehydrated state. Swellable hydrogel systems are produced by cross-linking water-soluble polymers, such as those listed in Section 1.2. The degree of swelling is easily modified by copolymerizing with a more or less hydrophilic monomer or by increasing the cross-linking density. A reversible swelling polymer can function as a mechanical piston in a novel delivery system, in which a triggered swelling response delivers the protein.

The rate-controlling step for osmotic delivery systems is the absorption of water into the device by osmosis. The solute is delivered through an orifice owing to the osmotic pressure gradient across the semipermeable membrane. The Alzet™ osmotic minipump has been used extensively in animal models to provide controlled protein delivery for 1–2 weeks. Thissen *et al.* (1991) infused rats with insulin-like growth factor-I (IGF-I) for 1 week using the osmotic minipump. Resistance to the effects of the IGF-I growth hormone were observed in protein-restricted young rats. More recently, applications of the minipump have been in the field of cancer research. Arteaga *et al.* (1994) implanted the osmotic system into mice to deliver tumor growth factor-α-ΔCys-Pseudomonas exotoxin recombinant fusion protein over 7 days. Their data suggest that antitumor agents selectively target epidermal growth factor receptors in breast cancer cells. A major advantage of osmotically controlled delivery is that release is independent of external conditions (pH or agitation), and good *in vivo/in vitro* correlations are obtained (Theeuwes, 1975).

The fourth mechanism controlling the release of proteins from a hydrogel system is limited by the diffusion of the protein through the hydrated polymer caused by concentration gradients. Such delivery systems can be modeled according to Fick's law, based on the concentration driving force. This chapter will focus on diffusion-controlled release systems for delivery of proteins.

1.2. Structure of Hydrophilic Polymers

The transport pathway of a protein through a hydrogel is complex owing to the varied structural properties of both components depending on environmental conditions. The hydrophilic polymers used in controlled release applications may be classified among three main types: nonporous, microporous, and macroporous (Peppas and Meadows, 1983). Nonporous hydrogels consist of covalently or ionically cross-linked macromolecular chains which form a three-dimensional network, characterized by its mesh size, ξ. The best definition of this correlation length has been given by de Gennes (1979). The typical size of this correlation length is of the order of 10 to 100 angstroms. The correlation length is equivalent to the mesh size that can be experimentally determined by quasi-elastic light scattering (QELS), as shown by Munch *et al.* (1976). For swollen polymeric networks, an additional parameter is usually necessary to determine the mesh size from experimental studies. Canal and Peppas (1989) described the procedure for evaluating the equilibrium polymer volume fraction.

In general, protein transport in nonporous polymers proceeds through a densely packed mesh of polymer chains. In this case, the transport process is strongly dependent on the polymer chain mobility, and the protein diffusion occurs essentially through a solid (Fig. 3). One method of producing nonporous hydrogels involves polymerizing with solvent levels below the equilibrium content. Nonporous membranes are employed as a barrier so protein release proceeds from another portion of the device or as a control device to impede the transport of highly water-soluble proteins.

Microporous hydrogels contain pores larger than the macromolecular correlation length. The pores are usually filled with water; their characteristic size is described by an average pore radius, r_P. Protein transport occurs by a combination of diffusion and convection whereby the water is transported through the pores, carrying solute with it. Microporous gels may have pores sizes between 0.01 and 0.1 μm. Macroporous hydrogel networks contain large pores, usually of diameter greater than 0.1 μm. The mass-transfer mechanism is by diffusion through water-filled pores and convection. In such highly swollen systems, the polymer structure has little to do with solute transport. Transport models describing protein diffusion in porous systems are commonly based on classical hydrodynamics (Satterfield

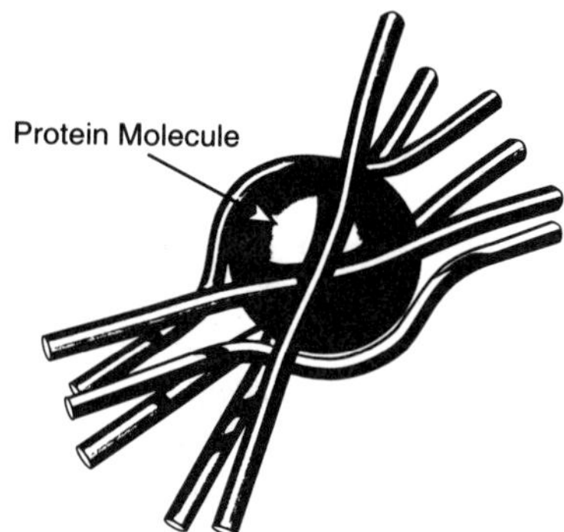

Figure 3. A view of the internal structure of a hydrogel on a molecular level.

et al., 1973). Domb *et al.* (1990) studied the permeation and transport of luteinizing hormone-releasing hormone (LHRH) analogs through poly(2-hydroxyethyl methacrylate) (PHEMA) membranes. Their findings suggest that decapeptide diffusion through hydrogels proceeds through the water-filled pores. Sato and Kim (1984) conducted an investigation to compare the transport mechanism for proteins to that for smaller compounds in porous and nonporous hydrogels. Insulin, cytochrome *c*, and albumin diffused mainly through the polymer dense region by a partitioning mechanism, with contributions from the water-filled pore regions.

Macroporous hydrogels can be synthesized with solvent in excess of the equilibrium value, so that phase separation creates pools of solvent that eventually leave behind large pores and channels. An alternative procedure that produces large mesh sizes (microporous) involves carrying out the polymerization in a nonsolvent, so that coiled polymer chains can fully extend upon hydration.

A hydrogel may be prepared in any of the three forms described above by altering the conditions of polymerization. PHEMA can be prepared as a nonporous, microporous, or macroporous gel by changing the quantity of water, cross-linking agent, and comonomer present during the co-polymerization/cross-linking of monomers. A broad spectrum of hydrophilic polymers has been synthesized for use in controlled release systems. In addition to PHEMA, some of the other commonly used hydrogels include poly(acrylic acid) (PAA), polyacrylamide (PAAm), poly(vinyl alcohol) (PVA), poly(methacrylic acid) (PMAA), poly(methyl methacrylate) (PMMA), poly-(*N*-vinylpyrrolidone) (PVP), poly(ethylene oxide) (PEO), poly(ethylene glycol) (PEG), and polysaccharides (Korsmeyer, 1991).

1.3. Methods for Loading Proteins into Hydrogels

1.3.1. ELECTROPHORETIC LOADING

A novel approach for loading proteins into hydrogel systems based on electrophoresis, a protein separation technique, was investigated by Shalaby *et al.* (1993). Bovine serum albumin (BSA) loading was increased 6- to 9-fold in PAAm by applying a current of 0.3 A across the hydrogel. The protein completely penetrated the hydrogel following only 30 min of electrophoretic operation with a current of 1.1 A. A combination of freeze drying and electrophoresis was proposed to produce uniformly distributed proteins in hydrogel delivery systems.

1.3.2. ENTRAPMENT

Jeffery *et al.* (1993) demonstrated the protein entrapment procedure using a (water-in-oil)-in-water (W/O/W) emulsion technique. Ovalbumin was dissolved in water and emulsified with a polymer solution in dichloromethane (oil phase) using a homogenizer at high speeds. The next step involved emulsifying in a PVA aqueous solution, to create a ternary W/O/W emulsion, and evaporation of the organic solvent, to form entrapped ovalbumin microspheres. The protein was stabilized in the organic solvent in reverse micelles. McGee *et al.* (1995) reported zero-order release of proteins in the range of 15 to 40 days from poly(DL-lactide-*co*-glycolide) microparticles loaded using a similar phase-separation technique. Other variations of the entrapment procedure are continually being developed (Hayashi *et al.*, 1994).

1.3.3. SOLUTION LOADING

Proteins can be incorporated into hydrogels after polymerization by use of the appropriate solvent. The selection of the solvent is based on its ability to act as a swelling agent for the hydrogel and a solubilizing agent for the protein, keeping in mind the implications of toxicity and denaturation. The hydrogel is immersed in a solution of the protein, allowing time for the protein to partition and penetrate into the hydrogel. After equilibrium is attained, the protein-loaded hydrogel is removed from the solution and dried slowly to prevent skin layer formation. This technique is commonly used to achieve loading concentrations below the solubility limit, as described by Carelli *et al.* (1989).

Penetration into the network requires that the dimensions of the "pores" exceed the protein dimensions; otherwise, the predominate process is adsorption. The adsorption of proteins onto hydrogels is an important area of research not discussed here (Dillman and Miller, 1973; Braatz *et al.*, 1992; Smith and Sefton, 1992; Strzinar and Sefton, 1992).

2. DIFFUSION-CONTROLLED DELIVERY SYSTEMS

2.1. Reservoir Systems

A reservoir delivery system consists of a protein core enclosed in a membrane, as shown in Fig. 4. The rate-limiting mechanism is diffusion of the protein through the polymer membrane. Zero-order release kinetics are obtained for saturated core systems, because the concentration gradient is constant across the hydrogel membrane. The simplicity of these systems makes them attractive as a predictable dosage formulation, although the encapsulating process is often quite complicated. The main advantage of the reservoir system is that the kinetic mechanism of release is independent of the molecular weight or molecular size of the active ingredient, providing zero-order release for proteins as well as drugs. However, it should be noted that reservoir systems are subject to catastrophic release of their contents in the event of membrane failure. The absence of thin spots, pinholes, or other membrane defects is critical for effective performance of a reservoir system.

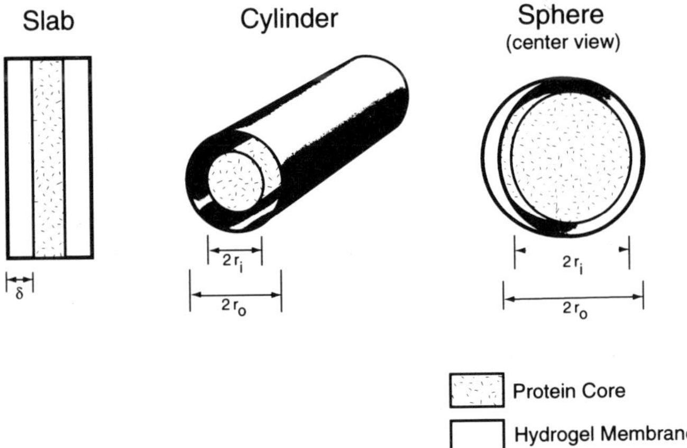

Figure 4. Reservoir controlled release devices with an inner protein core enclosed by a membrane of thickness δ (slab) and $r_o - r_i$ (cylinder and sphere).

Quality control and reproducibility are especially important for devices that contain an amount of active ingredient in excess of an acutely tolerated dose.

2.1.1. HYDROPHILIC POLYMER MEMBRANES

Reservoir delivery systems have been developed in a variety of styles, ranging from microcapsules to hollow fibers to liposomes. Hayashi *et al.* (1994) produced delivery systems by encapsulating proteins and hormones in poly-L-lactide microspheres by a solvent evaporation method. The release mechanism for hormones entrapped in liposomes was studied by Ho *et al.* (1986). Progesterone and hydrocortisone skin permeation was enhanced by the presence of the liposomes; no penetration of the liposomes was observed. Examples of the most common hydrogels employed in reservoir systems are crystalline–rubbery PEG, PAAm, celluloses, PAA, and PHEMA.

2.1.2. EFFECT OF FORMULATION VARIABLES ON PROTEIN RELEASE

The reservoir device geometry is an important variable that affects the release profile, because the amount of protein released is proportional to the surface area. The release rate (dM_t/dt) is inversely related to the hydrogel membrane thickness of a slab device, δ:

$$\frac{dM_t}{dt} = \frac{ADKc_s}{\delta} \tag{1}$$

The surface area is represented by A, D is the diffusion coefficient, K is the partition coefficient, and c_s is the solubility limit of the protein. Increased membrane thickness of a slab dramatically affects the release rate. The release rate for the cylindrical shape is proportional to the length, L, of the cylinder (Eq. 2). An increase in membrane thickness does not significantly affect the release rate from a cylindrical device. Spherical devices are affected the least by increased membrane thickness ($r_o - r_i$), as shown in Eq. (3).

$$\frac{dM_t}{dt} = \frac{2\pi LDKc_s}{\ln\left(\frac{r_o}{r_i}\right)} \tag{2}$$

$$\frac{dM_t}{dt} = \frac{4\pi DKc_s r_o r_i}{(r_o - r_i)} \tag{3}$$

Another important variable that affects the initial portion of the release is the storage time of the device. A freshly prepared reservoir system may show a lag time in reaching steady-state release, as a result of the reservoir not having equilibrated with the membrane before the beginning of the release test. Storage to ensure consistent equilibration may overcome this. However, a fully equilibrated system may give an initial burst effect (Baker, 1987). The concentration driving force remains constant in reservoir systems during the release, while the protein concentration exceeds the solubility limit.

2.1.3. APPLICATIONS

The most well known commercial reservoir controlled release systems deliver hormones for contraception from hydrophobic polymers. The Norplant™ subcutaneous device controls the release of levonorgestrel with silicone rubber, and the Progestasert™ intrauterine device (IUD) releases progesterone from reservoir devices of ethylene vinyl acetate. In the field of insecticides, reservoir dispensers called BioLure™ were developed to provide zero-order release of insect pheromones to disrupt mating (Smith *et al.*, 1983). The dispenser consists of a slab configuration with a rate-controlling membrane, with constant release described by Eq. 1.

2.2. Matrix Systems

Monolithic systems encompass several types of dispersed delivery devices, where the polymer is a rate-controlling membrane. A monolithic solution device contains the protein as a solution within the polymer. A monolithic dispersion contains dispersed solid protein in a rate-limiting polymer matrix, referred to as a matrix system (see Fig. 5). A variety of other systems are created by a combination of reservoir and matrix delivery devices (Baker, 1987). Matrix tablets are manufactured by compression of the mixture of protein and polymer powders. The controlling mechanism of release is the dissolution of the protein particles and diffusion through the water-filled pores and between polymer chains. Zero-order release can be achieved by proper design of the delivery system geometry and loading concentrations (Hsieh *et al.*, 1983). The novel developments of Langer and Folkman (1976) were instrumental in the field of sustained protein delivery from noninflammatory polymers. These delivery devices were copolymerized HEMA or PVA with a hydrophobic comonomer, ethylene vinyl acetate, or multilaminated layers.

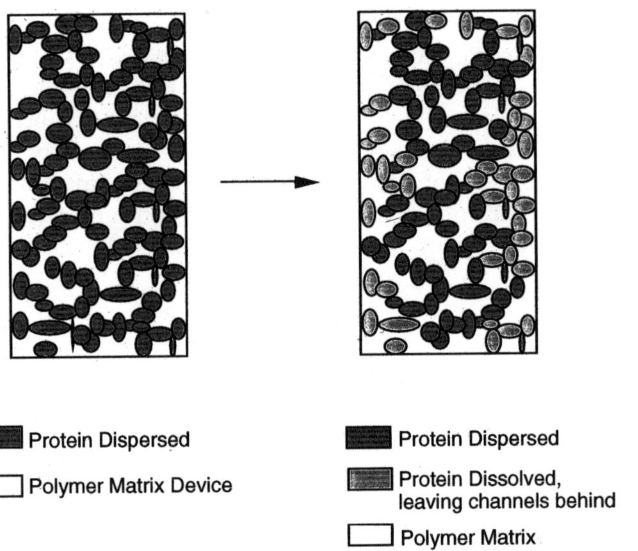

Figure 5. Matrix delivery device with protein particles dispersed throughout a polymer matrix.

2.2.1. HYDROPHILIC POLYMERS USED

The nonbiodegradable polymers used in matrix systems include hydroxyethyl cellulose, hydroxypropyl cellulose, hydroxypropyl methyl cellulose, PEO, PAA, and various PHEMA derivatives. The most common biodegradable copolymer used is PGLA.

2.2.2. EFFECT OF FORMULATION VARIABLES ON PROTEIN RELEASE

The loading concentration of the protein generally affects the release profile for monolithic systems, which function by diffusion, because the driving force is different for different protein loadings. A protein loading of 15–20 vol% for matrix devices ensures that all protein particles are in contact. Water penetration into the device acts to release the protein particles by dissolution, creating tortuous channels throughout the matrix. The device geometry is important in the release, similarly to the reservoir systems. The release rate is proportional to the square root of time:

$$\frac{dM_t}{dt} = 4\left(\frac{Dt}{\pi\delta}\right)^{1/2} \tag{4}$$

The mathematical equations describing the release of the protein and determination of the diffusion coefficient are discussed further in Section 4.

2.2.3. APPLICATIONS

Matrix systems for controlled release applications are predominately for delivery of soluble substances. Siegel and Langer (1984) revealed that large molecules, such as insulin, heparin, and albumin, were released over an extended period of time owing to the slow diffusion through the interconnected pore structure.

2.3. Biodegradable Hydrogels

The initial portion of the release of proteins from biodegradable systems is attributed to a diffusion-controlled mechanism (Shah *et al.*, 1992). Controlled release of proteins can also be achieved with biodegradable hydrogel systems. The hydrogel degradation mechanism may involve the degradation of polymer backbone, cross-linking agent, or pendant chains. Hydrogels with a degradable polymer backbone or cross-linking agent are eventually solubilized, whereas those with degradable pendant chains remain insoluble. The mechanism of protein delivery may depend on both hydrogel degradation and protein diffusion.

2.3.1. DEGRADABLE POLYMER BACKBONE

The degradation products of the polymer backbone are low-molecular-weight, water-soluble compounds that are eliminated from the body. Release is primarily controlled by protein diffusion through the hydrogel. As the backbone is degraded, the protein diffusion coefficient increases, which results in increased protein release rates. However, hydrogel degradation and protein release may be on separate time scales. For hydrogels that degrade quickly, a burst effect will dominate delivery. For slowly degrading hydrogels, the hydrogel will behave essentially as a matrix system, and the protein release will be diffusion-controlled.

Biodegradable hydrogels based on poly(ethylene glycol) block copolymers with poly(α-hydroxy acids) were developed (Zhu *et al.*, 1990; Sawhney *et al.*, 1993). Poly(ethylene glycol) is water-soluble; however, the incorporation of degradable monomer units along the polymer backbone renders the copolymer degradable. The hydrogel degradation rate may vary from 1 day to 4 months depending on the type of repeating unit (such as lactic acid or glycolic acid) and the copolymer ratio. The protein release rate is also dependent on the hydrogel composition. Sustained release rates of albumin and norethisterone were observed for over a month.

2.3.2. DEGRADABLE CROSS-LINKING AGENTS

Water-soluble polymers that are cross-linked with biodegradable cross-linking agents are known as degradable delivery systems. The polymer backbone chains remain intact while the cross-linking agents degrade. Final degradation products are usually of high molecular weight and are water-soluble. Degradation of the cross-links in such systems enhances hydrogel swelling, thus providing additional volume for diffusion. High-molecular-weight proteins are released only after much of the cross-linking has been eliminated. At this point, the mesh size has increased sufficiently to allow diffusional release of proteins and peptides.

Brondsted and Kopecek (1991) synthesized biodegradable hydrogels for site-specific delivery of protein drugs to the colon. The hydrogels contained acidic comonomers and enzymatically degradable azo-aromatic cross-links. In the low-pH environment of the stomach, the low degree of hydrogel swelling protects the proteins against degradation by digestive enzymes. As the hydrophilic polymer system passes down the gastrointestinal tract, the degree of swelling increases as a result of the increased ionization and repulsion of charges on the hydrogels with increasing pH. Upon arrival in the colon, the gels have reached a degree of swelling that allows the cross-links to be accessible to enzymes. Protein release is controlled by cross-link degradation induced enzymatically. The degradation rate of hydrogels containing degradable cross-links depends on the structure and length of the cross-linking agent as well as the degree of swelling (Kopecek *et al.*, 1992).

Functionalized albumin was also utilized as a degradable cross-linking agent for preparing poly(acrylic acid) (Park, 1988) and poly-*N*-vinylpyrrolidone biodegradable hydrogels (Shalaby and Park, 1990).

2.3.3. DEGRADABLE PENDANT CHAINS

Low-molecular-weight proteins are quickly released by bulk diffusion, presenting difficulties for a controlled release application. However, pendant chain degradable hydrogels represent a unique method for controlled delivery of these low-molecular-weight compounds. The protein is attached to the hydrogel polymer backbone via a biodegradable polymer link and is only released after the link is degraded. In this case, the hydrophilic polymer is not water-soluble.

3. FACTORS AFFECTING THE DIFFUSION OF PROTEINS

There are a variety of properties that can affect the transport of proteins through hydrophilic materials. These include the effects of pH, temperature, ionic strength, and solvent. The characteristics of the solute (size, shape, and ionizability) and the polymer are also important in determining the transport behavior.

The concentration of protein located within the polymeric delivery device is an important parameter, since it establishes the driving force for transport. The tranport process can be different for each type of experimental setup. For example, a diffusion study may involve the transport of proteins from a high concentration (donor cell) through a hydrogel into a compartment of lower concentration (receptor cell), where the driving force consists of the concentration gradient across the hydrogel. Another example is the driving force for a preloaded hydrogel due to the protein concentration difference between the hydrogel interior and the external solution, which is related to the flux, J, of that protein by Ficks law, as shown below.

$$J = -D\left(\frac{dc}{dx}\right) \tag{5}$$

It can easily be seen from Eq. (5) that changing the loading concentration and thus the driving force $[dc/(dx)]$, up the solubility limit can affect the flux for a given diffusion coefficient, D. Another important consideration is that the diffusion coefficient can be a function of the concentration, especially in highly concentrated systems (Crank, 1975). For dilute systems, the concentration dependence of D can be assumed to be negligible.

3.1. Environmental Conditions

The surrounding environment in contact with the delivery system can affect the polypeptide or protein, as well as the hydrogel. Insulin release from PMAA was studied *in vitro* and *in vivo* by Greenley *et al.* (1990). This system was investigated for possible applications in oral delivery of the polypeptide. The release of insulin following exposure to synthetic intestinal fluid demonstrated the usefulness of a pH-sensitive delivery device for insulin that prevents release in the acidic environment of the stomach. Dong and Hoffman (1991) synthesized pH-sensitive and thermally responsive hydrogels. Their results indicated that this class of hydrogels responded to changes in pH to a much greater extent than pure ionic hydrogels.

Zentner *et al.* (1979) examined the permeation of progesterone through PHEMA polymerized in water or ethanol and interpreted the results as verification of protein transport through the water-filled pores.

An important and practical consideration in delivering proteins from hydrophilic polymers lies in the potential for polymer–protein interactions that cause irreversible adsorption. This is a topic that has been the subject of considerable research, especially in the area of contact lens development. The effect of ionic strength and pH on the adsorption of BSA onto PMMA surfaces was studied by Suzawa and Murakami (1992). They concluded that the adsorption was maximized at the isoelectric point of BSA, under conditions that favor a compact protein conformation. Horbett (1982) identified the general properties of adsorbed proteins, including γ-globulin, fibrinogen, and albumin. This investigation hypothesized that interactions in response to the presence of a foreign material (polymer) were controlled by the plasma proteins. A thorough examination of polymer–protein interactions must be conducted during dosage form development.

3.2. Hydrogel Structure

The external environment in contact with the delivery device can have a large influence on the hydrogel structure and, hence, the diffusion process. The structure of hydrogels can be altered by a change in the external solution pH, ionic strength, temperature, or solvent or by the presence of an electric potential. Since there are many other factors that affect the swelling of the polymer, which in turn affect the structure, the environmental conditions must be controlled with great care. Many research groups have utilized this responsive characteristic to control the rate at which proteins can transport through such polymeric systems.

The dynamics of swelling and release must be considered concurrently in order to understand the effect of hydrogel structure. Kim and Lee (1991) observed the appearance of a swelling front for the case of ionizable hydrogels. The rate of penetration of the swelling front in P(MMA-*co*-MAA) increased with ionic strength due to the Donnan effect. The rate of diffusion was found to be controlled by the rate of swelling in P(MMA-*co*-MAA).

The polymer structure can be altered by copolymerizing to produce a network which alters the internal morphology. Caliceti *et al.* (1992) investigated the affect of copolymerizing HEMA in the presence of poly(ethylene glycol) methyl ether (MPEG). It was demonstrated by scanning electron microscopy (SEM) analysis that the hydrogel porosity was enhanced with the incorporation of higher-molecular-weight MPEG chains, ultimately producing a system able to deliver polypeptides and proteins.

The existence of multiple phases of water within hydrogels is a controversial issue, considered by one group of researchers to restrict the solute diffusional pathway. Thermal analysis of hydrated polymers has been conducted, and the multiple melting endotherms that these polymers exhibit when heated from subzero temperatures are attributed to "bound" versus "free" water. McNeil and Graham (1993) proposed that the water associated with PEO backbone chains (bound) was in the form of a trihydrate. Roorda (1994) argued that the phase transitions observed using differential scanning calorimetry (DSC) are due to the glass-transition temperature for the freezing hydrogel. Measurements of rapidly interchanging water molecules in PHEMA hydrogels using NMR provide supportive evidence that different water phases are nonexistent. Self-diffusion coefficients of water in PHEMA were measured with pulsed field-gradient NMR by Peschier *et al.*, 1993. Their data revealed only one diffusion coefficent, which was interpreted as indicating a homogeneous water phase. The NMR studies were carried out at equilibrium conditions and 25 °C, considered to more closely replicate *in vivo* conditions. The melting endotherms have been attributed to the nonequilibrium conditions of the calorimetric experiments (subzero temperatures).

4. TECHNIQUES FOR MEASUREMENT OF THE DIFFUSION COEFFICIENT

The protein diffusion coefficient in hydrophilic polymers can be measured by a variety of experimental techniques. Since the type of diffusion coefficient measured by each technique can be different, the method of choice will depend on the desired information and variable. A summary of the advantages and disadvantages of techniques available for diffusivity measurements is presented in Table I.

Diffusion describes the random motion that transports matter from one part of a system at high concentration to another at low concentration. Mathematically, this process relates the mass-transfer rate of a substance through unit area to the concentration gradient normal to the section by a proportionality constant, D (cm^2/sec), also referred to as the diffusion coefficient (Crank, 1975). Factors that affect protein diffusion in polymers include properties that alter polymer chain segmental mobility (degree of crystallinity, chain stiffness, degree of cross-linking), deformations that alter the free volume, and factors that can immobilize or denature the protein (Rabek, 1980).

Table I
Techniques for Measuring the Diffusion Coefficient of Proteins in Hydrogels

Technique	Advantages	Disadvantages
Membrane permeation	Easy to use Polymer in hydrogel state UV detection possible	Long times required to reach steady state Potential to rupture membrane
Absorption/desorption (dissolution)	Easy method Options for data analysis UV-detectable groups are abundant in proteins	Protein adsorption onto hydrogel is possible Polymer swelling complicates diffusion process
Scanning electron microscopy (SEM)	Concentration of protein within hydrogel is evaluated in two dimensions (x, y)	X-ray emitting atom must be present in protein Destructive method Dehydrated polymer is required
Infrared (IR) spectroscopy	Several assessories FTIR microscopy studies provide two-dimensional concentration profiles	Protein molecule must have unique IR band Microscopy requires microtomed samples
Quasi-elastic light scattering (QELS)	Measures translational, rotational, self, and mutual diffusion coefficients	Experimental artifacts observed at low concentration Protein interactions can interfere at high concentration

Diffusion coefficients for proteins (hormones, enzymes, or polypeptides) in polymer systems typically fall in the range from 10^{-5} to $10^{-9}\,\mathrm{cm}^2/\mathrm{sec}$. The experimental techniques described below have approximate time scales for measuring the corresponding variations in concentration for these systems, assuming that the diffusion of a polymer chain within the hydrogel is negligible compared to that of the protein. There are several experimental techniques that have been used to determine the diffusion coefficient, although the exact value determined will depend on the system and can be the mutual, D^v, intrinsic, D_A, multicomponent, D_A^B, or self diffusion coefficient, D_A^*. The mutual diffusion coefficient, D^v, describes the solute in a two-component system with a fixed reference, assuming that no volume change occurs on mixing. The intrinsic diffusion coefficient, D_A, describes the transfer of a component when no mass flow occurs. For constant partial volumes for a two-component system, v_A and v_B, the intrinsic diffusion coefficients are related to the mutual diffusion coefficient, D^v, by the following equation:

$$D^v = v_A c_A (D_B - D_A) + D_A \tag{6}$$

The concentrations of A and B are denoted by c_A and c_B, respectively. For the case in which there is a volume change on mixing, the system is fixed with respect to one component, B, and, therefore, the diffusivity is represented by

$$D_A^B = D^v (v_B)^2 \tag{7}$$

When radioactive labels are employed as tracers to measure the diffusion coefficient, the associated value is referred to as the self diffusion coefficient, D_A^*. In polymer–solvent systems, the intrinsic diffusivity of the polymer is negligible compared to that of the solvent, i.e. $D_B = 0$. In this situation, the diffusion coefficient can be specified by either the intrinsic or the mutual diffusion coefficient in accordance with the following relation:

$$D^v = D_A (1 - v_A c_A) = D_A v_B \tag{8}$$

Crank and Park (1968) reviewed experimental techniques for measuring diffusion of solutes through polymers, which generally apply to proteins as well as drug molecules. These techniques are based on two principles: permeation of the protein through a membrane, and sorption/desorption kinetics for the protein/polymer system.

Radioactive tracers were utilized by Bueche (1962) to measure self diffusion coefficients for polymer systems above their glass-transition temperature, T_g. Price *et al.* (1978) described a novel approach that used scanning electron microscopy (SEM) and dispersive energy X-ray fluorescence analysis to measure the interdiffusion ($D \approx 10^{-12}\,\mathrm{cm^2/sec}$) of compatible polymer/polymer systems. Quasi-elastic light scattering (QELS) is an unusual technique due to its ability to measure both the mutual and self diffusion coefficients. Patterson *et al.* (1981) and Amis *et al.* (1983) have demonstrated the application of this technique to polymeric gels.

4.1. Membrane Permeation Method

The membrane permeation method is used to study protein diffusion through a thin membrane, from a reservoir at high concentration (donor cell) through the hydrogel membrane and into another reservoir at lower concentration (receptor cell). For steady-state conditions, Colton *et al.* (1971) derived the following set of equations to determine the solute permeability:

$$\ln\left[\left(\frac{2c_t}{c_0}\right) - 1\right] = -UA\left(\frac{1}{V_1} + \frac{1}{V_2}\right)t \tag{9}$$

Here, c_t is the concentration of protein at any time t, c_o is the initial protein concentration, V_1 is the donor cell volume, V_2 is the receptor cell volume, U is the overall mass-transfer coefficient, and A is the surface area of the membrane exposed to solution.

The overall mass-transfer coefficient, U, is related to the local mass-transfer coefficients on either side of the membrane, k_1 and k_2, as follows:

$$\frac{1}{U} = \left(\frac{1}{k_1} + \frac{1}{P} + \frac{1}{k_2}\right) \tag{10}$$

Finally, the permeability, P, can be described by the partition coefficient, K, the diffusion coefficient, D, and the hydrogel thickness, δ:

$$P = \frac{KD}{\delta} \tag{11}$$

The partition coefficient, K, is the ratio of protein concentration located in the hydrogel, c_m^s, to the concentration of protein in solution at equilibrium, c_s^s.

$$K = \frac{c_m^s}{c_s^s} \tag{12}$$

The concentration of protein in the membrane at equilibrium is evaluated according to

$$c_m^s = \frac{(c_0 - c_s^s)V_0}{V_s} \tag{13}$$

Here, c_0 is the initial concentration of protein in solution, c_s^s is the protein concentration at equilibrium, V_0 is the initial volume of solution, and V_s is the volume of the swollen membrane.

The membrane permeation (or diffusion) technique is simple to setup, and UV analysis of the protein makes data acquisition and evaluation relatively easy. An additional advantage of this method is that the polymer membrane is maintained in the hydrated state throughout the experiment. The major disadvantages of the permeability technique are the long times required for steady state to be attained for large-molecular-weight polypeptides and proteins and the difficulty of maintaining mechanical integrity of swollen hydrophilic networks. High agitation rate in the donor and receptor cells is required to eliminate boundary layer effects, but this agitation may cause denaturation of the proteins.

4.2. Absorption/Desorption Method

A simple method for determination of the intrinsic diffusion coefficient of the protein (neglecting the diffusion of the polymer) is to measure the initial rate of protein absorption or desorption as a function of time according to Fick's law, assuming that D is constant (Crank, 1975).

$$\frac{\partial c}{\partial t} = D\frac{\partial^2 c}{\partial x^2} \tag{14}$$

This partial differential equation is solved according to the boundary conditions defined by the experiment. It is recommended that diffusion studies be carried out under sink conditions, and using a thin hydrogel membrane. Then, the absorption in or desorption from a thin hydrogel film of thickness δ can be described by the following initial and boundary conditions (Ritger and Peppas, 1987):

$$t = 0 \qquad -\frac{\delta}{2} < x < \frac{\delta}{2} \qquad c = c_0 \tag{15}$$

$$t > 0 \qquad x = \frac{\delta}{2}, -\frac{\delta}{2} \qquad c = c_s^s \tag{16}$$

Here, c_0 is the initial concentration of protein in the hydrogel, and c_s^s is the bulk concentration at equilibrium. From the solution of Eq. (14) with these boundary conditions, the mass of protein absorbed or desorbed at any time, M_t, normalized with respect to the amount at infinite times, M_∞, can be determined from

$$\frac{M_t}{M_\infty} = 1 - \sum_{n=0}^{\infty} \frac{8}{(2n+1)^2\pi^2} \exp\left[-D\frac{(2n+1)^2\pi^2}{\delta^2}t\right] \tag{17}$$

For short times, an alternative solution of Eq. (14) is given in the form of an error function series:

$$\frac{M_t}{M_\infty} = 4\left(\frac{Dt}{\delta^2}\right)^{1/2}\left[\frac{1}{\pi^{1/2}} + 2\sum_{n=1}^{\infty}(-1)^n \operatorname{ierfc}\frac{n\delta}{2(Dt)^{1/2}}\right] \tag{18}$$

Here, ierfc x is the integrated complementary error function of x (Ritger and Peppas, 1987). For short times, Eq. (18) can be approximated by the following equation:

$$\frac{M_t}{M_\infty} = 4\left(\frac{Dt}{\pi\delta^2}\right)^{1/2} \tag{19}$$

By plotting the fractional amount of absorbed or desorbed protein, M_t/M_∞, versus time, the diffusivity can be evaluated from the slope of the initial 60% of the total release, with an accuracy of 1%. This analysis is valid for systems that behave in a Fickian manner, such as an equilibrium swollen hydrogel with evenly dispersed protein. However, non-Fickian transport phenomena are typically observed for swelling hydrophilic polymers initially in the glassy state, as described by Peppas and Lustig (1985).

Ritger and Peppas (1987) proposed a simple semiempirical expression that represents a more general diffusion mechanism:

$$\frac{M_t}{M_\infty} = kt^n \tag{20}$$

The constants, k and n, are characteristic of the polymer/protein system, where n is the diffusional exponent. The classification of the type of protein transport through a slab geometry on the basis of the value of the exponent n is given in Table II. The value of n defining Fickian diffusion changes with geometry from 0.5, 0.45, to 0.43 for a slab, a cylinder, and a sphere (Ritger and Peppas, 1986).

Another approach for treating dynamic swelling hydrogels is to couple the Fickian diffusion with relaxation. One such model, by Berens and

Table II
Transport Mechanisms of Protein through a Hydrophilic Polymer Slab Geometry

Diffusional exponent, n	Type of transport	Time dependence
0.5	Fickian diffusion	$t^{1/2}$
$0.5 < n < 1.0$	Non-Fickian (anomalous) transport	t^{n-1}
1	Case II transport	Independent of time
$n > 1$	Super case II transport	t^{n-1}

Hofenberg (1978), expresses the protein release from the hydrogel as follows:

$$\frac{M_t}{M_\infty} = k_1 t + k_2 t^{1/2} \tag{21}$$

In this case, the protein uptake or release is written in terms of a relaxation-induced portion (first term) and a Fickian portion (second term). The previous two models are valid for the initial 60% of equilibrium, or M_∞.

The Higuchi equation (Eq. 22) has been applied to systems in which the initial protein concentration per unit volume (A) is above its solubility in that hydrogel matrix, c_s, as found in a reservoir system.

$$M_t = [c_s(2A - c_s)Dt]^{1/2} \tag{22}$$

A hydrogel delivery device can also be prepared by polymerizing in the presence of the protein (Korsmeyer, 1991). The protein must be stable under the polymerization conditions (temperature, solvent, pH) and not react into the hydrogel matrix. The pseudo-steady-state assumptions require that the protein concentration per unit volume exceed the saturation limit (i.e., $A \gg c_s$). Lee (1988) used an integral method for moving-boundary problems to reduce the problem as follows:

$$M_t = c_s(1 + H)\left(\frac{Dt}{3} + 1\right)^{1/2} \tag{23}$$

$$H = c_s^{-1}[5A - (A^2 - c_s^2)^{1/2}] - 4 \tag{24}$$

The major advantage of this analysis is that the analytical solution is valid over all A/c_s values, with an accuracy within 1% of the exact solution. This is especially important for hydrogels, where A/c_s values are small.

The absorption/desorption method is more involved experimentally compared to the membrane permeation technique, since the protein must be incorporated into the hydrogel, but not adsorbed, and subsequently dried. The dissolution study is typically carried out in USP no. 2 apparatus, with media continuously being monitored by UV analysis. Penetration of the dissolution media into the dehydrated polymer complicates the diffusion process, commonly producing a lag time prior to protein release.

4.3. Scanning Electron Microscopy (SEM)

Scanning electron microscopy (SEM) has been used to measure drug or protein concentration profiles as a function of position within the hydrogel sample. The effect of loading time on the concentration profile can be determined by analyzing replicate samples held under the same conditions, but at longer loading times. An X-ray microanalyzer counts the number of X-rays emitted from the sample for a particular atom, such as chlorine. This X-ray-emitting atom must be present in the protein in order to distinguish it from the dehydrated hydrogel. From the concentration profile, the diffusion coefficient is evaluated according to an appropriate model. For constant-density systems, the diffusion coefficient is determined from the Boltzmann solution for diffusion in infinite media, shown in Eq. (25).

$$D_{|c=c_1} = \frac{d\eta}{dc} \int \eta \, dc \tag{25}$$

Price *et al.* (1978) developed this technique for measurement of the interdiffusion of poly(vinyl chloride)–poly(ε-caprolactone) at 70 °C. The major disadvantage of this technique is the radiation damage of the hydrophilic polymer during the SEM scan and the necessity of dehydrating the delivery device (am Ende, 1993).

4.4. Fourier Transform Infrared (FTIR) Spectroscopy

The development of Fourier transform infrared (FTIR) spectroscopy has facilitated the expansion of the scope of applications for IR spectroscopy to new fields of research, from forensics to pharmaceutical on-line analysis. The most important requirement of this technique is that the protein contain a distinctive IR-absorbing functional group not present in the hydrogel. An advantage of FTIR spectroscopy is the ability to obtain spectral information for a polymer in the hydrated state, with the use of attenuated total reflectance (ATR)-FTIR spectroscopy. Use of FTIR microscopy to study protein concentration profiles within a hydrophilic polymer requires a cryotomed sample of the preloaded hydrogel with a thickness of approximately 4 μm.

Jabbari *et al.* (1993) used ATR-FTIR spectroscopy to evaluate the concentration of mucin (glycoprotein) that diffused into a PAA hydrogel. The mucin spectrum contained an IR band at 1550 cm^{-1}, representing the dimeric C=O stretching vibration, not found in the hydrogel. The diffusion of mucin into PAA was monitored by the increased intensity of the 1550-cm^{-1} band with time. The diffusion coefficient was determined according to Fick's law with the appropriate boundary conditions.

An innovative method for investigating the diffusion of polymer molecules across a hydrogel interface was developed by Sahlin (1992). This near-field FTIR microscopy technique has improved spatial resolution compared to standard FTIR microscopy. The concentration of the diffusing polymer was monitored across the gel/gel interface at discrete locations along the x-axis. The concentration profile was evaluated to determine the diffusion coefficient. FTIR microscopy was utilized by am Ende (1993) to examine the two-dimensional concentration profiles of solutes throughout the interior of the microtomed hydrogel. The basis of this technique can be applied to diffusing proteins through hydrogels.

4.5. Quasi-Elastic Light Scattering (QELS) Method

Light scattering techniques are useful for analyzing inhomogeneous media, such as macromolecules in solution or as part of a gel. The random motion of the macromolecules forms regions with different concentrations and dielectric constants, as well as different refractive indices, compared to those of the bulk material. Elastic light scattering refers to the scattered light being of the same frequency as the incident beam. Movement of the scattering centers produces a Doppler shift from the incident frequency, and, therefore, the scattered light is no longer monochromatic. This phenomenon is referred to as either quasi-elastic light scattering (QELS) or Rayleigh line width (Rabek, 1980).

The QELS technique measures the translational, rotational, self, and mutual diffusion coefficients, of macromolecules, such as proteins (Stock and Ray, 1985). The translational diffusion constant, D_0, for dilute solutions was shown to be proportional to the molecular weight of the protein, M_w:

$$D_0 = 2.37 \times 10^{-4} M_w^{-0.54} \tag{26}$$

Amis *et al.* (1983) studied the dynamic scattering in gels with a single decay process, which allowed them to determine the mutual diffusion coefficient. Here, D^v was found to be proportional to the solvent viscosity. The self diffusion coefficient D_A, evaluated from the slow decay process as $(0.7–39) \times 10^{-9}\,cm^2/sec$, was found to decrease drastically with increased concentration. Light scattering techniques have also been applied to evaluate the hydrogel mesh size, ξ, which describes the space available for the protein diffusion. Adam *et al.* (1976) analyzed semidilute polymer solutions to evaluate scaling laws of de Gennes (1979). The scaling laws correlate the mesh size to the polymer volume fraction, $v_{2,s}$, according to the following expression:

$$\xi \approx v_{2,s}^{-3/4} \tag{27}$$

At high concentrations, the mesh size (or correlation length) scales to the $-\frac{1}{2}$ power instead of the $-\frac{3}{4}$ power.

The major disadvantages of this technique for determination of the diffusion coefficient include the following: (a) at low concentrations, experimental artifacts may occur such as the detection of both the scattered and incident light (heterodyning); (b) at high concentrations, interactions between the protein molecules can cause departure from pure Brownian motion; and (c) structural defects, scratches, and trapped dust within the hydrogel sample can cause a strong scattering of light.

4.6. Other Techniques

There are several other experimental methods for determining diffusion coefficients for proteins in polymeric systems, such as interference microscopy and NMR spectroscopy. For a discussion of these methods, the reader is referred to Crank and Park (1968).

REFERENCES

Adam, M., Delsanti, M., and Jannink, G., 1976, Light scattering by cooperative diffusion in semi-dilute polymer solutions, *J. Phys., Lett.* **37**:153–156.

am Ende, M. T., 1993, Transport and Interaction of Ionizable Drugs and Proteins in Hydrophilic Polymers, Ph.D. Thesis, Purdue University.

Amis, E. J., Janmey, P. A., Ferry, J. D., and Yu, J., 1983, Quasi-elastic light scattering of gelatin solutions and gels, *Macromolecules* **16**:441–446.

Arteaga, C. L., Hurd, S. D., Dugger, R. C., Winnier, A. R., and Robertson, J. B., 1994, Epidermal growth factor receptors in human breast carcinoma cells: A potential selective target for transforming growth factor α-Pseudomonas exotoxin 40 fusion protein, *Cancer Res.* **54**:4703–4709.

Baker, R., 1987, *Controlled Release of Biologically Active Agents*, John Wiley & Sons, New York.

Berens, A. R., and Hofenberg, H. B., 1978, Diffusion and relaxation in glassy polymer powders: 2. Separation of diffusion and relaxation parameters, *Polymer* **19**:490–496.

Braatz, J. A., Heifetz, A. H., and Kehr, C. L., 1992, A new hydrophilic polymer for biomaterial coatings with low protein adsorption, *J. Biomater. Sci. Polym. Ed.* **3**(6):451–462.

Brondsted, H., and Kopecek, J., 1991, Hydrogels for site-specific oral drug delivery: Synthesis and characterization, *Biomaterials* **12**:584–592.

Bueche, F., 1962, *Physical Properties of Polymers*, Interscience Publishers, New York.

Caliceti, P., Veronese, F., Schiavon, O., Lora, S., and Carenza, M., 1992, Controlled release of proteins and peptides from hydrogels synthesized by gamma ray-induced polymerization, *Il Farmaco* **47**(3):275–286.

Canal, T., and Peppas, N. A., 1989, Correlation between mesh size and equilibrium degree of swelling of polymeric networks, *J. Biomed. Mater. Res.* **23**:1183–1193.

Carelli, V., Di Colo, G., Nannipieri, E., and Serafini, M.F., 1989, Evaluation of solution impregnation method for loading drugs into suspension-type polymer matrices: A study of factors determining the patterns of solid drug distribution in matrix and drug release from matrix, *Int. J. Pharm.* **55**:199–207.

Colton, C. K., Smith, K. A., Merrill, E. W., and Farrell, P. C., 1971, Permeability studies with cellulosic membranes, *J. Biomed. Mater. Res.* **5**:459–488.

Crank, J., 1975, *The Mathematics of Diffusion*, Academic Press, New York.

Crank, J., and Park, G. S., 1968, *Diffusion in Polymers*, Academic Press, New York.

de Gennes, P. G., 1979, *Scaling Concepts in Polymer Physics,* Cornell University Press, Ithaca, New York.

Dillman, W. J., Jr., and Miller, I. F., 1973, On the adsorption of serum proteins on polymer membrane surfaces, *J. Colloid. Interface Sci.* **44**:221–241.

Domb, A., Davidson, G. W. R., III, and Sanders, L. M., 1990, Diffusion of peptides through hydrogel membranes, *J. Controlled Release* **14**:133–144.

Dong, L. C., and Hoffman, A. S., 1991, A novel approach for preparation of pH sensitive hydrogels for enteric drug delivery, *J. Controlled Release* **15**:141–152.

Greenley, R. Z., Brown, R. M., Garbow, J., Vogt, C. E., Zia, H., Rodgers, R. L., Christie, M., and Luzzi, L. A., 1990, Polymer matrices for oral delivery, *Polym. Prepr. (Am. Chem. Soc., Div. Polym. Chem.)* **31**(2):182–183.

Hayashi, Y., Yoshioka, S., Aso, Y., Po, A. L. W., and Terao, T., 1994, Entrapment of proteins in poly(L-lactide) microspheres using reversed micelle solvent evaporation, *Pharm. Res.* **11**(2):337–340.

Heller, J., 1988, Chemically self-regulated drug delivery systems, *J. Controlled Release* **8**:111–125.

Ho, N. F. H., Ganesan, M. G., Weiner, N. D., and Flynn, G. L., 1986, Mechanisms of topical delivery of liposomally entrapped drugs, in: *Advances in Drug Delivery Systems* (J. M. Anderson and S. W. Kim, eds.), Controlled Release Series No. 1, Elsevier, New York, pp. 61–65.

Horbett, T. A., 1982, Protein adsorption on biomaterials, in: *Biomaterials: Interfacial Phenomena and Applications* (S. L. Cooper, N. A. Peppas, A. S. Hoffman, and B. D. Ratner, eds.), Advances in Chemistry Series No. 199, American Chemical Society, Washington, D.C., pp. 233–244.

Hsieh, D. S. T., Rhine, W. D., and Langer, R., 1983, Zero-order controlled release polymer matrices for micro- and macromolecules, *J. Pharm. Sci.* **72**(1):17–22.

Jabbari, E., Wisniewski, N., and Peppas, N. A., 1993, Evidence of mucoadhesion by chain interpenetration at a poly(acrylic acid)/mucin interface using ATR-FTIR spectroscopy, *J. Controlled Release* **26:**99–108.

Jeffery, H., Davis, S. S., and O'Hagen, D. T., 1993, The preparation and characterization of poly(lactide-*co*-glycolide) microparticles. II. The entrapment of a model protein using a (water-in-oil)-in-water emulsion solvent evaporation technique, *Pharm. Res.* **10:**362–368.

Kim, C. J., and Lee, P. I., 1991, Swelling-controlled drug release from anionic polyelectrolyte gel beads, *Proc. Int. Symp. Control. Rel. Bioact. Mater.* **18:**451–452.

Kopecek, J., Kopeckova, P., Brondsted, H., Rathi, R., Rihova, B., Yen, P.- Y., and Ikesue, K., 1992, Polymers for colon-specific drug delivery, *J. Controlled Release* **19:**121–130.

Korsmeyer, R. W., 1991, Diffusion controlled systems: Hydrogels, in: *Polymers for Controlled Drug Delivery* (P. J. Tarcha, ed.), CRC Press, Boca Raton, Florida, pp. 15–37.

Langer, R., 1990, New methods of drug delivery, *Science* **249:**1527–1533.

Langer, R., and Folkman, J., 1976, Polymers for the sustained release of proteins and other macromolecules, *Nature* **263:**797–800.

Langer, R., and Peppas, N., 1983, Chemical and physical structure of polymers as carriers for controlled release of bioactive agents: A review, *J. Mater. Sci.: Rev. Macromol. Chem. Phys.* **C23(1):**61–126.

Lee, P. I., 1988, Synthetic hydrogels for drug delivery: Preparation, characterization, and release kinetics, in: *Controlled Release Systems: Fabrication Technology*, Vol. II (D. S. T. Hsieh, ed.), CRC Press, Boca Raton, Florida, pp. 61–82.

Lee, V. H. L., 1992, *Peptide and Protein Drug Delivery*, Marcel Dekker, New York.

McGee, J. P., Davis, S. S., and O'Hagan, D. T., 1995, Zero order release of protein from poly(D,L-lactide-*co*-glycolide) microparticles prepared using a modified phase separation technique, *J. Controlled Release* **34:**77–86.

McNeill, M. E., and Graham, N. B., 1993, Properties controlling the diffusion and release of water-soluble solutes from poly(ethylene oxide) hydrogels 1. Polymer composition, *J. Biomater. Sci. Polym. Ed.* **4:**305–322.

Munch, J. P., Candau, S., Duplexxis, R., Picot, C., Herz, J., and Benoit, H., 1976, Spectrum of light scattered from viscoelastic gels, *J. Polym. Sci.: Polym. Phys. Ed.* **14:**1097–1109.

Park, K., 1988, Enzyme-digestible swelling hydrogels as platforms for long-term oral drug delivery: Synthesis and characterization, *Biomaterials* **9:**435–441.

Patterson, G. D., 1981, Light scattering from concentrated polymer solutions and gels, *Polym. Prepr. (Am. Chem. Soc., Div. Polym. Chem.)* **22:**709–713.

Peppas, N. A., and Lustig, S. R., 1985, The role of cross-links, entanglements, and relaxations of macromolecular carriers in the diffusional release of biologically active materials: Conceptual and scaling relationships, *Ann. N.Y. Acad. Sci.* **446:**26–41.

Peppas, N. A., and Meadows, D. L., 1983, Macromolecular structure and solute diffusion in membranes: An overview of recent theories, *J. Membr. Sci.* **16:**361–377.

Peschier, L. J. C., Bouwstra, J. A., de Bleyser, J., Junginger, H. E., and Leyte, J. C., 1993, Water mobility and structure in poly[2-hydroxyethylmethacrylate] hydrogels by means of the pulsed field gradient NMR technique, *Biomaterials* **14:**945–952.

Price, F. P., Gilmore, P. T., Thomas, E. L., and Laurence, R. L., 1978, Polymer/polymer diffusion, *J. Polym. Sci.: Polym. Symp.* **63:**33–44.

Rabek, J. F., 1980, *Experimental Methods in Polymer Chemistry: Physical Principles and Applications*, John Wiley & Sons, New York.

Ritger, P. L., and Peppas, N. A., 1987, A simple equation for description of solute release I. Fickian and non-Fickian release from non-swellable devices in the form of slabs, spheres, cylinders, or discs, *J. Controlled Release* **5**:23–36.

Roorda, W., 1994, Do hydrogels contain different classes of water?, *J. Biomater. Sci. Polym. Ed.* **5**:383–395.

Sahlin, J. J., 1992, Polymer Chain Interdiffusion in Gel/Gel Adhesion, Ph.D. thesis, Purdue University.

Sato, S., and Kim, S. W., 1984, Macromolecular diffusion through polymer membranes, *Int. J. Pharm.* **22**:229–255.

Satterfield, C. N., Colton, C. K., and Pitcher, W. H., Jr., 1973, Restricted diffusion of liquids within fine pores, AIChE J. **19**:628–635.

Sawhney, A. S., Pathak, C. P., and Hubbell, J. A., 1993, Bioerodible hydrogels based on photopolymerized poly(ethylene glycol)-*co*-(α-hydroxy acid) diacrylate macromers, *Macromolecules* **26**:581–587.

Shah, S. S., Cha, Y., and Pitt, C. G., 1992, Poly(glycolic acid-*co*-DL-lactic acid): Diffusion or degradation controlled drug delivery?, *J. Controlled Release* **18**:261–270.

Shalaby, W. S., and Park, K., 1990, Biochemical and mechanical characterization of enzyme-digestible hydrogels, *Pharm. Res.* **7**:816–823.

Shalaby, W. S. W., Abdallah, A. A., Park, H., and Park, K., 1993, Loading of bovine serum albumin into hydrogels by an electrophoretic process and its potential application to protein drugs, *Pharm. Res.* **10**(3):457–460.

Siegel, R. A., and Langer, R., 1984, Controlled release of polypeptides and other macromolecules, *Pharm. Res.* **1**:2–10.

Smith, B. A., and Sefton, M. V., 1992, Thrombin and albumin adsorption to PVA and heparin–PVA hydrogels. I. Single protein isotherms, *J. Biomed. Mater. Res.* **26**:947–958.

Smith, K. L., Baker, R. W., and Ninomiya, Y., 1983, Development of BioLure controlled release pheromone products, in: *Controlled Release Delivery Systems* (T. J. Roseman and S. Z. Mansdorf, eds.), Marcel Dekker, New York, pp. 325–335.

Stock, R. S., and Ray, W. H., 1985, Interpretation of photon correlation spectroscopy data: A comparison of analysis methods, *J. Polym. Sci.: Polym. Phys. Ed.* **23**:1393–1447.

Strzinar, I., and Sefton, M. V., 1992, Preparation and thrombogenicity of alkylated polyvinyl alcohol coated tubing, *J. Biomed. Mater. Res.* **26**:577–592.

Suzawa, T., and Murakami, T., 1980, Adsorption of bovine serum albumin on synthetic polymer latices, *J. Colloid Interface Sci.* **78**(1):267–268.

Theeuwes, F., 1975, Elementary osmotic pump, *J. Pharm. Sci.* **64**:1987–1991.

Thissen, J. P., Underwood, L. E., Maiter, D., Maes, M., Clemmons, D. R., and Ketelslegers, J. M., 1991, Failure of insulin-like growth factor-I (IGF-I) infusion to promote growth in protein-restricted rats despite normalization of serum IGF-I concentrations, *Endocrinology* **128**:885–890.

Zentner, G. M., Cardinal, J. R., and Gregonis, D. E., 1979, Progestin permeation through polymer membranes III: Polymerization solvent effect on progesterone permeation through hydrogel membranes, *J. Pharm. Sci.* **68**(6):794–797.

Zhu, K. J., Xiangzhou, L., and Shilin, Y., 1990, Preparation, characterization, and properties of polylactide (PLA)–poly(ethylene glycol) (PEG) copolymers: A potential drug carrier, *J. Appl. Polym. Sci.* **39**:1–9.

Chapter 6

Poly(ethylene glycol)-Coated Nanospheres: Potential Carriers for Intravenous Drug Administration

Ruxandra Gref, Yoshiharu Minamitake, Maria Teresa Peracchia, Avi Domb, Vladimir Trubetskoy, Vladimir Torchilin, and Robert Langer

I. Introduction

The development of injectable drug carriers, small enough to freely circulate through capillaries and with a long enough blood half-life to continuously deliver the encapsulated active compounds, could bring numerous advantages in the area of parenteral drug administration, such as:

- protection of labile drugs against degradation;
- reduction of toxic side effects associated with some highly active drugs such as those used for cancer treatment;

Ruxandra Gref • Laboratoire de Chimie-Physique Macromoléculaire (URA CNRS 494), ENSIC, 54001 Nancy Cedex, France. *Yoshiharu Minamitake* • Suntory Limited, Ohra-Gun, Gunma-Ken 370-05, Japan. *Maria Teresa Peracchia* • Dipartimento Farmaceutico, Universita degli Studi di Parma, 43100 Parma, Italy. *Avi Domb* • Department of Pharmaceutical Chemistry, The Hebrew University, Jerusalem, Israel 91120. *Vladimir Trubetskoy and Vladimir Torchilin* • Department of Radiology, Massachusetts General Hospital-East, Charlestown, Massachusetts 02129. *Robert Langer* • Department of Chemical Engineering, Massachusetts Institute of Technology, Cambridge, Massachusetts 02139.

Protein Delivery: Physical Systems, Sanders and Hendren, eds., Plenum Press, New York, 1997.

- more favorable drug pharmacokinetic profiles; and
- increase of patient comfort by avoiding repetitive injections or perfusion pumps.

Long-circulating particles provide numerous opportunities for the formulation of controlled release and site-specific (targeted) drug delivery systems. For example, by coupling a monoclonal antibody (or a fragment of it) to the carriers, specific cells with corresponding surface epitopes can potentially be targeted. We could imagine such particles acting as "magic missiles" directed toward a diseased tissue and liberating locally the active drugs (Goldberg, 1983).

However, when administered into the body, the particulate carriers are inevitably exposed to an attack from the immune system, which protects the body from invasion by foreign products. Thus, injected particles are removed from blood via recognition by phagocytic cells [essentially macrophages located in the mononuclear phagocyte system (MPS), mainly in the liver and spleen]. For example, 50–90% of polystyrene latex particles disappear from blood within the first five minutes after injection (Illum *et al.*, 1987). The same fate is observed in the case of particles made of other materials, such as poly(lactic acid) (PLA) (Bazile *et al.*, 1992), polycyanoacrylate (Kreuter *et al.*, 1979), polyacryl starch (Laakso *et al.*, 1986), or albumin (Gottlleb *et al.*, 1990).

It is generally assumed that the rapid carrier uptake by the MPS system is mediated by the adsorption of certain serum proteins and blood components (opsonins) onto the carrier surface. These components [such as fibrinogen, fibronectin, IgG, complement, and glycoproteins (Munthe-Kaas and Kaplan, 1980)] rapidly adsorb on the surface of the particles, depending on the physicochemical properties of the particle surface. The complement fraction C3, after deposition on the particle surface, plays a crucial role in particle recognition by the macrophages, resulting in their rapid phagocytosis (Artursson and Sjöholm, 1986; Leroux *et al.*, 1995). Detailed reviews describe the complex aspects of opsonization (Porter *et al.*, 1994; Müller, 1991; Donbrow, 1992).

Many strategies have been proposed to circumvent MPS capture. After a description of various types of systems (liposomes, emulsions, micelles, and nanoparticles) designed to achieve long blood circulation half-lives, this chapter will focus on newly developed injectable biodegradable nanospheres with a prolonged blood half-life due to a coating with poly(ethylene glycol) (PEG) and able to encapsulate and continuously release drugs (Spenlehauer *et al.*, 1992; Gref *et al.*, 1993a, 1994; Domb *et al.*, 1994; Stolnik *et al.*, 1995; Hagan *et al.*, 1995; Bazile *et al.*, 1995).

1.1. Approaches to Increase Particle Blood Circulation Time

Because of the rapid MPS clearance, site-specific targeting of intravenously administered particulate drug carriers to tissues other than liver and spleen is extremely difficult. To maintain the particles in the bloodstream, one possibility is to block the MPS phagocytic activity prior to particle injection; another possibility is to design MPS-avoiding particles.

The phagocytic potential of the liver and the spleen could be temporarily blocked by using different substances (such as rare-earth metal salts or carbon colloids). For example, injected liposome blood clearance can be reduced after MPS blockade with dextran sulfate or carbon (Souhami *et al.*, 1981). MPS blockade was also achieved by progressively increasing intravenous (i.v.) dosages of liposomes: liver saturation was first observed, followed by localization in the spleen and eventually accumulation in the bone marrow (Poste, 1983).

Although reversible, MPS blockade weakens the body's immune system and therefore cannot be used in therapy. As an alternative to hepatic blockade, hepatic avoidance is preferred. For this, the injected "stealth" particle should satisfy criteria related to its size and surface characteristics (such as hydrophilicity and charge). For example, it was established that large particles are rapidly cleared from blood; e.g., carriers larger than 5–7 μm are mechanically cleared by capillary filtration, mainly in the lungs (Kanke *et al.*, 1980), whereas particles smaller than 5 μm can circulate through capillaries but are readily captured by the MPS cells (Illum and Davis, 1983). Generally, the MPS uptake was found to increase with an increase in particle size, irrespective of the particle structure—e.g., nanospheres (Davis, 1981), liposomes (Gregoriadis *et al.*, 1977; Sato *et al.*, 1986), or fat emulsion droplets (Takino *et al.*, 1994)—although a small size is not sufficient to prolong blood half-life [as shown, for example, in the case of polystyrene particles with a mean diameter of only 60 nm, which are phagocytosed within minutes (Illum *et al.*, 1987)]. MPS-avoiding particles should also possess optimal surface characteristics to minimize the interactions with opsonins, which lead to MPS phagocytosis. The main parameters that govern these interactions are the surface charge (Petrak, 1993; Tabata and Ikada, 1988; Porter *et al.*, 1994), the nature of the surfactant used in the carrier preparation (Davis and Hansrani, 1985), and the surface hydrophobicity (Illum and Davis, 1983). The total effect on MPS uptake is due to all of these factors. However, investigations of the respective effects of these parameters is difficult, because of the interdependence between them (i.e., changing the charge may change other equally important factors such as surface hydrophobicity and particle size).

1.2. PEG Hydrophilic Coatings: Mechanism of Protein Rejection

It has been demonstrated that the creation of a hydrophilic neutral surface coating reduces particle blood clearance; surfaces with either no net charge or shielded charges generally give a low adsorption of serum proteins of importance for opsonization, while those containing either unprotected charges or hydrophobic domains result in a high opsonin adsorption followed by blood removal (Petrak, 1993; Malmstein, 1995). Steric barriers against opsonization can be obtained by the use of appropriate coatings, such as PEG coatings, which is a recent research trend. Hydrophilic and uncharged, commercially available with various end groups and well-defined molecular weights, PEG is considered as nontoxic and was approved by the U.S. Food and Drug Administration for internal use in the human body (Harris, 1985). A great deal of interest has been focused on PEG-modified proteins that have dramatically reduced immunogenicity and antigenicity (Abuchowski *et al.*, 1977). After i.v. administration, the macromolecular conjugates circulate longer in the bloodstream than the native proteins (Burnham, 1994).

PEG surface modification was used to increase biocompatibility (Sawhney *et al.*, 1993) or to obtain blood-compatible materials (Han *et al.*, 1993). PEG grafting on various substrates was shown to reduce the adsorption of various proteins (Prime and Whitesides, 1993; Llanos and Sefton, 1993) and fibrinogen (Han *et al.*, 1993) to the surface and to reduce complement activation (Kishida *et al.*, 1992).

Various models were proposed to explain the molecular interactions between PEG-coated surfaces and plasma proteins. It was proposed that, owing to the flexibility of PEG chains, a steric barrier is produced at the hydrophobic substrate coated with PEG and that protein adsorption (and thus opsonization) are thus avoided (Torchilin and Papisov, 1994). Dextran, another hydrophilic polymer with more rigid chains than PEG, was also used to coat polystyrene surfaces (Österberg *et al.*, 1995); the lightly packed coatings obtained by attaching dextran at one chain end to the surface were relatively ineffective compared to PEG at reducing protein adsorption.

The conditions that lead to protein repulsion from hydrophobic flat surfaces coated with PEG, attached to the substrate at only one chain end, were studied by Jeon *et al.* (1991) and Jeon and Andrade (1991). These authors consider a model in which different forces are taken into account:

- Hydrophobic attraction between the substrate and the hydrophobic regions of the protein
- van der Waals attractions between the hydrophobic substrate and the protein and between the PEG chains and the protein

- Steric repulsive forces between PEG and proteins, which are the main forces in the system; the tendency of blood proteins to approach hydrophobic substrates results in an elastic PEG restoring component and an osmotic pressure as a consequence of PEG chain crowding at the surface.

Using this model, Jeon and Andrade (1991) suggested that a high PEG surface density and a long PEG chain length are necessary for low protein adsorption and that the effect of surface density predominates over the effect of chain length.

2. PEG-COATED LONG-CIRCULATING DRUG CARRIERS

PEG-coated liposomes were the first supramolecular drug carriers to show an increased blood circulation time due to their PEG coating (Klibanov *et al.*, 1990; Blume and Cevc, 1990; Allen, 1994; Woodle *et al.*, 1992; Maruyama *et al.*, 1991). The enhancement of blood half-lives was attributed to the presence of "brush" PEG chains (Lasic *et al.*, 1991; Torchilin and Papisov, 1994). However, due to their nature (vesicles consisting of lipid bilayer membranes), liposomes present some inconveniences such as a limited physical stability and a difficulty to be freeze-dried. The drug entrapment yield strongly depends upon the physicochemical properties of the drug (in large amounts, hydrophobic drugs, because of direct interactions with the liposome membrane, are likely to change the physical characteristics of the liposomes) (Kulkarni *et al.*, 1995)

Sometimes, too high a PEG level in the dipalmitoylphosphatidylcholine membrane may lead to vesicle fusion or membrane rupture (Massenburg and Lentz, 1994). Studies with liposomes show that the bilayer rigidity is one of the most important contributing factors to prolonged circulatory half-life, by increasing liposome resistance to destabilizing effects of serum components such as high-density lipoproteins (Scherphof *et al.*, 1978). Cholesterol, a compound having a high glass transition temperature, was included in liposome bilayers, leading to an increase in stability and subsequently to an increase in blood circulation time (Gregoriadis and Senior, 1980). One of the possible approaches to increasing the stability of liposomes is the use of polymerizing lipids for their production (Ringsdorf *et al.*, 1988). On the other hand, too high a stability might be sometimes inconvenient, by hindering drug release and liposome removal from the organism.

To overcome the stability problems, another approach in designing PEG-coated long-circulating carriers is the grafting or the adsorption of

PEG on polymeric cores in the solid state. For example, PEG was covalently attached to the surface of polycyanoacrylate nanoparticles during the polymerization process (Vauthier *et al.*, 1994) and to the surface of chemically cross-linked albumin nanospheres (Müller and Kissel, 1993). In the latter case, the uptake of the particles by cell culture macrophages was significantly reduced, depending on PEG chain length and surface density. However, by covalently attaching polymers to the particle surface, nonhomogeneous surface coverings were sometimes obtained (Artursson *et al.*, 1990), presumably due to a nonhomogeneous initial surface distribution of reactive sites and the increasing steric hindrance due to the already attached polymer chains. Moreover, practical difficulties such as drug denaturation or particle destruction can be encountered when attempts are made to attach polymers to ready-made particles containing drugs.

Surface modification by polymer adsorption is an alternative to surface modification by polymer grafting. For example, polystyrene nanospheres coated by Poloxamer or Poloxamine (Illum and Davis, 1983; Müller, 1991) or poly(methyl methacrylate) colloidal carriers coated by Poloxamer (Tröster and Kreuter, 1988) circulate longer in blood. This family of surfactants consist of poly(propylene glycol) (PPG) blocks, which adsorb on the hydrophobic polystyrene surface, and of more hydrophilic PEG blocks, which stick out of the surface in aqueous solutions and prevent opsonin adsorption. In spite of the increase in blood circulation time, particle coating by polymer adsorption was found to have several drawbacks (Petrak, 1993):

- The surface coverage was not complete and homogeneous.
- Hydrophilic polymer chain desorption may occur by replacement with blood compounds.
- Nondegradable "model" polymers such as polystyrene are not realistic therapeutic systems. This approach is difficult to generalize to hydrophilic polymer chain adsorption on biodegradable materials such as poly(lactic acid) (Müller, 1991).

PEG-containing micelles are another type of carriers that circulate longer in the blood. For example, adriamycin was chemically attached to a poly(aspartic acid) backbone with a pendant PEG block (Kataoka *et al.*, 1993). These polymers are endowed with self-assembling micelle-forming properties due to the presence of the hydrophobic drug attached to them. In spite of increased blood residence and antitumor activity improvement, these molecular aggregates have several inconveniences. For example, some drugs cannot be attached to polymers without substantial chemical modifications. From a practical point of view, each time a drug is covalently linked to a polymer, a new chemical entity is created. This then requires preclinical studies, independent of those on the parent drug, for regulatory approval to test clinically. Indeed, drug might be released together with a polymer fragment, and its therapeutic activity is thus considerably modified. Therefore, drug physical entrapment in long-circulating micelles, nanoparticles, or liposomes would be preferred.

3. PEG-COATED BIODEGRADABLE NANOSPHERES: POTENTIAL LONG-CIRCULATING DRUG CARRIERS

One recent approach to designing long-circulating drug carriers is to form compact biodegradable polymeric cores, stable in biological fluids and able to physically entrap and release drugs in a continuous and controlled manner. According to the models in the literature (Jeon and Andrade, 1991; Jeon *et al.*, 1991), PEG chains attached at one chain end to a nanosphere surface (Fig. 1) should inhibit MPS clearance. Ideally, these particles should degrade, after drug release, into nontoxic elements, in order to avoid accumulation of empty particles in the body. The size of the nanospheres should be as small as possible (ideally less than 200 nm), in order to avoid particle removal by a filtering effect. The particles should be able to encapsulate drugs, avoiding the requirement for chemical linking. The drug loading (weight fraction of encapsulated agent in the particles) should be reasonably high (e.g., more than 30%), and high entrapment efficiencies should be obtained (e.g., more than 80% of the amount of agent to be used

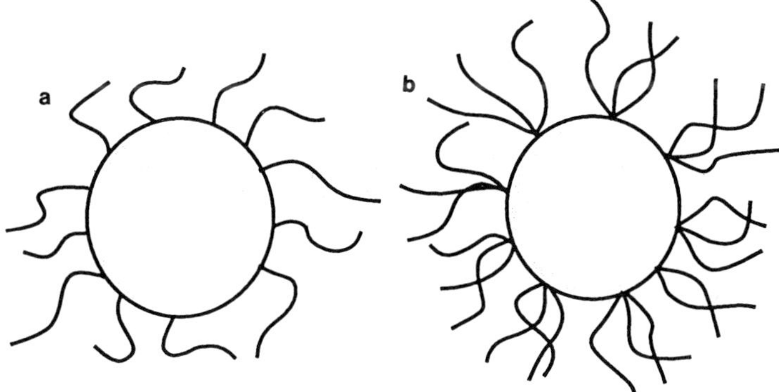

Figure 1. Schematic representation of PEG-coated nanospheres. The coating steric hindrance should avoid blood protein (P) adsorption on the particle core, formed in a hydrophobic biodegradable polymer R (PLA, PLGA, PCL, or PSA). (a) Nanosphere prepared from diblock PEG–R polymer; (b) nanosphere prepared from multiblock PEG_n–R polymer.

in the first step of the encapsulation process should be incorporated into the final carrier).

An important requirement of particulate drug carriers is their lack of toxicity. It is preferred that the delivery system degrades, after or during drug liberation, into harmless, excretable components that naturally occur in the body. Ideally, the core should consist of biodegradable polymers shown to be safe in the human body, such as (PLA) or poly(lactic-*co*-glycolic acid) (PLGA). Under physiological conditions, these polyesters undergo bulk erosion (Li *et al.*, 1990). The degradation kinetics are strongly dependent upon polymer physicochemical properties (Vert *et al.*, 1991).

To prepare PEG-coated nanospheres as depicted in Fig. 1, first, amphiphilic bioerodible polymers (symbolized as PEG–R), composed of a PEG block and a hydrophobic biodegradable block (R = PLA or PLGA), were synthesized. PEG-coated nanospheres were then formed by taking advantage of the different solubilities of R and PEG in aqueous and organic solutions (Spenlehauer et al., 1992; Gref et al., 1993a). Nanospheres were also formed with diblock PEG–poly(ε-caprolactone) (PEG–PCL) and PEG–poly(sebacic acid) (PEG–PSA) (Peracchia *et al.*, 1996).

In order to further increase the PEG density on the surface (Fig. 1b), nonlinear multiblock PEG_n–R polymers were synthesized (Domb *et al.*, 1994, 1996). They are composed of several (*n*) PEG blocks, attached together at one chain end to one hydrophobic R block.

PEG–PLA diblock copolymers were also used to prepare micelles (Hagan *et al.*, 1995; Piskin *et al.*, 1995) able to entrap hydrophobic drugs such as adriamycin (Piskin *et al.*, 1995). Stable micelles were formed when the PEG content in the polymer was between 25 and 80%. However, the maximum drug loading in the micelles was relatively low (about 12 mg/g) (Piskin *et al.*, 1995).

PLGA nanospheres were coated by adsorption of water-soluble PEG–PLA diblock copolymers to obtain long-circulating particles (Stolnik *et al.*, 1994). The advantage of the PEG–PLA over Poloxamers and Poloxamine lies in replacing the poly(propylene oxide) (PPO) moiety of the latter polymers with the biodegradable PLA chain. However, it is often stated that a polymer coating obtained by polymer adsorption is not homogeneous and might be replaced by blood proteins (and in particular opsonins) on the surface of the particles (Bazile *et al.*, 1995).

3.1. Biodegradable Polymers Containing PEG Blocks

Biodegradable polymers containing PEG blocks are a recent research trend. They consist of rigid and often crystalline polyester blocks with soft

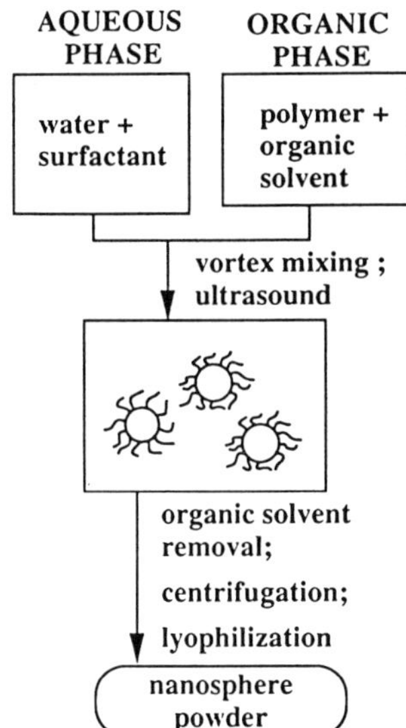

Figure 2. Schematic representation of the nanosphere preparation procedure.

and flexible PEG blocks. Thus, various mechanical and degradation behaviors can be achieved, opening up to these new polymers a wide range of applications such as temporary scaffolds or drug delivery devices. For example, triblock and diblock copolymers of the type R–PEG–R and PEG–R were prepared, where R is a hydrophobic biodegradable polymer as in the following nonexhaustive list:

- Polycaprolactone (PCL) (Perret and Skoulios, 1972; Wang and Qiu, 1993; Nair, 1993)
- Poly(L-lactic acid) (Younes and Cohn, 1987; Jedlinski *et al.*, 1993; Liu and Hu, 1993; Pitt *et al.*, 1993; Kricheldorf and Meier-Haack, 1993; Li and Kissel, 1993; Cerrai and Tricoli, 1994)
- Poly(DL-lactic acid) (Churchill and Hutchinson, 1986; Zhu *et al.*, 1990; Deng *et al.*, 1990; Stolnik *et al.*, 1995; Piskin *et al.*, 1995)
- Poly(lactic-*co*-glycolic acid) (Li and Kissel, 1993; Minamitake *et al.*, 1996)

The degradation behavior of these polymers was studied *in vitro* (Hu and Liu, 1993, 1994) and *in vivo* (Younes *et al.*, 1988; Ronneberger *et al.*, 1995). Biodegradable PLA-PEG-PLA triblock copolymers showed faster *in vivo* degradation kinetics than PLA, but the overall biological response to implants was comparable in both cases (Younes *et al.*, 1988). Similarly, PLGA–PEG–PLGA and PLGA polymers were equally well tolerated in cage implants over a 21-day implantation period (Ronneberger *et al.*, 1995).

Starlike multiblock copolymers were also formed. PLA copolymers were coupled to three or four PEG blocks by polymerization of lactide in the presence of stannous octanoate as a catalyst (Zhu *et al.*, 1989). Multiblock linear copolymers were also formed by reacting PEG with α-hydroxy-carboxylic acids (Cohn *et al.*, 1988). PEG_n–R polymers were formed by first attaching two or more $PEG\text{-}NH_2$ chains to citric, mucic, or tartaric acid or other natural polyfunctional molecules, leaving one or more active hydroxyl end groups for the ring opening or condensation polymerization of bioerodible R polymers (Domb *et al.*, 1994).

3.2. Preparation of PEG-Coated Nanospheres

Nanospheres can be produced by various methods (Kreuter, 1994). By using diblock PEG–R copolymers with different PEG molecular weight (2, 5, 12, and 20 kDa), nanospheres were formed by an emulsion/evaporation procedure (Fig. 2); (Gref *et al.*, 1994). First, PEG–R was dissolved in an appropriate organic solvent, such as ethyl acetate or methylene chloride. Then, an oil-in-water (O/W) emulsion was formed in an aqueous phase, by vortexing and sonicating. Sonication can be replaced by microfluidization. In the latter case, the O/W emulsion formed by vortexing is divided into two jet streams which connect in a ceramic chamber under high pressure and then pass through a microchannel. Owing to the resulting high turbulence, the emulsion breaks down into a very fine one. The main advantage of microfluidization over sonication is the possibility of large-scale production.

After the fine O/W emulsion was produced, the organic solvent was slowly evaporated. PEG is water-soluble but is practically insoluble in most organic solvents, such as ethyl acetate. For the hydrophobic part R of the copolymers, the situation is the opposite: R is insoluble in water but very soluble in ethyl acetate or methylene chloride. Consequently, there should be a tendency for block rearrangement inside the droplets, leading to a migration of PEG chains toward the interface with water, whereas R chains concentrate inside the droplets, forming a phase-separated structure as indicated in Fig.1. The organic solvent was then removed by evaporation at room temperature, under gentle stirring or by using a rotavapor. The R chains inside the core form an entangled structure. The solidified particles were recovered by centrifugation and lyophilization.

PEG-coated nanospheres were also prepared by the solvent diffusion method by using water-miscible organic solvents, such as acetone (Bazile *et al.*, 1995). Immediately after the organic phase containing PEG–PLA is poured into water, the polymer precipitates, leading to the formation of tiny particles (usually, less than 200 nm).

One of the advantages of this type of nanospheres is that, owing to the amphiphilic nature of the block copolymers PEG–R and PEG_n–R, the use of other surfactants (not always biocompatible) can be avoided. This is not the case with PLA (Bazile *et al.*, 1995) or PLGA (Gref *et al.*, 1994) polymers, whatever the nanosphere preparation method. The use of surfactants such as poly(vinyl alcohol) (PVA) or Pluronic is necessary for the formation of PLA and PLGA nanoparticles, and thus additional washing steps are required to remove adsorbed surfactant.

Water-soluble PLA-PEG copolymers with PLA:PEG ratios of 2:5 and 3:4 (PEG chains of 5 and 2 kDa, respectively) were used to coat PLGA nanospheres and thus prepare sterically stabilized particles (Stolnik *et al.*, 1995). For this, the PLGA nanospheres were either prepared in the presence of PEG–PLA copolymers by the precipitation–solvent evaporation method or were prepared without any stabilizing agent and subsequently coated by incubation in aqueous solutions of PEG–PLA.

4. NANOSPHERE CHARACTERIZATION

Various techniques can be used for the characterization of surface-modified nanoparticles (Müller, 1991). In the case of PEG-covered nanospheres, atomic force microscopy and freeze-fracture have been used to determine their morphology, light scattering has been employed to measure their size, and surface analysis techniques have been applied to detect the presence of PEG on the surface and the stability of this coating, etc. Some of these techniques are presented in this section.

4.1. Morphology Studies

Scanning electron microscopy and atomic force microscopy (AFM), a nondestructive technique particularly useful for temperature-sensitive samples and allowing a high resolution, were used for the observation of PEG-coated nanospheres (Gref *et al.*, 1994). In the AFM technique, the probe, a very thin cantilever, is placed directly in contact with the sample. It oscillates vertically (tapping mode), with a high frequency (hundreds of kilohertz). When it approaches the sample, interactive repulsive forces at the atomic level take place. During horizontal sweeping of the sample, the movement of the probe is detected by a photoreceptor on the basis of the

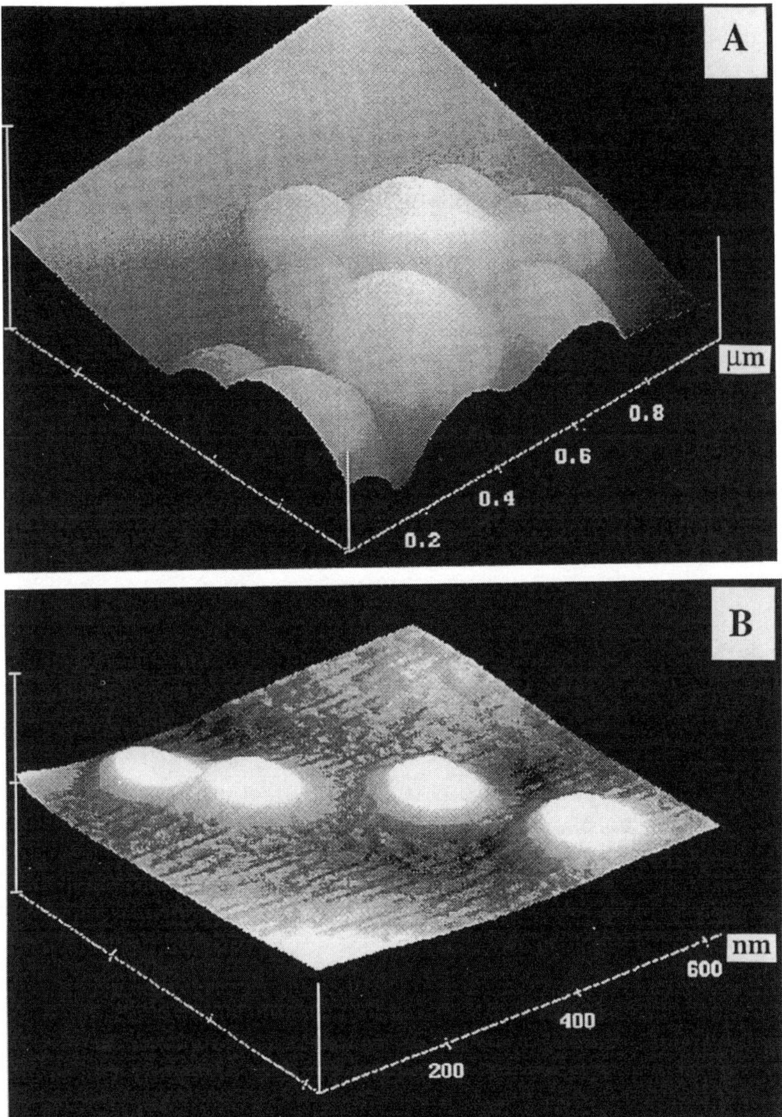

Figure 3. Spherical shape of nanospheres evidenced by images of PLGA (A) and PEG5K–PLGA5K (B) nanospheres taken with an atomic force microscopic (Nanoscope III, Digital Instruments). (Reprinted, with permission, from Gref *et al.*, 1994; Copyright (1995) AAAS.) (C) Picture of freeze-fractured $(PEG20K)_3$–PLA nanospheres taken by using transmission electron microscopy.

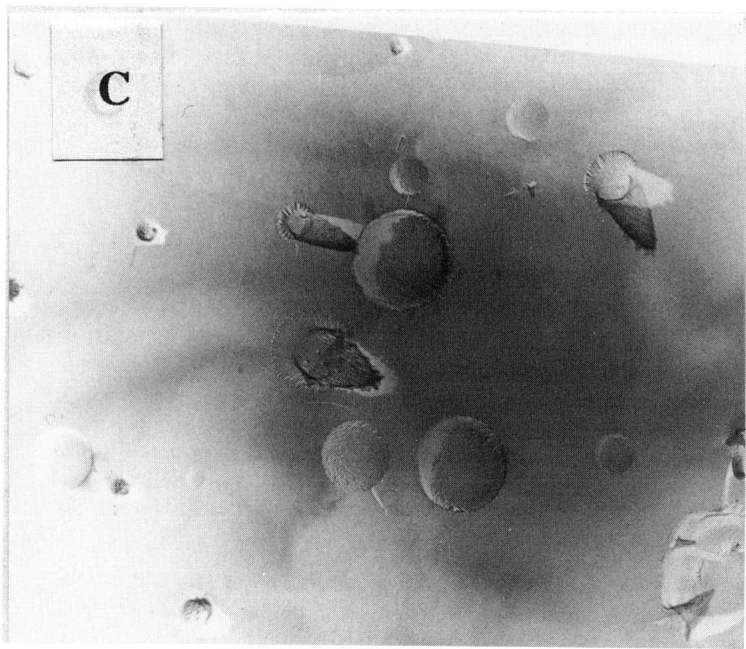

Figure 3. (*Continued*)

deviation of a laser beam reflected on the probe. Thus, the topological surface structure is detected. Figure 3A and B represent typical images of nanosphere samples, showing a spherical shape and allowing an estimation of particle size (for size distribution measurement, quasi-elastic light scattering was used). The same spherical shape of nanospheres prepared by using $(PEG20K)_3$–PLA polymers can be observed on pictures after freeze-fracture of particle suspensions (Fig. 3C).

4.2. Size Distribution Measurement

Quasi-elastic light scattering (QELS), also called photon correlation spectroscopy (PCS), was used to determine the mean diameter of nanospheres and to follow their aggregation and deaggregation behavior (Gref *et al.*, 1994). It was thus established that the emulsion/evaporation procedure leads to the formation of nanospheres with a very narrow size distribution. The influence of different parameters involved in the nanosphere preparation procedure by the emulsion/evaporation method was studied, in order to obtain <200-nm particles with a narrow unimodal size distribution (Minamitake *et al.*, 1996). For this optimization study, nanospheres were prepared following the emulsion/evaporation procedure. For a given PEG–R copolymer, the following were the main factors influencing nanosphere size:

- Polymer concentration in the organic solvent and type of organic solvent
- Volume ratio of organic and aqueous phases
- Nature and concentration of the surfactant in the aqueous phase
- Sonication time and intensity

It appeared that the lower the PVA molecular weight, the smaller was the nanosphere size. A similar observation was drawn from size-optimization studies of nanospheres prepared from PLGA (Scholes *et al.*, 1993). The lower the polymer concentration was, the lower the viscosity of the organic phase and thus the smaller the resulting nanospheres. However, reducing the polymer concentration leads to lower particle recovery.

The diameter of PEG–PLA prepared by the solvent diffusion method depended on the chain length of both PEG and PLA blocks (Bazile *et al.*, 1995). For the same length of the PEG block, the longer the PLA chains in the diblock PEG–PLA copolymers were, the higher the nanosphere diameter. For the same chain length of PLA chains, the longer the PEG blocks were, the smaller the PEG–PLA nanospheres. Blends of PLA60K and PEG–PLA polymers were also studied, and in all cases, the particle diameter increased whereas the polydispersity decreased with increasing weight percent of PLA in the blends.

PLGA nanospheres were prepared by the precipitation–solvent evaporation method in the presence of PEG–PLA polymers (Stolnik *et al.*, 1995). The resulting PEG-coated particles had an average particle size between 120 and 140 nm and a polydispersity index between 0.08 and 0.13, indicative of a relatively narrow size distribution.

4.3. Detection and Stability of the PEG Coating

The location of PEG chains on the surface of PEG–PLA nanospheres was checked by a degradation test (Bazile *et al.*, 1995). The particles were incubated in sodium hydroxide solutions for different incubation times until total destruction, and the solutions were then neutralized by addition of phosphate buffer. The optical densities of these solutions (OD1) were measured after addition of iodine complexation agent. Control experiments were performed with water instead of sodium hydroxide (OD2) and water instead of the nanoparticle suspensions (OD3). OD1–OD3 is a measure of the amount of PEG detached during the degradation process. Encapsulated PEG chains that are not on the surface are not accessible to the iodine complexation agent. As no evolution of OD1–OD3 was observed during the degradation process of PEG2K–PLA10K nanospheres, the authors concluded that all the PEG chains were located at the surface of these particles. It was therefore calculated that one PEG chain would occupy about 1.9 nm^2.

X-ray photoelectron spectroscopy (XPS), also known as electron spectroscopy for chemical analysis (ESCA), was used to detect the PEG coating on the nanosphere surface (Gref *et al.*, 1994). XPS is based on the irradiation of a surface with a beam of "soft" X rays, resulting in the emission of photoelectrons. Only photoelectrons having a sufficient kinetic energy (emitted from the sample's immediate surface) can pass through the studied material and be detected. The binding energy of the emitted electrons is recorded. It depends on the atomic environment of the atoms irradiated. Sampling depth, typically in the range 1–10 nm (Clark, 1982; Hayward *et al.*, 1986), depends on electron kinetic energy, the nature of the material, and the angle of the analyzer with respect to the surface. However, XPS experiments are conducted in vacuum, so the PEG chain configuration at the analyzed surface is different from that in aqueous solutions. Therefore, XPS only enables semiquantitative conclusions to be made; it allows, for example, for the selection among different preparation techniques of the one that maximizes the PEG relative density on the surface or for the study of PEG detachment after various incubation times, from which conclusions on the coating stability may be drawn. For example, it was established that in the case of nanospheres prepared from multiblock PEG_n–R copolymers without the use of any surfactant, the PEG surface density was significantly increased compared to that of nanospheres prepared from PEG–R copolymers (Peracchia *et al.*, 1996).

Further, by using XPS it was possible to study the stability of the coating PEG layer during degradation studies (incubation of nanospheres in water, at 37 °C). It appeared that only a very low fraction of PEG (usually less than 10% and depending upon PEG molecular weight) was detached during 24 hr of incubation (Gref *et al.*, 1993b), and this observation was confirmed by a quantitative determination of the amount of PEG released by using an iodine complexation method (described by Baleux, 1972).

4.4. Surface Hydrophobicity and Charge Determination

Other techniques can be used to measure and compare surface hydrophobicity and charge, two of the most prominent factors which determine the biodistribution of nanospheres. For example, particle surface hydrophobicity can be determined by Rose Bengal binding methods, hydrophobic interaction chromatography (HIC), aqueous two-phase partitioning, or the critical flocculation temperature method (Müller, 1991).

HIC was carried out using alkyl agarose gels in order to assess the ability of PEG5K–PLA4.5K and PEG2K–PLA3.4K copolymers to alter the surface hydrophobicity of polystyrene and PLGA nanospheres (Stolnik *et al.*, 1995). The chromatograms recorded showed a heterogeneity of the surface hydrophobicity, indicating the possibility of a nonhomogeneous surface coverage. Surface charge of PLGA nanospheres coated with PEG–PLA was determined by measuring the zeta potential (Stolnik *et al.*, 1995; Table I). Incubation of PLGA nanospheres with PEG–PLA copolymers leads to an increase in their size (indicating the formation of an adsorbed coating layer) and to an increase in the zeta potential from −40.3 mV (PLGA) to −18.5 mV (PLGA coated with PEG5K–PLA4.5K) (Table I). This surface charge reduction was attributed to the formation of a coating layer which shifts the plane of shear to the boundary of the layer.

The stability of PEG-coated nanosphere suspensions was determined by measuring the critical coagulation/flocculation concentration as a function of electrolyte concentration (Stolnik *et al.*, 1995). In the presence of

Table I

Effect of the Adsorption of PEG–PLA Copolymers and of Poloxamine 908 on Polystyrene and PLGA Nanospheres on the Particle Size and Zeta Potential[a]

System[b]	Particle size ± SD (nm) [Coating layer thickness (nm)]	Zeta potential ± SD (mV)
Polystyrene		
Uncoated	164 ± 3.2	−44.7 ± 1.7
PLA:PEG 2:5	178 ± 3.5 (7.2)	−21.3 ± 1.6
PLA:PEG 3:4	168 ± 3.2 (1.9)	−32.2 ± 1.5
Poloxamine 908	188 ± 2.6 (9.2)	−15.8 ± 2.2
PLGA		
Uncoated	140 ± 2.7	−40.3 ± 2.3
PLA:PEG 2:5	161 ± 3.7 (10.5)	−18.5 ± 1.4
PLA:PEG 3:4	147 ± 3.6 (3.3)	−26.9 ± 1.3
Poloxamine 908	160 ± 3.8 (9.8)	−14.9 ± 1.8

[a]Reproduced with permission from Stolnik *et al.* (1995).
[b]Uncoated and nanospheres coated with the copolymers.

PEG-based coatings, an electrosteric stabilization was achieved by steric forces in combination with electrostatic repulsion.

PEG–PLA-coated PLGA nanospheres were studied with regard to their ability to provide a repulsive barrier against protein adsorption (Stolnik *et al.*, 1995). Incubation of uncoated PLGA particles with albumin solutions resulted in a significant increase in the nanosphere size, indicating the formation of an adsorbed layer around 10 nm in thickness. The size of the same particles coated with PEG–PLA did not increase, showing that this coating can produce a surface that significantly decreases albumin adsorption.

Two-dimensional polyacrylamide gel electrophoresis (2D-PAGE), which has been successfully used to establish a correlation between adsorbed proteins and *in vivo* behavior of polystyrene carriers coated with Poloxamers (Blunk *et al.*, 1994) could be a helpful tool to study competitive plasma protein adsorption on PEG-coated nanospheres as a function of PEG chain length and surface density. Indeed, the relevance of the adsorbed plasma proteins (and their ratios and conformations) with respect to the *in vivo* fate of the particles remains to be uncovered.

5. DRUG ENCAPSULATION IN PEG-COATED NANOSPHERES

5.1. Drug Encapsulation and Release Properties

To study the encapsulation properties of PEG–R and PEG_n–R nanospheres, lidocaine, an antiarrhythmic compound easily detectable by UV spectroscopy, was used as a model lipophilic drug (Gref *et al.*, 1994; Peracchia *et al.*, 1996). Drug-loaded nanospheres were prepared by the same procedure as depicted in Fig. 2, in which lidocaine was dissolved together with the block copolymers in methylene chloride (Peracchia *et al.*, 1996). High loadings (up to 33% by nanosphere weight) and entrapment efficiencies (more than 95%) were achieved. This is presumably due to the good solubility of lidocaine in the organic solvents used for nanosphere fabrication and its low water solubility. Similar results were obtained with other poorly water-soluble drugs, such as prednisolone [using methylene chloride/chloroform (1:1, v/v) as solvent] or carmustine, an antitumoral drug.

Verrecchia *et al.* (1993, 1995) entrapped another water-insoluble drug, [^{14}C]ibuprofen ([^{14}C]-IBP), in PEG2K–PLA30K nanospheres. The *in vivo* data after i.v. administration of the particles in rats showed an increase of IBP plasma half-life, from a few minutes when encapsulated in "non-stealth" PLGA nanospheres up to two and a half hours in the case of "stealth"

PEG–PLA nanospheres (Verrecchia *et al.*, 1993). Consequently, after 24 hr, the area under the curve (AUC) of [^{14}C]-IBP pharmacokinetics is increased by a factor of 12 (Verrecchia *et al.*, 1995).

Tissue distribution studies 6 hr after injection of PEG–PLA nanospheres revealed that 53% of the injected dose was in the intestines, 11.3% in the liver, and 9.1% in plasma; in the case of PLGA nanospheres, 52.2% of the dose was in the intestines, 20% in the liver, and 0.39% in plasma (Verrecchia *et al.*, 1995). IBP is a drug with a fast hepatic metabolism and an intestinal and fecal elimination, and the tissue distribution of the drug is not modified in both cases (PEG–PLA and PLGA nanoparticles), except for the plasma residence time.

These data were confirmed by autoradiographs of i.v.-treated rats, 6 hr after particle administration. In the case of PEG–PLA nanospheres, a high amount of radioactivity was located in the heart and the blood vessels, confirming that nanoparticles still circulated 6 hr after the injection. Conversely, in the case of PLGA nanospheres, [^{14}C]-IBP associated radioactivity was essentially located in the liver, spleen, and intestines. PEG–PLA nanospheres slowly released biologically active IBP from the polymer matrix, whereas PLGA particles are removed from blood before drug release (Verrecchia *et al.*, 1995). Thus, the "stealth" PEG–PLA nanoparticles can be considered as a sustained release parenteral dosage form.

5.2. Parameters Influencing Drug Release

To design PEG-coated nanospheres for therapeutic applications, a comprehensive study of the parameters which govern drug release would be necessary. By using lidocaine as a model drug, Peracchia *et al.* (1996) showed that drug release is a function of diblock PEG–R copolymer composition and molecular weight, nanosphere size, and drug loading. Lidocaine-loaded nanospheres were prepared from various diblock and multiblock PEG_n–R copolymers, where R was a polyester (PLA, PLGA, or PCL) or a polyanhydride (PSA). The entrapment efficiency and the nanosphere mean diameter were both affected by the polymer structure (Tables II and III). The encapsulation efficiency depends mainly upon the R chemical composition and slightly decreases with an increase in the amount of PEG in the polymers. The nanosphere size is practically independent of the drug loading. Lidocaine was entrapped in PEG–R nanospheres with high entrapment efficiencies (Table II). In the case of PEG–PLGA polymers, the release was continuous; an initial fast release ("burst effect") was observed in the first 2 hr, and afterward the drug was released at a

Table II
Encapsulation Properties of PEG–PLGA, PEG_3–PLA, and PEG–PSA Polymers: Influence of the Theoretical Loading on the Mean Diameter (Measured by QELS) and Actual Loading

Polymer structure	Mean diameter (nm) for a theoretical drug loading of		Actual drug loading (wt. %) for a theoretical drug loading of	
	20 wt. %	33 wt. %	20 wt. %	33 wt. %
PLGA	150	135	18	29.7
PEG5K–PLGA	165	170	18	28.1
PEG12–KPLGA	181	175	18.4	29.7
PEG20–KPLGA	180	181	17.8	25.7
$(PEG5K)_3$–PLA	167	167	15	23.1
$(PEG12K)_3$–PLA	183	140	14	23.1
$(PEG5K)_3$–PLA	162	171	15.6	28.4
PSA	144	211	18	30.4
PEG5K–PSA	164	190	17.8	29.7
PEG12K–PSA	166	188	18.4	29.4
PEG20K–PSA	153	198	19.6	32.7

practically constant rate over 8 hr (Peracchia *et al.*, 1996). In the case of uncoated PLGA particles, the fast initial drug release was explained as being due to drug located near the surface of the nanospheres (Niwa *et al.*, 1993). This might be the case with PEG–R particles also. Presumably, release kinetics from PEG–R nanospheres are related to the physicochemical

Table III
Encapsulation Properties of PEG_3–PLA Polymers: Influence of the Theoretical Loading on the Mean Diameter (Measured by QELS) and Actual Loading

Copolymer structure	Mean diameter (nm) for a theoretical drug loading of				Actual drug loading (wt. %) for a theoretical drug loading of			
	10 wt. %	20 wt. %	33 wt. %	50 wt. %	10 wt. %	20 wt. %	33 wt. %	50 wt. %
PLA	118	108	114	116	9.8	20	32	45
$(PEG5K)_3$–PLA	161	184	200	192	7.1	12	25.7	35
$(PEG12K)_3$–PLA	205	200	206	220	6.7	13.6	23.1	34
$(PEG20K)_3$–PLA	238	234	238	240	6.5	12.6	22.3	30

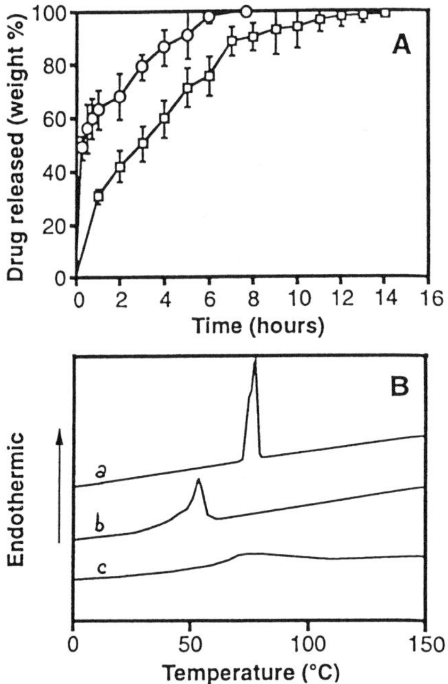

Figure 4. (A) Comparison of lidocaine release kinetics from PEG20K–PLGA80K nanospheres at 10% (○) and 33% (□) loading. (B) Differential scanning calorimetry patterns (heating rate, 10 °C/min) of lidocaine crystals (a) and PEG20K–PLGA80K nanospheres at 33% (b) and 10% (c) loading. (Reprinted, with permission, from Gref *et al.*, 1994; copyright (1995) AAAS.)

properties of each polymer, core density, water uptake, and interactions between drug and polymer.

Figure 4A shows the influence of lidocaine loading on its release from PEG–PLGA nanospheres. Whereas nanospheres containing 33 wt. % lidocaine constantly release lidocaine over 12 hr, particles containing only 10 wt. % drug complete their release within 6 hr. To better understand this phenomenon, the physical state of the encapsulated drug was taken into account. The possibility of drug crystallization was investigated by calorimetric (DSC) and X-ray studies. At low loading, lidocaine is present as a molecular dispersion inside the polymer matrix; above a certain content in the polymer, part of it crystallizes (Fig. 4B). Surface analysis (XPS) studies excluded the location of these crystals outside or on the surface of the nanospheres. Presumably, the crystallized drug should dissolve and diffuse out of the nanospheres more slowly, and this could explain a longer release period at a higher loading.

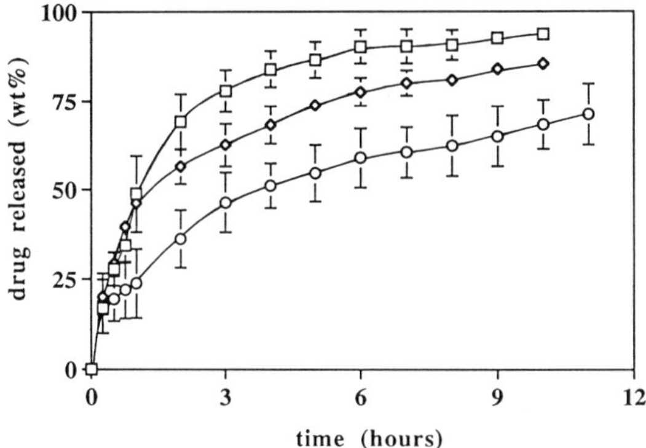

Figure 5. Lidocaine release from nanospheres prepared from PLA (□), $(PEG5K)_3$–PLA (◇), and $(PEG20K)_3$–PLA (○), at a theoretical drug loading of 20%.

Lidocaine was released from multiblock PEG_3–PLA nanospheres over more than 10 hr (Fig. 5; Peracchia *et al.*, 1996). For these polymers, lower encapsulation efficiencies compared to those of diblock PEG–PLA particles were observed (Table II), probably due to the more hydrophilic nature of polymers containing several PEG blocks. The nanosphere size (measured by QELS) increases with the PEG molecular weight, presumably due to an increase in the thickness of the "brush" PEG coating with the PEG chain length.

According to these results, different release patterns can be achieved by an appropriate choice of polymer composition and drug loading.

6. *EX VIVO* STUDIES (PHAGOCYTOSIS ASSAY)

To study the ability of the PEG coating to prevent macrophage recognition, Bazile *et al.* (1995) performed kinetic studies regarding the uptake of PLA and PEG–PLA nanospheres by cultured THP-1 macrophage cells (Fig. 6). PLA and PEG–PLA nanospheres were core-labeled by entrapping a radioactive polymer (^{14}C-labeled PLA18K) during the preparation procedure. These *ex vivo* experiments showed that after two hours, about nine times less PEG5K–PLA30K nanospheres than PLA50K ones were phagocytosed. A comparative study with particles prepared from PEG2K–PLA30K and PEG5K–PLA30K showed that the higher the PEG molecular weight, the lower the PEG–PLA nanosphere uptake by macrophages (Fig. 6). The study clearly shows the efficacy of a PEG coating in preventing macrophage uptake in culture.

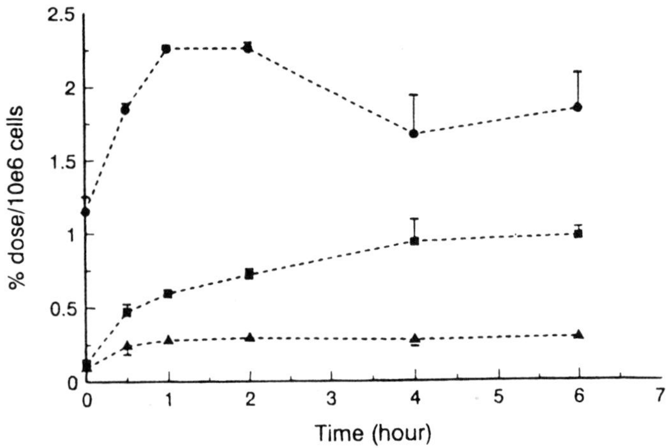

Figure 6. Uptake kinetics of ^{14}C-labeled PLA18K (●), PEG2K–PLA30K (■), and PEG5K–PLA30K (▲) nanospheres by cultured THP-1 cells. (Reprinted, with permission, from Bazile *et al.*, 1995.)

7. BLOOD HALF-LIFE AND ORGAN DISTRIBUTION OF PEG-COATED NANOSPHERES

To determine the ability of the PEG coating to prolong blood half-life and alter organ distribution of the particles, PEG-coated and noncoated nanospheres were injected into BALB/c mice (Gref *et al.*, 1994). For this, the particles were core-labeled with a radioactive gamma-emitting isotope, indium 111 (^{111}In). Indium-111 was chelated with diethylenetriaminepentaacetic acid stearyl amide (DTPA-SA) and then added to the polymer dissolved in dichloromethane, and the nanospheres were prepared by the emulsion/evaporation procedure (Fig. 2). The excess of nonentrapped ^{111}In was removed by repetitive washing. Nanospheres were stable when incubated at 37 °C in phosphate buffer solution (PBS) or horse serum for 4 hr, presumably due to the firm anchoring of the hydrophobic stearyl moiety to the core of the particles. It was concluded that this labeling method could be effective for *in vivo* studies (direct counting of radioactivity in blood or organ samples, or gamma-scintigraphy).

Figure 7 shows that blood circulation time after i.v. administration in mice increases as the molecular weight of PEG increases, presumably due to an increase in the protective effect of the coating PEG layer. Five minutes after injection, 66% of noncoated particles were accumulated in the liver,

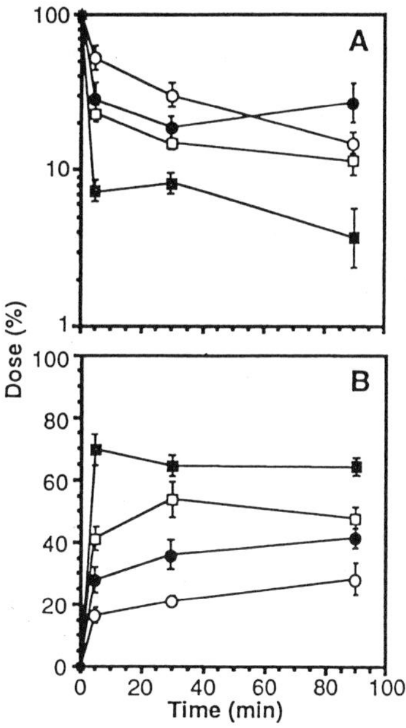

Figure 7. Comparative blood circulation time (A) and liver uptake (B) after injection of ^{111}In-labeled nanospheres in the tail vein of BALB/c mice. The nanospheres were prepared from PLGA (■); PEG5K–PLGA45K (□), PEG12K–PLGA100K (●), and PEG20K–PLGA180K (○) polymers, with a lactic acid:glycolic acid molar ratio in PLGA of 3:1. (Reprinted, with permission, from Gref *et al.*, 1994; copyright (1995) AAAS.)

while after 5 hr, the amount of 20K PEG-coated nanospheres in the liver did not exceed 30%.

The increase in blood circulation time with the increase in PEG molecular weight or surface density was confirmed by gamma-scintigraphy studies with ^{111}In-labeled nanospheres (Gref *et al.*, 1994). Fifteen minutes after injection in mice, unmodified PLGA nanospheres were completely eliminated from the blood, and only liver- and spleen-associated radioactivity was observed. Within this time, PEG20K–PLGA nanospheres circulate well, and the radioactivity in the blood pool (heart and lung) was detected.

To further improve blood circulation time, it would be necessary to further increase the molecular weight of the PEG coating, but when one does so, PEG loses its elimination properties. An alternative is to use shorter PEG chains but maximize the surface density.

To study the influence of PEG surface density, a series of nanospheres were prepared with PEG5K–PLGA copolymers with different PLGA chain

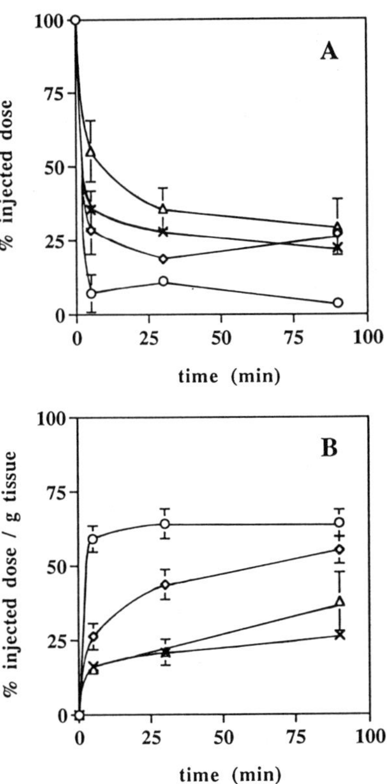

Figure 8. Comparative blood circulation time (A) and liver uptake (B) after injection of ^{111}In-labeled nanospheres in the tail vein of BALB/c mice. The nanospheres were prepared from PLGA (○), PEG5K–PLGA45K (◇), PEG5K–PLGA25K (×), and PEG5K–PLGA20K (Δ) polymers, with a lactic acid:glycolic acid molar ratio in PLGA of 3:1.

lengths. If we assume that all the PEG chains migrate to the surface and all the PLGA chains form the core, then we should obtain particles with various PEG surface densities. This value can be obtained from the particle density and mean diameter (measured by QELS). For example, for nanospheres prepared from PEG5K–PLGA45K, one PEG chain would occupy about 2 nm^2; the corresponding values are 1 nm^2 for PEG5K–PLGA25K and 0.7 nm^2 for PEG5K–PLGA20K. This last value approaches the optimal values calculated by Jeon *et al.* (1991) in their theoretical studies to minimize protein surface adsorption. *In vivo* experiments (Fig. 8) show that increasing surface density has a similar effect as increasing PEG molecular weight.

Autoradiography of a rat sagittal section 6 hr after intravenous administration of [^{14}C]-PLA core-labeled PEG5K–PLA30K nanospheres showed that a high amount of radioactivity was located in the heart and blood vessels but that capture by MPS organs such as liver and bone marrow also took place (Bazile *et al.*, 1995). Phagocytosis was delayed due to the PEG coating, but the final destination of the PEG-coated particles is always the MPS, as observed with other types of particles. In the case of PEG-coated liposomes, it has been suggested (Senior *et al.*, 1991) that the PEG barrier ensures only a gradual plasma protein adsorption onto the surface, whereas unprotected particles are instantaneously coated with plasma proteins, and this could explain why the particles are eventually captured by the MPS.

The blood half-life in rats of the [^{14}C]-PLA core-labeled PEG5K-PLA30K nanoparticles was 6 hr (Bazile *et al.*, 1995). Biodistribution studies confirmed that after 6 hr the PEG5K–PLA30K nanospheres were predominantly captured by liver (9% of the injected dose) and spleen (11% of the injected dose). Interestingly, a coating with Pluronic F68, a triblock copolymer of the type PEG–PPG–PEG, did not allow [^{14}C]-PLA nanospheres to keep circulating in the vascular compartment (Bazile *et al.*, 1995). These authors suggested that adsorbed polymers containing PEG blocks (such as Pluronic) are, at the molecular level, heterogeneously dispersed on the surface of the particles, because they cannot rearrange during the adsorption process and thus cannot offer an efficient coating against opsonization.

The coating of PLGA nanospheres with PEG–PLA copolymers also resulted in extended circulation times. For example, 3 hr post i.v. injection, the best results were obtained with PLGA nanospheres coated with PEG5K–PLA4.5K (28.5% of the injected radioactivity was in the blood, 22.7% in the liver, and 10.7% in the spleen). Good results were also found for PLGA coated with PEG2K–PLA3.4K (17.5% of the injected radioactivity was in the blood, 20.9% in the liver and 8.4% in the spleen). Whereas Poloxamine 908 ensured a prolonged blood circulation time for PLGA particles (20.3% were still circulating after 3 hr) as in the case of polystyrene ones (38.6% remained in the blood after 3 hr), PEG–PLA polymers were ineffective in protecting the polystyrene nanospheres against MPS uptake. The authors suggested that there might be a difference in stability of the PEG–PLA coatings on polystyrene and PLGA surfaces. The PLA part of PEG–PLA would be a better anchor group on a surface of a similar structure (PLGA) and would therefore provide a coating stable enough toward desorption and effective against gradual adsorption of plasma proteins.

The sum of the biodistribution results obtained with two animal models (mice and rats) and on different PEG-coated nanoparticles (PEG–R and

PLGA coated with PEG–PLA, labeled with ^{111}In or [^{14}C]-PLA) have shown that PEG–R copolymers can be successfully used to produce or coat biodegradable nanospheres and thus obtain sterically stabilized particles with dramatically increased blood half-lives and decreased hepatic uptake as compared to naked PLA or PLGA nanospheres.

8. CONCLUSION

Although at a relatively early stage of development, PEG-coated long-circulating nanospheres can be considered as a new sustained release intravenous dosage form. These particles, of a mean diameter generally less than 200 nm, are composed of a PEG coating and a biocompatible core, which has been formed by using a variety of polymers such as polyesters (PLA, PLGA, or PCL) or polyanhydrides. They have been shown to have blood half-lives in rats and mice of up to several hours, as compared to a few minutes for uncoated particles, and the hepatic uptake was drastically reduced compared to that of uncoated particles.

Various drugs (such as lidocaine, prednisolone, carmustine, and ibuprofen) have been encapsulated inside the hydrophobic core and released in a continuous manner. These poorly water soluble drugs have also a low solubility in biological fluids and thus a low bioavailability. High loadings were achieved, together with a good entrapping efficiency, depending on the physicochemical properties of the drug and the polymers used. According to *in vitro* studies, drug release can be controlled by appropriately choosing the polymer nature (polyester or polyanhydride) and physicochemical properties (such as chemical composition, molecular weight, or crystallinity) (Peracchia *et al.*, 1996). In *in vivo* studies, biologically active ibuprofen was released from PEG–PLA nanoparticles, and the tissue distribution of the released drug was not modified except for the residence time in plasma (Verecchia *et al.*, 1995).

Generally, active substances with toxic side effects or very labile drugs, such as those used to fight cancer, for treatment of heart failure, or for prolongation of anesthesia, can be potential candidates for encapsulation into PEG-coated nanospheres. In this way, the compounds could be continuously released directly in the blood via passive targeting with PEG-coated nanospheres. Moreover, as previously shown with long-circulating liposomes (Gabizon and Papahadjopoulos, 1988), PEG-coated long-circulating nanospheres could enhance the localization of drugs in tumors or in inflammatory or infected tissues.

With further studies, the PEG-coated nanospheres could be used in a variety of applications, in drug delivery, medical imaging, or gene therapy.

For example, by encapsulating NMR contrast agents, the long-circulating nanospheres have potential applications for medical imaging. Target-specific superparamagnetic contrast agents could be further designed by encapsulating magnetite particles in the core of the nanospheres bearing antibodies on their surface. Polymorphic (cationic, fusogenic) liposomes provide a promising new approach to gene therapy, because they greatly improve transfection by exogenous DNA (Lasic and Papahadjopoulos, 1995). Similarly, by the attachment of appropriate proteins such as transferrin to the PEG end group (Wagner *et al.*, 1990) on the surface of DNA-containing particles, nanosphere endocytosis for gene therapy could possibly be achieved.

ACKNOWLEDGMENT

We thank Professor N. Lotan, Professor H. Brem, and Dr. S. Cohen for helpful discussions, E. Shaw for technical assistance with AFM studies, Dr. A. Milshteyn for performing the *in vivo* experiments on mice, the French Foreign Affairs Ministry for the awarding of a Lavoisier grant to Dr. R. Gref, and the National Institutes of Health for grants U01 CA52857 and GM26698.

REFERENCES

Abuchowski, A., Van Es, T., Palczuk, N., and Davis, F., 1977, Alteration of immunological properties of bovine serum albumin by covalent attachment of polyethylene glycol, *J. Biol. Chem.,* **252:**3578–3581.

Allen, T., 1994, The use of glycolipids and hydrophilic polymers in avoiding rapid uptake of liposomes by the mononuclear phagocyte system, *Adv. Drug Delivery Rev.* **13:**285–309.

Artursson, P., and Sjöholm, I., 1986, Effect of opsonins on the macrophage uptake of polyacrylstarch microparticles, *Int. J. Pharm.* **32:**165–170.

Artursson, P., Brown, L., Dix, J., Goddard, P., and Petrak, K., 1990, Preparation of sterically stabilized nanoparticles by desolvation from graft copolymers, *J. Polym. Sci., Part A: Polym. Chem.* **28:**2651–2663.

Baleux, B., 1972, Chimie analytique. Dosage colorimétrique d'agents de surface non ioniques polyoxyéthylènes à l'aide d'une solution iodo-iodurée, *C. R. Acad. Sci., Paris, Seri. C* **1972:**1617–1620.

Bazile, D.V., Ropert, C., Huve, P., Verrecchia, T., Marlard, M., Frydman, A., Veillard, M., and Spenlehauer, G., 1992, Body distribution of fully biodegradable ^{14}C-poly(lactic acid) nanoparticles coated with albumin after parenteral administration to rats, *Biomaterials* **13:**1093–1102.

Bazile, D.V., Prudhomme, C., Bassoullet, M.-T., Marlard, M., Spenlehauer, G., and Veillard, M., 1995, Stealth MePEG-PLA nanoparticles avoid uptake by the mononuclear phagocytes system, *J. Pharm. Sci.* **84:**493–498.

Blume, G., and Cevc, G., 1990, Liposomes for the sustained drug release *in vivo*, *Biochim. Biophys. Acta* **1066:**91–97.

Blunk, T., Hochstrasser, D., Sanchez, J., Müller, B., and Müller, R, 1994, Colloidal carriers for intravenous drug targeting—Plasma protein adsorption patterns on surface-modified latex particles evaluated by two-dimensional polyacrylamide gel electrophoresis, *Electrophoresis* **14:**1382–1387.

Burnham, N. L., 1994, Polymers for delivering peptides and proteins, *Am. J. Hosp. Pharm.* **51**(2)**:**210–218.

Cerrai, P., and Tricoli, M., 1994, Block copolymers from L-lactide and poly(ethylene glycol) through a non-catalyzed route, *Makromol. Chem. Rapid Commun.* **14**(9)**:**529–538.

Churchill, J. R., and Hutchinson, F. G., c/o Imperial Chemical Industries, 1986, European Patent 0 166 596 A3.

Clark, D. T., 1982, Advances in ESCA applied to polymer characterization, *J. Appl. Chem.* **54**(2)**:**415–438.

Cohn, D., Younes, H., and Gideon, U., 1988, European Patent 0 295 055 A2.

Davis, S. S., 1981, Colloids as drug delivery systems, *Pharm. Technol.* **5:**71–88.

Davis, S. S., and Hansrani, P., 1985, The influence of emulsifying agents on the phagocytosis of lipid emulsions by macrophages, *Int. J. Pharm.* **23:**69–77.

Deng, X., Xiong, C., Cheng, L., and Xu, R., 1990, Synthesis and characterization of block copolymers from D,L-lactide and poly(ethylene glycol) with stannous chloride, *J. Polym. Sci., Part C: Polym. Lett.* **28:**411–416.

Domb, A., Gref, R., Minamitake, Y., Peracchia, M. T., and Langer, R., 1994, Nanoparticles and microparticles of non-linear hydrophilic–hydrophobic multiblock copolymers, U.S. Serial No. 08/265,440.

Domb, A., Elmalek, O., Gref, R., Minamitake, Y., Peracchia, M., and Langer, R., 1996, Synthesis and applications of non-linear block copolymers of PEG and poly(hydroxy acid) esters, Paper in preparation.

Donbrow, M. (ed.), 1992, *Microcapsules and Nanoparticles in Medicine and Pharmacy,* CRC Press, Boca Raton, Florida.

Gabizon, A., and Papahadjopoulos, D., 1988, Liposome formulations with prolonged circulation time in blood and enhanced uptake by tumors, *Proc. Natl. Acad. Sci. USA* **85:**6949–6953.

Goldberg, E. (ed.), 1983, *Targeted Drugs*, John Wiley & Sons, New York, 296 pp.

Gottlleb, S., Ernst, A., Litt, L., Schwarz, K.Q., and Meltzer, R.S., 1990, *J. Am. Soc. Echocardiography* **3:**238.

Gref, R., Minamitake, Y., and Langer, R., 1993a, U.S. Serial No. 08/096,370 and U.S. Serial No. 08/210,677.

Gref, R., Minamitake, Y., Peracchia, M.T., Trubetskoy, V., Milshteyn, A., Sinkule, J., Torchilin, V., and Langer, R., 1993b, Biodegradable PEG-coated stealth nanospheres, *Proc. Int. Symp. Control. Rel. Bioact. Mater.* **20:**131–132.

Gref, R., Minamitake, Y., Peracchia, M.T., Trubetskoy, V., Torchilin, V. and Langer, R., 1994, Biodegradable long-circulating nanospheres, *Science* **263:**1600–1603.

Gregoriadis, G., and Senior, J., 1980, The phospholipid component of small unilamellar liposomes controls the rate of clearance of entrapped solutes from the circulation, *FEBS Lett.* **119**(1)**:**43–46.

Gregoriadis, G., Neerunjun, D. E., and Hunt, R., 1977, Fate of liposome-associated agent injected into normal and tumour-bearing rodents. Attempts to improve localization in tumour tissues, *Life Sci.* **21:**357–370.

Hagan, S. A., Dunn, S. E., Garnett, M. C., Davies, M. C., Illum, L., and Davis, S. S., 1995, Pla Peg micelles—a novel drug delivery system, *Proc. Int. Symp. Control. Rel. Bioact. Mater.* **22:**194–195.

Han, D., Jeong, S., Ahn, K., Kim., Y., and Min, B., 1993, Preparation and surface properties of POE-sulfonate grafted polyurethanes for enhanced blood compatibility, *J. Biomater. Sci. Polym. Ed.* **4**(6):579–589.

Harris, J. M., 1985, Laboratory synthesis of polyethylene glycol derivatives, *J. Macromol. Sci.-Rev. Macromol. Chem. Phys.* **C25**(3):325–373.

Hayward, J. A., Durrani, A., Lu Y., Clayton C., and Chapman D., 1986, Biomembranes as models for polymer surfaces. IV. ESCA analyses of a phosphorylcholine surface covalently bound to hydroxylated substrates, *Biomaterials* **7**:252–258.

Hu, D., and Liu, H., 1993, Effect of soft segment on degradation kinetics in polyethylene glycol/poly(L-lactide) block copolymers, *Polym. Bull.* **30**:669–676.

Hu, D., and Liu, H., 1994, Structural analysis and degradation behaviour in polyethylene glycol/poly(L-lactide) copolymers, *J. Appl. Polym. Sci.* **51**:473–482.

Illum, L., and Davis, S. S., 1983, Effect of the nonionic surfactant poloxamer 338 on the fate and deposition of polystyrene microspheres following intravenous administration, *J. Pharm. Sci.* **72**:1086–1089.

Illum, L., Davis, S. S., Müller, R. H., Mak, E., and West, P., 1987, The organ distribution and circulation time of intravenously injected colloidal carriers sterically stabilized with a block copolymer-Poloxamine 908, *Life Sci.* **40**:367–374.

Jedlinski, Z., Kurcok, P., Walach, W., Janeczek, H., and Radecka, I., 1993, Polymerization of lactones, 17a) Synthesis of ethylene glycol–L-lactide block copolymers, *Makromol. Chem.* **194**:1681–1689.

Jeon, S. I., and Andrade, J. D., 1991, Protein–surface interactions in the presence of polyethylene oxide. II. Effect of protein size, *J. Colloid Interface Sci.* **142**(1):159–166.

Jeon, S. I., Lee, J. H., Andrade, J. D., and De Gennes, P. G., 1991, Protein–surface interactions in the presence of polyethylene oxide. I. Simplified theory, *J. Colloid Interface Sci.* **142**(1):149–158.

Kanke, M., Simmons, G. H., Weiss, D. L., Bivins, B. A., and DeLuca, P. P., 1980, Clearance of ^{141}Ce labelled microspheres from blood and distribution in specific organs following intravenous and intraarterial administration in Beagle dogs, *J. Pharm. Sci.* **69**:755–762.

Kataoka, K., Kwon, G. S., Yokoyama, M., Okano, T., and Sakurai, Y., 1993, Block copolymer micelles as vehicles for drug delivery, *J. Controlled Release* **24**:119–132.

Kishida, A., Mishima, K., Corretge, E., Konishi, H., and Ikada, Y., 1992, Interactions of poly(ethylene glycol)-grafted cellulose membranes with proteins and platelets, *Biomaterials* **13**(2):113–118.

Klibanov, A., Maruyama, K., Torchilin, V., and Huang, L., 1990, Amphiphatic polyethylene glycols effectively prolong the circulation time of liposomes, *FEBS Lett.* **268**:235–238.

Kreuter, J., 1994, Nanoparticles, in: *Colloidal Drug Delivery Systems* (J. Kreuter, ed.), Marcel Dekker, New York, pp. 219–342.

Kreuter, J., Täuber, U., and Illi, V., 1979, Distribution and elimination of poly(methyl-2-^{14}C-methacrylate) nanoparticles radioactivity after injection in rats and mice, *J. Pharm. Sci.* **68**:1443–1447.

Kricheldorf, H., and Meier-Haack, J., 1993, Polylactones, 22a) ABA triblock copolymers of L-lactide and poly(ethylene glycol), *Makromol. Chem.* **194**:715–725.

Kulkarni, S. B., Betageri, G. V., and Singh, M., 1995, Factors affecting microencapsulation of drugs in liposomes, *J. Microencapsulation* **12**:229–246.

Laakso, T., Artursson, P., and Sjöholm, G., 1986, Biodegradable microspheres. IV. Factors affecting the distribution and degradation of polyacryl starch microparticles, *J. Pharm. Sci.* **75**:962–967.

Lasic, D., and Papahadjopoulos, D., 1995, Liposomes revisited, *Science* **267**:1275–1276.

Lasic, D., Martin, F., Gabizon, A., Huang, S., and Papahadjopoulos, D., 1991, Sterically stabilized liposomes: A hypothesis on the molecular origin of the extended circulation times, *Biochim. Biophys. Acta* **1070:**187–192.

Leroux, J. C., DeJaeghere, F., Anner, B., Doelker, E., and Gurny, R., 1995, An investigation of the role of plasma and serum opsonins on the internalization of biodegradable poly(D,L-lactic acid) nanoparticles by human monocytes, *Life Sci.* **57**(7)**:**695–703.

Li, S. M., Garreau, H., and Vert, M., 1990, Structure-property relationships in the case of degradation of massive poly(α-hydroxy acids) in aqueous media, *J. Mater. Sci.: Mater. Med.* **1:**131–139.

Li, Y., and Kissel, T., 1993, Synthesis and characterization of biodegradable ABA triblock copolymers consisting of poly(L-lactic acid) or poly(L-lactic-*co*-glycolic acid) attached to central poly(oxyethylene) B blocks, *J. Controlled Release* **27:**247–257.

Liu, H., and Hu, D., 1993, Melting behaviour of hydrolyzable poly(oxyethylene)/poly(L-lactide) copolymers, *Makromol. Chem.* **194:**3393–3403.

Llanos, G., and Sefton, M., 1993, Immobilization of poly(ethylene glycol) onto poly(vinyl alcohol) hydrogel. 2. Evaluation of thrombogenicity, *J. Biomed. Mater. Res.* **27:**1383–1391.

Malmstein, M., 1995, Protein adsorption at phospholipid surfaces, *J. Colloid Interface Sci.* **172:**106–115.

Maruyama, K., Yuda, T., Okamoto, A., Ishikura, C., Kojima, S., and Iwatsuru, M., 1991, Effect of molecular weight in amphiphatic polyethyleneglycol on prolonging the circulation time of large unilamellar liposomes, *Chem. Pharm. Bull.* **39:**1620–1622.

Massenburg, D., and Lentz, B., 1994, Poly(ethylene glycol)-induced fusion and rupture of dipalmitoylphosphatidylcholine large, unilamellar extruded vesicles, *Biochemistry* **32:** 9172–9180.

Minamitake, Y., Gref, R., Hrkach, J., Peracchia, M. T., Domb, A., and Langer, R., 1994, Injectable nanospheres made of biodegradable polyethylene glycol–poly(lactic-*co*-glycolic) acid diblock copolymers, Paper in preparation.

Müller, B. G., and Kissel, T., 1993, Camouflage nanospheres: A new approach to bypassing phagocytic blood clearance by surface modified particulate carriers, *Pharm. Pharmacol. Lett.* **3:**67–70.

Müller, R. H., 1991, *Colloidal Carriers for Controlled Drug Delivery and Targeting*, CRC Press, Boca Raton, Florida.

Munthe-Kaas, A. C., and Kaplan, G., 1980, Endocytosis by macrophages, in: *The Reticuloendothelial System, a Comprehensive Treatise, Vol 1, Morphology* (H. Friedman, M. Escobar, and S. M. Reichard, eds.), Plenum Press, New York, pp. 19–55.

Nair, M. c/o Eastman Kodak Co., 1993, European Patent 0 552 802 A2.

Niwa, T., Takeuchi, H., Hino, T., Kunou, N., and Kawashima, Y., 1993, Preparations of biodegradable nanospheres of water-soluble and insoluble drugs with D,L-lactide/glycolide copolymer by a novel spontaneous emulsification solvent diffusion method, and the drug release behavior, *J. Controlled Release* **25:**89–98.

Österberg, E., Bergström, K., Holmberg, K., Schuman, T. P., Riggs, J. A., Burns, N. L., VanAlstine, J. M., and Harris, J. M., 1995, Protein-rejecting ability of surface-bound dextran in end-on and side-on configurations: Comparison to PEG, *J. Biomed. Mater. Res.* **29:**741–747.

Peracchia, M. T., Gref, R., Minamitake, Y., Lotan, N., Domb, A., and Langer, R., 1996, PEG-coated nanospheres from amphiphilic diblock and multiblock copolymers: Investgigation of their drug encapsulation and release characteristics, submitted to *J. Controlled Release.*

Perret, R., and Skoulios, A., 1972, Synthèse et caractérisation de copolymères séquencés polyoxyéthylène/poly-ε-caprolactone, *Makromol. Chem.* **156:**143–156.

Petrak, K., 1993, Design and properties of particulate carriers for intravascular administration, in: *Pharmaceutical Particulate Carriers* (A. Rolland, ed.), Marcel Dekker, New York.

Piskin, E., Kaitian, X., Denkbas, E. B., and Küçükyavuz, Z., 1995, Novel PDLLA/PEG copolymer micelles as drug carriers, *J. Biomater. Sci. Polymer Ed.* **7**(4):359–373.

Pitt, C., Wang, J., Shah, S., Sik, R., and Chignell, C., 1993, ESR spectroscopy as a probe of the morphology of hydrogels and polymer–polymer blends, *Macromolecules* **26**:2159–2164.

Porter, C. J., Davies, M. C., Davis, S. S., and Illum, L., 1994, Microparticulate systems for site-specific therapy—bone marrow targeting, in: *Site-Specific Pharmacotherapy* (A. Domb, ed.), John Wiley & Sons, New York.

Poste, G., 1983, Liposome targeting *in vivo*: Problems and opportunities, *Biol. Cell.*, **47**(1):19–37.

Prime, K., and Whitesides, G., 1993 Adsorption of proteins onto surfaces containing end-attached oligo(ethylene oxide)—a model system using self-assembled monolayers, *J. Am. Chem. Soc.* **115**:10714–10721.

Ringsdorf, H., Schlarb, B., and Venzmer, J., 1988, Molecular architecture and function of polymeric oriented systems: Models for the study of organization, surface recognition and dynamics of biomembranes, *Angew. Chem.* **27**:113–158.

Ronneberger, B., Kao, W. J., Anderson, J. M., and Kissel, T., 1996, *In vivo* biocompatibility study of ABA triblock copolymers consisting of poly(L-lactic-*co*-glycolic acid) A blocks attached to central poly(oxyethylene) B blocks, *J. Biomed. Mater. Res.* **30**:31–40.

Sato, Y., Kiwada, H., and Kato, Y., 1986, Effects of dose and vehicle size on the pharmacokinetics of liposomes, *Chem. Pharm. Bull.* **34**:4244–4252.

Sawhney, A., Pathak, C., and Hubbell, J., 1993, Interfacial photopolymerization of poly(ethylene glycol)-based hydrogels upon alginate poly(L-lysine) microcapsules for enhanced biocompatibility, *Biomaterials* **14**:1008–1016.

Scherphof, G., Roerdink, E., Waite, M., and Parks, J., 1978, Disintegration of phosphatidylcholine liposomes in plasma as a result of interaction with high-density lipoproteins, *Biochim. Biophys. Acta* **542**:296–307.

Scholes, P., Coombes, A., Illum, L., Davis, S., Vert, M., and Davies, M., 1993, The preparation of sub-200 nm poly(lactide-*co*-glycolide) microspheres for site-specific drug delivery, *J. Controlled Release* **25**:145–153.

Senior, J., Delgado, C., Fisher, D., Tilcock, C., and Gregoriadis, G., 1991, Influence of surface hydrophilicity of liposomes on their interaction with plasma protein and clearance from the circulation: Studies with poly(ethylene glycol)-coated vesicles, *Biochim. Biophy. Acta* **1062**:77–82.

Souhami, R. L., Patel, H. M., and Ryman, B. E., 1981, The effect of reticuloendothelial blockade on the blood clearance and tissue distribution of liposomes, *Biochim. Biophys. Acta* **674**:354–371.

Spenlehauer, G., Bazile, D., Veillard, M., Prud'homme, C., and Michalon, J.P., 1992, European Patent 0520888A1 and European Patent 0520889A1.

Stolnik, S., Dunn, S. E., Garnett, M. C., Davies, M. C., Coombes, A. G. A., Taylor, D. C., Irving, M. P., Purkiss, S. C., Tadros, T. F., Davis, S. S., and Illum, L., 1995, Surface modification of poly(lactide-*co*-glycolide) nanospheres by degradable poly(lactide)-poly(ethylene glycol) copolymers, *Pharm. Res.* **11**:1800–1808.

Tabata, Y., and Ikada, Y., 1988, Macrophage phagocytosis of biodegradable microspheres composed of L-lactic acid/glycolic acid homo- and copolymers, *J. Biomed. Mater. Res.* **22**:837–858.

Takino, T., Konishi, K., Takakura, Y., and Hashida, M., 1994, Long circulating emulsion carrier systems for highly lipophilic drugs, *Biol. Pharm. Bull.* **17**(1):121–125.

Torchilin, V. P., and Papisov, M. I., 1994, Hypothesis: Why do polyethylene glycol-coated liposmes circulate so long?, *J. Liposome Res.* **4**(1)**:**725–739.

Tröster, S. D., and Kreuter, J., 1988, Contact angles of sufactants with a potential to alter the body distribution of colloidal drug carriers on poly(methyl methacrylate) surfaces, *Int. J. Pharm.* **45:**91–100.

Vauthier, C., Popa, M., Puisieux, F., and Couvreur, P., 1994, Modification de la surface des nanoparticules de poly(cyanoacrylate dalkyle) par greffage de poly(éthylène glycol), IXèmes Journées Scientifiques du GTRV, Paris, France.

Verrecchia, T., Bazile, D. V., Archimbaud, Y., Marlard, M., Spenlehauer, G., and Veillard, M., 1993, Compared bioavailability of IBP-5823 administered by the i.v. route in (i) stealth PLAPEG/cholate and (ii) non stealth PLAGA/albumin nanoparticles, VIIIèmes Journées Scientifiques du GTRV, Nancy, France.

Verrecchia, T., Splenlehauer, G., Bazile, D. V., Murry-Brelier, A., Archimbaud, Y., and Veillard, M. 1995, Non-stealth (poly(lactic acid/albumin)) and stealth (poly(lactic acid–polyethylene glycol)) nanoparticles as injectable drug carriers, *J. Controlled Release* **36:**49–61.

Vert, M., Li, S. M., and Garreau, H., 1991, More about the degradation of LA/GA-derived matrices in aqueous media, *J. Controlled Release* **16:**15–26.

Wagner, E., Zenke, M., Cotten, M., Beug, H., and Birnstiel, M. L., 1990, Transferrin/polycation conjugates as carriers for DNA uptake into cells, *Proc. Natl. Acad. Sci. USA* **87:**3410–3414.

Wang, S., and Qiu, B., 1993, Polycaprolactone-poly(ethylene glycol) block copolymer, I: Synthesis and degradability *in vitro*, *Polym. Adv. Technol.* **4:**363–366.

Woodle, M. C., Newman, M. S., and Martin, F. J., 1992, Liposome leakage and blood circulation: Comparison of adsorbed block copolymers with covalent attachment of PEG, *Int. J. Pharm.* **88:**327–334.

Younes, H., and Cohn, D., 1987, Morphological study of biodegradable PEO/PLA block copolymers, *J. Biomed. Mater. Res.* **21:**1301–1316.

Younes, H., Nataf, P., Cohn, D., Appelbaum, Y., Pizov, G., and Uretzky, G., 1988, Biodegradable PELA copolymers: *In vitro* degradation and tissue reaction, *Biomater., Artif. Cells, Artif. Org.* **16:**705–719.

Zhu, K., Bihai, S., and Shilin, Y., 1989, "Super microcapsules" (SMC). I. Preparation and characterization of star polyethylene oxide (PEO)–polylactide (PLA) copolymers, *J. Poly. Sci., Part A: Polym. Chem.* **27:**2151–2159.

Zhu, K., Xiangzhou, L., and Shilin, Y., 1990, Preparation, characterization and properties of polylactide (PLA)–poly(ethylene glycol) (PEG) copolymers: A potential drug carrier, *J. Appl. Polym. Sci.* **39:**1–9.

Chapter 7

Multiple Emulsions for the Delivery of Proteins

Merrick L. Shively

1. INTRODUCTION

Due to the well-known enzymatic lability or environmental sensitivity of proteins, a common characteristic among protein delivery systems is the ability to protect the protein from the external environment. Modes of protection may be chemical or physical. A well-recognized system that provides physical protection of proteins is microcapsules, in which a solid membrane separates the solid contents of the microcapsule from the external environment. An alternative approach would be to replace a solid membrane with a liquid membrane, i.e., multiphase or multiple emulsions. Multiple emulsions may be prepared as either oil-in-water-in-oil (O/W/O) or water-in-oil-in-water (W/O/W) systems. Multiple-emulsion systems, for the purpose of delivering proteins, would be comprised of an interior aqueous phase, containing the water-soluble protein, separated from the external aqueous phase by an oil phase, i.e., W/O/W emulsions (Fig. 1). Multiple emulsions, therefore, provide an alternative technique for the encapsulation of proteins and other materials that would otherwise be metabolized, rapidly cleared, or toxic to the patient. Multiple emulsions have been utilized for parenteral and oral administration (Brodin *et al.*, 1978). Although there is a physical resemblance to microcapsules, multiple

---

Merrick L. Shively • NeXagen, Inc., Boulder, Colorado 80301; *current address:* Drug Delivery Solutions, LLC, Louisville, Colorado 80027.

Protein Delivery: Physical Systems, Sanders and Hendren, eds., Plenum Press, New York, 1997.

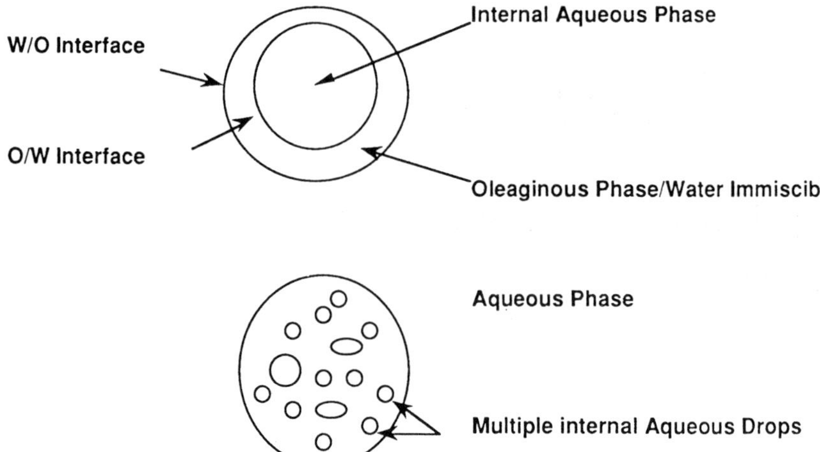

Figure 1. Schematic representation of two types of multiple-emulsion systems. The upper droplet contains a singular aqueous droplet whereas the lower droplet contains numerous aqueous droplets.

emulsions have historically suffered from technical obstacles that have prevented their commercialization. This chapter will review the physical stability issues, methods of preparation, modes of release, and specific applications of conventional systems as well as present the preparation and use of a novel multiple-emulsion system.

2. METHODS OF PREPARATION

The initial step in the fabrication of a multiple emulsion (W/O/W) is to prepare a primary emulsion (W/O). It is generally agreed that the surfactant for the primary emulsion should have an HLB value of 3–6 (in the hydrophilic–lipophilic balance system of surfactant classification). Surfactants that have successfully been utilized include Span 80 (sorbitan oleate; Nianxi *et al.*, 1992; Zheng *et al.*, 1993; Omotosho *et al.*, 1990), E644 (polyamine; Nianxi *et al.*, 1992), N205 (polyamine; Nianxi *et al.*, 1992), TX-4 (polyoxyethylene alkylphenol ether; Nianxi *et al.*, 1992), MOA3 (polyoxyethylene aliphatic alcohol ether; Nianxi *et al.*, 1992), Brij 93 (Nianxi *et al.*, 1992), polyoxamers [poly(ethylene oxide)–poly(propylene oxide)–poly(ethylene oxide) block copolymer; Law *et al.*, 1986], and egg lecithin. The surfactant or combination of surfactants is then dissolved in the oil phase. Oil phases that have been utilized to prepare multiple emulsions include isopropyl myristate (Omotosho *et al.*, 1990; Florence and Whitehill, 1981; Law *et al.*, 1986), octane (Omotosho *et al.*, 1990), 0.3 M palmitic acid in octadecyl triglyceride (Shichiri *et al.*, 1974), and mineral oil (Zheng *et al.*, 1993). The aqueous and oil phases are then dispersed using either a homogenizer (Zheng *et al.*, 1993), a microfluidizer (Microfluidics Corp., Newton, MA; Zheng *et al.*, 1993), mechanical stirring (Law *et al.*, 1986), or sonication (Shichiri *et al.*, 1974; Engel *et al.*, 1968) to prepare a W/O primary emulsion.

The second stage of the process disperses the primary emulsion in an aqueous phase. A high-HLB (9–11) surfactant is dissolved in the external phase. The primary emulsion is then added to the surfactant solution and mixed. The method and intensity of mixing are chosen to achieve maximum dispersion with minimal rupture of multiple-emulsion droplets. The resultant multiple-emulsion preparation usually contains two types of droplets (Fig. 1). Although preparations of multiple emulsions usually contain both types of droplets, the relative amount of each type is dependent on the method of preparation and the surfactant at the O/W interface (Florence and Whitehill, 1981).

3. STABILITY ISSUES

3.1. Background

The physical stability of multiple emulsions may be better understood after a brief review of methods to stabilize liquid/liquid dispersions, i.e., oil-in-water emulsions. The physical stabilization of emulsions is accomplished through a combination of droplet size reduction, homogenization, electrostatic repulsion, and/or the generation of physical barriers (Weiner, 1986; Washington, 1990). Electrostatic repulsion of dispersed droplets and physical barriers to coalescence may be obtained through the appropriate choice of surfactants, i.e., surface-active agents comprised of distinct hydrophobic/hydrophilic regions. Surfactants, by their inherent nature, are partially soluble in the oil phase as well as the aqueous phase, hence the HLB (hydrophilic–lipophilic balance) system of surfactant classification. Cationic or anionic surfactants, at the appropriate pH, provide electrostatic repulsion while nonionic surfactants provide an aqueous hydration barrier to droplet coalescence. Despite the fact that the theories and rationales to improve the physical stability of emulsions are well understood (Weiner, 1986; Washington, 1990), the physical stability of emulsions is a major technical hurdle. As a result, only a few emulsion (O/W) pharmaceutical products (e.g., Intralipid®) have been commercialized.

3.2. Surfactant Migration

Multiple emulsions (W/O/W), like the simpler O/W emulsions, suffer from physical stability problems. However, owing to the additional interface, i.e., the W/O interface, the physical stability issue is made more complicated. Conventional multiple emulsions therefore require at least two surfactants. At least one surfactant is required to stabilize each interface (i.e., the W/O and O/W interfaces; Fig. 1). Surfactants of different HLBs are utilized to separately stabilize each phase. The W/O interface is stabilized with relatively low HLB, hydrophobic surfactants while the O/W interface is stabilized with relatively high HLB, hydrophilic surfactants. Owing to the partial solubility of surfactants in aqueous and oil phases, surfactants migrate from one interface to another (Magdassi *et al.*, 1984; Lin *et al.*, 1973). This surfactant migration results in a nonoptimal surfactant HLB mixture at both interfaces, leading to significant physical instability (Nianxi *et al.*, 1992). Several approaches have been evaluated to minimize the effects of surfactant migration such as gelation of internal phase (Florence and Whitehill, 1980), polymerization of internal phase (Florence and Whitehill, 1982), interfacial complexation (Law *et al.*, 1986), and the use of a W/O microemulsion (Chatenay *et al.*, 1985).

3.3. Osmotic Gradients

The presence of a membrane between the internal and the external aqueous phase provides for generation of osmotic gradients. For multiple emulsions, any osmotic pressure difference between the internal and external phases has the potential to significantly reduce the physical stability of these systems (Matsumoto and Kohda, 1980).

3.4. Process Denaturation of Protein

The denaturation of protein during the preparation of the primary (W/O) emulsion is a concern. Techniques utilized to minimize physical denaturation of proteins or in which denaturation does not appear to be a problem include processing of the oil and aqueous phase at low temperatures (e.g., 4 °C), relatively high surfactant concentrations (15 and 25% for Span 80 and Brij 93, respectively), sonication (Engel *et al.*, 1968), and low-shear mixing methods (Zheng *et al.*, 1993). Although not specifically addressed, the use of high surfactant concentrations probably reduces the surface free energy and therefore the amount of work (i.e., mixing) that is required to achieve a given reduction in particle size. Due to lower yields, compared to that observed for low-shear methods, high-shear methods (e.g., Microfluidizer) are not extensively utilized.

Chemical degradation of the protein within the internal aqueous phase may be controlled through incorporation of antioxidants in the aqueous and oil phases, e.g., catalase and α-tocopherol, respectively (Zheng *et al.*, 1993). Although samples should be prepared under aseptic conditions, antibiotics are commonly incorporated within the internal aqueous phase to reduce bacterial growth. A reported mixture of antibiotics includes 50,000 U of penicillin, 25,000 U of polymixin, 50 mg of streptomycin, and 40 mg of gentamicin per liter of aqueous phase (Zheng *et al.*, 1993).

3.5. Methods to Determine Physical Stability

In addition to the traditional particle size methods to characterize the stability of W/O or O/W emulsions [e.g., Coulter Counters (Davis *et al.*, 1976), electron microscopy (Davis and Burbage, 1977), photomicroscopy (Law *et al.*, 1986), and light scattering (Washington, 1990)], additional methods such as measurement of viscosity (Kita *et al.*, 1977a) and conductivity (Kita *et al.*, 1977b) may be utilized for multiple emulsions. Short-term stability of 20 weeks has been shown for one system (Florence and Whitehill, 1981). Owing to the inherent instability of multiple emulsions, storage is normally limited to less than a week (Zheng *et al.*, 1993). It would therefore be necessary to administer conventional multiple emulsions soon after their preparation. These short-term storage properties represent significant limitations to commercialization.

4. APPLICATIONS

4.1. Parenteral Administration

Although parenteral administration is currently the most widely utilized route of administration for proteins, there are situations in which the protection or microenvironment provided by multiple-emulsion systems may be desirable, for instance, situations in which the circulating half-life may be increased or, possibly, situations in which the optimal storage conditions significantly differ from the conditions of the biological milieu. Owing to the limited availability of proteins, most studies evaluating the feasibility of multiple emulsions have been limited to nonproteinaceous materials (Brodin *et al.*, 1978; Omotosho *et al.*, 1990). With the improving ability to produce larger quantities of proteins and the resultant interest in their administration, reports involving the development of protein delivery systems are becoming more frequent.

The preparation of a hemoglobin-containing multiple-emulsion system as a blood substitute is one such example (Zheng *et al.*, 1993; Davis *et al.*, 1984). Since free hemoglobin is quickly phagocytosed by reticuloendothelial cells throughout the body (Guyton, 1981), a delivery system whereby the hemoglobin could serve to carry oxygen for a longer period of time was desired. Options may include the incorporation of hemoglobin within microcapsules, liposomes, or multiple emulsions. Once again, multiple emulsions are unique in that the interior aqueous phase may be customized to achieve optimal activity and stability of the encapsulated protein. For example, Zheng *et al.* (1993) modified the environment of the internal aqueous phase to maintain an oxygen affinity similar to that of red blood cells by incorporating pyridoxal 5-phosphate within phosphate-buffered saline at pH 7.4. In addition, a water-soluble antioxidant was added to protect the hemoglobin from free-radical oxidation (i.e., catalase). In order to achieve even greater protection against oxidation of hemoglobin, an oil-soluble antioxidant was incorporated within the oil phase [i.e., α-tocopherol (Szebeni *et al.*, 1984) and cholesterol (Szebeni *et al.*, 1988)]. By appropriate attention to the specific requirements of hemoglobin, Zheng *et al.* (1993) achieved a multiple-emulsion delivery system, with greater than 90% encapsulation efficiency, that had equivalent shear stability (leakage of hemoglobin) and oxygen-carrying capacity to whole blood.

4.2. Oral Administration

Multiple emulsions have also been proposed for the oral delivery of proteins. Because of the relatively large quantities of insulin historically available and the well-known interest in alternative insulin delivery systems (see Chapter 13), the incorporation of insulin within W/O/W multiple emulsions for oral administration has served as the model system (Engel *et al.*, 1968; Shichiri *et al.*, 1974). Engel *et al.* (1968) reported on the intraduodenal administration of insulin multiple emulsions to Wistar rats and gerbils. The internal aqueous phase of the primary W/O emulsion contained 100 U of insulin per milliliter of 0.003 M zinc chloride, pH 6.5. Zinc chloride was added to stabilize the W/O emulsion. Loss of insulin activity as the insulin concentration was increased was thought to be due to the formation of inactive zinc–insulin complexes. At doses from 80 to 150 U/kg, a significant ($P < 0.05$) decrease in blood glucose was, however, observed for each animal, e.g., 35 and 57% (compared to baseline) for rats and gerbils, respectively. Shichiri *et al.* (1974) improved upon the work of Engel and co-workers by decreasing the pH of the internal phase to 2.2, thus preventing the formation of inactive zinc–insulin complexes. Shichiri and co-workers reported significant ($P < 0.05$) decreases in blood glucose and increases in plasma insulin at doses as low as 10 U/kg. Although the ability of insulin to be orally absorbed is enhanced by multiple emulsions, the interanimal variability was considered too high for this delivery system to be applicable for the human administration of insulin.

The parenteral and oral studies presented indicate that multiple emulsions may be suitable delivery vehicles for certain biologically active molecules. A major attribute of multiple-emulsion systems is the ability to provide an individually tailored aqueous microenvironment for a protein. As long as no osmotic pressure gradient is generated between the interior and exterior aqueous phases, numerous formulation variables may be utilized. Although the feasibility of the use of multiple emulsions for the delivery of proteins has been shown, the long-term storage of these systems remains a serious technical obstacle to further commercialization. In lieu of waiting for the development of new surfactants to improve the physical stability of multiple emulsions, we have investigated self-emulsifying systems (Myers and Shively, 1992), as described in the following section.

5. SOLID-STATE EMULSIONS

Solid-state emulsions are solids that self-emulsify *into a liquid system* upon contact with an aqueous phase. Depending on the materials and processing utilized, O/W or W/O/W emulsions may be fabricated (Myers and Shively, 1992). The ability to store these systems as solids and to reconstitute them when needed for administration is attractive. Since the rationale for delivering proteins and peptides by multiple emulsions has been shown and the long-term stability of aqueous W/O/W systems is not critical, the majority of the work in the area of solid-state emulsions has focused on the solid-state stability of these systems. The physical properties of these systems will briefly be reviewed and some *in vivo* data obtained for a model peptide presented.

5.1. Method of Preparation

Solid-state emulsions are prepared using similar techniques (Chiou and Reigelman, 1971) to those used to prepare solid-state dispersions (suspensions), hence the name (Shively, 1993a). As for conventional multiple emulsions, a primary (W/O) emulsion is prepared (Myers and Shively, 1992). The same opportunities therefore exist to custom-tailor the contents of the interior aqueous phase and oil phase to the particular requirements of the protein. An aliquot of the primary emulsion is added to a rotary vacuum flask. Sucrose, or another suitable lattice material, is dissolved in sufficient water and added to the flask. A typical ratio of sucrose to primary emulsion is 3:1 (sucrose:primary emulsion, w/w). Vacuum is applied until a brittle foam is produced.

The resulting solid (or foam) is stored at a temperature below the glass-transition temperature of sucrose in a desiccator until required (Shively and Myers, 1993). The effects of excipients on aging (Myers and Shively, 1993), process and storage conditions (Shively and Myers, 1993), and the particle size (Shively, 1993a, b) have also been reported. The addition of buffer to the prepared solid-state emulsion results in a significant majority of single internal aqueous droplets within an oil membrane (Myers and Shively, 1992).

5.2. Physical Properties of Solid-State Emulsions

Based on recent carbon-13 and proton solid-state NMR experiments (Shively and Dec, 1994), W/O/W solid-state emulsions are thought to be comprised of lattice materials in which the guest molecule is actually a W/O droplet. The lattice is comprised of intermolecularly hydrogen-bonded sucrose or similar molecules (Shively, 1993a). Owing to the complexity of such a system, compared to other multimolecular lattice systems (Takemoto and Sonoda, 1984), the precise arrangement of lattice molecules may never be elucidated. If this theory is true, the apparent stability of these systems may be the result of the oil–water interface being a function of the guest–host relationship, i.e., not dependent on modification or protection of the surface by a surfactant.

5.3. Oral Administration of Vancomycin Solid-State Emulsion

A preliminary study in rats was initiated with W/O/W solid-state emulsions to evaluate the intestinal absorption of a model peptide, vancomycin (Geary and Schlameus, 1993). In contrast to past studies in which multiple emulsions were directly administered to the small intestine, formulations in this study were orally administered. Therefore, this study was designed to evaluate the physical stability of multiple-emulsion solid-state emulsions in the gastric environment and intestinal absorption. The primary emulsion was comprised of 73.5% (w/w) sesame oil, 1.5% (w/w) monoglycerol stearate, and 25% aqueous phase. The isotonic aqueous phase contained 200 mg of vancomycin per milliliter and was maintained at pH 4.5 with a phosphate buffer. The final solid-state emulsion contained 11.1 mg of vancomycin per gram of solid. Conscious rats were orally administered a 50-mg/kg dose of a vancomycin solid-state emulsion formulation. Blood samples were withdrawn from previously surgically placed catheters, and

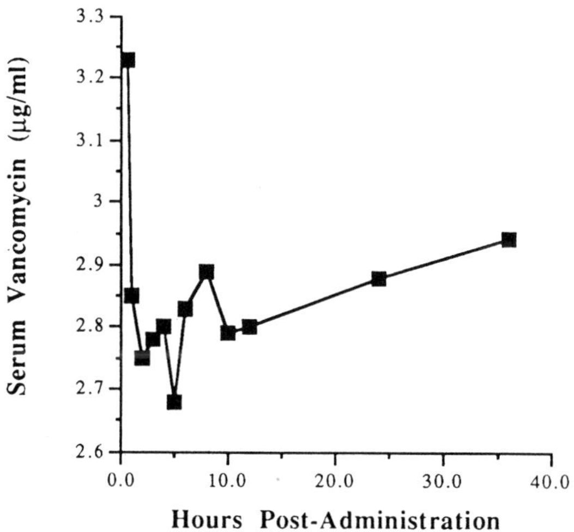

Figure 2. Serum concentrations of vancomycin in conscious rats following administration of vancomycin within W/O/W solid-state emulsion. A 50-mg/kg dose of a vancomycin solid-state emulsion formulation was orally administered to rats. Serum levels of vancomycin were measured immediately prior to administration ($t = 0$) and 0.5, 1, 2, 3, 4, 5, 6, 8, 10, 12, 24, and 36 hr after administration. Data represent the mean $\pm$ SEM obtained in 3 animals. Sensitivity limit of assay was 2.0 μg/ml (95% confidence greater than zero). Rats orally administered the equivalent dose of the commercial oral product resulted in no detectable serum levels (data not shown). (After Shively and Thompson, 1995.)

serum was isolated. Serum samples were analyzed by fluorescence polarization immunoassay (TDx, Abbott Laboratories) having a sensitivity level of 2.0μg/ml (95% confidence from zero). Physical mixtures and a commercial suspension were also administered as controls. Analysis of the results ($N = 4$ for each treatment group) indicated that, as expected, there was no detectable vancomycin in the serum of control treatment groups. Serum levels of vancomycin in 3 animals treated with the multiple-emulsion solid-state emulsion, representative of the treatment group, are shown in Fig. 2. The intestinal absorption of vancomycin, for solid-state-emulsion-treated animals, was highly reproducible over 36 hr (AUC of 101.6 ± 1.0 mg/ml $\times$ hr). These results corresponded, based on intravenous administration, to an absolute absorption of approximately 30%. Analysis of Fig. 2 indicates that vancomycin was not only absorbed but was detected for a sustained time period. Although untested, a rationale for such a concentration–time profile may be that multiple-emulsion droplets are sequestered and vancomycin is released as the droplets break down or diffuses out of the droplet. Vancomycin could not be detected in serum (data not shown) after the oral administration of the commercially available vancomycin suspension (Lilly Laboratories, Inc.) at an equivalent dose.

6. MISCELLANEOUS APPLICATIONS

6.1. Vaccine Adjuvants

Multiple emulsions have been utilized as adjuvants for the delivery of vaccines (Herbert, 1965; Aitken, 1973; Blackall *et al.*, 1992). The incorporation of ovalbumin or other foreign protein has been found to result in higher and more sustained antibody titers compared to those obtained with administration of vaccines in saline solutions.

6.2. Enzyme Immobilization

Although not associated with the delivery of proteins, multiple emulsions or liquid membranes have been utilized for enzyme immobilization in place of solid-membrane methods or conventional methods (Scheper *et al.*, 1987; May and Li, 1972). Multiple-emulsion systems have proved useful because there is no membrane fouling, they can be utilized in cell-free fermentation broths, and enzyme inhibitors can be excluded.

7. SUMMARY

Multiple emulsions are unique in that a true liquid phase is maintained separate from an external aqueous phase. This may be especially important for bioactive molecules that cannot be appropriately stabilized in the solid state. In addition, the separation of aqueous phases enables highly specialized environments, conducive to protein activity, to be prepared. The physical instability of conventional systems remains a major factor limiting their wider application. Attempts to improve the physical stability of the aqueous dispersions through interfacial complexation and the use of microemulsions are improving the short-term stability. As an alternative approach, solid-state emulsions attempt to store the multiple emulsion as a solid. Although solid-state emulsions appear to have the potential to be useful protein delivery systems, a substantial experimental data base has yet to be generated.

REFERENCES

Aitken, I. D., 1973, The serological response of the chicken to a protein antigen in multiple emulsion oil adjuvant, *Immunology* **25:**957–966.

Blackall, P. J., Eaves, L. E., Rogers, D. G., and Firth, G., 1992, An evaluation of inactivated infectious Coryza vaccines containing a double-emulsion adjuvant system, *Avian Dis.* **36:**632–636.

Brodin, A. F., Kavaliunas, D. R., and Frank, S. G., 1978, Prolonged release from multiple emulsions, *Acta Pharm. Suec.* **15:**1–12.

Chatenay, D., Urbach, W., Cazabat, A. M., Vacher, M., and Waks, M., 1985, Proteins in membrane mimetic systems. Insertion of myelin basic protein into microemulsion droplets, *Biophys. J.* **48:**893–898.

Chiou, W. L., and Reigelman, S., 1971, Pharmaceutical applications of solid dispersion systems, *J. Pharm. Sci.* **60:**1281–1302.

Davis, S. S., and Burbage, A. S., 1977, Electron microsopy of water-in-oil-in-water emulsions, *J. Colloid Interface Sci.* **62:**361–363.

Davis, S. S., Purewal, T. S., and Burbage, A. S., 1976, The particle size analysis of multiple emulsions, *J. Pharm. Pharmacol. Suppl.* **28:**60P.

Davis, T. A., Asher, W. J., and Wallace, H. W., 1984, Artificial red blood cells with crosslinked hemoglobin membranes, *Appl. Biochem. Biotechnol.* **10:**123–132.

Engel, R. H., Riggi, S. J., and Fahrenbach, M. J., 1968, Insulin: Intestinal absorption as water-in-oil-in-water emulsions, *Nature* **219:**856–857.

Florence, A. T., and Whitehill, D., 1980, Multiple W/O/W emulsions stabilized with poloxamer and acrylamide gels, *J. Pharm. Pharmacol.* **32:**64P.

Florence, A. T., and Whitehill, D., 1981, Some features of breakdown in water-in-oil-in-water multiple emulsions, *J. Colloid Interface Sci.* **79:**243–256.

Florence, A. T., and Whitehill, D., 1982, Stabilization of water/oil/water multiple emulsions by polymerization of aqueous phase, *J. Pharm. Pharmacol.* **34:**687–691.

Geary, R. S., and Schlameus, H. W., 1993, Vancomycin and insulin used as models for oral delivery of peptides, *J. Controlled Release* **23:**65–74.

Guyton, A. C., 1981, *Textbook of Medical Physiology,* Sixth ed., W. B. Saunders Co., Philidelphia, pp. 872–873.

Herbert, W. J., 1965, Multiple emulsions. A new form of mineral-oil antigen adjuvant, *Lancet* **1965:**771.

Kita, Y., Masumoto, S., and Yonezawa, D., 1977a, Viscometric method for estimating the stability of w/o/w multiple-phase emulsions, *J. Colloid Interface Sci.* **62:**745–751.

Kita, Y., Masumoto, S., and Yonezawa, D., 1977b, An attempt at measuring the stability of w/o/w-type multiple phase emulsions by analyzing the concentration of anions, *Nippon Kagaku Kaishi* **6:**748.

Law, T. K., Whateley, T. L., and Florence, A. T., 1986, Stabilization of W/O/W multiple emulsions by interfacial complexation of macromolecules and nonionic surfactants, *J. Controlled Release* **3:**279–290.

Lin, T. J., Kurihara, H., and Ohta, H., 1973, Effect of surfactant migration on the stability of emulsions, *J. Soc. Cosmet. Chem.* **24:**797–814.

Magdassi, S., Frenkel, M., Garti, N., and Kasan, R., 1984, Multiple emulsions II: HLB shift caused by emulsifier migration to external interface, *J. Colloid Interface Sci.* **97:**374–379.

Matsumoto, S., and Kohda, M., 1980, The viscosity of water-in-oil-in-water emulsions: An attempt to estimate the water permeation coefficient of the oil layer from the viscosity changes in diluted systems on ageing under osmotic pressure gradients, *J. Colloid Interface Sci.* **73:**13–20.

May, S. W., and Li, N. N., 1972, The immobilization of urease using liquid-surfactant membranes, *Biochem. Biophys. Res. Commun.* **47:**1179–1185.

Myers, S. L., and Shively, M. L., 1992, Preparation and characterization of emulsifiable glasses: Oil-in-water and water-in-oil-in-water emulsions, *J. Colloid Interface Sci.* **149:**271–278.

Myers, S. L., and Shively, M. L., 1993, Solid state emulsions: The effects of maltodextrin on microcrystalline aging, *Pharm. Res.* **10:**1389–1391.

Nianxi, Y., Mingzu, Z., and Peihong, N., 1992, A study of the stability of W/O/W multiple emulsions, *J. Microencapsul.* **9:**143–151.

Omotosho, J. A., Florence, A. T., and Whateley, T. L., 1990, Absorption and lymphatic uptake of 5-fluorouracil in the rat following oral administration of W/O/W multiple emulsions, *Int. J. Pharm.* **61:**51–56.

Scheper, T., Makryaleas, K., Nowottny, C., Likidis, Z., Tsikas, D., and Schugerl, K., 1987, Liquid surfactant membrane emulsions. A new technique for enzyme immobilization, *Ann. N.Y. Acad. Sci.* **501:**165–170.

Shichiri, M., Shimizu, Y., Yoshida, Y., Kawamori, R., Fukuchi, M., Shigeta, Y., and Abe, H., 1974, Enteral absorption of water-in-oil-in-water insulin emulsions in rabbits, *Diabetologia* **10:**317–321.

Shively, M. L., 1993a, Characterization of oil-in-water emulsions prepared from solid-state emulsions: Effect of matrix and oil phase, *Pharm. Res.* **10:**1153–1156.

Shively, M. L., 1993b, Droplet size distribution within oil-in-water emulsions prepared from solid-state dispersions, *J. Coll. Interface Sci.* **155:**66–69.

Shively, M. L., and Dec, S. F., 1994, Solid state emulsions: Evaluation by ^{1}H and ^{13}C solid-state nuclear magnetic resonance, *Pharm. Res.* **11:**1301–1305.

Shively, M. L., and Myers, S. L., 1993, Solid state emulsions: The effects of process and storage conditions, *Pharm. Res.* **10:**1071–1075.

Shively, M. L., and Thompson, D. C., 1995, Oral bioavailability of vancomycin solid-state emulsions, *Int. J. Pharm.* **117:**119–122.

Szebeni, J. Winterbourn, C. C., and Carrell, R. W., 1984, Oxidative interactions between hemoglobin and membrane lipid. A liposome model, *Biochem. J.* **220:**685–692.

Szebeni, J., Hauser, H., Eskelson, C. D., Waston, R.R., and Winterhalter, K. H., 1988, Interaction of hemoglobin derivatives with liposomes. Membrane cholesterol protects against the changes of hemoglobin, *Biochemistry* **27:**6425–6434.

Takemoto, K., and Sonoda, N., 1984, Inclusion compounds in urea, thiourea and selenourea, in: *Inclusion Compounds: Structural Aspects of Inclusion Compounds Formed by Organic Host Lattices,* Vol. 2 (J. L. Atwood, J. E. D. Davies, and D. D. MacNicol, eds.), Academic Press, New York, pp. 47–67.

Washington, C., 1990, The stability of intravenous fat emulsions in total parenteral nutrition mixtures, *Int. J. Pharm.,* **66:**1–21.

Weiner, N., 1986, Strategies for formulation and evaluation of emulsions and suspensions: Some thermodynamic considerations, *Drug Dev. Ind. Pharm.* **12:**933–951.

Zheng, S., Zheng, Y., Beissinger, R., Wasan, D. T., and McCormick, D. L., 1993, Hemoglobin multiple emulsion as an oxygen delivery system, *Biochim. Biophys. Acta* **1158:**65–74.

Chapter 8

Transdermal Peptide Delivery Using Electroporation

Russell O. Potts, D. Bommannan, Ooi Wong, Janet A. Tamada, Jim E. Riviere, and Nancy A. Monteiro-Riviere

1. INTRODUCTION

Transdermal drug delivery has become an attractive dosage form as demonstrated by the success of many recent products such as nicotine patches for smoking cessation, as well as estradiol and other hormone replacement transdermal systems. The feasibility of passive transdermal delivery, however, is limited by the size, charge, and dose of the drug to be administered. Owing to these limitations, viable candidates for passive transdermal delivery remain few in number and are restricted to small-molecular-weight, lipophilic, uncharged, and potent drugs. Peptides and proteins, owing to their large size and ionic character, do not readily pass through the skin, and effective transport often requires enhancement techniques. Iontophoresis, the electromigrational movement of charged molecules through the skin under a low-voltage and continuous electrical driving force, is one such enhancement method. The iontophoretic delivery of large-molecular-weight compounds such as luteinizing hormone-releasing

Russell O. Potts, D. Bommannan, Ooi Wong, and Janet A. Tamada • Cygnus, Inc., Redwood City, California 94063. *Jim E. Riviere and Nancy A. Monteiro-Riviere* • Cutaneous Pharmacology and Toxicology Center, College of Veterinary Medicine, North Carolina State University, Raleigh, North Carolina 27606.

Protein Delivery: Physical Systems, Sanders and Hendren, eds., Plenum Press, New York, 1997.

hormone (LHRH) or analogs (Meyer *et al.*, 1988; Miller *et al.*, 1990; Heit *et al.*, 1993; Srinivasan *et al.*, 1990), thyrotropin-releasing hormone (Burnette and Marrero, 1986), and insulin (Meyer *et al.*, 1989; Siddiqui *et al.*, 1987; Srinivasan *et al.*, 1989; Chien *et al.*, 1989) as well as smaller-molecular-weight compounds such as lidocaine (Riviere *et al.*, 1992a; Singh and Roberts, 1989) has been reported. However, the success of iontophoretic delivery of large-molecular-weight compounds such as peptides remains elusive, primarily owing to the impermeable nature of skin and the consequent inability to deliver therapeutically meaningful doses in humans.

Large-molecular-weight (or, more correctly, larger molecular-volume) compounds can be introduced into cells via a process known as electroporation or electropermeabilization (Neumann, 1992). This technique involves the application of short, transient (microsecond–millisecond) electrical pulses of high magnitude ($\sim$1 kV/cm) which induce a short-lived (up to seconds) and reversible, high-permeability state in the membrane lipids. Electroporation has found particular application in DNA transfection, where large nucleic acid molecules are transported into cells following the application of a series of electrical pulses (Neumann, 1992). Transfection has also been achieved in tissue where an electroporative pulse was applied to the skin following the subcutaneous injection of plasmid DNA (Titomirov *et al.*, 1991). In addition, recent evidence shows that other compounds of therapeutic interest may be introduced into viable red blood cells [see review by Tsong (1990)]. Finally, these techniques have found therapeutic use in chemotherapy, the application of electrical pulses to superficial tumors resulting in the selective uptake by the tumor of a systematically administered drug (Mir *et al.*, 1991; Belehradek *et al.*, 1991). Thus, electroporation has significant potential to enhance drug uptake by cells and tissue.

The mechanism underlying electroporation has been studied using artificial lipid bilayer membranes and liposomes. It has been hypothesized that the lipid bilayers are reversibly permeabilized by the formation of transient pores, shown schematically in Fig. 1 (Kinosita *et al.*, 1992; Chernomordik *et al.*, 1983; Chang and Reese, 1990; Tsong, 1990; Chernomordik *et al.*, 1987). Although tantalizing data exist indicating the formation of transient "pores" in cell membranes following electroporation (Chang and Reese, 1990), the definitive demonstration of pores remains elusive. Regardless of the precise mechanism, however, high membrane permeability is induced by an electrical pulse of appropriate amplitude and duration. Such changes in permeability are also accompanied by a large increase in membrane electrical conductivity due to the enhanced transport of small ions such as sodium and chloride (Chernomordik and Chizmadzhev, 1989; Kinosita *et al.*, 1992; Chernomordik *et al.*, 1987).

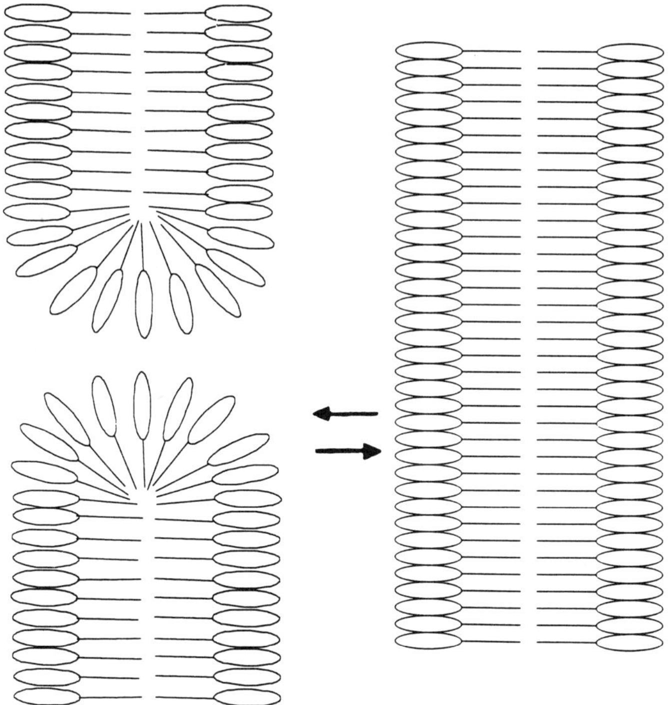

Figure 1. A schematic representation of changes in lipid bilayer structure induced by electroporation.

Mammalian skin is perhaps the most formidable transport barrier found in nature. The lipids of the stratum corneum, the outermost skin layer, form the primary barrier to transport of many compounds of therapeutic interest (Scheuplein, 1965, 1978; Potts and Guy, 1992; Blank and Scheuplein, 1969; Elias, 1983, 1987, 1991). As shown schematically in Fig. 2, these lipids form broad multilamellar arrays in the extracellular space surrounding the remains of epidermal cells known as corneocytes. The lipids have a unique composition (fatty acids, cholesterol, and ceramides; no phospholipids are present) and form the only continuous domain within the stratum corneum (Elias, 1983, 1987, 1991). Despite profound differences between stratum corneum lipids and those of the phospholipid bilayers more commonly found in other biomembranes, direct comparison of passive transport through each suggests a common mechanism involving free-volume fluctuations in the lipid alkyl chains. Transport within the lipid hydrocarbon domain substantially restricts the permeability of large mol-

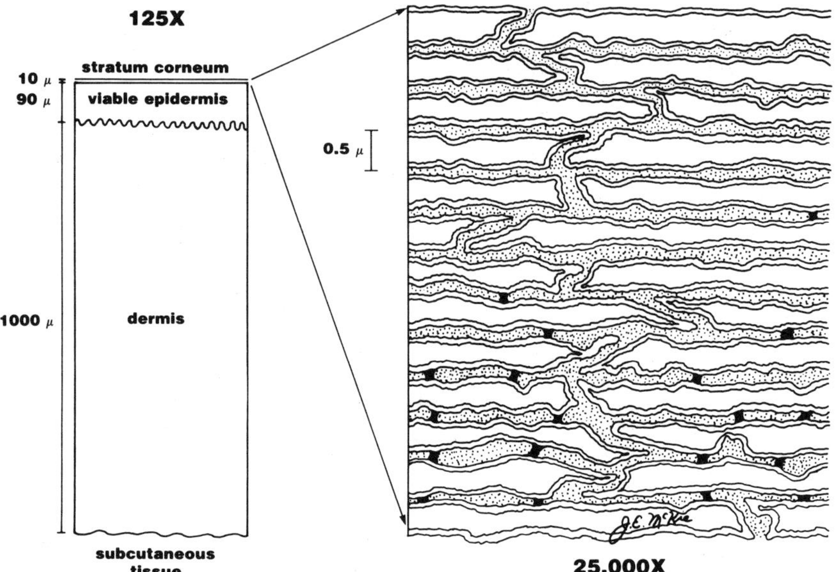

Figure 2. A schematic representation of stratum corneum structure.

ecules in both stratum corneum lipids and phospholipid bilayers (Potts and Guy, 1992). However, the absolute magnitude of permeability is about 10^5 greater in bilayer lipids relative to that in the stratum corneum, primarily owing to the multilamellar and highly tortuous lipid domain in the latter (Potts and Guy, 1992). Consequently, passive transdermal drug delivery, while successful for a number of small and lipid-soluble molecules such as nitroglycerin, nicotine, and estradiol, has yet to realize the full therapeutic potential for larger compounds.

As discussed above, (a) pulsed electric fields reversibly permeabilize cell lipid membranes, and (b) the multilamellar lipid domains of the stratum corneum act as the rate-limiting moiety for transdermal delivery. Thus, it is reasonable to hypothesize enhanced transport through skin following electroporation.

In this chapter, we compare the iontophoretic transport of peptides (LHRH, neurotensin, vasopressin, calcitonin, and insulin) through electroporated and nonelectroporated skin. The results demonstrate enhanced transport of peptides following electroporation and a return to baseline values within a few hours after treatment. These results are consistent with an intrinsic alteration in skin transport properties due to electroporation.

2. RESULTS AND DISCUSSION

2.1. *In Vitro* Transport

The delivery of LHRH through human skin *in vitro* was measured before, during, and after the application of direct current (0.5 mA/cm^2 for 30 min), either without or with (Fig. 3) a single exponential electrical pulse, applied at the initiation of the direct-current treatment (Bommannan *et al.*, 1994). Prior to electrical treatment, the passive flux of LHRH was near 0.05 μg/(cm$^2 \cdot$ hr) for both pulsed and nonpulsed samples. The application of

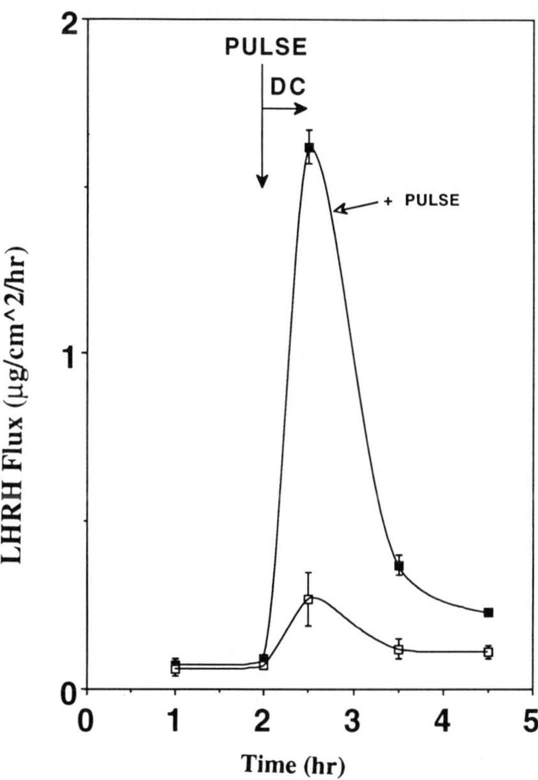

Figure 3. The delivery of LHRH through human skin *in vitro* before, during, and after the application of an iontophoretic current (0.5 mA/cm^2 for 30 min), either in the presence or absence of an exponential pulse (1000-V initial amplitude, 5-msec time constant).

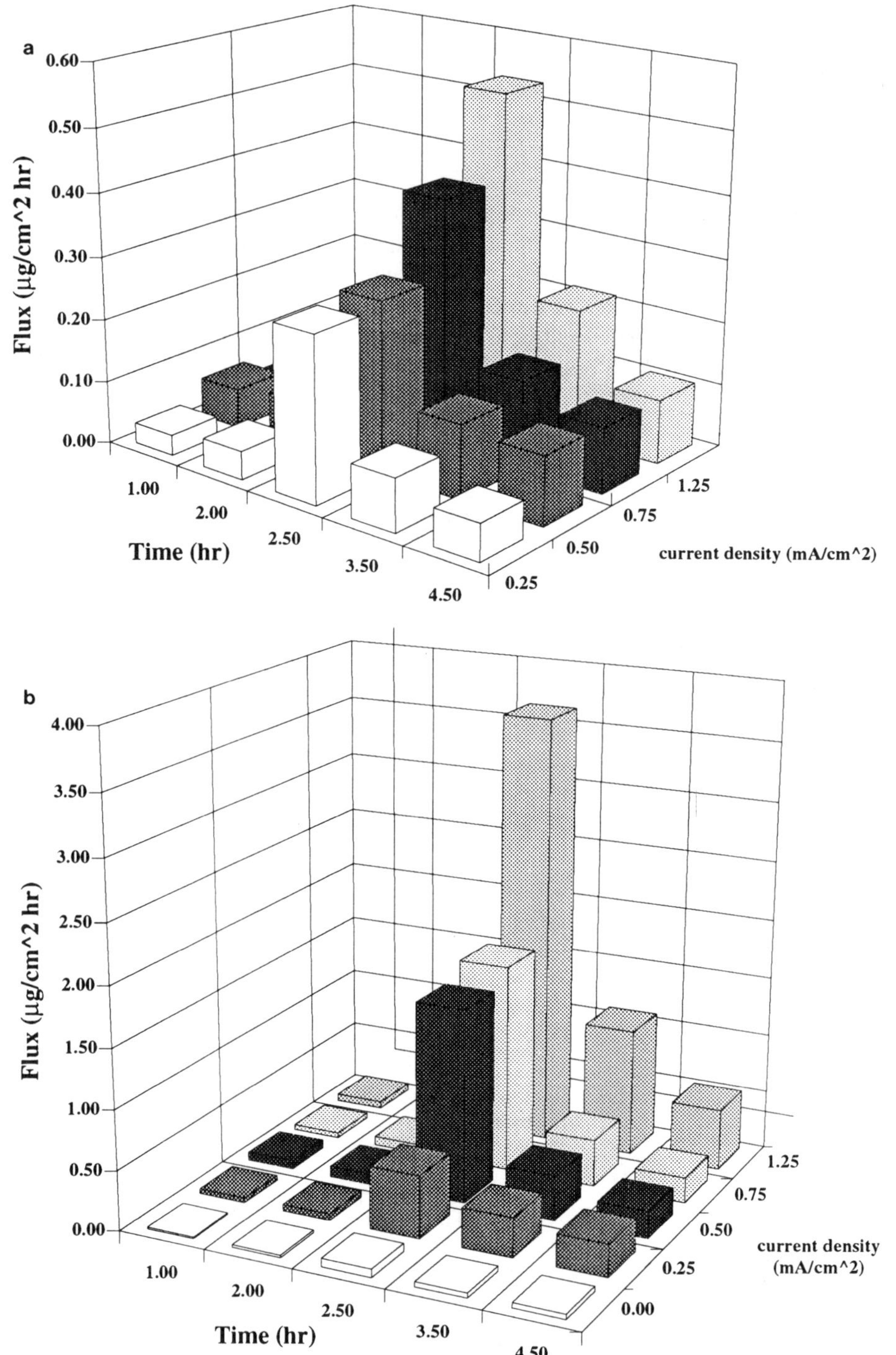
a
0.60
0.50
0.40
0.30
0.20
0.10
0.00
Flux (μg/cm^2 hr)
1.00
2.00
2.50
3.50
4.50
Time (hr)
1.25
0.75
0.50
0.25
current density (mA/cm^2)
b
4.00
3.50
3.00
2.50
2.00
1.50
1.00
0.50
0.00
Flux (μg/cm^2 hr)
1.00
2.00
2.50
3.50
4.50
Time (hr)
1.25
0.75
0.50
0.25
0.00
current density
(mA/cm^2)

direct current resulted in a significant increase in LHRH flux [0.27 μg/($cm^2 \cdot hr$) (SD = 0.08; $n = 3$)] relative to passive delivery. The concentration of LHRH in the donor was 2.5 mg/ml, and therefore the apparent permeability coefficient (defined as the flux divided by donor concentration) is 1×10^{-4} cm/hr. Similar permeability coefficient values have been reported for the iontophoretic transport of the LHRH analog leuprolide through human skin (Lu *et al.*, 1992) and of LHRH through porcine skin flap skin (Heit *et al.*, 1993). These results demonstrate that, although iontophoretic conditions were applied for only 30 min in our studies, a permeability coefficient was obtained which is comparable to steady-state values reported by others.

The application of a single, exponentially decaying pulse (5-msec time constant, 1000-V initial amplitude) just prior to iontophoresis resulted in an average flux of 1.62 μg/($cm^2 \cdot hr$) (SD = 0.05; $n = 3$). Two hours after the cessation of iontophoresis, the passive flux of LHRH decreased to average values of 0.23 (SD = 0.01; $n = 3$) and 0.10 (SD = 0.03; $n = 3$) μg/($cm^2 \cdot hr$) with and without a pulse, respectively. In all experiments, analysis of the contents of the receiver chamber by high-pressure liquid chromatography (HPLC) and thin-layer chromatography (TLC) techniques showed that intact LHRH passed through the skin. Thus, the application of a single electroporative pulse resulted in the reversibly enhanced iontophoretic transport of intact LHRH through human skin at a rate more than 30-fold greater than that for passive delivery and more than 5-fold greater than that for iontophoresis alone.

The same experimental protocol was repeated over a range of current densities. The results obtained at current densities from 0 to 1.25 mA/cm^2, either in the absence (Fig. 4a) or presence (Fig. 4b) of an electrical pulse, are qualitatively the same as the results obtained at 0.5 mA/cm^2 (Fig. 3). An average passive flux (prior to electrical treatment) of 0.057 μg/($cm^2 \cdot hr$) (SD = 0.038; $n = 76$) was obtained for all samples evaluated, yielding a permeability coefficient of 2×10^{-5} cm/hr. The passive flux of the LHRH analog leuprolide acetate was measured through human skin by several investigators. Lu *et al.* (1992) found that transport was below the limits of their HPLC detection. The value increased to 2.2×10^{-4} cm/hr, however, when 10% urea was used as a penetration enhancer. Srinivasan *et al.* (1990) also found that the passive transport of LHRH was below their limits of detection, suggesting a permeability coefficient value of less than

Figure 4. (a) The delivery of LHRH through human skin *in vitro* before, during, and after the application of an iontophoretic current as a function of the applied current. (b) The delivery of LHRH through human skin *in vitro* before, during, and after the application of an electroporative pulse plus iontophoretic current as a function of the applied current.

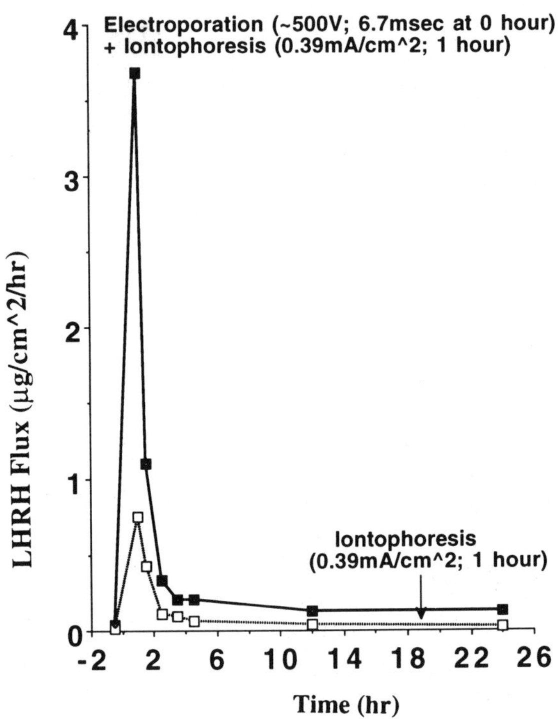

Figure 5. The delivery of LHRH through human skin *in vitro* before, during, and after the application of an iontophoretic current (0.39 mA/cm^2 for 60 min), either in the presence or absence of an exponential pulse (500-V initial amplitude, 6.7-msec time constant).

4×10^{-6} cm/hr. When the skin was pretreated with ethanol for 2 hr, however, these investigators measured a value of 6×10^{-4} cm/hr. Thus, the passive transport of LHRH through human skin is very small, with a permeability coefficient near 10^{-5} cm/hr in the absence of a penetration enhancer.

In a separate experiment, the LHRH flux was measured for 24 hr during and following electrotreatment. The results (Fig. 5) were obtained using a direct current of 0.39 mA/cm^2 for 1 h, with or without an exponential pulse (500-V initial amplitude and 6.7-msec time constant). The results again showed a significant increase in LHRH flux following the application of an electrical pulse. Moreover, the flux decreased after treatment to values near that obtained prior to electrotreatment, or in the absence of a pulse.

These results show a dramatic increase in the iontophoretic flux of LHRH through human skin following the application of a single electrical pulse. The increased flux due to pulsation cannot be explained by increased current since the charge introduced by the pulse ($\sim 1\,\text{mA}\cdot\text{sec}$) is negligible compared to the total charge introduced throughout the electrotreatment ($\sim 1000\,\text{mA}\cdot\text{sec}$). Furthermore, electrical pulsation resulted in increased LHRH flux, even in the absence of a constant-current, electromigrational driving force (data not shown). Thus, it is likely that the transport properties of human skin are transiently altered by electrical pulsation.

A plot of the LHRH transport data (Fig. 4a, b) shows a linear dependence of flux upon current density for both treatment protocols (Fig. 6). The flux of an ionic peptide (J_{pep}) of charge Z_i through skin is related to the applied current (i) by

$$J_{\text{pep}} = t_{\text{pep}} \cdot i / Z_i \cdot F \tag{1}$$

where F is the Faraday constant. The ionic mobility of a peptide in the skin (u_{pep}) can be determined from the transport number (t_{pep}), defined by

$$t_{\text{pep}} = c_{\text{pep}} \cdot u_{\text{pep}} \cdot Z_{\text{pep}} / \sum c_i \cdot u_i \cdot Z_i \tag{2}$$

In this equation, c_{pep} and Z_{pep} are the peptide concentration and charge, respectively, in the donor solution, whereas c_i, u_i, and Z_i are the concentration, mobility, and charge, respectively, of all other ionic species. Thus, the slope of a plot of J_{pep} versus i is proportional to t_{pep}, the peptide transport number. The transport number represents the fraction of the total charge transferred by the ion in question. The transport number of LHRH was determined from a linear regression analysis of the data in Fig. 6, yielding values of 4.3×10^{-5} and 4.5×10^{-6} in the presence and absence of a pulse, respectively. These results show that the application of a single pulse immediately prior to constant-current treatment resulted in a 10-fold increase in the LHRH transport number. Since the concentration and charge of all other species were fixed in these experiments, these results demonstrate that the application of an electrical pulse resulted in a dramatic increase in the mobility of LHRH relative to the other ionic species present. In other words, the application of a single pulse, of about 5-msec duration, resulted in a 10-fold increase in LHRH transport through human skin averaged over the iontophoretic treatment period of 30 min. Furthermore, as discussed above, the increased transport could not be caused by additional charge introduced by the pulse, as this charge was significant relative to the total charge passing through the skin during 30 min of direct-current application.

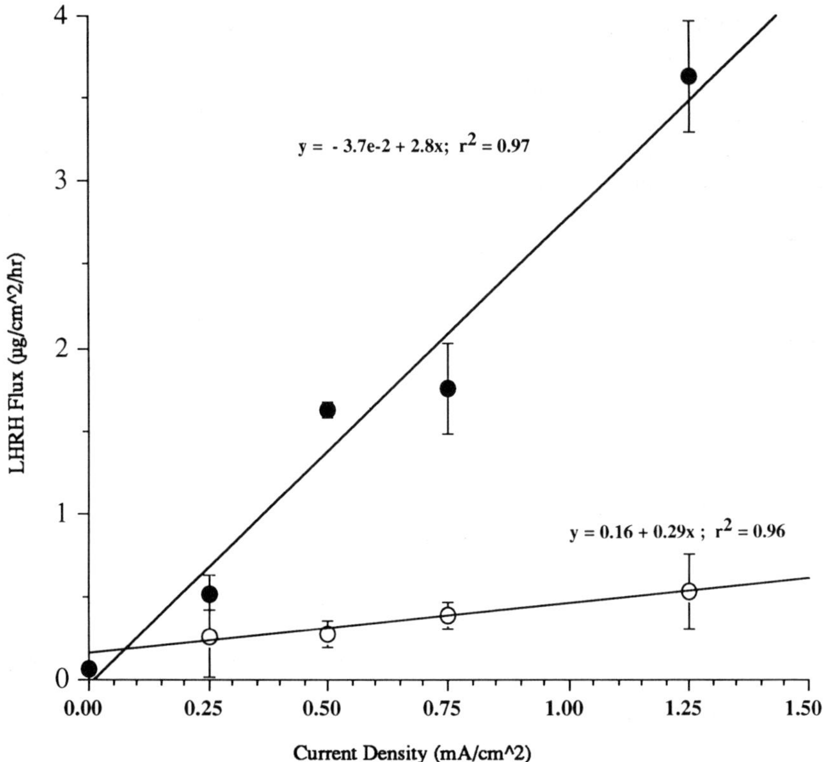

Figure 6. The iontophoretic LHRH flux through human skin as a function of current density in the absence (○) or presence (●) of an electroporative pulse.

Similar results were obtained for calcitonin at a constant current density (0.5 mA/cm^2 for 1 hr) in the presence or absence of a single exponential pulse of 500-V initial potential and 10-msec time constant applied at the onset of the iontophoresis. These results (data not shown) reveal an approximate 2-fold increase in calcitonin flux through human skin due to electrical pulsation. Furthermore, the high-permeability state induced by electroporation returned to pretreatment values within 24 hr. Finally, it should be noted that calcitonin concentration was determined using ELISA, which measures only intact peptide.

Similar to the results obtained with LHRH, calcitonin flux increased linearly with applied current in the presence or absence of an electical pulse (data not shown). Electrical pulsation, however, resulted in about 2-fold greater flux at each current density tested. From a linear regression analysis of these data using Eq. (1), the calcitonin transport number was determined to be 0.50×10^{-6} and 0.19×10^{-6}, respectively, in the presence and absence of a pulse. Thus, as with the results obtained with LHRH, electrical pulsation resulted in a reversible increase in skin permeability, albeit of smaller magnitude for the larger calcitonin molecule. Once again, the increased transport cannot be explained by incremental charge due to the pulse, suggesting that electrical pulsation results in a transient, reversible high-permeability state in human skin.

Similar experiments were performed with neurotensin, vasopressin, and insulin. In the case of neurotensin and vasopressin, the application of a single pulse resulted in an increased flux relative to that obtained in the absence of a pulse, and the flux in both treatment protocols increased with current density (data not shown). In contrast, a single pulse did not increase the transport of insulin, although the flux did increase with applied current (data not shown). The transport number for each peptide was calculated from the flux versus current data, and the values obtained are shown in Table I. These results are important from various points of view. Most important, for all peptides tested that are smaller than insulin, the application of a single pulse resulted in a substantial increase in transport through skin, associated with an increase in ionic mobility. Thus, electrical pulsation alters the transport properties of human skin. Although an electric field was used as the driving force, flux enhancement was also achieved via passive transport following pulsation. Hence, delivery enhancement is independent of the driving force (e.g., electrical or chemical).

The peptide transport data also provide insight into iontophoretic transport through the skin. The measure of intrinsic transport through skin

Table I

Transport Numbers for the Iontophoretic Flux of Various Peptides with Differing Molecular Weights through Human Skin in the Presence and Absence of an Electrical Pulse

		Transport number ($\times 10^6$)	
Peptide	MW	Pulse	No Pulse
Vasopressin	1084	39	8.3
LHRH	1182	43	4.5
Neurotensin	1693	12	5.8
Calcitonin	3430	0.50	0.19
Insulin	6000	0.035	0.035

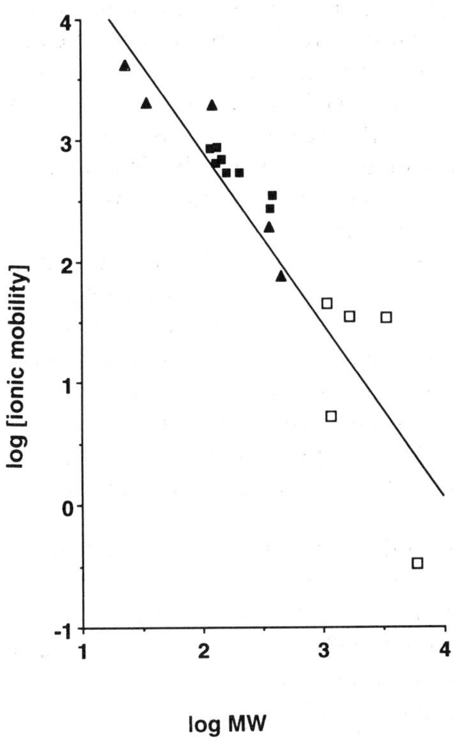

Figure 7. The ionic mobility [defined in Eq. (3)] vs. molecular weight for the iontophoretic delivery of peptides through human skin. The data obtained in the present study (□) are compared with data from Yoshida and Roberts (1992) (■) and Green *et al.* (1991) (▲).

is the ionic mobility, defined by

$$u_{\text{pep}} = J_{\text{pep}} \cdot F\left(\sum c_i \cdot u_i \cdot Z_i\right) \Big/ C_{\text{pep}} \cdot i \tag{3}$$

Since the same buffer was used in all experiments presented here, and the peptide concentration is negligible compared to that of the buffer, the value of $\Sigma c_i \cdot u_i \cdot Z_i$ is constant, and hence u_{pep} is proportional to $J_{\text{pep}}/(C_{\text{pep}} \cdot i)$. The values of $J_{\text{pep}}/(C_{\text{pep}} \cdot i)$ for the iontophoretic transport of the peptides studied here are plotted as function of their molecular weight in Fig. 7 along wth values derived from experimental results for the iontophoretic delivery of anionic peptides (Green *et al.*, 1991), carboxylic acids (Miller and Smith,

1989), vasopressin (Lelawongs *et al.*, 1989), and an LHRH analog (Heit *et al.*, 1994). These results show decreased u_{pep} with increased molecular weight which is nearly identical with small-molecule results obtained for anion and cation transport through human skin (Yoshida and Roberts, 1992), and anion transport through hairless mouse skin (Green *et al.*, 1991). While those results were obtained in slightly different buffer systems than employed here, all were isotonic and neutral pH, minimizing differences in the $\Sigma c_i \cdot u_i \cdot Z_i$ term. In fact, some of the data scatter may well reflect the small differences in buffer composition. The close agreement among these results may reflect the fact that iontophoretic transport through human (Yoshida and Roberts, 1992) and mouse (Green *et al.*, 1991) skin is likely to occur through existing pores of similar size. Regardless, these results show that the ionic mobility of peptides through skin decreases dramatically with increased molecular weight.

In contrast with the results obtained with small peptides, experimental results obtained with insulin showed no transport enhancement following a single pulse. Interestingly, insulin delivery through skin was enhanced by the application of 12 equally spaced pulses (1 pulse/5 min) during 60 min of constant current application, showing that a multipulse protocol can augment the effect. Multiple pulses, therefore, appear to create more and/or larger pathways which allow the passage of more and/or larger molecules through the skin. These results are consistent with a model whereby electrical pulsation creates transport pathways through the skin whose size, number, and duration are a function of the pulse frequency.

Passive peptide flux values measured following the cessation of electrical treatment show that LHRH (Figs. 3 and 5) transport decreased, reaching a constant, pretreatment value after 12–24 hr. Similar results were obtained with all peptides evaluated. It is important to note that the flux measured during electrotreatment was averaged over the entire treatment period, suggesting that significantly higher flux was achieved immediately after pulsing. Regardless of the magnitude of the initial flux, however, the high-permeability state induced in human skin by electrical pulsation reversed within 12–24 hr. Similar results have been reported for human skin by Prausnitz *et al.* (1993). Results obtained with other membranes, however, show that the reversal of an electro-induced high-permeability state occurs on a much shorter time scale (Neumann *et al.*, 1992). While this difference could be associated with longer-lived "electropores" in skin than those found in cell membranes, it is more likely that the results reflect LHRH binding to skin during iontophoresis. The subsequent desorption of peptide may then take many hours. Skin has a demonstrated reservoir capacity due to iontophoretic delivery (Wearley and Chien, 1990; Heit *et al.*, 1994), consistent with this hypothesis. The rate of flux relaxation is not substan-

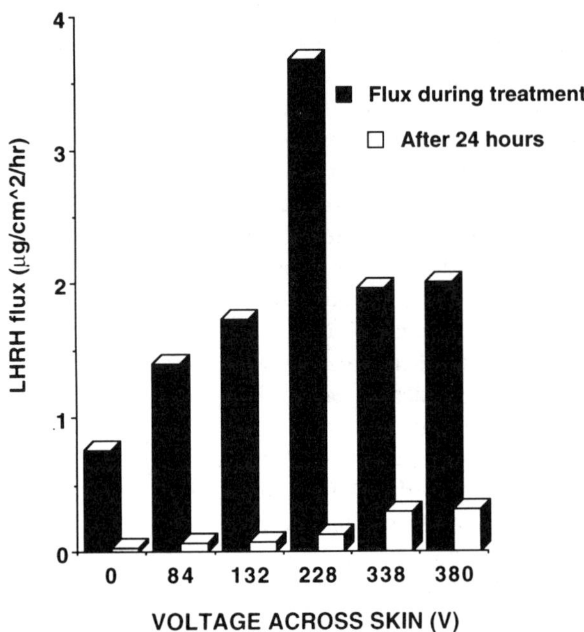

Figure 8. The effect of pulse voltage on the iontophoretic delivery of LHRH through human skin. For each pulse voltage, a single pulse (6–10 msec) was applied, followed by 60 min of iontophoresis (0.39 mA/cm^2).

tially different between pulsed and nonpulsed experiments, again suggesting that pore resealing is not rate-limiting. Finally, a mechanism-based analysis of skin electroporation data suggests that the actual lipid resealing time is on the order of 10 min (Chizmadzhev *et al.*, 1995).

Peptide flux through human skin was also measured at constant current as a function of the pulse voltage from about 250 to 1000 V. Results obtained with LHRH (Fig. 8) show that flux increased up to about 228 V but decreased somewhat at higher potential. Results obtained with calcitonin (data not shown) showed a similar trend, with maximal flux at 250–500 V but somewhat lower flux at higher voltage. Irreversible effects, however, as measured by the peptide flux 24 hr following pulsation, increased steadily with applied voltage. Similar results were obtained by Prausnitz *et al.* (1993), who measured calcein transport through skin using a multiple-pulse protocol. Their results showed that calcein flux increased by about 10^4 as the pulse voltage was increased up to about 300 V, and flux returned to pretreatment values 24 hr later. As the pulse voltage was increased to near 1000 V, however, the calcein flux increased only slightly, and the effect became progressively less reversible. Thus, there appears to be an optimal pulsation voltage which results in enhanced skin delivery yet minimizes irreversible effects.

The results presented here show that electrical pulsation causes reversible enhancement of peptide transport through human skin *in vitro*. In all cases, only intact peptide was measured. Moreover, the enhanced flux cannot be accounted for by increased current, and it is most likely due to increased ionic mobility of the peptide within the skin. Thus, electrical pulsation reversibly alters the ion-transport properties of skin. The rate-limiting barrier to skin transport of ionized compounds is the lipids of the stratum corneum. Alteration of stratum corneum lipids results in a significant increase in skin transport (Golden *et al.*, 1987; Potts and Francoeur, 1990). For example, heating the skin to temperatures just above the stratum corneum lipid phase-transition temperature results in a 100-fold increase in sodium-ion conductivity. The sodium-ion conductivity returns to pretreatment values when the skin is cooled (Oh *et al.*, 1993). Similarly, alteration of stratum corneum lipid structure with chemical perturbants such as oleic acid increases ion conductivity (Potts *et al.*, 1992). The application of a transient electrical pulse to other lipid-based membranes creates a high-permeability state associated with the reversible formation of pores within the membrane (electroporation). Thus, it seems likely that electrical pulsation of human skin results in the formation of transient pores within stratum corneum lipids.

2.2. Isolated Perfused Porcine Skin Flap

In order to assess the efficacy of electroporation in a model system that closely resembles human clinical use, the transdermal delivery of LHRH was studied using the isolated perfused porcine skin flap (IPPSF) model. The IPPSF model is a viable and vascularized *in vitro* system that was developed to quantify percutaneous absorption and cutaneous toxicity (Riviere *et al.*, 1994; Riviere and Monteiro-Riviere, 1991). The model has been used successfully to predict human iontophoretic delivery of arbutamine (Riviere *et al.*, 1992b), to assess iontophoretic-induced skin irritation (Monteiro-Riviere, 1990), and to characterize the pathway of compound delivery across the skin (Monteiro-Riviere *et al.*, 1994). Of particular relevance to the studies described here are our studies, *in vivo* and using the IPPSF model, of the iontophoretic delivery of LHRH (Heit *et al.*, 1993, 1994). These studies have shown that LHRH is an excellent model peptide with which to assess electrically assisted transdermal delivery and that iontophoresis allows biologically effective transdermal delivery of LHRH using a small electrode area ($10\,cm^2$) and low current density (less than $0.2\,mA/cm^2$), producing minimal tissue irritation. While therapeutically relevant amounts of LHRH were delivered, the "onset time" (defined as the time after current initiation required to achieve a therapeutic dose) was often unacceptably long. As demonstrated by the *in vivo* results shown above, electroporation prior to iontophoresis can significantly decrease the "onset time" as well as increase the efficacy of transdermal delivery.

The objective of these IPPSF studies was to assess the effect of electroporation on the iontophoretic delivery of LHRH. The experimental design and Porex® electrodes have been fully described elsewhere (Heit *et al.*, 1993; 1994; Monteiro-Riviere, 1990). Briefly, a 1.0-mg/ml solution of LHRH in 10 mM 2-(*N*-morpholino)ethanesulfonic acid buffer with 154 mM NaCl was placed in 4.5-cm^2 Ag/AgCl electrodes. Direct current (positive polarity, anode in the donor) was applied at a current density of $0.4\,mA/cm^2$ for 30 min. The venous effluent of the IPPSF was collected and assayed for LHRH using a ^{125}I radioimmunoassay. An electroporative pulse of 500 V and 5-msec exponential time constant was applied immediately prior to iontophoresis. In some experiments, the electroporation/iontophoresis protocol was repeated for various intervals to assess the effect of repeated applications.

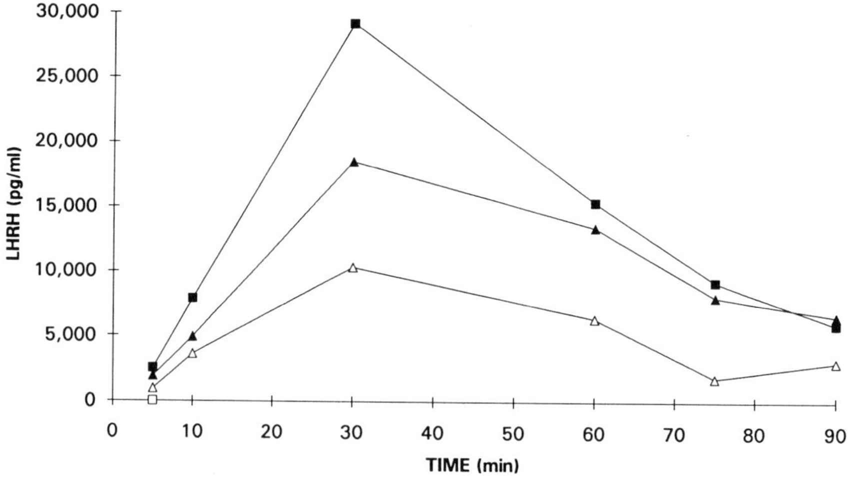

Figure 9. The mean LHRH concentration in the IPPSF perfusate resulting from one 500-V pulse followed by 30 min of iontophoresis ($0.2\,mA/cm^2$) (▲), three 500-V pulses, each followed by 10 min of iontophoresis (■), or 30 min of iontophoresis alone (△).

Figure 9 compares the mean LHRH flux seen after control iontophoresis to that seen after a single pulse followed by 30 min of iontophoresis or after three successive applications of a single pulse, each followed by 10 min of iontophoretic current. The LHRH flux of the iontophoretic control is comparable to reported values (Heit *et al.*, 1993, 1994). In contrast, the application of a single pulse to initiate the experiment resulted in a nearly twofold increase in LHRH concentration at the end of 30 min of iontophoresis. Furthermore, the application of a pulse every 10 min resulted in a nearly threefold increase in LHRH transport after 30 min of iontophoresis. The incremental current supplied by the pulse(s) was negligible compared to the iontophoretic current, again demonstrating that enhanced transport was not due to increased current. Moreover, the LHRH delivery increased with increasing number of pulses, suggesting that pulsing altered the intrinsic transport properties of skin. Note also that electroporation resulted in greater LHRH flux within 5 min. The overall effect of electroporation on LHRH flux can be seen from the results in Table II, which shows an increase in the area under the IPPSF efflux profiles due to the application of electroporative pulse(s).

Interindividual variability is a common problem in transdermal research. Since two IPPSFs are created from each individual pig, experiments were conducted with pairs of flaps, where one flap was used for the iontophoretic control while the other flap received the combined electroporation plus iontophoresis protocol. This experimental design was optimal since it substantially reduces the interindividual pig variability that confounds treatment differences seen in Table II. Figure 10 shows the results obtained for pairs of IPPSFs in which LHRH was delivered by either one

Table II

Electrically Enhanced Transdermal Delivery of LHRH in Isolated Perfused Porcine Skin Flaps: Comparison of Iontophoresis Alone vs. Electroporation plus Iontophoresis

	AUC[a] [(ng/(ml·min)]		
Treatment	Mean	Standard error	Number of flaps
Iontophoresis	557	266	6
Electroporation plus iontophoresis			
1 pulse	1054	436	5
2 pulses	1514	831	2

[a]AUC, Area under the curve.

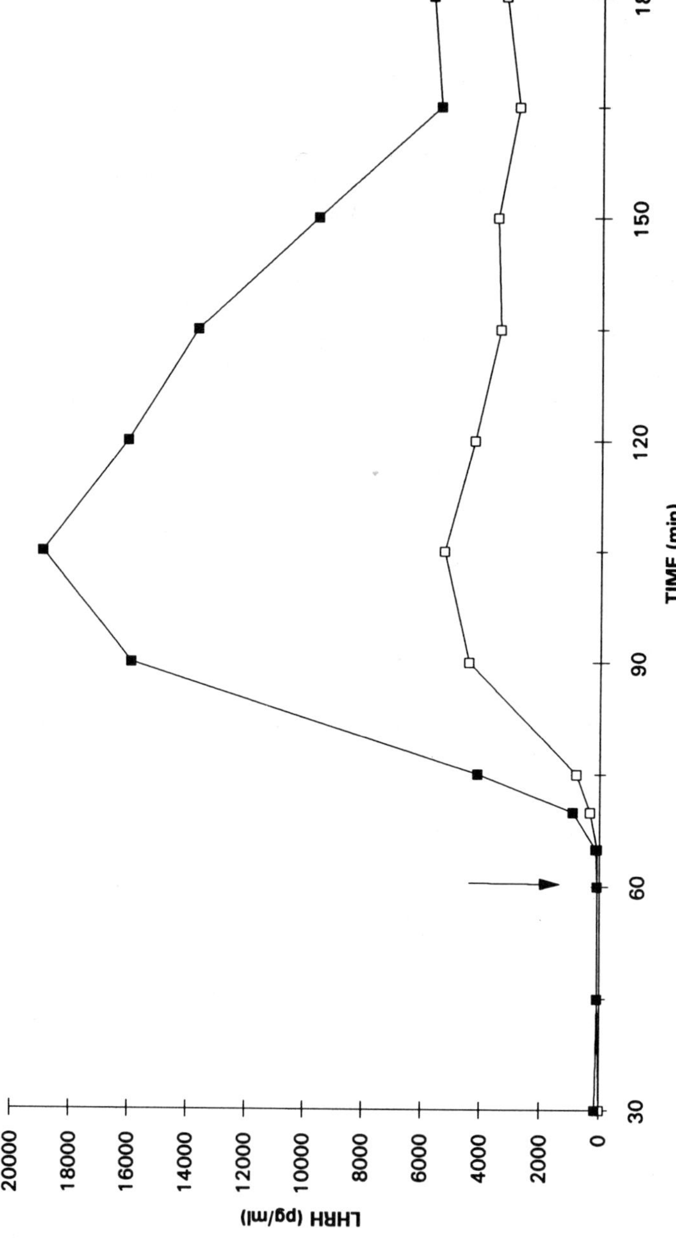

Figure 10. Representation pairs of IPPSFs in which LHRH was delivered by either one 500-V pulse followed by 30 min of iontophoresis (■) or 30 min of iontophoresis alone (□). The arrow indicates the beginning of treatment.

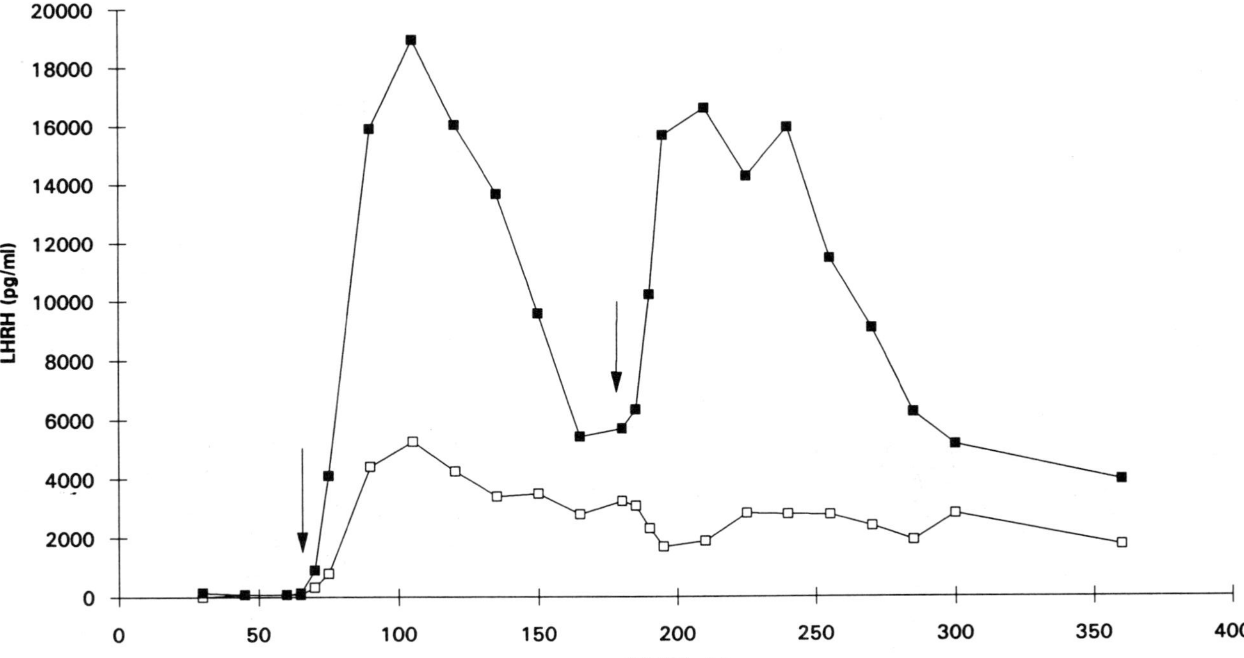

Figure 11. A representative pair of IPPSFs in which LHRH was delivered in two separate episodes consisting of either one 500-V pulse followed by 30 min of iontophoresis (■) or 30 min of iontophoresis alone (□). The arrows indicate the beginning of each treatment period.

pulse (500 V, ~5-msec exponential decay) followed by 30 min of iontophoresis or iontophoresis alone. The results show a nearly fourfold increase in maximal LHRH concentration in the perfusate following electroporation. In addition, rapid onset of LHRH delivery is seen within 10 min after the electroporative pulse.

The results obtained following the repeated application of the pulse/iontophoresis protocol are shown in Fig. 11. In these experiments the same treatment protocol as described above was applied at 60 and 180 min. These results show negligible LHRH delivery prior to electrical treatment (e.g., passive transport). During the application of iontophoretic current for 30 min, the LHRH concentrations in the venous effluent were 6 and 20 μg/ml, respectively, in the absence and the presence of a pulse. During the next 90 min the LHRH flux decreased for both treatment protocols, at which time a subsequent pulse/iontophoresis episode was initiated. The second application of iontophoretic current alone resulted in no significant change in LHRH flux. In contrast, however, the application of a pulse immediately prior to iontophoresis resulted in a second, dramatic increase in LHRH flux. These results demonstrate the ability of electroporation to repeatedly enhance LHRH transport in a pulsatile manner relative to iontophoresis. The pulsatile delivery of hormone peptides such as LHRH may be crucial to their therapeutic success. Thus, these results suggest that electroporation may provide a novel means of achieving the programmable delivery of peptides.

2.3. Skin Toxicology following Electroporation

A major factor in the clinical acceptability of electrically enhanced transdermal delivery is its effect on the skin. The pig is a widely accepted animal model for assessing electrically assisted transdermal delivery (Monteiro-Riviere, 1990; Riviere and Monteiro-Riviere, 1991). Preliminary studies using electroporation (Riviere *et al.*, 1995) conducted with pigs had two objectives. The first was to identify any unique skin changes associated with electroporation and to determine the effect of pulses on iontophoresis-induced irritation. The second objective was to define a pulse/iontophoresis protocol for drug delivery that was minimally irritating.

The initial study involved eight pigs and utilized phosphate-buffered-saline-containing glass wells with Ag/AgCl electrodes. The second experiment, conducted with six animals, used Porex® electrodes containing LHRH as described in the IPPSF experiments above. Gross and histological end points were scored as previously described (Monteiro-Riviere,

1990), and the degree of erythema, edema, and the presence of petechiae were evaluated on a scale of 0–4. Routinely processed sections were examined for any epidermal changes and the presence of previously documented skin alteration due to iontophoresis (Monteiro-Riviere *et al.*, 1994). Because changes may be localized, two sections were cut from each site (5 and 75 μm into the tissue block), so that multiple sections could be examined.

The results of the first study defined the extremes of iontophoresis and electroporation. Pulses of 0, 250, 500, and 1000 V were applied followed by constant-current anodal iontophoresis of 0, 0.2, and 2.0 mA/cm^2 for 30 min or 10 mA/cm^2 for 10 min. The results of gross evaluation observed either immediately after treatment or 4 hr after treatment are tabulated in Table III. Immediately following treatment, erythema was directly proportional to increased pulse voltage, although the response usually dissipated within 5 min. The pulse voltage alone had no impact on edema or petechiae. Erythema, edema, and petechiae all increased significantly with increased current in the absence of a pulse. The application of an electroporative pulse did not increase the iontophoretic-induced irritation at any current tested. All changes tended to decrease by 4 hr after treatment. Most importantly, note that the erythema, edema, and petechiae observed under conditions of electroporative LHRH delivery (500–1000 V, 0.2 mA/cm^2) were comparable to the results obtained with iontophoresis alone. Thus, an electroporative pulse has no adverse effects on skin irration.

Microscopic changes, consisting primarily of focal intraepidermal edema and/or vacuolization, increased with higher voltage pulses. Similar to the gross observations, the most severe changes were primarily associated with higher current densities, although this was not reproducibly seen at all sites as in the previously reported lidocaine study (Monteiro-Riviere, 1990). The random occurrence of this same lesion was observed in other LHRH studies involving only iontophoresis (Heit *et al.*, 1994).

Based upon this study and the transport results described above, a second study was designed to examine the gross and histological alterations observed at voltage and current values which bracketed the conditions used in the IPPSF transport studies. In this study, a single 0-, 500-, or 1000-V pulse was administered before a current density of either 0.4 or 0.8 mA/cm^2 was applied for 30 min. Similarly to the previous studies, erythema and edema were most severe at the 0.8-mA/cm^2 current density, regardless of pulse amplitude. Furthermore, histological evaluation showed that epidermal vacuolization was associated with all treatments and increased with increased current density. At 0.4-mA/cm^2 current density and all three pulse voltages studied, minimal alteration was noted. The results of these studies show that, at both the gross and light microscopic level, electroporation

Table III
Gross Observations of the Skin under the Active Electrode[a]

Pulse (V)	Current (mA/cm^2)	Erythema		Edema		Petechiae	
		Immediately after treatment	4 hr after treatment	Immediately after treatment	4 hr after treatment	Immediately after treatment	4 hr after treatment
0	0.0	0.00	0.00	0.00	0.00	0.00	0.00
	0.2	1.75	0.50	1.75	1.00	0.75	0.00
	2.0	1.25	0.50	1.25	1.00	0.75	0.00
	10.0	1.75	0.50	1.75	1.50	1.50	0.50
250	0.0	0.00	N/A[b]	0.00	N/A	0.00	N/A
	0.2	0.75	0.00	0.75	0.50	0.00	0.00
	2.0	1.25	0.50	1.50	1.00	0.50	0.50
	10.0	2.00	0.00	1.67	1.00	1.67	1.00
500	0.0	1.00	N/A	0.00	N/A	0.00	N/A
	0.2	1.25	0.00	1.00	1.50	0.00	0.00
	2.0	1.75	0.50	1.50	1.00	1.50	0.00
	10.0	2.30	0.00	1.33	1.00	1.33	0.00
1000	0.0	2.60	1.67	0.80	0.67	0.00	0.00
	0.2	1.25	1.00	1.25	0.50	0.00	0.00
	2.0	1.50	0.00	1.00	1.00	0.25	0.50
	10.0	2.00	1.00	1.75	1.00	1.50	0.50

[a] All iontophoretic treatment was applied for 30 min except that at 10 mA/cm^2, which was applied for 10 min. The degree of erythema, edema, and the presence of petechiae were evaluated on a scale of 0–4.
[b] N/A: Not available.

does not result in any skin change not previously seen with iontophoresis alone. The only dermal alterations seen with electroporation were mild intraepidermal vacuolization and a transient erythema.

3. CONCLUSION

The results presented here demonstrated significantly enhanced delivery of peptides through the skin following the application of an electrical pulse. Furthermore, enhanced delivery was reversible and not associated with any untoward dermal irritation. These results have clear therapeutic significance with respect to drug delivery as the vast majority of peptides and proteins cannot be successfully administered by traditional means. The therapeutic utility of this enhanced transport is obvious from the data in Fig. 6. The therapeutically acceptable limit of current for iontophoretic delivery is about $0.5\,mA/cm^2$. The iontophoretic delivery of LHRH under the experimental conditions described here is less than $0.3\,\mu g/(cm^2 \cdot hr)$ at this current. By contrast, an LHRH flux greater than $1.5\,\mu g/(cm^2 \cdot hr)$ was achieved by the application of a single pulse to initiate iontophoretic delivery. An unacceptable current of greater than $2.0\,mA/cm^2$ would be required to deliver the equivalent mass of drug iontophoretically (see Fig. 6). Similar results have been obtained with other peptides and proteins. Furthermore, multiple pulsing can further enhance delivery (Prausnitz *et al.*, 1993). Hence, electroporation allows the enhanced delivery of peptides which cannot be effectively delivered by other transdermal means. Finally, the results serve to underscore the profound similarities of electrical pulse-enhanced transport in skin to that seen in a variety of other lipid-based membranes.

REFERENCES

Belehradek, J. B., Jr., Orlowski, S., Poddevin, B., Paoletti, C., and Mir, L. M., 1991, Electrochemistry of spontaneous mammary tumors in mice, *Eur. J. Cancer* **27**:73–76.

Blank, I. H., and Scheuplein, R. J., 1969, Transport into and within the skin, *Br. J. Dermatol.* **81**:4–10.

Bommannan, D. B., Tamada, J., Leung, L., and Potts, R. O., 1994, Effect of electroporation on transdermal iontophoretic delivery of luteinizing hormone releasing hormone (LHRH) *in vitro*, *Pharm. Res.* **11**:1809–1814.

Burnette, R. R., and Marrero, D., 1986, Comparison between the iontophoretic and passive transport of thyrotropin releasing hormone across excised nude mouse skin, *J. Pharm. Sci.* **75**:738–743.

Chang, D. C., and Reese, T. C., 1990, Changes in membrane structure induced by electroporation as revealed by rapid-freezing electron microscopy, *Biophys. J.* **58:**1–12.

Chernomordik, L. V., and Chizmadzhev, Y. A., 1989, in: *Electroporation and Electrofusion in Cell Biology* (E. Neumann, A. E. Sowers, and C. A. Jordan, eds.), Plenum Press, New York, pp. 83–95.

Chernomordik, L. V., Sukharev, S. I., Abidor, I. G., and Chizmadzhev, Y. A., 1983, Breakdown of lipid bilayer membranes in an electric field, *Biochim. Biophys. Acta* **736:**203–213.

Chernomordik, L. V., Sukharev, S. I., Popov, S. V., Pastushenko, V. F., Sokiro, A. G., Abidor, I. G., and Chizmadzhev, Y. A., 1987, The electrical breakdown of cell and lipid membranes: The similarity of phenomenologies, *Biochim. Biophys. Acta* **902:**360–373.

Chien, Y. W., Siddiqui, O., Shi, W.-M., Lelawongs, P., and Liu, J.-C., 1989, Direct current iontophoretic transdermal delivery of peptide and protein drugs, *J. Pharm. Sci.* **78:**376–383.

Chizmadzhev, Y. A., Zarnytsin, V. G., Weaver, J. C., and Potts, R. O. 1995, Mechanism of electroinduced ionic species transport through a multilamellar lipid system, *Biophys. J.*, **68:**749–765.

Elias, P. M., 1983, Epidermal lipids, barrier function, and desquamation, *J. Invest. Dermatol.* **80:**44s–49s.

Elias, P. M., 1987, Plastic wrap revisited, *Arch. Dermatol.* **123:**1405–1406.

Elias, P. M., 1991, Epidermal barrier function: Intercellular lamellar lipid structures, origin, composition and metabolism, *J. Controlled Release* **15:**199–208.

Golden, G. M., Guzek, D. B., Kennedy, A. H., McKie, J. E., and Potts, R. O., 1987, Stratum corneum lipid phase transitions and water barrier properties, *Biochemistry* **26:**2382–2388.

Green, P. G., Hinz, R. S., Cullander, C., Yamane, G. M., and Guy, R. H., 1991, Iontophoretic delivery of amino acids and amino acid derivatives across skin *in vitro*, *Pharm. Res.* **9:**1113–1120.

Heit, M., Williams, P., Jayes, P., Chang, S., and Riviere, J., 1993, Transdermal iontophoretic peptide delivery: *In vitro* and *in vivo* studies with luteinizing hormone release hormone, *J. Pharm. Sci.* **82:**240–243.

Heit, M. C., Monteiro-Riviere, N. A., Jayes, F. L., and Riviere, J. E., 1994, Transdermal iontophoretic delivery of luteinizing hormone releasing hormone (LHRH): Effect of repeated administration, *Pharm. Res.* **11:**1000–1003.

Kinosita, K., Hibino, M., Itoh, H., Shigemori, M., Hirano, K., Kirino, Y., and Hayakawa, T., 1992, in: *Guide to Electroporation and Electrofusion* (D. C. Chang, B. M. Chassy, J. A. Saunders, and A. E. Sowers, eds.), Academic Press, New York, pp. 29–46.

Lelawongs, P., Liu, J.-C., Siddiqui, O., Chien, Y.-C., 1989, Transdermal iontophoretic delivery of arginine–vasopressin: Physicochemical considerations, *Int. J. Pharm.* **56:**13–22.

Lu, M. F., Lee, D., and Rao, G. S., 1992, Percutaneous absorption enhancement of leuprolide, *Pharm. Res.* **9:**1575–1579.

Meyer, B. R., Kreis, W., Eschbach, J., O'Mara, V., Rosen, S., and Sibalis, D., 1988, Successful transdermal administration of therapeutic doses of a polypeptide to normal human volunteers, *Clin. Pharmacol. Ther.* **44:**607–612.

Meyer, R., Katzeff, H., Eschbach, J., Trimmer, J., Zacharias, S., Rosen, S., and Sibalis, D., 1989, Transdermal delivery of human insulin to albino rabbits using electrical current, *Am. J. Med. Sci.* **297:**321–325.

Miller, L. L., and Smith, G. A., 1989, Iontophoretic transport of acetate and carboxylate ions through hairless mouse skin. A cation exchange membrane model, *Int. J. Pharm.* **49:**15–22.

Miller, L., Kolaskie, C., Smith, G., and Rivier, J., 1990, Transdermal iontophoresis of gonadotropin releasing hormone (LHRH) and two analogues, *J. Pharm. Sci.* **79:**490–493.

Mir, L. M., Orlowski, S., Belehradek, J. B., Jr., and Paoletti, C., 1991, Electrochemotherapy potentiation of antitumor effect of bleomycin by local electric pulses, *Eur. J. Cancer* **27**:68–72.

Monteiro-Riviere, N. A., 1990, Altered epidermal morphology secondary to lidocaine iontophoresis: *In vitro* and *in vivo* studies in porcine skin, *Fundam. Appl. Toxicol.* **15**:174–185.

Monteiro-Riviere, N. A., Inman, A. O., and Riviere, J. E., 1994, Identification of the pathway of transdermal iontophoretic drug delivery: Light and ultrastructural studies using mercuric chloride in pigs, *Pharm. Res.* **11**:251–256.

Neumann, E., 1992, Membrane electroporation and direct gene transfer, *Bioelectrochem. Bioenerg.* **28**:247–267.

Neumann, E., Sprafke, A., Boldt, E., and Wolf, H., 1992, in: *Guide to Electroporation and Electrofusion* (D. C. Chang, B. M. Chassy, J. A. Saunders, and A. E. Sowers, eds.), Academic Press, New York, pp. 77–90.

Oh, S. Y., Leung, L., Bommannan, D. B., Guy, R. H., and Potts, R. O., 1993, Effect of current, ionic strength and temperature on the electrical properties of skin, *J. Controlled Release* **27**:115–125.

Potts, R. O., and Francoeur, M. L., 1990, Lipid biophysics of water loss through the skin, *Proc. Natl. Acad. Sci. USA* **87**:3871–3873.

Potts, R. O., and Guy, R. H., 1992, Predicting skin permeability, *Pharm. Res.* **9**:663–669.

Potts, R. O., Guy, R. H., and Francoeur, M. L., 1992, Routes of ionic permeability through mammalian skin, *Solid State Ionics* **53**:165–169.

Prausnitz, M. R., Bose, V. G., Langer, R., and Weaver, J. C., 1993, Electroporation of mammalian skin: A mechanism to enhance transdermal drug delivery, *Proc. Natl. Acad. Sci. USA* **90**:10504–10508.

Riviere, J. E., and Monteiro-Riviere, N. A., 1991, The isolated perfused porcine skin flap as an *in vitro* model for percutaneous toxicology, *Crit. Rev. Toxicol.* **21**:329–344.

Riviere, J. E., Bowman, K. F., Monteiro-Riviere, N. A., Carver, M. P., and Dix, L. P., 1994. The isolated perfused porcine skin flat (IPPSF). I. A novel *in vitro* model for percutaneous absorption and cutaneous toxicology studies, *Fundam. Appl. Toxicol.* **7**:444–453.

Riviere, J. E., Monteiro-Riviere, N. A., and Inman, A. O., 1992a, Determination of lidocaine concentrations in skin after transdermal iontophoresis: Effects of vasoactive drugs, *Pharm. Res.* **9**:211–214.

Riviere, J. E. Williams, P. L., Hilman, R., and Mishky, L., 1992b, Quantitative prediction of transdermal iontophoretic delivery of arbutamine in humans using the *in vitro* isolated perfused porcine skin flap (IPPSF), *J. Pharm. Sci.* **81**:504–507.

Riviere, J. E., Monteiro-Riviere, N. A., Rogers, R. A., Bommannan, D. B., Tamada, J. A., and Potts, R. O., 1995, Pulsatile transdermal delivery of LHRH using electroporation: Drug delivery and skin toxicology, *J. Controlled Release* **36**:221–228.

Scheuplein, R. J., 1965, Mechanism of percutaneous adsorption. I. Routes of penetration and the influence of solubility, *J. Invest. Dermatol.* **45**:334–346.

Scheuplein, R. J., 1978, Permeability of the skin: A review of major concepts, *Curr. Probl. Dermatol.* **7**:172–186.

Siddiqui, O., Sun, Y., Liu, J.-C., and Chien, Y. W., 1987, Facilitated transdermal transport of insulin, *J. Pharm. Sci.* **76**:341–345.

Singh, J., and Roberts, M., 1989, Transdermal delivery of drugs by iontophoresis: A review, *Drug Des. Del.* **4**:1–12.

Srinivasan, V., Higuchi, W. I., Sims, S. M., Ghanem, A. H., and Behl, C. R., 1989, Transdermal iontophoretic drug delivery: Mechanistic analysis and application to polypeptide delivery, *J. Pharm. Sci.* **78**:370–375.

Srinivasan, V., Su, M.-H., Higuchi, W. I., and Behl, C. R., 1990, Iontophoresis of polypeptides: Effect of ethanol pretreatment of human skin, *J. Pharm. Sci.* **79**:588–591.

Titomirov, A. V., Sukharev, S., and Kistanova, E., 1991, *In vivo* electroporation and stable transformation of skin cells of newborn mice by plasmind DNA, *Biochim. Biophys. Acta* **1088**:131–134.

Tsong, T. Y., 1990, On electroporation of cell membranes and some related phenomena, *Bioelectrochem. Bioenerg.* **24**:271–295.

Wearley, L. L., and Chien, Y. W., 1990, Iontophoretic transdermal permeation of verapamil (III): Effect of binding and concentration gradient on reversibility of skin permeation rate, *Int. J. Pharm.* **59**:87–94.

Yoshida, N. H., and Roberts, M. S., 1992, Structure–transport relationships in transdermal iontophoresis, *Adv. Drug Delivery Rev.* **9**:239–264.

Chapter 9

Protein Delivery with Infusion Pumps

Ulrike Bremer, C. Russell Horres, and Michael L. Francoeur

1. INTRODUCTION

1.1. Rationale for Infusion Therapy

Endogenous proteins or peptides are intimately involved in numerous physiological processes, ranging from the control of cell growth to the defense of the body from infection by viruses or bacteria. It was on the basis of this wide range of biological activities that the therapeutic potential of these agents was originally recognized. Despite this awareness, efforts to utilize existing biomolecules or to design new drugs based on peptides or proteins has been historically slow owing to a lack of analytical and synthetic methodologies. However, with the advent of new molecular biological techniques such as cloning, the commercial production of many such molecules can be fairly routine, which led directly to the emergence of the biotechnology industry in the 1980s.

The medicinal use of proteins and peptides, nevertheless, has imposed many challenging new problems on the pharmaceutical industry. In general, these molecules are difficult to stabilize as they often undergo a variety of physical and chemical transformations including precipitation, aggregation,

---

Ulrike Bremer • Pharmetrix, Inc., Menlo Park, California 94025. *C. Russell Horres* • CyberRx, Inc., San Diego, California 92130. *Michael L. Francoeur* • Pharmetrix, Inc., Menlo Park, California 94025; *current address:* De Novo, Inc., Menlo Park, California 94025.

Protein Delivery: Physical Systems, Sanders and Hendren, eds., Plenum Press, New York, 1997.

oxidation, racemization, and deamidation. Further, they are poorly absorbed after oral administration (Kennedy, 1991) as a consequence of enzymatic and acid-catalyzed degradation as well as by extensive first-pass liver metabolism (Chien, 1992). Thus, biotechnology products have been produced mainly in parenteral dosage forms which permit repetitive and acute dosing by injection. Multiple daily injections with a needle, however, are not attractive to patients or physicians and decrease effectiveness through loss of compliance as well as an increased risk of infection. While other noninvasive methods of delivery such as iontophoresis and nasal or buccal administration are under investigation, the continuous infusion of biotechnology products by pumps offers many advantages, particularly if the device is small and portable.

The first and foremost advantage is the opportunity to avoid repeated injections and to achieve a constant or controlled rate of delivery of the drug. Banerjee *et al.* (1991) have pointed out that many proteins or peptides are rapidly excreted or degraded by the body, exhibiting, in effect, an extremely short biological half-life. Many sophisticated electronic infusion devices allow complex patterns of dosing which are customized to patients' needs and would not require repeated injections in order to maintain a constant level of protein in the blood. The second major advantage of infusion therapy over repeated injections is cost-effectiveness. It is quite simply, less time-consuming and less costly for a clinician or nurse to administer a single dose instead of multiple injections given over a period of time. While most of the more than 20 approved biotechnology products are currently given on an intermittent basis, health-care providers are beginning to appreciate the obvious advantages and value of continuous administration, from both a cost and an effectiveness point of view.

Most proteins that are currently used or under development must be given parenterally because of poor oral bioavailability and in order to ensure a rapid onset of action, with the main routes being intravenous, intramuscular, or subcutaneous. Intravenous administration results in the fastest, intramuscular injection in a more sustained, and subcutaneous injection in the slowest onset of action (Kompella *et al.*, 1991). Table I shows pharmacokinetic parameters of Neupogen®. Actimmune®, and Proleukin® and the manufacturer's recommended administration (Physicians' Desk Reference, 1993) for these biotechnology drugs, indicating that a sustained drug release is desired but made difficult by relatively short half-lives or high clearance rates. By administrating the proteins subcutaneously, absorption into the systemic circulation is slowed, and thus an effective plasma level can be sustained for a prolonged period of time. If, however, the drug needs to be maintained over hours or even days at the exact same concentration, constant-rate infusion would be the preferred mode of delivery. This type of

Table I
Pharmacokinetic Parameters and Recommended Administration of Neupogen®, Actimmune®, and Proleukin®[a]

Drug	Administration[b]	Dose (μg/kg)	Plasma level (μg/ml)	Volume of distribution	Elimination half-life	Clearance	Recommended administration
Neupogen® (Filgrastim; granulocyte colony-stimulating factor)	Continuous IV infusion	20	48 (mean)				Single daily SC injection or continuous SC or IV infusion
	SC injection	3.5	4 (max.)	150 ml/kg	3.5 hr	0.5–0.7 ml/(min · kg)	
Actimmune® (Interferon gamma-1b)	IV injection	100			38 min	1.4 l/min	Three times weekly SC injection
	IM injection	100	1.5 (4-hr max.)		2.9 hr		
	SC injection	100	0.6 (7-hr max.)		5.9 hr		
Proleukin® (Aldesleukin; interleukin-2 product)	5 min IV infusion			High	85 min	268 ml/min	15-min *N* infusion every 8 hr

[a]Source: Physicians' Desk Reference, 1993.
[b]Abbreviations: IV, Intravenous; SC, subcutaneous; IM, intramuscular.

dosing can be very tightly controlled when infusion pumps are used, which is important for proteins having a high potential for toxicity or side effects and a small therapeutic window. Even if the drugs have a high therapeutic index, almost all protein/biotech drugs are expensive, making the delivery efficiency and accuracy achieved with infusion pumps quite desirable. Further, some proteins are most efficacious when administered in a pulse mode or a circadian pattern, which can be done with several of today's programmable infusion devices. The pituitary gonadotropins, for example, require a pulsatile delivery mode to achieve certain therapeutic responses. When infused in pulses every 90 to 120 minutes, pituitary gonadotropins induce ovulation, whereas they are given continuously, the resulting constant plasma levels will inhibit ovulation.

1.2. Limitations of Infusion Therapy

The primary disadvantage of infusion therapy is its impact on patient lifestyle. The physical size of the devices and the precautions associated with parenteral therapy are both factors in limiting patient lifestyle. Portable, electromechanical devices must be carefully handled to avoid breakage and are typically not waterproof. Patients will tolerate these restrictions on an active lifestyle for a few months, but long-term use of the devices can be problematic. In addition to the precautions associated with using the devices, there are logistical issues of battery charging or replacement and dedicated pump accessories, which contribute to the overall complexity of infusion therapy. Many of these problems can be improved through the availability of small, comfortable-to-wear, disposable systems.

The route of administration for infusion therapy is also limiting. The least restrictive mode is subcutaneous infusion, pioneered in the insulin delivery studies by Champion *et al.* (1980) and Pickup *et al.* (1980). Small-gauge (30 G), short needles ($\frac{1}{2}$ inch) with soft plastic wings for handling and taping to the body have proved to be relatively comfortable to the continuously infused diabetic patient. Attempts to leave the needles in place for longer than 72 hours, however, have been met with numerous difficulties. Although some patients were able to leave the needle in place for two weeks without complications, statistics proved that changing the site every two to three days minimized site complications. Site preparation and care are also important in preventing potentially serious infections.

Continuous intravenous (i.v.) delivery presents another level of complications. Infusion into peripheral veins can become problematic after only a few days and typically requires moving the needle to a new location. Unlike the self-administered subcutaneous infusion of the diabetic, self-cannulation is difficult, and nursing intervention is usually required. New, soft, peripheral cannulas are now available, and so called long-line catheters that can be threaded into larger veins have been developed. If the patient is likely to require more than a few weeks of i.v. therapy, an indwelling central venous line is the most practical means of accessing the circulation. The effectiveness of this route as a long-term access has been one of the most dramatic advances in medicine of this century. Patients have been supported with parenteral nutrition through such catheters for up to 15 years, with catheter changes as infrequent as every two to three years, and in one case, a single venous catheter was used for 10 years (Heimburger, 1992). Thus, a safe and effective route to continuous parenteral drug therapy is available. Central venous therapy for an intermediate term can be conducted with a percutaneous catheter; for a longer term therapy, drug ports are implanted beneath the skin where they can be readily accessed with needles. Flushing and care of the central venous lines are important to the long-term success. This type of therapy places additional demands on the patient and caregiver.

Implantable pumps with drug reservoirs have had some success as a means of long-term drug therapy. The difficulties of such systems are noteworthy, including drug instability, the necessity for surgical intervention, problems of system reliability, and the lack of control after the device is implanted. These factors combine to impact significantly on the cost of such systems, which can run to tens of thousands of dollars. With engineering and pharmaceutical advances, implantable infusion systems may become more useful for chronic therapies. In spite of these limitations, infusion therapy remains an important component of drug delivery. Especially with biotech drugs that are aimed at very serious diseases like cancer or AIDS, the medical rationale will outweigh the concerns as long as there are no alternative ways of administration with the same therapeutic effectiveness.

2. HISTORY OF INFUSION THERAPY

From the early approaches to venous cannulation in the 18th century to the introduction of gravity infusion of physiological solutions in the early 1900s, the therapeutic rationale for infusion had its beginnings. The introduction of parenteral pharmaceutical agents with either direct or indirect effects on physiologic systems provided further need for regulation of the rate and amount of infusate. The first practical means of flow control, which is still widely used today in gravity systems, is the tubing clamp. Modern variants of this simple flow regulator are known as roller clamps because the mechanism used to progressively reduce the diameter of the intravenous administration tubing is a small roller. Since flow in a tube cannot be visualized, air chambers, known as drip chambers, were added to infusion sets to allow the flow to be viewed as a series of drops. Nurses then found themselves involved with frequently monitoring the number of drops per minute as a means of quantifying delivery rates. The inherent creep in the clamped tubing combines with changing fluid levels and physiological downstream pressures and resistances to demand constant adjusting of the clamp in rate-sensitive infusions. It should be noted that, in the early days of parenteral administration, it was common to infuse directly into the subcutaneous space, from which fluids would be absorbed into the bloodstream. This process, known as clisis, could absorb 120–300 ml per day without harm. This delivery mechanism is rarely used today because it is not suitable for poorly absorbed agents or for agents that may cause local tissue damage.

In the mid 1960s, devices were developed to measure and regulate drop rate in i.v. infusions. These devices, termed i.v. controllers, enjoyed significant popularity for the next 30 years, whereupon they were gradually displaced by true pumps, capable of producing flow without the assistance of gravity, which were introduced in the early 1970s. Large-volume pumps and controllers dominate hospital-based infusion therapy. They are capable of delivering in excess of 50 ml per day and often deliver a liter or more of fluid. Proteins are delivered by such devices, but they are diluted in a carrier vehicle. As the practice of i.v. therapy expanded, more interest was placed on highly accurate and precise systems for infusing directly from syringes. Large, noisy, multi-gear ratio syringe drivers were common in the research laboratories of the 1950s and began to increase in sophistication and find more clinical applications in the 1970s. In Europe, syringe pumps are immensely popular because of their economy, accuracy, and ability to deliver therapeutic agents without adding unnecessary and potentially stressing diluting fluid.

Microelectronics technology has had a major impact on infusion devices in the last two decades, resulting in smaller and programmable delivery devices. A major driver of programmability was a flurry of activity around the concept of continuous subcutaneous insulin delivery in the early 1980s. Believing that tight control of blood glucose was the means to reduce long-term complications, leading diabetologists were experimenting with delivering insulin at a patient-specific basal rate supplemented with patient-controlled pre-meal boluses. Additional research, focused on the chronobiology of insulin needs, sought to prevent the early-morning rise in glucose levels by adjusting the basal rate for certain time periods. With conventional delivery systems, such sophistication in therapy would require a nurse around the clock to manually input changes in the delivery rate. The extreme potency of insulin makes it ideally suited for delivery by the accurate and precise syringe pumps. Several manufacturers responded by developing portable, battery-powered versions which were about the size of a videocassette. Within 10 years, programmable versions, the size of an audiocassette, were available which could adjust basal infusion rates up to 50 times per day. Unfortunately, the results of improved outcomes were long in coming, and most companies abandoned their marketing efforts by the early 1990s.

An important by-product of the race to miniaturize infusion devices has been the development of truly portable and wearable systems for infusions of up to four liters per day. Such devices have been enabling technology for the home-health-care market and are employed today in a wide variety of infusions treating resistant infections or providing parenteral nutrition, oncotherapy, or pain control. Such devices are also having an impact upon the hospital-based infusion practice as they have made possible the miniaturization and combination of two to four pumps in a space no larger than that occupied by conventional pump of the 1980s. The most recent advance in infusion therapy has been the concept of a disposable pump, to be used once and discarded. This concept is particularly attractive in the outpatient and home therapy arenas, where the logistics, costs, and complexity of a programmable device may not be required. Baxter Travenol has been a pioneer in this market, and now several companies are offering these convenient cost-effective systems.

3. STATIONARY AND PORTABLE INFUSION PUMPS

There are various ways to classify today's infusion devices based upon their functionality, driving mechanism, size, or ability to be disconnected periodically from the infusion apparatus. In order to simplify the classification, we have chosen to differentiate only between stationary and portable devices. Stationary devices are used for patients who remain in bed during the treatment, either at home or in a hospital environment. In general, these devices are large, programmable systems that remain near the bedside and require close supervision by medical personnel. Portable infusion pumps, on the other hand, are used by patients who can freely move during their therapy. They include implantable and externally worn pumps, which range from somewhat bulky forms to small and concealable devices as shown in Fig. 1.

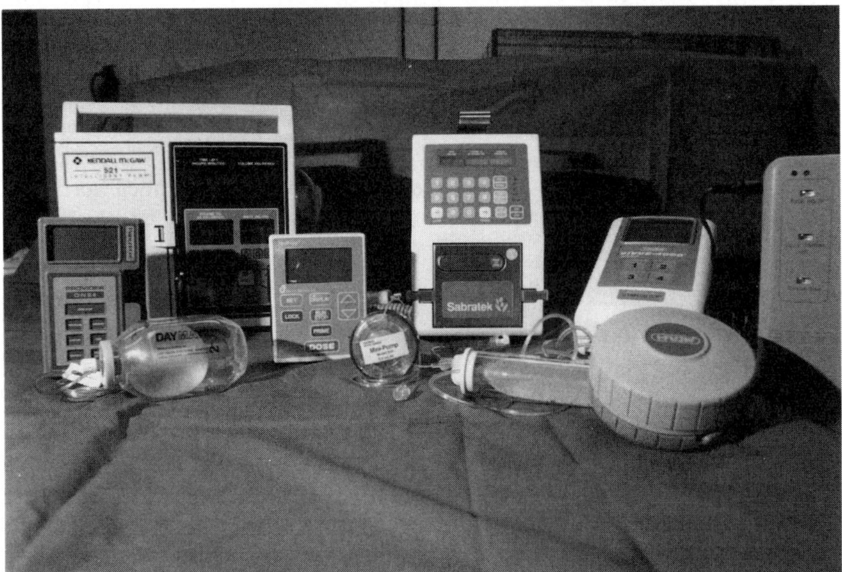

Figure 1. Portable infusion pumps. Background (left to right): KENDALL McGRAW 521 INTELLIGENT PUMP, Sabratek, I-Flow Vivus 4000, EZ-Flow 80™. Foreground (left to right): Pancretec PROVIDER ONE, DAYMATE™ Infusion Device 2ml/hr, MedMate 1100, Mini-Pump, Model 524, Multiday Infusor 0.5 ml/hr, I-FLOW Model PARAGON. (Picture on file, Pharmetrix Corp., Menlo Park, California; pumps supplied by Bay Area I. V. Therapy Inc., Los Altos, California.)

3.1. Stationary Infusion Pumps

Today's stationary infusion pumps are highly sophisticated electromechanical devices that allow the user to program the desired flow rate and infusion volume for one or more i.v. lines. They are equipped with several alarm features to indicate potentially hazardous conditions such as changes in the flow rate, air in the line, or occlusion of the catheter. The pump mechanism is either a peristaltic compression and release of the administration line or motor-controlled piston movement. The price of these devices ranges from $2,000 to $8,000, which excludes the disposable i.v. sets. The majority of these pumps are used in a hospital setting, where they are operated and supervised during operation by the medical staff. Health-care providers also lease stationary infusion pumps to chronically ill patients for home use. A flow-rate accuracy of 3–10% makes these pumps highly desirable for administration of proteins with a narrow therapeutic index.

Pumps must have the ability to actually deliver fluids against gravity and most physiological pressures. The installed base of these devices is very large in the United States and is gaining rapidly throughout the world. There are approximately 200,000 electromechanical devices in use in the United States, and the market is dominated by four large suppliers, Baxter, Abbot, IVAC, and IMED. As an industry, infusion device manufacturing is relatively young, having begun about 25 years ago. Large-volume pumps and i.v. controllers typically can be thought of as 10- to 15-pound boxes, approximately 8 to 10 cubic inches in volume. They can be moved, and most have internal batteries that allow limited excursions away from AC power. The recent trend has been toward smaller, multiline units. Syringe pumps are the oldest devices in the category and have benefited significantly from advances in miniaturization of electronics. Recent trends in patient-controlled analgesia and newer i.v. anesthetic agents have produced a resurgence of syringe pumps in the hospital market.

The use of large-volume pumps and the other i.v. controllers for the delivery of proteins presents unique challenges. Since these systems work best when delivering several hundred milliliters of fluid per day, dilution of the active agent is often an issue. Considerable concern was raised in the late 1970s regarding the use of these systems to deliver insulin. It was discovered that significant adsorption of the polypeptide on to glass containers was reducing the effective dose to the patient. Plastic bags are gaining favor as the primary solution reservoir. These range in volume from 50 to 1000 milliliters. They also need to be tested for their compatibility with biopeptides and proteins. Abbott has a system that allows delivery directly from drug vials, and several adapters exist for venting syringes to allow them to be used as a reservoir. It is interesting to note that even though pumps can deliver against gravity, they are commonly located above the patient, with the fluid reservoirs suspended above the pump.

Because of the small volume of the drug reservoir, the low residual volume in the tubing, and the increased accuracy inherent in the device design, syringe pumps are very suitable for protein delivery. Most syringe pumps accommodate a range of syringe sizes and are capable of accurate or precise delivery at low flow rates. The one major drawback appears to be the smoothness of travel of the syringe plunger; at extremely slow rates, a stick-slip phenomenon can occur, which can degrade precision. Accuracy in a syringe pump is primarily determined by the syringe tolerances. A relatively recent addition to the syringe pump market is a class of devices that deliver at constant rate over a present time interval. These simple-to-use, battery-powered pumps have found some favor in antibiotic delivery applications where prefilled syringes serve as the drug container. It should be recognized that unless the syringe pump is directly hooked to a dedicated line to the patient, it is possible to deliver the medication in a retrograde manner into other open drug lines sharing the same catheter access. Check valves are frequently employed to prevent this inadvertent misadministration. Baxter Travenol, C. R. Bard, and Becton Dickinson are major U.S. suppliers of hospital-based syringe infusion devices. This mode of drug delivery is much more prevalent in Europe than in the United States, reflecting the European concern for cost-effective and accurate means of delivering drugs. Welmed (U.K.), Vial (France), and B. Braun (Germany) are major European suppliers of hospital syringe pumps.

Stationary infusion devices can often be programmed or controlled by external computers to achieve sophisticated infusion patterns. An example of drug plasma levels after pulsatile delivery using a stationary device is shown in Fig. 2. These results were obtained in a study by Paolisso *et al.* (1989) comparing the efficacy of pulsatile and continuous infusion of insulin and glucagon in normal humans. The Biostator (Life Science Instruments, Miles Laboratories) used in this study is a closed-loop system. One intravenous line is used for continuous blood withdrawal and determination of the blood glucose levels. The second line is used for insulin (or glucose) infusion at a rate based on the measured glucose levels. The study concluded

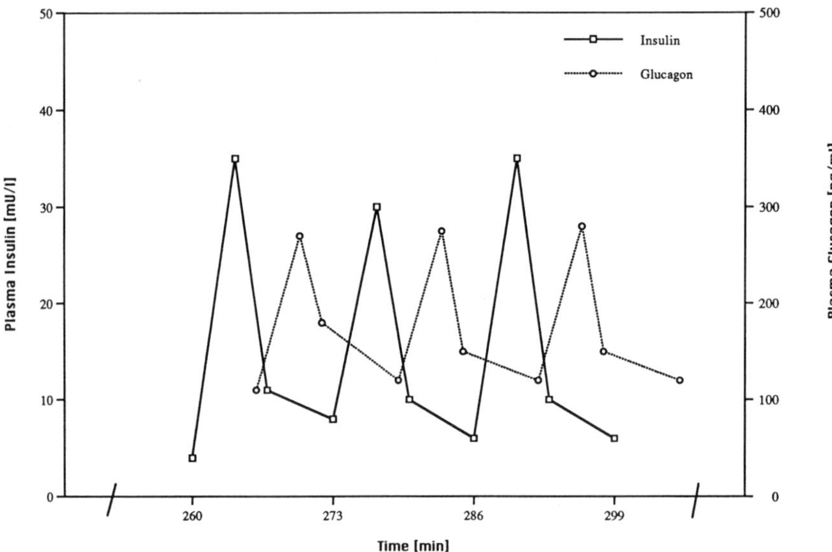

Figure 2. Plasma levels after pulsatile infusion of insulin and glucagon with Biostator in 13-min intervals with a 6-min phase displacement between the two hormones ($n = 6$ each).

that the stimulatory effect of glucagon on endogenous glucose production dominates over the inhibitory effect of insulin when both hormones are delivered in a pulsatile and out-of-phase mode.

3.2. Implantable Infusion Pumps

The inconvenience of being connected to a stationary infusion device for prolonged periods of time has been long recognized and a significant amount of private and government support has been directed toward the development of small implantable devices capable of delivering 10–50 milliliters of infusate over days to months. The first commercial implantable pump was designed and manufactured by Infusaid Corporation. This device is a two chambered pump that is driven by the vapor pressure of a liquid contained within an expandable bellows acting upon the drug compartment. The infusion rate is controlled by selecting the resistance of the drug outlet. Once implanted, the near constancy of the body temperature ensures reasonably constant delivery. A drug loading port is built directly into the center of the hockey-puck-shaped device, which is usually implanted subcutaneously on the chest wall where it can be accessed for reloading. The pressure generated by the filling operation compresses the driving gas to a liquid state and allows the device to be recycled indefinitely. More recent models have incorporated electronically controlled valves to allow rate control and bolus delivery patterns. Applications for the device have included chemotherapy and treatment of diabetes. Considerable effort has been directed toward developing a stabilized form of U-500 Insulin that can survive the rigors of prolonged storage at elevated temperatures with the accompanying agitation of being stored in an implantable device.

Medtronic, a pacemaker manufacturer, has also pioneered the development of implantable infusion devices. Their device is a miniaturized, battery operated, peristaltic pump which is actually smaller than the Infusaid device. Using technology similar to that used to program pacemakers, Medtronic can remotely program and control their device. Although these devices have clearly shown that it is technologically possible to produce implantable delivery systems, the cost per device, the need for surgical intervention for implantation, and the extreme requirements for drug stability are still limitations to their widespread use. With polypeptides, particular attention must be paid to the stability of the molecules in implantable systems as the reservoirs are never completely purged of liquid, and thus some amount of residual material may be present for many months.

3.3. External Infusion Pumps

Some portable infusion pumps are battery-driven miniature versions of the stationary devices. They are equipped with the same programming and alarm features, but the size is reduced so that the pumps can be attached to a belt or worn in a harness. Abbott, Baxter, and Pharmacia Deltec offer devices of this type. Other pumps are completely or partially disposable, and the principle of operation is based on mechanical or osmotic mechanisms. I-Flow, for example, pressurizes a collapsible infusate bag with a spring. The housing with the spring mechanism is reusable while the infusate bag is disposed of after use. Examples of single-use devices are Baxter's, Imed's, and McGaw's balloon pumps, which consist of an elastomeric bag that contains the infusate under pressure. These balloon pumps apply pressure to the infusate and then control the output by using restrictors or filters, which allow a defined amount of drug to pass. When proteins are delivered through any kind of restrictors, however, there is always a risk that the molecules may be denatured by the high-shear fluid forces experienced within the device. An alternative control means was employed by Pharmetrix in the design of the osmotically driven Mini-Pump. The delivery rate is

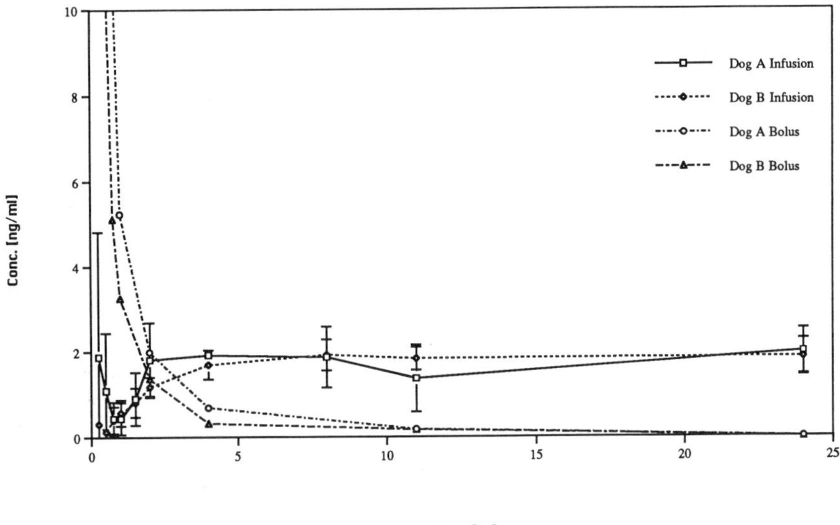

Figure 3. Plasma levels of somatostatin analog in beagles after subcutaneous infusion with Harvard syringe pump ($n = 6$ each) and after i.v. bolus injection.

regulated not by a restriction element for the infusate but by the osmotic membrane permeability for the driving fluid. This driving fluid passes a semipermeable membrane and expands an elastomeric diaphragm, which then expels the infusate at a continuous rate that is independent of the infusate viscosity or molecular weight.

The performance of this osmotic pump (Pharmetrix, 1992) has been compared to that of an electrically driven Harvard syringe pump by following subcutaneous infusion of a somatostatin analog in beagle dogs over 24 h. After an i.v. bolus injection, the drug is eliminated rapidly from the body with a total clearance of 16–19 l/hr. The subcutaneous infusion with either the syringe pump (Fig. 3) or the osmotic pump (Fig. 4) results in a relatively constant plasma level after the first two hours. The variability between disposable osmotic pumps ($n = 6$ for each dog) is actually less than the variability between the syringe pumps. Comparable performance has thus been demonstrated between the portable, lightweight Pharmetrix pump and the stationary, electromechanical syringe pump. These studies also confirm the compatibility of proteins with the components of the device, supporting its potential for delivering biotech products in humans. The desired flow rate for use of the osmotic pump in conditions that differ from those for human use can be accommodated in device design, by

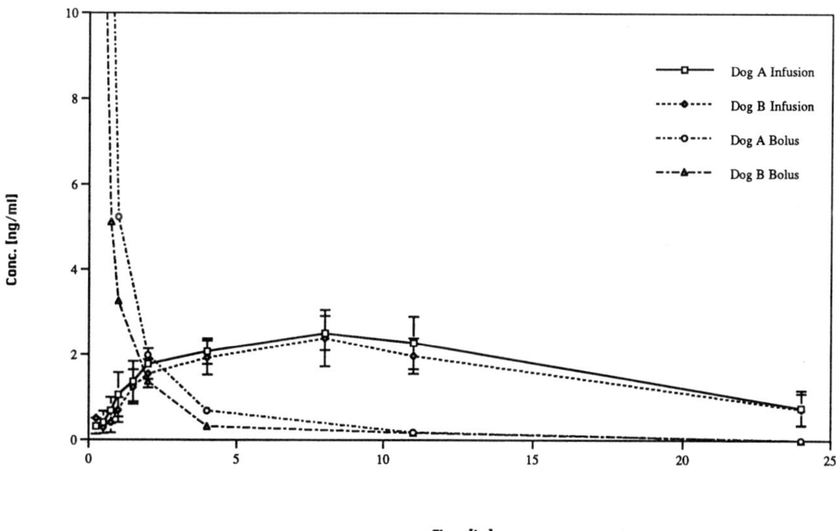

Figure 4. Plasma levels of somatostatin analog in beagles after subcutaneous infusion with Pharmetrix osmotic pump ($n = 6$ each) and after i.v. bolus injection.

simply changing the area of the semipermeable membrane of the osmotic pump.

4. SUMMARY

When a therapeutic effect is optimized by precise control of specific temporal patterns of plasma levels, infusion offers distinct advantages over oral administration, bolus injection, or depot delivery of polypeptides. The limitations of oral delivery are well known, and although research is under way into development of carrier systems that prevent degradation of labile agents, it is unlikely that the variances in absorption will meet the need for precise control. Depot delivery from subcutaneous or intramuscular implants presents a difficult situation when local tissue reactions to the agent sometimes occur. Removal of a depot system in the event of adverse reactions presents additional difficulties. Bolus injections are unable to sustain constant plasma levels unless the drug half-life is long or the injections are frequently administered. Insulin injections, for example, would be required every 30–60 minutes to approximate the plasma levels provided by a continuous infusion; such frequent injections would not be practical on a 24-hour basis. For the developer of new polypeptides, parenteral administration offers the most direct route to the marketplace. The step from periodic injections to tightly controlled infusion is a logical progression as compared with modification of the molecules or vehicles to obtain equivalent profiles. In Table II several different types of devices that can be used for infusion of proteins are compared.

Microelectronics have played a major role in the miniaturization of infusion devices and undoubtedly will continue to do so. Micromachining, a spin-off technology of integrated circuit manufacture, will also find application in small infusion devices. In the future, we will have cost-effective disposable devices (Saaman *et al.*, 1994) built on this technology that are programmable and thus can be adapted to meet each individual therapeutic need (Horres, 1994). We can also expect to see more closed-loop drug delivery systems where biosensors and infusion devices are combined to optimize a particular therapy. Recent positive results obtained in diabetics by a decade on tight glucose control may forecast a resurgence of popularity of insulin pumps. At the other end of the spectrum, low-cost, small, and simple-to-use osmotically powered systems are close to being marketed; these systems will make infusion almost as convenient as transdermal patches. We will also see major advances in how drugs and devices are interfaced. Prefilled and ready-to-use drug cartridges have proven to be

Table II
Comparison of Infusion Devices with Different Driving Mechanisms

Device	Volume range (ml)	Flow rate range (ml/hr)	Accuracy	Flow consistency	Price($)
Gravity controller	100–2000	5–250	15–30%	Poor at low rates	75–150
Peristaltic pumps	25–4000	1–2000	5–10%	Poor at low rates	1200–2500
Piston pumps	25–4000	1–2000	3–5%	Very poor at low rates	1000–3000
Syringe pumps	1–50	0.04–50	3–5%	Fair at low rates	750–2000
Balloon/spring pumps	10–200	0.2–200	10–25%[a]	Excellent	10–25[b]
Osmotic pumps	5–25	0.04–2	10–15%	Excellent	NA[c]

[a]Accuracy depends on infusate characteristics.
[b]Per use.
[c]Not available.

efficient in surgical and emergency medicine and can greatly improve most infusion applications. It is anticipated that coded, prefilled cartridges or pouches will be automatically recognized by preprogrammed pumps to reduce operator labor and entry errors.

REFERENCES

Banerjee, P. S., Hosny, E. A., and Robinson, J. R., 1991, Parenteral delivery of peptide and protein, in: *Peptide and Protein Drug Delivery* (V. H. L. Lee, ed.), Marcel Dekker, New York, p. 487–543.

Champion, M. C., Shepherd, G. A., Rodger, N. W., and Dupre J., 1980. Continuous subcutaneous infusion of insulin in the management of diabetes mellitus, *Diabetes* **29:**206–212.

Chien, Yie W., 1992, Parenteral drug delivery and delivery systems, in: *Novel Drug Delivery Systems,* 2nd ed. (J. Swarbrick, ed.), Marcel Dekker, New York, p. 381–528.

Heimburger, D. C., 1992, Ten-year survival of a Broviac catheter, *Nutr. Clin. Pract.* **7**(2)**:**74–76.

Horres, C. R., 1994, U.S. Patent 5,290,240; Electrochemical controlled dispensing assembly and method for selective and controlled delivery of a dispensing fluid, Pharmetrix Corp., Menlo Park, California.

Kennedy, F. P., 1991, Recent developments in insulin delivery techniques, *Drugs* **42**(2)**:**219.

Kompella, U. B., and Lee, V. H. L., 1991, Pharmacokinetics of peptide and protein drugs in: *Peptide and Protein Drug Delivery* (V. H. L. Lee, ed.), Marcel Dekker, New York, p. 391–484.

Paolisso, G., Scheen, A. J., Albert, A., and Lefebvre, P. J., 1989, Effects of pulsatile delivery of insulin and glucagon in humans, *Am. J. Physiol.* **257**:E686–E696.

Pharmetrix, 1992, Data on file, Pharmetrix Corp., Menlo Park, California.

Physicians' Desk Reference, 1993, 47th ed., Medical Economics Data, Montvale, New Jersey, pp. 606, 875, 1078.

Pickup, J., Keen, H., Vibert, G. C., White, M. C., and Kohner, E. M., 1980, Continuous subcutaneous insulin infusion in the treatment of diabetes mellitus, *Diabetes Care* **3**:290–300.

Saaman, A., Aubert, C., and Neftel, F., 1994, First disposable micropump for peptide delivery, *Proc. Int. Symp. Control. Rel. Bioact. Mater.* **21**:359–360.

Chapter 10

Oral Delivery of Microencapsulated Proteins

Mary D. DiBiase and Eric M. Morrel

1. INTRODUCTION

While the neonatal mammalian small intestine is able to absorb macromolecules (Walker and Isselbacher, 1974; Walker, 1979), in the mature gut the oral bioavailability of polypeptide drugs is generally poor (Lee *et al.*, 1991; Humphrey and Ringrose, 1986; Pusztai, 1989). This low bioavailability is not surprising given the large number of barriers in the gastrointestinal tract to the uptake of intact peptides and proteins. Proteolytic enzymes in the stomach, the intestinal lumen, and the brush border of the enterocytes digest proteins by hydrolysis. Most digestion in the small intestine is due to the proteolytic enzymes trypsin, chymotrypsin, and carboxypeptidase, all from the pancreas (Castro, 1981; Matthews *et al.*, 1968). Any remaining peptides larger than three amino acids are further hydrolyzed in the brush border. Tripeptides are hydrolyzed extracellularly, by the brush border enzymes, or, if absorbed, by cytoplasmic amino peptidases (Nicholson and Peters, 1977). Besides enzymatic barriers, there are significant physical barriers to peptide drug absorption in the intestinal wall. Within the intestinal mucosa, the membrane of the microvilli consists of the typical trilaminar arrangement observed in biological membranes: two molecular layers of lipid with the hydrocarbon tails oriented inward and the hydrophilic heads on the outer part of the protein-coated membrane. The

Mary D. DiBiase • Biogen, Cambridge, Massachusetts 02142. *Eric M. Morrel* • Ascent Pediatrics, Inc., Wilmington, Massachusetts 01887.

Protein Delivery: Physical Systems, Sanders and Hendren, eds., Plenum Press, New York, 1997.

outer membranes of adjacent cells are fused at the basal membrane, forming a tight junction—a significant barrier between the lumen and the intercellular space (Humphrey, 1986; Csaky, 1984). The glycocalyx, a uniform layer of filamentous glycoproteins on the surface of the microvilli (Egberts *et al.*, 1984), possesses a negative charge at physiologic pH due to the presence of sialic acids at the terminal portion of the carbohydrate chain. The glycocalyx, furthermore, lies within a layer of mucus, about 1–5 μm thick, upon which lies an unstirred water layer about 100–400 μm in depth (Humphrey, 1986; Thomson and Dietschy, 1984). The unstirred water layer can act as a barrier for both water-soluble (Alpers, 1987) and hydrophobic substances (Humphrey, 1986), while the combination of mucus and glycocalyx may be a barrier to polar molecules due to both the viscosity and the electronegativity of that layer (Smithson *et al.*, 1981; Esposito *et al.*, 1983).

Despite all of the above obstacles to peptide and protein absorption, there is both clinical and experimental evidence that large molecules may penetrate the intestinal mucosa (Gruber *et al.*, 1987). The degree of this macromolecular absorption is certainly not of any nutritional significance but still is often high enough to be of biological importance (Alpers and Isselbacher, 1967; Bernstein and Ovary, 1968; Bockman and Winborn, 1966; Casley-Smith, 1967; Chisui, 1968; Cornell *et al.*, 1971; Danforth and Moore, 1959; Walker *et al.*, 1972; Warshaw *et al.*, 1971), and such macromolecular absorption has been observed to occur under normal physiologic conditions (Gruskay and Cooke, 1955; Korenblat *et al.*, 1968; Wilson and Walzer, 1935). Adults, for example, develop precipitins in serum following ingestion of milk proteins (Korenblat *et al.*, 1968). The oral absorption of β-lactam antibiotics, angiotensin-converting enzyme inhibitors, and cyclosporins further reveals that enteral uptake of peptides or peptide analogs is possible. In experimental systems, even large proteins such as insulin (Danforth and Moore, 1959) and horseradish peroxidase (Warshaw *et al.*, 1971) in rats and large-molecular-weight antigens in guinea pigs (Bernstein and Ovary, 1968) have been noted to cross the intestinal wall. The permeability, reactivity (chemical or enzymatic), transit time (Amidon *et al.*, 1988; Sinko *et al.*, 1991), and solubility and dissolution rates (Oh *et al.*, 1993) will affect the degree of oral absorption for each peptide or protein. Factors such as molecular weight (Bernstein and Ovary, 1968), charge, structure, and configuration all will determine those characteristics which will influence intestinal permeability.

One way to potentially enhance the uptake of proteins and peptides is to microencapsulate them prior to administration. Besides providing protection from intestinal enzymes, microcapsules could be formulated with characteristics such as charge and hydrophobicity, which might exploit the above-summarized intestinal wall features and result in increased systemic uptake as compared to that of the native, nonencapsulated peptide. This chapter is intended to provide, first, an overview of the specific mechanisms that result in the transport of proteins and peptides across the intestinal wall. A summary of the observed and theorized mechanisms of microparticulate intestinal absorption will then be presented. The final section will focus on various case studies wherein microspheres have been used as oral protein and peptide drug delivery systems.

2. MECHANISMS OF INTESTINAL ABSORPTION OF PROTEINS AND PEPTIDES

The intestinal absorption of intact peptides or proteins occurs to varying extents (Gardner, 1984; Silk *et al.*, 1985; Humphrey and Ringrose, 1986) and generally involves either a passive mechanism, some type of carrier-mediated transport, or receptor-mediated or non-receptor-mediated endocytotic transport. Each mechanism will contribute to the absorption of a peptide or protein to a different degree, depending upon such characteristics as the size, charge, or lipophilicity of the peptide or protein.

2.1. Passive Diffusion

Most peptide drugs—those larger than three amino acids—that are passively absorbed, are absorbed by diffusion through the lipid membrane of enterocytes. Although the diffusivity of the drug in the membrane and the membrane thickness are both important factors affecting permeability, the principal determinant of the degree of absorption of a drug across the lipid membrane is the partition coefficient between the membrane and the lumen (Matthews, 1991; Houston and Wood, 1980). Lipid-soluble molecules are taken up readily when compared to molecules taken up by other passive-transport mechanisms, such as diffusion through aqueous pores (Menzies, 1984). In one experimental system, the relatively nonpolar rat jejunum exhibits increased absorption of lipid molecules with decreasing polarity of those molecules (Westergaard, 1987). This phenomenon is also demonstrated in the absorption of endotoxins from gram-negative bacteria due to the interaction between the lipid A portion of the endotoxin and the apolar epithelial membrane (Kabir *et al.*, 1978). With regard to drug design, increasing the lipophilicity of a drug can result in increased intestinal absorption by either passive mechanisms or the carrier pathways that will be discussed below (Clayton *et al.*, 1975; Wyvratt and Patchett, 1985).

Increasing lipophilicity can, however, result in a decrease in solubility, which, in turn, can result in solubility/dissolution problems and ultimately adversely affect intestinal absorption (Oh, 1991).

Besides diffusion across the lipid bilayer of intestinal epithelial cells, another mechanism by which peptides and proteins can be absorbed is by diffusion through aqueous pores in the cell membrane. Aqueous pore diffusion in undamaged cells, though, is probably dominated by smaller-molecular-weight species, particularly those that are water-soluble and are not taken up by carrier-mediated transport mechanisms (Pusztai, 1989).

The effect of molecular weight on drug absorption by either of the diffusion pathways discussed above has not been fully explored. Cyclosporin, with a molecular weight of 1200, exhibits a low, but acceptable, extent of oral absorption of 23% (Wood *et al.*, 1983). Drugs of molecular weight 2000–3000, therefore, might be expected to exhibit bioavailabilities of 10–20% (Amidon and Lee, 1994). A number of significant drugs, e.g. labetolol, morphine, and propanolol, exhibit low systemic availabilities (Benet and Williams, 1990). Even with an upper molecular-weight limit of 1000–2000 for peptide drugs which might exhibit some reasonable bioavailability, the number of possible peptide drugs that might be synthesized from natural amino acids is quite large (Amidon and Lee, 1994).

Passive diffusion can also occur by paracellular routes. Horseradish peroxidase (HRP) has been observed to cross the mucosal barrier intact in both normal rat jejunum (Heyman *et al.*, 1982) and surgically traumatized guinea pig intestinal tissue (Rhodes and Karnovsky, 1971). The actual route of transport of the HRP in the normal intestinal tissue was not known but was concluded to be either by an intracellular pathway or intercellular, through tight junctions (Heyman *et al.*, 1982). Transport of the HRP in the guinea pig intestine was observed to be through the epithelial tight junctions but required bovine serum albumin (BSA) as a macromolecular cofactor to alter the tight-junction barrier. While the use of BSA, which has sometimes been found to be contaminated with endotoxins or phospholipases (Dvorak and Bast, 1970), complicates analysis of these observations, the result still suggests the possibility of intercellular diffusion. Similarly, paracellular transport of macromolecules has been observed when the brush border has been damaged through disease or experimentally, as with hypertonic solutions (Menzies, 1984; Wheeler *et al.*, 1978). Paracellular transport of macromolecular food proteins also has been observed, albeit under the more natural circumstance of passage across the villus tip, where epithelial cells are routinely shed into the lumen. In that extrusion zone, where the intercellular junctions have been naturally disrupted, macromolecular transport has been observed in both normal and diseased small intestine (Jackson *et al.*, 1983).

2.2. Carrier-Mediated Transport

Carrier-mediated transport represents another pathway for intestinal absorption, particularly for di- and tripeptides (Adibi and Phillips, 1968; Matthews and Payne, 1980; Humphrey, 1986; Matthews, 1975, 1983, 1991) and their analogs (Kramer *et al.*, 1990; Kimura *et al.*, 1983; Nakishima *et al.*, 1984; Allen *et al.*, 1979; Yokohama *et al.*, 1984a, b; Friedman and Amidon, 1989a; Hu and Amidon, 1988; Sinko and Amidon, 1988, 1989; Tsuji *et al.*, 1987). Di- and tripeptides, as well as amino acids, are produced naturally by the enzymatic degradation of polypeptide and protein fragments remaining after gastric and luminal digestion of ingested macromolecules. Intraluminal hydrolysis accounts for only a small portion of the protein digestion (Newey and Smith, 1962), with the exo- and endopeptidases of the brush border of the microvilli being the principal producers of absorbable peptide fragments (Adibi and Kim, 1981).

Carrier-mediated transport involves cotransport of the absorbable species with a proton. The required proton gradient is hypothesized to be maintained by a Na^+–H^+ exchanger. The lumen of the intestine is acidic relative to the epithelial cell cytosol. The low cytosolic sodium concentration, required to produce the transporter driving force, is maintained by the Na^+–K^+ ATPase in the basolateral membrane. The sodium/proton exchanger working in concert with the sodium/potassium ATPase, therefore, results in a transport mechanism for the uptake of di- and tripeptides into the intestinal wall (Ganapthy and Leibach, 1985).

The carriers for free amino acids are not the same as those for di- and tripeptides, and the number of different carriers and the specificity of those pathways are not known (Matthews *et al.*, 1968). The large number of natural di- and tripeptides, furthermore, limits the ability to completely describe all existing mechanisms of carrier transport (Matthews, 1975, 1983). Through extensive study of these mechanisms (Matthews, 1991), however, a number of general guidelines have been developed.

The carrier transport pathway is stereospecific (Matthews, 1975, 1983; Boyd and Ward, 1982; Asatoor *et al.*, 1973; Cheeseman and Smyth, 1972); peptides of the D-configuration are handled by the transporter (Boyd and Ward, 1982; Asatoor *et al.*, 1973; Cheeseman and Smyth, 1972) but are poorly taken up and slowly hydrolyzed (Matthews, 1975, 1983). An α-peptide bond is preferred (Matthews, 1975, 1983), though not required (Bai *et al.*, 1991) for carrier transport, whereas methylation, acetylation, or other modification of the N-terminal α-amino group (Addison *et al.*, 1974; Das and Radhakrishnan, 1975; Rubino *et al.*, 1971; Addison *et al.*, 1975), as well as modification of the C-terminal carboxyl (Matthews, 1975, 1983), decreases or eliminates affinity for the transport carrier. The presence of a β-amino acid as part of a dipeptide is compatible with carrier transport (Addison *et al.*, 1974, 1975; Tomita *et al.*, 1990), but the presence of a γ-amino acid eliminates affinity for the carrier (Das and Radhakrishnan, 1975; Tomita *et al.*, 1990). There is little evidence for carrier transport of tetrapeptides (Adibi and Morse, 1977; Burston *et al.*, 1979; Kerchner and Geary, 1983); when observed, tetrapeptide uptake has been concluded to be by passive mechanisms (Matthews and Payne, 1980; Boyd and Ward, 1982; Addison *et al.*, 1975). Similarly, penta- and hexapeptides are rarely absorbed intact and are generally digested by the brush border peptidases, down to molecules that can be absorbed by carrier pathways (Gruber, *et al.*, 1987).

Peptide analogs such as β-lactam antibiotics (Kramer *et al.*, 1990; Quay, 1972; Quay and Foster, 1970; Sinko and Amidon, 1988; Kimura *et al.*, 1983; Nakashima *et al.*, 1984), angiotensin-converting enzyme (ACE) inhibitors (Kramer *et al.*, 1990; Friedman and Amidon, 1989a, b; Hu and Amidon, 1988; Yee and Amidon, 1990), alafosfalin (Allen *et al.*, 1979), and thyrotropin releasing hormone (TRH) (Yokohama *et al.*, 1984a, b) are also absorbed via the carrier transport system, as has been demonstrated in studies of competitive inhibition with di- and tripeptides, though no competitive inhibition is seen with free amino acids (Sinko and Amidon, 1988; Oh *et al.*, 1989). The study of the transport of these peptide analogs has reflected some of the specificities of the carrier pathway that were outlined above and revealed other characteristics of this route of absorption. Work with TRH, which is absorbed by the carrier though it lacks both a free N-terminal α-amino group and a C-terminal carboxyl group (Yokohama *et al.*, 1984b; Humphrey and Ringrose, 1986), has shown that the carrier pathway is saturable with increasing dose (Yokohama *et al.*, 1984b). Substitution in TRH of a pyroglutamyl residue with carboxybutyrolactone, creating the analog DN 1417, results, however, in a molecule taken up only by passive mechanisms (Addison *et al.*, 1975). The lack of requirement for a free N-terminal α-amino group for carrier transport also has been proven in studies of such β-lactam antibiotics as cefixime and ceftibuten (Tsuji *et al.*, 1987; Oh *et al.*, 1990; Tsuji *et al.*, 1986) as well as with ACE inhibitors such as captopril, enalapril, and lisinopril (Hu and Amidon, 1988; Friedman and Amidon, 1989a, b; Yee and Amidon, 1990). Investigations of ACE inhibitors have shown that the ester prodrugs such as enalapril are absorbed by the transporter whereas the diacids, such as enalaprilat, are poorly absorbed. Lisinopril, which is somewhat similar in structure to enalaprilat, exhibits an affinity for the transporter, but with low carrier permeability (Friedman and Amidon, 1989a, b; Tsuji *et al.*, 1986; Yee and Amidon, 1990). The underlying reasons for these specificities are not entirely understood.

2.3. Receptor-Mediated and Non-Receptor-Mediated Endocytosis

With the carrier transport pathway restricted almost exclusively to uptake of amino acids, dipeptides, and tripeptides, and passive diffusional uptake also limited mostly to smaller, and generally more lipophilic, species, the principal pathway for uptake of macromolecules in the intestine is by endocytosis. Similarly to passive absorption, endocytotic uptake is truly appreciable only in the immature intestine, decreasing significantly after gut closure (Pusztai, 1989). The capacity for uptake of macromolecules in the mature intestine, however, is still significant enough to allow antigen sampling for the development of mucosal immunity or permit appreciable absorption of various toxins.

Endocytotic uptake in the intestine can occur by either receptor-mediated or non-receptor-mediated pathways. For both pathways, endocytotic uptake follows the same general process, one that is similar to macrophage phagocytosis. Endocytotic uptake starts with either nonspecific adsorption or receptor binding of large molecules to the absorptive cell. Fluid-phase markers have also been observed to be internalized during receptor-mediated endocytosis (Doxsey *et al.*, 1987; Sandvig *et al.*, 1987). Once enough molecules are adsorbed or bound, invagination of the membrane causes the formation of a membrane-bound vesicle (endosome or phagosome). The phagosome will migrate across the cell and, in most cases, depending on the type of absorptive cell, fuse with a lysosome to form large vacuoles (phagolysosome). The majority, though not all, of the contents of the phagolysosome are then digested (Heyman *et al.*, 1982), after which the phagolysosome fuses with the basal surface of the absorptive cell and releases the contents of the vacuole into the interstitium (O'Hagan *et al.*, 1987). Since there is not always fusion of the endocytotic vesicle with a lysosome, and since lysosomal digestion is not always complete, macromolecules can be absorbed intact into the intestinal wall.

The efficiency of macromolecular uptake by endocytosis, as well as the pathway of transcellular transport of the resulting endosomes, is related to whether fluid-phase or absorptive, receptor-mediated or non-receptor-mediated, endocytosis is involved (Mayorga *et al.*, 1989; Roederer *et al.*, 1987; Wessling-Resnick and Braell, 1990; Abrahamson and Rodewald, 1981; Tartakoff, 1987). The ability of the endosomes to discriminate between soluble and membrane-associated species has been demonstrated in neonatal rodents with soluble HRP and immunoglobulin. Immunoglobulin is taken up in the neonatal rodent by receptor-mediated transport across the intestinal epithelial cells. When the apical side of the intestinal epithelial cells is exposed to both HRP and immunoglobulin, both are taken up within the same vesicles and found within endosomes. The HRP, though, is directed toward lysosomes, whereas the immunoglobulin is found in more deeply situated tubules and vesicles (Abrahamson and Rodewald, 1981).

Factors such as charge (Shen and Ryser, 1978), or the specific characteristics of the ligands and receptors (Salzman and Maxfield, 1989; Dunn *et al.*, 1989; Tartakoff, 1987) will also influence endocytosis and the fate of endosomes. The conjugation of poly(L-lysine) to HRP, for example, has been observed to result in a 200-fold increase in uptake of the HRP by cultured mouse fibroblasts. This was concluded to be a result of the addition of a strong cationic charge to the HRP (Shen and Ryser, 1978).

Intestinal absorption of macromolecules by endocytosis occurs primarily through two types of cells: the enterocytes in the epithelium of villi and the so-called M cells (membranous or microfold cells), which are scattered throughout the intestine but are mostly concentrated in the Peyer's patches overlying the gut-associated lymphoid tissues (GALT). The principal function of the M cells is, in fact, to sample and present protein antigens (Fara, 1985; Bockman and Cooper, 1973; Owen, 1977; Owen and Jones, 1974) or microorganisms (Wolf *et al.*, 1981; Owen *et al.*, 1986b) to the underlying lymphoid tissue. That the M cells lack a well-developed lysosomal apparatus (Owen *et al.*, 1986a) assists in this antigen-presenting function, because the low number of lysosomes improves the chances for presentation of intact absorbed proteins to the underlying lymphoid cells. When the concentration of an absorbable antigen is low, transport through the M cells dominates, whereas once concentrations are high, endocytotic uptake is seen in other epithelial cells, both in the Peyer's patch and in the villi (Owen, 1977; Walker, 1982). Once an antigen binds to the apical membrane of an M cell, endocytosis, transcytosis, and exocytosis occur fairly rapidly (Bockman and Cooper, 1973; Owen, 1977; Danforth and Moore, 1959).

Much of the work on non-receptor-mediated endocytosis has involved the use of HRP as a tracer. In one study in rats (Cornell *et al.*, 1971), HRP was injected into the lumen of ligated portions of the jejunum and ileum. Uptake of the HRP was comparable in the jejunum and ileum and was greatest near the apical portion of the villi, with less in the base and none in the crypts. Besides being absorbed at the apical surface membrane, HRP was also present within membrane-bound cytoplasmic cannicular vesicles and vacuolar structures. HRP was also absorbed into extracellular spaces between adjacent absorptive cells and in the lamina propria. In a study of HRP infusion into ileal segments of mice (Owen, 1977), HRP was observed to be taken up more readily by M cells than by columnar cells. HRP absorbed by M cells was released into the extracellular space, for uptake by lymphoid cells, rather than being sequestered into lysosomes, as was seen with HRP absorption by epithelial cells in other locations (Straus, 1969). In a study in rabbits (Heyman *et al.*, 1982), an estimated 67–97% of the HRP that was endocytosed by epithelium in villi was degraded during transcytosis.

A related study by Heyman (1990) with CaCo-2 cells confirmed HRP transport by endocytosis, primarily along a degradative pathway most likely associated with the lysosomal system. Other work with CaCo-2 cells and HRP (Hidalgo *et al.*, 1989) found that HRP is endocytosed by a fluid-phase mechanism; this proposed uptake mechanism was supported by the Heyman study, in which monensin was found to have no effect on HRP transcytosis. The degree of HRP degradation also varied depending on whether the marker was presented at the apical or the basal membrane of the CaCo-2 cells. Different cell types of intestinal epithelium, therefore, could process endocytosed proteins in disparate ways (Heyman *et al.*, 1990).

Within the small intestine, there are several types of ligand–receptor pairs for receptor-mediated endocytosis. Lectins are one such class of receptors or ligands that are often associated with receptor-mediated endocytosis and have been seriously considered as conjugates to enhance oral delivery of protein and peptide drugs. M-cell luminal membranes display a variety of lectin-binding specificities (Owen and Bhalla, 1983; Roy, 1987). Lectin–antigen conjugates that attach to M-cell apical membranes are absorbed more efficiently than conjugates that do not adhere (Neutra *et al.*, 1987). Most microbial and plant lectins or toxins interact with the epithelium of the small intestines, though many of those lectins/toxins in binding, cause some damage to the cells, with subsequent endocytosis of the lectin/toxin (Pusztai, 1989). Lectins can be bound to toxins, and act to provide a mechanism for uptake of the toxin by the cell, though the lectin itself is not toxic (Blaustein *et al.*, 1987; Pappenheimer, 1977; Gill, 1978; Dallas and Falkow, 1980; McDonel, 1980; Pusztai, 1986a). Alternatively, lectins may be unassociated with a toxin but simply consist of multiple, and varied, lectin subunits. Even with no associated toxin, these lectins still may have some toxic effect, though far less an effect than is seen with lectins conjugated to toxins. Once bound, these lectins can cause extensive morphological changes in the cell membrane and interfere with the cell metabolism. Often, these types of lectins can be found in dietary plants and microorganisms of the intestinal tract (Goldstein and Poretz, 1986; Pusztai, 1986b; Pusztai *et al.*, 1986; Pusztai, 1988). PHA, from kidney beans (*Phaseolus vulgaris*), is one of the most studied lectins (Greer *et al.*, 1985; King *et al.*, 1980, 1982). When ingested, PHA is mostly excreted intact, but a relatively high percentage, 5–10%, is taken up systemically (Pusztai, 1988). This degree of uptake is particularly significant when compared to the 0.1%

uptake of tomato lectins, nontoxic lectins that also bind strongly to intestinal epithelium (Kilpatrick *et al.*, 1985). Furthermore, though most of the endocytosed PHA interacts with lysosomes (King *et al.*, 1986), most of the lectin is exocytosed intact.

Two other related systems of receptor-mediated transport are those associated with iron and vitamin B_{12} uptake. In both systems, the specific binding protein is released into the intestine and binds to the ligand in the lumen of the gut. In the case of iron, transferrin is released in the stomach, and binds to the iron, and then this complex binds to receptors in the duodenal mucosa. Vitamin B_{12} forms a stable complex (Grasback *et al.*, 1959) with intrinsic factor (IF) released in the stomach, and this B_{12}–IF complex in turn interacts with enterocyte membrane-bound receptors for IF (Mathan *et al.*, 1974), which are located in the terminal ileum. The actual uptake of the B_{12}–IF complex is not completely understood, though it is known to be slow, taking over 3 hr, is energy-dependent (Chanarin *et al.*, 1978), and, like the uptake of the iron–transferrin complex, involves endocytosis (Russell-Jones and de Aizpurua, 1988; Castle, 1953; Fox and Castle, 1942; Allen and Majerus, 1972a, b; Jani *et al.*, 1989).

Immunoglobulins, specifically IgA and IgG, have also been shown to adhere selectively to M-cell luminal membranes in several experimental models. Mouse monoclonal IgA antibodies were observed to adhere to the M-cell membranes and to be transported across the epithelium, independent of the antigen specificities of the antibodies. IgG was found to inhibit this binding of IgA by attaching to the M-cell surface. These results suggest that there exists a common immunoglobulin domain which, regardless of species specificity, mediates adherence to M cells (Weltzin *et al.*, 1989).

3. MECHANISMS OF INTESTINAL ABSORPTION OF MICROPARTICULATES

Given the susceptibility of orally administered peptide and protein drugs to enzymatic degradation in the lumen and brush border of the intestine, there has been increasing interest in the microencapsulation of those drugs and in understanding the uptake of microparticulates in the intestine. Most particulate uptake involves transcellular pathways across the intestinal epithelium, though there have been some observations of paracellular transport of particles. The focus of the discussion below will be on these pathways and on how factors such as size, charge, and hydrophobicity affect particulate absorption in the intestine.

3.1. Transcellular Pathway

Based on the majority of observations of particulate uptake in the intestine, transcellular transport, specifically through the epithelium of the Peyer's patch, appears to be the primary pathway of microparticulate absorption (LeFevre and Joel, 1984; Pappo *et al.*, 1991; Eldridge *et al.*, 1990; Damge *et al.*, 1990; Pappo and Ermak, 1989). Furthermore, most of the transcellular transport involves M cells (Bockman and Cooper, 1973; Joel *et al.*, 1978, 1970; LeFevre *et al.*, 1978a), though there have been a few select instances in which other cell types appeared to be involved (Wells *et al.*, 1988; Landsverk, 1988).

The size of a microparticle may be the most significant parameter with regard to absorption into the Peyer's patch and the ultimate fate of that particle once inside the gut-associated lymphoid tissue (GALT). Particles ranging from 27-nm carbon (LeFevre and Joel, 1986; Joel *et al.*, 1978; Hammer *et al.*, 1983; LeFevre *et al.*, 1985b) to 10-μm polymer microspheres (Eldridge *et al.*, 1990) have been observed within Peyer's patches following oral administration in test animals. Polymer microspheres greater than 10 μm have specifically been observed not to be absorbed (Eldridge *et al.*, 1990; LeFevre *et al.*, 1980), so 10 μm is often considered the cutoff for particulate absorption. Smaller particles also appear to be absorbed more avidly than larger ones. Particles 1 μm in diameter have been observed to be taken up less efficiently than 100-nm and 500-nm particles (Jani *et al.*, 1989), and 2.65 μm latex particles are more readily absorbed than 9.13-μm particles in the Peyer's patches of mice (Ebel, 1990). Even after transepithelial transport, size can still affect the destiny of an absorbed particle. In one study, polymer particles less than 5 μm in diameter were observed to be transported through efferent lymphatics in macrophages, whereas most particles of the same polymer but greater than 5 μm in diameter remained within the Peyer's patch (Eldridge *et al.*, 1990). In another study, latex particles less than 5.7 μm in diameter were absorbed into the blood, whereas larger particles were not seen in the blood but were observed in the Peyer's patch and mesenteric lymph node (LeFevre *et al.*, 1980).

The hydrophobicity or charge of a particle has also been observed to affect uptake (LeFevre *et al.*, 1978a, 1979, 1980, 1985b; Pappo and Ermak, 1989; Eldridge *et al.*, 1990; Eldridge, 1988). Microspheres consisting of polymers exhibiting relatively high hydrophobicity have been demonstrated to be more readily absorbed into Peyer's patches than those made of less hydrophobic polymers (Eldridge *et al.*, 1990; LeFevre *et al.*, 1985b). Carboxylated latex particles, carrying a negative charge, exhibited decreased uptake relative to uncharged latex microspheres (Jani *et al.*, 1989).

There still are some general observations regarding this pathway for particulate absorption which are somewhat independent of the specific characteristics of a particle. Absorption of microparticulates into the Peyer's patch is most likely via intracellular vesicles (O'Hagan *et al.*, 1987; Jani *et al.*, 1989), but generally without formation of a phagolysosome because of the relative deficiency of lysosomes in M cells (Owen *et al.*, 1986a). This absorption process is quite rapid (Wolf *et al.*, 1981; Landsverk, 1988; Pappo and Ermak, 1989; Damge *et al.*, 1990), with transport of microspheres across the M cells occurring within 10 minutes (Pappo and Ermak, 1989; Damge *et al*, 1990). In one investigation, the rate of transport of 650- to 750-nm latex particles was comparable to that observed with soluble antigens (Bockman and Cooper, 1973; Owen, 1977), lectin–ferritin conjugates (Neutra *et al.*, 1987), and microorganisms (Wolf *et al.*, 1981; Owen *et al.*, 1986). This rapid transport suggests that there is some form of membrane binding, based on studies showing that transport across M cells is generally faster with binding (Neutra *et al.*, 1987). The transport of microparticulates through the M cells into the Peyer's patch dome may be facilitated by fenestrations in the basal lamina supporting the follicle epithelium (McCluggage *et al.*, 1986; Pappo *et al.*, 1988).

Assuming a high concentration of microparticles in the intestinal lumen, as time progresses, an increasingly larger fraction of microparticles are transported across the follicle epithelium, with a distinct directionality of particles being taken up from the lumen and discharged into the M-cell pocket (Pappo and Ermak, 1989). This creates a concentration gradient from the lumen to the subepithelial dome, with localization of particles on the M-cell apical membrane, in the M-cell pocket, and in the subepithelial dome (Pappo *et al.*, 1991). The uptake of particles, however, is not always continuous. One study revealed what appeared to be an 80-min pause in uptake, 10 min after infusion into an intestinal loop, despite a high concentration of latex microspheres in the lumen (Pappo and Ermak, 1989). Once absorbed into the GALT, microparticulates generally are transported into the lymphatic system, often after being taken up by macrophages (LeFevre and Joel, 1984; Jani *et al.*, 1989; LeFevre *et al.*, 1978b, 1989).

Though dependent upon the size of the microparticulates, and not large on an absolute scale, the fraction of microparticulates taken up into the Peyer's patch following an intraluminally administered dose, estimated to be as high as 5%, is still quite significant (Pappo and Ermak, 1989). This can be compared to estimates of 0.01–0.02% for the fraction of antigen absorbed through non-Peyer's patch and non-receptor-mediated pathways following similar dosing (Gruskay and Cooke, 1955; Warshaw *et al.*, 1971). While the uptake of antigen into Peyer's patches is known to be more efficient than absorption elsewhere in the gut (Keljo and Hamilton, 1983), the observed high level of uptake for polystyrene nanoparticles (Pappo and Ermak, 1989) suggests that M-cell phagocytosis of those polymer particulates may be more proficient than that seen with biological particles (Wolf *et al.*, 1981; Owen *et al.*, 1986b). With regard to protein and peptide drug delivery, particulate uptake has been shown to be further enhanced when the particles are conjugated to anti-M-cell antibodies (Pappo *et al.*, 1991).

3.2. Paracellular Transport

While the previous discussion reveals that a large number of investigations have reported transcellular transport of particulates, the number of studies demonstrating paracellular transport is far smaller. Some of the earliest work on paracellular transport of particulates was performed by Volkheimer and co-workers (Volkheimer and Schulz, 1968, 1969; Volkheimer, 1975, 1977; Volkheimer *et al.*, 1968, 1969), who observed intestinal absorption into the blood, following oral administration in several animal models, including humans, of a wide variety of particles (e.g., starch grains, diatoms, pollens). *Candida albicans* has also been reported to be absorbed in the intestine by a paracellular pathway (Krause *et al.*, 1969; Stone *et al.*, 1974), and lipoidal loaded alkyl cyanoacrylate nanocapsules were observed in the intercellular space between enterocytes 10–15 min after injection into a jejunal loop (Damge *et al.*, 1990).

While, as discussed above, the maximum particle size generally reported for transcellular transport is 10 μm (Eldridge *et al.*, 1990; LeFevre *et al.*, 1980), Volkheimer observed the absorption of particles from 5 to 150 μm into the blood following oral administration. According to Volkheimer, the optimal size for particles absorbed by this process, which he assumed to involve paracellular transport and which he labeled "persorption," was 5–70 μm. The particle hardness was also reported to be a factor in the degree of absorption into the blood, with harder particles exhibiting more efficient absorption (Volkheimer and Schulz, 1968, 1969; Volkheimer *et al.*, 1968, 1969; Volkheimer, 1975, 1977).

Some form of local desquamation of intestinal epithelium seems to be the most common explanation of how particles, of any size, might be absorbed via the usually tight junctions between epithelial cells (Luckey, 1974; Csaky, 1984). One hypothesized mechanism for the extrusion of epithelial cells is a combination of the pressure from reproducing epithelial cells, which loosens the surrounding cells, and variations in lymphatic pressure in the lacteals acting on the epithelium. A decrease in lymphatic pressure could then, in theory, draw particulate matter into the site through this region of deepithelialization (Luckey, 1974). Persorption also may involve the "kneading" of large particles through the intestinal epithelial layer, and increased gastrointestinal motility, therefore, has been surmised to be important for increasing persorption (Volkheimer *et al.*, 1968). This hypothesis was supported by studies in which villi movement was either stimulated by drugs such as caffeine or neostigmine or deterred by atropine or barbituric acid and persorption rates of starch granules increased and decreased, respectively (Volkheimer, 1977; Volkheimer *et al.*, 1968).

3.3. Liposome Absorption

The uptake of liposomes into the intestinal wall appears to parallel the transcellular transport of particulates. All reports of liposome absorption indicate that the Peyer's patches are the principal pathway for transepithelial transport (Aramaki *et al.*, 1993; Michalek *et al.*, 1992; Childers *et al.*, 1990; Alpar *et al.*, 1992; Rowland and Woodley, 1980, 1981), though the specifics of this process are poorly characterized (Aramaki *et al.*, 1993). Similar to the observed uptake of latex particles by Peyer's patches (LeFevre *et al.*, 1985a), distearoylphophatidylcholine, phosphatidylcholine, phosphatidylserine, and cholesterol liposomes (DSPC-liposomes), 374 and 855 nm in diameter, were preferentially taken up by Peyer's patches in the lower ileum of rats (Aramaki *et al.*, 1993). Notably, these DSPC-liposomes have been found to be very stable in the acidity of the gastrointestinal tract (Alpar *et al.*, 1992; Rowland and Woodley, 1980, 1981), a characteristic considered critical for the potential transport of any encapsulated drug (Chiang and Weiener, 1987a, b). A negative charge may enhance uptake of liposomes, as was reported for somewhat larger, negatively charged phosphatidylserine liposomes in rat Peyer's patches (Tomizawa *et al.*, 1993). Also, similarly to polymer particulates, liposomes have been observed to be transported across the epithelium by endocytosis (Childers *et al.*, 1990; Aramaki *et al.*, 1993) and then exocytosed into the intercellular space of underlying lymphoid cells (Childers *et al.*, 1990). The ultimate fate of liposomes, once absorbed into the intestinal wall, is in some dispute. Some studies report uptake of intact liposomes into the portal vein (Das *et al.*, 1984) or even the systemic blood circulation (Dapergolas and Gregoriadis, 1976a). Other investigations dispute the possibility of intact liposome uptake into the systemic blood circulation (Patel and Ryman, 1977; Deshmukh *et al.*, 1981).

4. CASE STUDIES

4.1. Introduction

From the recognition and verification that labile therapeutic compounds can be protected from the harsh environment of the gastrointestinal system and that particles in the <10-μm range can cross the barrier of the intestinal mucosa have arisen many potential applications of microencapsulation for drug or vaccine delivery. Requirements for biocompatibility and stability have excluded many potential matrices for use although the potential for microspheres (monolithic systems) or microcapsules (reservoir systems) both to protect labile compounds and to aid in systemic delivery has inspired numerous companies and academic institutions to investigate such approaches for the oral delivery of peptides and proteins. Research and development in this area has progressed along many different pathways. Certain institutions have applied expert knowledge of microencapsulation techniques, whereas others have initiated projects based on an understanding of the physiological factors involved. Compounds under investigation for delivery range from established or novel systemic or targeted therapeutics to antigens for vaccine delivery. The systems themselves range from conventional aqueous insoluble microspheres of polymers with distinct properties to lipid systems, microencapsulated cells, and aqueous soluble bioadhesive microspheres. Although it is recognized that many of these systems will protect peptides and proteins from endogenous pH and enzymes, the main drive for the development of these types of systems is the panacea of significantly improving absorption and achieving systemic therapy, by oral administration, of compounds that are conventionally administered by the parenteral route.

The following case histories will elucidate the interest in not only therapeutic peptide and protein administration but also microencapsulated antigens for oral vaccine delivery. Oral immunization offers many practical advantages over parenteral. Not only is this route of administration more acceptable to patients, but the reduction in the need for highly trained personnel and refrigerated storage results in simpler logistics for mass immunization. In addition, oral immunization has been shown, in various systems, to induce a vigorous immune response at mucosal surfaces, the most common sites of entry of infectious agents.

4.2. Polyester Microspheres

Poly(lactic/glycolic acid) (PLGA) polymers and closely related analogs have been used for biomedical applications for many years, both as surgical sutures and as bone plates for internal fixation. In addition, PLGA polymers have been used for the preparation of parenteral controlled release drug delivery systems. Long experience with these compounds shows that they are completely biodegradable by hydrolysis of the ester linkages, to toxicologically acceptable products that are eliminated from the body. The polymer degradation rate is determined primarily by the ratio of lactide to glycolide present in the copolymer. Microspheres of PLGA are manufactured by solvent extraction and solvent evaporation techniques generally involving three inter-related processes: droplet formation, droplet stabilization, and droplet hardening (Arshady, 1991). Because of their well-documented history and their hydrophobicity, PLGA microspheres have also been investigated as potential antigen delivery systems for oral vaccines. Antigens in particulate form are generally effective oral immunogens, while soluble antigens are not (Cox and Taubman, 1984), possibly due to increased survival of the particulate antigen in the gastrointestinal tract and its enhanced adsorption and recognition in the immune inductive environment of the Peyer's patches. PLGA microspheres, with various ratios of lactide to glycolide, have been shown to be specifically taken up into the Peyer's patch lymphoid tissue of the gut. As previously discussed, the majority of the microspheres $<5\,\mu m$ in diameter are shown to leave the Peyer's patches and enter the mesenteric lymph nodes, whereas the majority of particles $>5\,\mu m$ in diameter remain in the Peyer's patches until digested, thereby stimulating a purely mucosal response (Eldridge *et al.*, 1990, 1989a). A number of preclinical studies on various antigens encapsulated into PLGA microspheres for oral delivery have been completed. The use of PLGA microspheres as an oral antigen delivery system for staphylococcal enterotoxin B (SEB) in mice has been described (Eldridge *et al.*, 1990, 1989b). The microparticulates induced circulating toxin-specific antibodies and a concurrent secret ory IgA antitoxin response in saliva, gut wash fluid, and bronchial-alveolar wash. Induction of the pulmonary antibody response by oral immunization with microencapsulated SEB has potential implications for the development of oral vaccines against resp iratory tract pathogens (Eldridge *et al.*, 1991). PLGA microspheres have also been investigated as oral delivery systems for entrapped influenza virus (Moldoveanu *et al.*, 1993), parainfluenza virus (Ray *et al.*, 1993), simian immunodeficiency virus, and purified enterotoxigenic *Escherichia coli* colonization factor antigens (Edelman *et al.*, 1993; Reid *et al.*, 1993).

Although it appears that the majority of work with PLGA microspheres for oral delivery is directed toward vaccine administration, PLGA nanospheres have also been investigated as a potential alternate oral delivery system for cyclosporin (Sanchez *et al.*, 1993).

4.3. Zein Microspheres

Prolamine is the name given to a characteristic class of proteins occurring specifically in cereals. Proteins of this class are rich in hydrophobic amino acids and soluble in aqueous alcohol. Zein, one of the typical prolamines and a major storage protein of corn (Larkins *et al.*, 1984; Swallen, 1941), possesses many applicable characteristics for designing an oral drug delivery matrix: zein is insoluble in aqueous media yet is degraded over time by protease enzymes to constituent peptides and amino acids; it is hydrophobic and exhibits mucoadhesive properties; it is also GRAS, is covered by a United States Pharmacopoeia monograph, and has been previously used in pharmaceutical coating methodology. Zein has also been investigated as a potential sustained release, direct compression tablet excipient (Katayama and Kanke, 1992).

Alkermes Inc. (Cambridge, Massachusetts) has developed a patented drug delivery system (Mathiowitz *et al.*, 1993) using zein as a microencapsulation matrix for therapeutic compounds, including peptides and proteins, for enteral delivery. The microspheres are formed by phase separation in a nonsolvent followed by solvent removal by extraction. The manufacturing process uses all GRAS materials and yields monolithic-type microspheres with greater than 90% therapeutic peptide or protein encapsulation efficiency and an average size of 1–5 μm. The size distribution of the microspheres is Gaussian and can be varied by modification of the manufacturing procedure.

The zein microsphere system (OraLease®) produced by Alkermes has been designed to protect drugs from the harsh environment of the stomach and the small intestine, increase the residence time of drugs targeted to the gastrointestinal tract, and enable or improve the transport of drugs from the lumen into the body. Oral peptide and protein delivery is one application of the microsphere system that is being investigated. Peptides and proteins successfully incorporated into the OraLease system include calcitonin, erythropoietin, desmopressin (dDAVP), vasopressin, and insulin. The company has also invested effort in the development of a solid oral dosage form for delivery of the microspheres as an appropriate pharmaceutical product. A Phase I clinical trial of an OraLease formulation of dDAVP delivered in capsules in normal volunteers has been completed and demonstrated that the OraLease formulation of dDAVP is safe and well tolerated and showed a dose-dependent physiological effect in humans (Alkermes press release, November 22, 1993).

Because of their hydrophobicity and biocompatibility, zein microspheres are also being investigated as potential drug delivery systems for oral vaccines (Alkermes, 1994). Potential oral vaccine targets include several infectious diseases such as pneumonia, meningitis and diarrhea. Applications of OraLease as a drug delivery systems may also be further advanced by the reported bioadhesive properties of zein microspheres (Mathiowitz *et al.*, 1994).

4.4. Proteinoid Microspheres

Proteinoids are man-made condensation polymers produced by random or directed assembly of natural or synthetic amino acids or small peptide chains. Following the discovery, in the late 1950s, that linear condensation polymers of mixed natural amino acids could interact with water to form hollow microspheres, proteinoids have been the subject of extensive investigations.

Emisphere Technologies Inc. (Hawthorne, New York) has developed a patented oral drug delivery system (Steiner *et al.*, 1990), citing the ability to encapsulate therapeutic agents, including peptides, proteins, and antigens, in microspheres composed of proteinoids. The system was developed to allow drugs to be absorbed unchanged into the bloodstream while protecting them from the harsh environment of the gastrointestinal tract. Reservoir-type microspheres, capable of carrying a cargo of drug, are formed by linear thermal condensation of amino acids at elevated temperatures in acidic medium. A spray-drying process can also produce uniform proteinoid material. The proteinoid microspheres assemble and disassemble purely on the basis of pH. Inclusion in the polymer mix of a stoichiometric excess of acidic dicarboxylic or polycarboxylic amino acid results in an acidic proteinoid, and inclusion of an excess of basic diamino or polyamino monomer results in a basic proteinoid. Ability to modify the system as such allows manipulation of the solubility of the microspheres to be dependent on the pH of the environment. It is reported that acidic proteinoids remain intact in the stomach but, when discharged into the small intestine, where a higher pH is encountered, undergo spontaneous disassociation to release drug (Robinson, 1993). However, direct information as to whether or not the proteinoids actively increase absorption of entrapped drug is not available at the time of writing. A variety of molecules have been incorporated into proteinoid microspheres, including insulin, heparin, glycoproteins, polio vaccine, monoclonal antibody IgG 2a, calcitonin, human growth hormone, and influenza vaccine (Milstein *et al.*, 1992; Emisphere Technologies, 1993). The size distribution of the microspheres tends to be Gaussian and can be varied by varying the ionic strength and the choice of amino acids to be incorporated in the polymer [a library of over 600 proteinoid compounds had been synthesized at Emisphere as of 1993 (Emisphere Technologies, 1993)]. The encapsulation efficiency of the process is reported to vary according to the type of proteinoid and the characteristics of the compound being encapsulated. Heparin has been incorporated at 2.1% in a hydrophobic amino acid proteinoid and at 8.9% in a positively charged amino acid proteinoid whereas insulin was incorporated at 43% and 13%, respectively (Milstein *et al.*, 1992).

The drug delivery system is reportedly nontoxic in rats, mice, guinea pigs, chickens, dogs, and monkeys (Milstein *et al.*, 1992; Steiner *et al.*, 1990). A potential for use of proteinoid microspheres for vaccine delivery has also been demonstrated by oral immunization of rats with proteinoid microspheres encapsulating HA-NA and M1 influenza virus antigens (Santiago *et al.*, 1993). The study reports that a single enteric dose of M1 entrapped in proteinoid microspheres was able to induce a significant IgG response as early as two weeks post dosing, while rats dosed orally with the same M1 total dose (no microspheres) showed no detectable antibody response. A single enteric dose of HA-NA spheres induced a response up to eight times higher than that observed in rats dosed with unencapsulated antigen. Studies in *cebus* and *cynomolgus* monkeys indicate that the oral dosing of proteinoid encapsulating low-molecular-weight heparin (LMWH) elicits a consistent clinical response (Milstein *et al.*, 1992). Initial clinical assessment of the system has also been conducted with a human safety and dose escalation study in the United Kingdom. A total of 14 subjects were fasted overnight and dosed with a microsphere suspension containing LMWH. Study results show that the system is effective and nontoxic in humans and that all doses administered were well tolerated. Emisphere plans to develop a proteinoid microsphere pharmaceutical product as an oral suspension, capsule, or tablet formulation.

4.5. Polycyanoacrylate Microspheres

The alkylcyanoacrylates are biodegradable polymers which have been used as tissue adhesives in surgery (Woodward *et al.*, 1965) and have been investigated as nanoparticles for controlled release of adsorbed drugs by parenteral administration in humans (Verdun *et al.*, 1986). Distribution of the polymer in the body is reported to be dependent on physicochemical properties of the particles such as particle size, surface charge, and rate of degradation. The degradation rate has been shown to be dependent on the molecular weight of the polymer. Polyalkylcyanoacrylate nanospheres have been thoroughly studied for applications ranging from ophthalmic delivery to use as carriers in cancer chemotherapy. They are generally prepared by polymerization of alkylcyanoacrylate in an acidic aqueous medium containing the drug (Chouinard *et al.*, 1994) although nanocapsules of polyisobutylcyanoacrylate, composed of an oily core surrounded by a polymeric film, have been developed to enable better delivery of lipophilic compounds (Chouinard *et al.*, 1991). It has been reported that polyalkylcyanoacrylate nanocapsules, defined as spherical vesicles less than 300 nm in diameter, can improve the intestinal absorption of lipophilic drugs and do pass through the intestinal epithelium (Damge *et al.*, 1988).

Insulin polybutylcyanoacrylate nanocapsules can be prepared by interfacial emulsion polymerization of isobutyl 2-cyanoacrylate. Insulin is added to a lipid phase containing miglyol and isobutyl 2-cyanoacrylate dissolved in ethanol. This lipid phase is then added to an aqueous phase containing nonionic surfactant. The suspension is concentrated by evaporation and then purified (Damge *et al.*, 1990). In corroborating studies, insulin nanocapsules, with a mean diameter of 0.22 μm and an encapsulation efficiency of 54.9%, administered orally to diabetic rats, induced a significant decrease in glycemia after 2 days, and the effect was maintained for up to 20 days (Damge *et al.*, 1988). Although an earlier study reported that alkylcyanoacrylate nanoparticles with adsorbed insulin were not effective following oral administration, the hypoglycemic effect of insulin nanocapsules was again reported in a further study in which the effect of site of administration in the gastrointestinal tract was assessed (O'Hagan, 1994). The hypoglycemic effect lasted from 11 to 16 days, depending on the site of administration.

Although insulin-loaded polybutylcyanoacrylate nanocapsules have been demonstrated to be effective at reducing the glucose-induced peak of hyperglycemia in both rats and dogs, other peptides investigated in this system, including secretin, cholecystokinin, and somatostatin, did not exert a prolonged biological effect after oral administration (Damge *et al.*, 1990).

Polybutylcyanoacrylate nanoparticles have also been investigated as potential antigen delivery systems. Particles with adsorbed ovalbumin and mean particles sizes of 0.1 and 3 μm were administered to rats by gastric intubation (O'Hagan *et al.*, 1989). Both groups of rats showed enhanced salivary IgA antibody responses in comparison to those shown by rats administered soluble ovalbumin. However, only the group receiving 0.1-μm particles showed an enhanced IgG antibody response. The authors of the study suggested that the particles may gain access to lymphoid tissue through the M cells of the Peyer's patches.

4.6. Lipid-Based Systems

In the following, lipid-based systems for oral drug delivery of peptides and proteins have been classified into liposomes and emulsions, although the distinction between these classifications is somewhat vague.

4.6.1. EMULSIONS

Cortecs International Limited (Middlesex, U.K.) has been developing a patented oral drug delivery system for peptides and proteins since 1987. The system, Macromol®, involves loosely associating the peptides or proteins to be delivered with two types of lipid which can cross the cell membranes in the gut and are taken into the systemic circulation through the lymphatic system. The emulsion system comprises a water-in-oil microemulsion in which the aqueous phase contains the therapeutic and the oil phase contains lecithin, nonesterified fatty acids, and cholesterol in critical proportions equivalent to those required by the cell for optimal secretion of chylomicrons (Cho and Flynn, 1989). Protease inhibitors have also been incorporated into the system to protect the therapeutic compound. The emulsion is converted to a solid oral dosage form by coating onto a solid core; during this process, the water is driven off, leaving the protein embedded in the oil. The delivery system is administered in hard gelatin capsules. Cortecs is focusing on the oral delivery of insulin and calcitonin by this method. Macromol® has been tested directly in preclinical studies in pigs with both insulin and calcitonin (New *et al.*, 1993); the oral availability of calcitonin was reported to be increased by an order of magnitude as judged by fall in plasma calcium levels (New *et al.*, 1994a). Human trials have also been completed, and results have been reported that demonstrate oral uptake of calcitonin at commercially viable doses as measured by appearance of collagen cross-links in the urine (New *et al.*, 1994b). A Phase II study of oral calcitonin is in progress at the time of writing (New *et al.*, 1994c).

Affinity Biotech. Inc. (Aston, Pennsylvania) is also investigating the oral delivery of peptides and proteins using a microemulsion delivery system. The company is reported to be running preclinical studies of the formulation with calcitonin and human growth hormone (Anonymous, 1994).

4.6.2. LIPOSOMES

Liposomes are microscopic closed vesicles composed of a bilayered phospholipid membrane. The ability of liposomes to transport drugs or other bioactive molecules to various tissues has been a subject of extensive investigation for some time. Among various routes of administration, oral dosing is the simplest and has obvious advantages, and a number of studies have been performed to investigate uptake and biodistribution of liposomes following oral delivery (Das *et al.*, 1984; Aramaki *et al.*, 1993).

Liposomes have been investigated for oral delivery of both peptide and protein therapeutics and vaccines. Liposomes have been used in rats and dogs as a means of preventing insulin degradation in the upper gastrointestinal tract and enhancing insulin absorption (Patel and Ryman, 1976; Patel *et al.* 1982; Dapergolas and Gregoriadis, 1976b). Liposomes of various fluid (e.g., phosphatidylcholine, cholesterol, and dicetyl phosphate) and solid (e.g., dipalmitoyl phosphatidylcholine, cholesterol, and dicetyl phosphate) formulations were produced in these studies by sonication processes. However, although liposomes do exhibit some ability to protect insulin and enhance absorption, massive amounts of peroral liposomal insulin are required to achieve modest reductions in blood glucose, and a significant amount of further study will likely be required before feasible dosing levels are attained (Spangler, 1990). Liposomes have also been investigated for their ability to deliver blood coagulation factor VIII. Oral administration of factor VIII in liposomes to a hemophiliac patient led to a higher rise in blood levels than did that of the free component (Hemker *et al.*, 1980).

Investigations to support the use of oral liposomes as delivery systems for vaccines have included Peyer's patch uptake studies in rats (Michalek *et al.*, 1992); homogeneous unilamellar liposomes of ~100 nm were reported to be taken up into M cells. *In vivo* studies in rats to develop a liposomal vaccine against *Streptococcus mutans* were also reported in the study; oral administration of antigen in liposomes resulted in a mucosal response that was higher than that obtained when the oral administration consisted of antigen alone. Studies on the effectiveness of liposomes as adjuvants of orally and nasally administered tetanus toxoid in guinea pigs (Alpar *et al.*, 1992) and soluble proteins in mice (Clarke and Stokes, 1992) have been reported. The distearoyl phosphatidylcholine and cholesterol tetanus toxoid liposome formulation significantly improved the immune response as compared to that obtained with the free antigen. However, Clarke and Stokes (1992) concluded from their *in vivo* studies that liposomes containing ovalbumin or keyhole limpet hemocyanin were ineffective at eliciting any significant increase in serum or intestinal antibody response as compared with the free antigen. In further investigations, *in vitro* studies performed by

Clarke and Stokes reported that the addition of bile caused a rapid and profound release of protein marker from the liposomes. These results led to the overall conclusion of the study that liposomes may be useful as carriers for orally administered compounds but are ineffective as adjuvants for the nonparticulate, naturally weak immunogens used in the investigation.

5. CONCLUSION

With the comprehensive research base that has developed in this area, if therapeutic peptides or proteins can be successfully and, to some extent, economically microencapsulated to produce a homogeneous and stable drug delivery system, the potential for a successful system is unchallengeable. However, although many systems are in the process of preclinical and even early clinical trials, there are still many issues in this field that need to be addressed and resolved. Certain studies are performed with compounds that, for a number of reasons, are not viable commercially for oral delivery, usually because of related issues such as dose, cost, and encapsulation efficiency. Although many such studies are cited as "proof of concept," it is extremely difficult to define a model peptide or protein with which to optimize a delivery system. Detailed scientific information on mechanisms of action is also lacking in many circumstances and, if available, would allow logical approaches to system optimization and ultimately potential widespread system application.

REFERENCES

Abrahamson, D., and Rodewald, R., 1981, Evidence for the sorting of endocytic vesicle contents during the receptor-mediated transport of IgG across the newborn rat intestine, *J. Cell Biol.* **91:**270–280.

Addison, J. M., Matthews, D. M., and Burston, D., 1974, Evidence for active transport of the dipeptide carnosine (β-alanyl-L-histidine) by hamster jejunum *in vitro, Clin. Sci. Mol. Med.* **46:**707–714.

Addison, J. M., Burston, D., Dalrymple, J. A., Matthews, D. M., Payne, J. W., Sleisenger, M. H., and Wilkinson, S., 1975, A common mechanism for transport of di- and tripeptides by hamster jejunum *in vitro, Clin Sci. Mol. Med.* **49:**313–322.

Adibi, S. A., and Kim, Y. S., 1981, Peptide absorption and hydrolysis, in: *Physiology of the Gastrointestinal Tract*, 1st ed. (L. R. Johnson, ed.), Raven Press, New York, p. 1073–1085.

Adibi, S. A., and Morse, E. L., 1977, The number of glycine residues which limit intact absorption of glycine oligopeptides in human jejunum, *J. Clin. Invest.* **60:**1008–1016.

Adibi, S. A., and Phillips, E., 1968, Evidence for greater absorption of amino acid from peptide than from free form by human intestine, *Clin. Res.* **16:**446–448.

Alkermes, Inc., 1994, Annual Report, Cambridge Massachusetts.

Allen, J. G., Havas, L., Leichtt, E., Lenox-Smith, J., and Nisbet, L. J., 1979, Phosphonopeptides as antibacterial agents: Metabolism and pharmacokinetics of alafosfalin in animals and humans, *Antimicrob. Agents, Chemother.* **16**:306–313.

Allen, R. H., and Majerus, P. W., 1972a, Isolation of vitamin B12-binding proteins using affinity chromatography II. Purification and properties of a human granulocyte vitamin B12-binding protein, *J. Biol. Chem.* **242**:7702–7708.

Allen, R. H., and Majerus, P. W., 1972b, Isolation of vitamin B12-binding proteins using affinity chromatography III. Purification and properties of human plasma transcobalamin II. *J. Biol. Chem.* **242**:7709–7717.

Alpar, H. O., Bowen, J. C., and Brown, M. R. W., 1992, Effectiveness of liposomes as adjuvants of orally and nasally administered tetanus toxoid, *Int. J. Pharm.* **88**:335–344.

Alpers, D. H., 1987, Digestion and absorption of carbohydrates and proteins, in: *Physiology of the Gastrointestinal Tract,* Vol. 2. (L. R. Johnson, ed.), Raven Press, New York, pp. 1469–1487.

Alpers, D. H., and Isselbacher, K. J., 1967, Protein synthesis by the rat intestinal mucosa: The role of ribonuclease, *J. Biol. Chem.* **242**:5617–5620.

Amidon, G. L., and Lee, H. J., 1994, Absorption of peptide and peptidomimetic drugs, *Annu. Rev. Pharmacol. Toxicol.* **34**:321–341.

Amidon, G. L., Sinko, P. J., and Fleisher, D., 1988, Estimating human oral fraction dose absorbed: A correlation using rat intestinal membrane permeability for passive and carrier-mediated compounds, *Pharm. Res.* **5**:651–654.

Anonymous, 1994, *BioScan* **8**(April).

Aramaki, Y., Tomizawa, H., Hara, T., Yachi, K., Kikuchi, H., and Tsuchiya, S., 1993, Stability of liposomes *in vitro* and their uptake by rat Peyer's patches following oral administration, *Pharm. Res.* **10**:1228–1231.

Arshady, R., 1991, Preparation of biodegradable microspheres and microcapsules: 2. Polyactides and related polyesters, *J. Controlled Release* **17**:1–22.

Asatoor, A. M., Chadha, A., Milne, M. D., and Prosser, D. I., 1973, Intestinal absorption of stereoisomers of dipeptides in the rat, *Clin. Sci. Mol. Mol. Med.* **45**:199–212.

Bai, P. F., Subramanian, P., Mosberg, H. I., and Amidon, G. L., 1991, Structural requirements for the intestinal mucosal cell peptide transporter: The need for N-terminal α-amino group, *Pharm. Res.* **8**:593–599.

Benet, L. Z., and Williams, R. L., 1990, Design and optimization of dosage regimens: Pharmacokinetic data, in: *The Pharmacological Basis of Therapeutics* (A. G. Gilman, T. W. Rall, A. S. Nies, and P. Taylor, eds.), Pergamon, New York, 1655.

Bernstein, I. D. and Ovary, Z., 1968, Absorption of antigens from the gastrointestinal tract, *Int. Arch. Allergy Appl. Immunol.* **33**:521–529.

Blaustein, R. O., Germann, W. J., Finkelstein, A., and DasGupta, B. R., 1987, The N-terminal half of the heavy chain of botulinum type A neurotoxin forms of channels in planar phospholipid bilayers, *FEBS Lett.* **226**:115–120.

Bockman, D. E., and Cooper, M. D., 1973, Pinocytosis by epithelium associated with lymphoid follicles in the bursa of Facricius appendix and Peyer's patches. An electron microscopic study, *Am. J. Anat.* **136**:455–477.

Bockman, D. E., and Winborn, W. B., 1966, Light and electron microscopy of intestinal ferritin absorption. Observations in sensitized and nonsensitized hamsters (*Mesocricetus auratus*), *Anat. Rec.* **155**:603–622.

Boyd, C. A. R., and Ward, M. R., 1982, A micro-electrode study of oligopeptide absorption by the small intestinal epithelium of *Necturus macuosus*, *J. Physiol.* **324**:411–428.

Burston, D., Taylor, E., and Matthews, D. M., 1979, Intestinal handling of two tetrapeptides by rodent small intestine *in vitro*, *Biochim. Biophy. Acta.* **553**:175–178.

Casley-Smith, J. R., 1967, The passage of ferritin into jejunal epithelial cells, *Experientia* **23**:370–371.

Castle, W. B., 1953, *N. Engl. J. Med.* **11**:603.

Castro, G., 1981, in: *Development of Mammalian Absorptive Processes*, Ciba Foundation Symposium 70, Excerpta Medica, Amsterdam, p. 201–225.

Chanarin, I., Muir, M., Hughes, A., and Hoffbrand, A. V., 1978, Evidence for intestinal origin of transcobalamin II during vitamin B_{12} absorption, *Br. Med. J.* **1**:1453–1455.

Cheeseman, C. I., and Smyth, D. H., 1972, Specific transfer process for intestinal absorption of peptides, *Proc. Physiol. Soc.* **1972**(November):45p–46p.

Chiang, C.-M., and Weiner, N., 1987a, Gastrointestinal uptake of liposomes. I: *In vitro* and *in situ* studies, *Int. J. Pharm.* **37**:75–85.

Chiang, C.-M., and Weiner, N. 1987b, Gastrointestinal uptake of liposomes. II: *In vivo* studies, *Int. J. Pharm.* **40**:143–150.

Childers, N. K., Denys, F. R., McGee, N. F., and Michalek, S. M., 1990, Ultrastructural study of liposome uptake by M cells of rat Peyer's patch: An oral vaccine system for delivery of purified antigen, *Regional Immunol.* **3**:8–16.

Chisui, N. S., 1968, Morphological aspects suggesting the transfer procedure absorption of some proteins in intestine of adult rats, *Rev. Roum. Med. Intern.* **5**:65–71.

Cho, Y. W., and Flynn, M., 1989, Oral delivery of insulin, *Lancet* **1989**:1518–1519.

Chouinard, F., Kan, F. W. K., Leroux, J.-C., Foucher, C., and Lenaerts, V., 1991, Preparation and purification of isohexylcyanoacrylate nanocapsules, *Int. J. Pharm.* **72**:211–217.

Chouinard, F., Buczkowski, S., and Lenaerts, V., 1994, Polyalkylcyanoacrylate nanocapsules: Physicochemical characterization and mechanism of formation, *Pharm. Res.* **11**:869–873.

Clarke, C. J., and Stokes, C. R., 1992, The intestinal and serum humoral immune response of mice to systemically and orally administered antigens in liposomes: I. The response to liposome-entrapped soluble proteins, *Vet. Immunol. Immunopathol.* **32**:125–138.

Clayton, J. P., Cole, M., Elson, S. W., Hardy, K. D., Mizen, L. W., and Sutherland, R., 1975, Preparation, hydrolysis and oral absorption of alpha-carboxy esters of carbenicillin, *J. Med. Chem.* **18**:172–177.

Cornell, R., Walker, W. A., and Isselbacher, K. J., 1971, Small intestinal absorption of horseradish peroxidase. A cytochemical study, *Lab. Invest.* **25**:42–48.

Cox, D. S., and Taubman, M. A., 1984, Oral induction of the secretory antibody response by soluble and particulate antigens, *Int. Arch. Appl. Immunol.* **75**:126–131.

Csaky, T. Z., 1984, Intestinal permeation and permeability: An overview, in: *Pharmacology of Intestinal Permeation I* (T. Z. Csaky, ed.), Spinger-Verlag, Berlin.

Dallas, W. S., and Falkow, S., 1980, Amino acid sequence homology between cholera toxin and *E. coli* heat-labile toxin, *Nature* **288**:499–201.

Damge, C., Michel, C., Aprahamian, M., and Couvreur, P., 1988, New approach for oral administration of insulin with polyalkylcyanoacrylate nanocapsules as drug carrier, *Diabetes* **37**:246–251.

Damge, C., Michel, C., Aprahamian, M., Couvreur, P., and Devissaguet, J. P., 1990, Nanocapsules as carriers for oral peptide delivery, *J. Controlled Release* **13**:233–239.

Danforth, E., and Moore, R. D., 1959, Intestinal absorption of insulin in the rat, *Endocrinology* **65**:118–126.

Dapergolas, G., and Gregoriadis, G., 1976a, *Lancet* **ii**:1054–1056.

Dapergolas, G., and Gregoriadis, G., 1976b, Hypoglycaemic effect of liposome entrapped insulin administered intragastrically into rats, *Lancet* **ii**:824–827.

Das, M., and Radhakrishnan, A. N., 1975, Studies on a wide-spectrum intestinal dipeptide uptake system in the monkey and in the human, *Biochem. J.* **146:**133–139.

Das, N., Das, M. K., and Bachhawat, B. K., 1984, Detection of liposomes in portal blood following oral administration, *J. Appl. Biochem.* **6:**346–352.

Deshmukh, D. S., Bear, W. D., and Brockerhoff, H., 1981, Can intact liposomes be absorbed in the gut? *Life Sci.* **28:**239–242.

Doxsey, S. J., Brodsky, F. M., Blank, G. S., and Helenius, A., 1987, Inhibition of endocytosis by anti-clathrins antibodies, *Cell* **50:**453–463.

Dunn, K. W., McGraw, T. E., and Maxfield, F. R., 1989, Iterative fractionation of recyclin receptor from lysosomally destined ligands in an early sorting endosome, *J. Cell Biol.* **109:**3303–3314.

Dvorak, A. F., and Bast, R. C., 1970, Nature of the immunogen in crystalline serum albumin, *Immunochemistry* **7:**118–124.

Ebel, J. P., 1990, A method for quantifying particle absorption form the small intestine of the mouse, *Pharm. Res.* **7:**848–851.

Edelman, R., Russell, R. G., Losonsky, G., Tall, B. D., Tacket, C. O., Levine, M. M., and Lewis, D. H., 1993, Immunization of rabbits with enterotoxigenic *E. coli* colonization factor antigen (CFA/I) encapsulated in biodegradable microspheres of poly(lactide-*co*-glycolide), *Vaccine* **11**(2):155–158.

Egberts, H. J., Koninkx, J. F., Dijk, J., and Mouwen, J. M., 1984, Biological and pathobiological aspects of the glycocalyx of the small intestinal epithelium—a review, *Vet. Quat.* **6**:186–199.

Eldridge, J., 1988, Paper presented at the Oral Immunization Symposium, Birmingham, Alabama.

Eldridge, J. H., Gilley, R. M., Staas, J. K., Moldoveanu, Z., Meulbroek, J. A., and Tice, T. R., 1989a, Biodegradable microspheres: Vaccine delivery system for oral immunization, *Curr. Top. Microbiol. Immunol.* **146:**59–65.

Eldridge, J. H., Meulbroek, J. A., Staas, J. K., Tice, T. R., and Gilley, R. M. 1989b, Vaccine-containing biodegradable microspheres specifically enter the gut-associated lymphoid tissue following oral administration and induce a disseminated mucosal immune response. *Adv. Exp. Med. Biol.* **251:**191–202.

Eldridge, J. H., Hammond, C. J., Meulbroek, J. A., Staas, J. K., Gilley, R. M., and Tice, T. R., 1990, Controlled vaccine release in the gut-associated lymphoid tissues. 1. Orally administered biodegradable microspheres target the Peyer's patches, *J. Controlled Release* **11:**205–214.

Eldridge, J. H., Staas, J. K., Meulbroek, J. A., McGhee, J. R., Tice, T. R. and Gilley, R. M., 1991, Biodegradable microspheres as a vaccine delivery system, *Mol. Immunol.* **28:**287–294.

Emisphere Technologies, 1993, Annual Reports, Emisphere Technologies, Inc., Hawthorne, New York.

Esposito, G., Faelli, A., Tosco, M., Orsenigo, M., De Gasperi, R., and Pacces, N., 1983, Influence of the enteric surface coat on the unidirectional flux of acetamide across the wall of the rat small intestine, *Experientia* **39**:149–151.

Fara, J. W., 1985, in: *Rate Control in Drug Therapy* (L. F. Prescott and W. S. Nimmo, eds.), Churchill Livingstone, New York, pp. 144–150.

Fox, H. J., and Castle, W. B., 1942, *Am. J. Med. Sci.* **203**:18–26.

Friedman, D. I., and Amidon, G. L., 1989a, Intestinal absorption mechanism of two prodrug ACE inhibitors in rats: Enalapril maleate and fosinopril sodium, *Pharm. Res.* **6**:1043–1047.

Friedman, D. I., and Amidon, G. L., 1989b, The intestinal absorption mechanism of dipeptide ACE inhibitors of the lysyl-proline type: Lisonopril and SQ 29,852, *Pharm. Sci.* **78**:995–998.

Ganapthy, V., and Leibach, F., 1985, Is intestinal peptide transport energized by a proton gradient? *Am. J. Physiol.* **249**:G153–G160.

Gardner, M. L., 1984., Intestinal assimilation of intact peptides and proteins from the diet—a neglected field, *Biol. Rev.,* **59**:289–331.

Gill, D. M., 1978, Seven toxic peptides that cross cell membranes. in: *Bacterial Toxins and Cell Membranes* (J. Jeljasewicz and T. Wadstrom, eds.), Academic Press, New York, pp. 291–332.

Goldstein, I. J., and Poretz, R. D., 1986, Isolation, physiochemical characterization and carbohydrate specificity of lectins, in: *The Lectins* (I. E. Leiner, N. Sharon, and I. J. Goldstein eds.), Academic Press, Orlando, Florida, pp. 33–47.

Grasback, R., Kantero, I., and Slurala, M., 1959, Influence of calcium ions on vitamin B_{12} absorption, steatorrheoea and pernicious anaemia, *Lancet* **i**:234–236.

Greer, F., Brewer, A. C., and Pusztai, A., 1985, Effect of kidney bean (*Phaseolus vulgaris*) toxin on tissue weight and composition and some metabolic functions of rats, *Br. J. Nutr.* **54**:95–103.

Gruber, P., Longer, M. A., and Robinson, J. R., 1987, Some biological issues in oral, controlled drug delivery, *Adv. Drug Delivery Rev.* **1:**1–18.

Gruskay, F. L., and Cooke, R. E., 1955, The gastrointestinal absorption of unaltered protein in normal infants and in infants recovering from diarrhea, *Pediatrics* **16**:763–768.

Hammer, R., Joel, D. D., and LeFevre, M. E., 1983, Ultrastructure of macrophages of the murine Peyer's patch dome, *Exp. Cell. Biol.* **51:**61–69.

Hemker, H. C., Muller, A. D., Hermens, W. T., and Zwaal, R. F. A., 1980, Oral treatment of haemophilia A by gastrointestinal absorption of factor VIII entrapped in liposomes, *Lancet* **1980:**70–71.

Heyman, M., Duroc, R., Desjeux, J.-F., and Morgat, J. L., 1982, Horseradish peroxidase transport across adult rabbit jejunum *in vitro*, *Am. J. Physiol.* **242:**G558–G564.

Heyman, M., Crain-DeNoyelle, A-M., Nath, S. M., and Desjeux, J-F., 1990, Quantification of protein transcytosis in the human colon carcinoma cell line CaCo-2, *J. Cell. Physiol.* **143:**391–395.

Hidalgo, I. J., Raub, T. J., and Borchardt, R. T., 1989, Characterization of the human colon carcinoma cell line (CaCo-2) as a model system for intestinal epithelial permeability, *Gastroenterology* **96:**736–749.

Houston, J. B., and Wood, S. G., 1980, Gastrointestinal absorption of drugs, in: *Progress in Drug Metabolism* (J. W. Bridges and L. F. Chasseaud, eds.), John Wiley & Sons, New York.

Hu, M., and Amidon, G. L., 1988, Passive and carrier–mediated intestinal absorption components of captopril, *J. Pharm. Sci.* **77:**1007–1011.

Humphrey, M. J., 1986, The oral bioavailability of peptides and related drugs in: *Delivery Systems for Peptide Drugs* (S. S. Davis, T. I. Illum, and E. Tomlinson, eds.), Plenum Press, New York, pp. 139–151.

Humphrey, M. J., and Ringrose, P. S., 1986, Peptides and related drugs: A review of their absorption, metabolism, and excretion, *Drug Metab. Rev.* **17:**283–310.

Jackson, D., Walker-Smith, J. A., and Philips, A. D., 1983, Macromolecular absorption by histological normal and abnormal small intestinal mucosa in childhood: An *in vitro* study using organ culture, *J. Pediatr. Gastroenterol. Nutr.* **2:**235–247.

Jani, P., Halbert, W., Langridge, J., and Florence, A. T., 1989, The uptake and translocation of latex nanospheres and microspheres after oral administration to rats, *J. Pharm. Pharmacol.* **41:**809–812.

Joel, D. D., Sordat, B., Hess, W., and Cotier, H., 1970, Uptake and retention of particles from the intestine by Peyer's patches in mice, *Experientia* **26:**694.

Joel, D. D., Laissue, J. A., and LeFevre, M. E., 1978, Distribution and fate of ingested carbon particles in mice, *J. Reticulendothel. Soc.* **24:**477–487.

Kabir, S., Rosenstreich, D. L., and Mergenhagen, S. E., 1978, Bacterial endotoxins and cell membranes, in: *Bacterial Toxins and Cell Membranes* (J. Jeljaszewicz and T. Wadstrom, eds.), Academic Press, New York, pp. 59–87.

Katayama, H., and Kanke, M., 1992, Drug release from directly compressed tablets containing zein, *Drug Dev. Ind. Pharm.* **18:**2173–2184.

Keljo, D. J., and Hamilton, J. R., 1983, Quantitative determination of macromolecular transport across intestinal Peyer's patches, *Am. J. Physiol.* **244:**G637–644.

Kerchner, G. A., and Geary, L. E., 1983, Studies on the transport of enkephalin-like oligopeptides in rat intestinal mucosa, *J. Pharmacol. Exp. Ther.* **226**:33–38.

Kilpatrick, D. C., Pusztai, A., Grant, G., Graham, C., and Ewen, S. W. B., 1985, Tomato lectin resists digestion in the mammalian alimentary canal and binds to intestinal villi without deleterious effects, *FEBS Lett.* **185:**299–305.

Kimura, T., Yamamoto, T., Mizuno, M., Suga, Y., Kitade, S., and Sezaki, H., 1983, Characterization of aminocephalosporin transport across rat small intestine, *J. Pharmacobiodyn.* **6:**246–253.

King, T. P., Pusztai, A., and Clarke, E. M. W., 1980, Kidney bean (*Phaseolus vulgaris*) lectin-induced lesion in rat small intestine. 1. Light microscope studies. *J. Comp. Pathol.* **90:**585–595.

King, T. P., Pusztai, A., and Clarke, E. M. W., 1982, Kidney bean (*Phaseolus vulgaris*) lectin-induced lesion in rat small intestine. 3. Ultrastructural studies, *J. Comp. Pathol.* **92:**357–373.

King, T. P., Pusztai, A., Grant, G., and Slater, D., 1986, Immunogold localization of ingested kidney bean (*Phaseolus vulgaris*) lectins in epithelial cells of the rat small intestine, *Histochem. J.* **18:**413–420.

Korenblat, R. E., Rothberg, R. M., and Minden, P., 1968, Immune response of human adults after oral and prenatal exposure to bovine serum albumin, *J. Allergy* **41:**226–235.

Kramer, W., Girbig, F., Gutjahr, U., Kleeman, H. W., Leipe, I., Urbach, H., and Wagner, A., 1990, Interaction of renin inhibitors with the intestinal uptake system for oligopeptides and beta-lactam antibiotics, *Biochim. Biophys. Acta* **1027:**25–30.

Krause, W., Matheis, H., and Wulf, K., 1969, Fungamie and funguria after oral administration of *Candida albicans, Lancet* **i:**598–599.

Landsverk, T., 1988, Phagocytosis and transcytosis by follicle-associated epithelium of the ileal Peyer's patch in calves, *Immunol. Cell. Biol.* **66:**261–268.

Larkins, B. A., Pedersen, K., Marks, M. D., and Wilson, D. R., 1984, The zein proteins of maize endosperm, *Trends Biochem. Sci.* **1984**(July)**:**306–308.

Lee, V. H. L., Dodda-Kashi, S., Grass, G. M., and Rubas, W., 1991, Oral route of peptide and protein drug delivery, in: *Peptide and Protein Drug Delivery* (V. H. L. Lee, ed.), Marcel Dekker, New York, pp. 691–738.

LeFevre, M. E., and Joel, D. D., 1984, Peyer's patch epithelium: An imperfect barrier to large particulates in mice, in: *Intestinal Toxicology* (C. M. Schiller, ed.), Raven Press, New York, p. 45–58.

LeFevre, M. E., and Joel, D. D., 1986, Distribution of label after intragastric administration of ^{7}Be-labeled carbon to weanling and aged mice, *Proc. Soc. Exp. Biol. Med.* **182:**112–119.

LeFevre, M. E., Olivo, R., Vanderhoff, J. W., and Joel, D. D., 1978a, Accumulation of latex in Peyer's patches and its subsequent appearance in villi and mesenteric lymph nodes, *Proc. Soc. Exp. Biol. Med.* **159:**298–302.

LeFevre, M. E., Vanderhoff, J. W., Lasissue, J. A., and Joel, D. D., 1978b, Accumulation of 2 μm latex particles in mouse Peyer's patches during chronic latex feeding, *Experientia* **34:**120–122.

LeFevre, M. E., Hammer, R., and Joel, D. D., 1979, Macrophages of the mammalian small intestine: A review, *J. Reticulendothel. Soc.* **26:**553–573.

LeFevre, M. E., Hancock, D. C., and Joel, D. D., 1980, Intestinal barrier to large particles in mice, *J. Toxicol. Environ. Health* **6:**691–704.

LeFevre, M. E., Joel, D. D., and Schidlovsky, G., 1985a, Retention of ingested latex particles in Peyer's patches of germfree and conventional mice, *Proc. Sci. Exp. Biol. Med.* **179:**522–528.

LeFevre, M. E., Warren, J. B., and Joel, D. D., 1985b, Particles and macrophages in murine Peyer's patches, *Exp. Cell. Biol.* **53:**121–129.

LeFevre, M. E., Boccio, A. M., and Joel, D. D., 1989, Intestinal uptake of fluorescent micropheres in young and aged mice, *Proc. Soc. Exp. Biol. Med.* **190:**23–27.

Luckey, T. D., 1974, The villus chemostat in man, *Am. J. Clin. Nutr.* **27:**1266–1276.

Mathan, V. I., Baboir, B. M., and Donaldson, R. M., 1974, Kinetics of the attachment of intrinsic factor bound cobamides to ileal receptors, *J. Clin. Invest.* **54:**598–608.

Mathiowitz, E., Bernstein, H., Morrel, E., and Schwaller, K., 1993, Method for producing protein microspheres, U.S. Patent 5,271,961, December 21, 1993.

Mathiowitz, E., Chickering, D., Jacob, J., DiBiase, M., Berstein, H., Gunn, K., and Sherman, M., 1994, GI transit studies of hydrophobic microspheres, *Proc. Int. Symp. Control. Rel. Bioact. Mater.* **21**.

Matthews, D. M., 1975, Intestinal absorption of peptides, *Physiol. Rev.* **55:**537–540.

Matthews, D. M., 1983, Intestinal absorption of peptides, *Biochem. Soc. Trans.* **11:**808–810.

Matthews, D. M., 1991, *Protein Absorption*, John Wiley & Sons, New York.

Matthews, D. M., and Payne, J. W., 1980, Transmembrane transport of small peptides, *Curr. Top. Membr. Transp.* **14:**331–425.

Matthews, D. M., Craft, I. L., Geddes, D. M., Wise, I. J., and Hyde, C. W., 1968, Absorption of glycine and glycine peptides from the small intestine of the rat, *Clin. Sci.* **35:**415–424.

Mayorga, L. S., Diax, R., and Stahl, P. D., 1989, Reconstitution of endosomal proteolysis in a cell-free system, *J. Biol. Chem.* **264:**5392–5399.

McClugage, S. G., Low, F. N., and Zimny, M. L., 1986, Porosity of the basement membrane overlying Peyer's patches in rats and monkeys, *Gastroenterology* **91:**1128–1133.

McDonel, J. L., 1980, Binding of *Clostridium perfringes* (^{125}I) enterotoxin to rabbit intestinal cells, *Biochemistry* **19:**4801–4807.

Menzies, I. S., 1984, Transmucosal passage of inert molecules in health and disease, in: *Intestinal Absorption and Secretion* (E. Skadhauge and K. Heintze, eds.), MTP Press, Lancaster, pp. 527–543.

Michalek, S. M., Childers, N. K., Katz, J., Dertzbaugh, M., Zhang, S., Russell, M. W., Macrina, F. L., Jackson, S., and Mestecky, J., 1992, Liposomes and conjugate vaccines for antigen delivery and induction of mucosal immune responses, in: *Genetically Engineered Vaccines* (J. E. Ciardi *et al.*, eds.), Plenum Press, New York, pp. 191–198.

Milstein, S., Baughman, R., Santiago, N., Rivera, T., Falzarano, L., Dewland, P., and Welch, S., 1992, Initial clinical assessment of the oral administration of low molecular weight heparin (LMWH) using the proteinoid oral delivery system, American Association of Pharmaceutical Scientists 1992 Annual Meeting and Exposition, November 15–19, San Antonio, Texas.

Moldoveanu, Z., Novak, M., Huang, W.-Q., Gilley, R. M., Staas, J. K., Schfer, D., Compans, R. W., and Mestecky, J., 1993, Oral immunization with influenza virus in biodegradable microspheres, *J. Infect. Dis.* **167:**84–90.

Nakishima, E., Tsuji, A., Kagatani, S., and Yamana, T., 1984, Intestinal absorption mechanism of amino-β-lactam antibiotics. III. Kinetics of carrier-mediated transport across the rat small intestine *in situ*, *J. Pharmacobio-Dyn.* **7:**452–464.

Neutra, M. R., Phillips, T. L., Mayer, E. L., and Fishkind, D. J., 1987, Transport of membrane-bound macromolecules by M cells in follicle-associated epithelium of rabbit Peyer's patch, *Cell Tissue Res.* **247**:537–546.

New, R. R. C., Guard, P., Hotten, P., Stevens, D., Harris, R., Shepherd, T., and Flynn, M. J., 1993, Use of a lipid carrier to deliver calcitonin and insulin via the small intestine, Sixth International Symposium on Recent Advances in Drug Delivery Systems, Salt Lake City, Utah, February 22–25.

New, R. R. C., Guard, P. W., Littlewood, G. M., and Flynn, M. J., 1994a, Oral delivery of calcitonin to animals and man in an emulsion-based carrier system, U.K. Association of Pharmaceutical Sciences, Leicester, March 29–31.

New, R. R. C., Guard, P. W., Littlewood, G. M., and Flynn, M. J., 1994b, Changes in urinary crosslinks after administration of calcitonin to humans by subcutaneous and oral routes, Controlled Release Society 21st International Symposium, Nice, France, June.

New, R. R. C., Guard, P. W., Littlewood, G. M., Sandbank, B. M., and Flynn, M. J., 1994c, Administration of calcitonin to humans: Comparison of intra-nasal and oral routes, Third European Symposium on Controlled Release Drug Delivery, Noordwijk aan Zee, The Netherlands, April 6–8.

Newey, H., and Smith, D. H., 1962, *J. Physiol.* **164**:527.

Nicholson, J. A., and Peters, T. J., 1977, Subcellular distribution of hydrolase activities for glycine and leucine homopeptides in human jejunum, *Clin. Sci. Mol. Med.* **54**:205–207.

Oh, D., 1991, Estimating Oral Drug Absorption in Humans, Ph.D. thesis, University of Michigan, Ann Arbor.

Oh, D., Curl, R. L., and Amidon, G. L., 1993, Estimating the fraction dose absorbed from suspensions of poorly soluble compounds in humans: A mathematical model, *Pharm. Res.* **10**:264–270.

Oh, D. M., Sinko, P. J., and Amidon, G. L., 1989, Characterization of oral absorption of some penicillins: Determination of intrinsic membrane absorption parameters in the intestine *in situ*, *Pharm. Res.* **6**:S-91.

Oh, D. M., Sinko, P. J., and Amidon, G. L., 1990, Peptide transport of β-lactam antibiotics: Structural requirements for an α-amino group, *Pharm. Res.* **7**:S-119.

O'Hagan, D. T., 1994, Microparticles as oral vaccines, in: *Novel Delivery Systems for Oral Vaccines* (D. T. O'Hagan, ed.), CRC Press, Boca Raton, Florida, p. 188–197.

O'Hagan, D. T., Palin, K. J., and Davis, S. S., 1987, Intestinal absorption of proteins and macromolecules and the immunological response, *CRC Crit. Rev. Ther. Drug Carrier Syst.* **4**:197–220.

O'Hagan, D. T., Palin, K. J., and Davis, S. S., 1989, Polybutylcyanoacrylate particles as adjuvants for oral immunization, *Vaccine* **7**:213–216.

Owen, R. L., 1977, Sequential uptake of horseradish peroxidase by lymphoid follicle epithelium of Peyer's patches in the normal unobstructed mouse intestine: An ultrastructural study, *Gastroenterology* **72**:440–451.

Owen, R. L., and Bhalla, D. K., 1983, Cytochemical analysis of alkaline phosphatase and sterase activities and of lectin-binding and anionic sites in rat and mouse Peyer's patch M cells, *Am. J. Anat.* **168**:199–212.

Owen, R. L., and Jones, A. I., 1974, Epithelial cell specialization within human Peyer's patches: An ultrastructural study of intestinal lymphoid follicles, *Gastroenterology* **66**:189–203.

Owen, R. L., Apple, R. T., and Bhalla, D. K., 1986a, Morphometric and cytochemical analysis of lysosomes in rat Peyer's patch follicle epithelium: Their reduction in volume fraction and acid phosphate content in M cells compared to adjacent enterocytes, *Anat. Rec.* **216**:521–527.

Owen, R. L., Pierce, N. F., Apple, R. T., and Cray, W. C., (1986b), M cell transport of *Vibrio cholerae* from the intestinal lumen into Peyer's patches: A mechanism for antigen sampling and for microbial transepithelial migration, *J. Infect. Dis.* **153**:1108–1118.

Pappenheimer, A. M., Jr., 1977, Diphtheria toxin, *Annu. Rev. Biochem.* **46:**69–94.

Pappo, J., and Ermak, T. H., 1989, Uptake and translocation of fluorescent latex particles by rabbit Peyer's patch follicle epithelium: A quantitative model for M cell uptake, *Clin. Exp. Immunol.* **76:**144–148.

Pappo, J., Steger, H. J., and Owen, R. L., 1988, Differential adherence of epithelium overlying gut-associated lymphoid tissue. An ultrastructural study, *Lab. Invest.* **58:**692–697.

Pappo, J., Ermak, T. H., and Steger, H. J., 1991, Monoclonal antibody-directed targeting of fluorescent polystyrene microspheres to Peyer's patch M cells, *Immunology* **73:**277–280.

Patel, H. M., and Ryman, B. E., 1976, Oral administration of insulin by encapsulation within liposomes, *FEBS Lett.* **62:**60–62.

Patel, H. M., and Ryman, B. E., 1977, The gastrointestinal absorption of liposomally entrapped insulin in normal rats (proceedings), *Biochem. Soc. Trans.* **5:**1054–1056.

Patel, H. M., Stevenson, R. W., Parsons, J. A., and Ryman, B. E., 1982, Use of liposomes to aid intestinal absorption of entrapped insulin in normal and diabetic dogs, *Biochim. Biophys. Acta* **716:**188–193.

Pusztai, A., 1986a, The role in food poisoning of toxins and allergens from higher plants, in: *Developments in Food Microbiology,* Vol. 2 (R. K. Robinson, ed.), Elsevier Applied Science Publishers, London, pp. 179–194.

Pusztai, A., 1986b, The biological effects of lectins in the diet of animals and man, in: *Lectins, Biology, Biochemistry, Clinical Biochemistry*, Vol. 5 (T. C. Bog-Hansen, and E. van Driessche, eds.), Walter de Gruyter, Berlin, pp. 317–327.

Pusztai, A., 1988, Lectins, in: *Toxicants of Plant Origin. III. Proteins and Amino Acids* (P. R. Cheeke, ed.), CRC Press, Boca Raton, Florida.

Pusztai, A., 1989, Transport of proteins through the membranes of the adult gastrointestinal tract—potential for drug delivery?, *Adv. Drug Delivery Rev.* **3:**215–228.

Pusztai, A., Grant, G., and De Olveira, J. T. A., 1986, Local (gut) and systemic responses to dietary lectins, *IRCS Med. Sci.* **14:**205–208.

Quay, J. F., 1972, Transport interaction of glycine and cephalexin in rat jejunum, *Physiologist* **13:**241.

Quay, J. F., and Foster, L., 1970, Cephalexin penetration of the surviving rat intestine, *Physiologist* **13:**287–288.

Ray, R., Novak, M., Duncan, J. D., Matsuoka, Y., and Compans, R. W., 1993, Microencapsulated human parainfluenza virus induces a protective immune response, *J. Infect. Dis.* **167:**752–755.

Reid, R. H., Boedecker, E. C., McQueen, C. E., Davis, D., Tseng, L-Y., Kodak, J., Sau, K., Wilhelmsen, C. L., Nellore, R., Dalal, P., and Bhagat, H. R., 1993, Preclinical evaluation of microencapsulated CFA/II oral vaccine against enterotoxigenic *E. coli*, *Vaccine* **11:**159–168.

Rhodes, R. S., and Karnovsky, M. J., 1971, Loss of macromolecular barrier function associated with surgical trauma in the intestine, *Lab. Invest.* **25:**220–229.

Robinson, J. R., 1993, Recent advances in formulation of poorly absorbed drugs. Current status on targeted drug delivery to the gastrointesintal tract, Capsugel Symposium, Short Hills, New Jersey, April 22.

Roederer, M., Bowser, R, and Murphy, R. F., 1987, Kinetics and temperature dependence of exposure of endocytosed material to proteolytic enzymes and low pH: Evidence for a maturation model for the formation of lysosomes, *J. Cell. Physiol.* **131:**200–209.

Rowland, R. N., and Woodley, J. F., 1980, The stability of liposomes *in vitro* to pH, bile salts and pancreatic lipase, *Biochim. Biophys. Acta* **620:**400–409.

Rowland, R. N., and Woodley, J. F., 1981, The uptake of distearoyl phosphatidylcholine/cholesterol liposomes by rat intestinal sacs *in vitro*, *Biochim. Biophys. Acta* **673:**217–223.

Roy, M. J., 1987, Precocious development of lectin (*Ulex europeaeus* agglutinin I) receptors in dome epithelium of gut-associated lymphoid tissues, *Cell Tissue Res.* **248:**483–489.

Rubino, A., Field, M., and Schwachman, H., 1971, Intestinal transport of amino acid residues of dipeptides. I. Influx of the glycine residue of glycyl-L-proline across mucosal border, *J. Biol. Chem.* **246:**3542–3548.

Salzman, N. H., and Maxfield, F. R., 1989, Fusion accessibility of endocytotic compartments along the recycling and endocytotic pathways in intact cells, *J. Cell. Biol.* **109:**2097–2104.

Sanchez, A., Vila-Jato, J. L., and Alonso, M. J., 1993, Development of biodegradable microspheres and nanospheres for the controlled release of cyclosporin A, *Int. J. Pharm.* **99:**263–273.

Sandvig, K., Olsnes, S., Petersen, O. W., and Deurs, B. B., 1978, Acidification of the cytosol inhibits endocytosis from coated pits, *J. Cell. Biol.* **105:**679–689.

Santiago, N., Milstein, S., Rivera, T., Garcia, E., Zaidi, T., Hong, H., and Bucher, D., 1993, Oral immunization of rats with proteinoid microspheres encapsulating influenza virus antigens, *Pharm. Res.* **10:**1243–1247.

Shen, W.-C., and Ryser, H. J.-P., 1978, Conjugation of poly-L-lysine to albumin and horseradish peroxidase: A novel method of enhancing the cellular uptake of proteins, *Proc. Natl. Acad. Sci. USA* **75:**1872–1876.

Silk, D. B., Grimble, G. K., and Rees, R. G., 1985, Protein digestion and amino acid and peptide absorption, *Proc. Nutr. Soc.* **44:**63–72.

Sinko, P. J., and Amidon, G. L., 1988, Characterization of the oral absorption of β-lactam antibiotics. I. Cephalosporins. Determination of intrinsic membrane absorption parameters in the rat intestine *in situ*, *Pharm. Res.* **5:**645–650.

Sinko, P. J., and Amidon, G. L., 1989, Characterization of the oral absorption of β-lactam antibiotics. II. Competitive absorption and peptide carrier specificity, *J. Pharm. Sci.* **78:**732–726.

Sinko, P. J., Leesman, G. D., and Amidon, G. L., 1991, Predicting fraction dose absorbed in humans using a macroscopic mass balance approach, *Pharm. Res.* **8:**979–988.

Smithson, K. W., Millar, D., Jacobs, L., and Gray, G., 1981, Intestinal diffusion barrier: Unstirred water layer or membrane surface mucous coat, *Science* **214:**1241–1244.

Spangler, R. S., 1990, Insulin administration via liposomes, *Diabetes Care* **13:**911–922.

Steiner, S., and Rosen, R., 1990, Delivery systems for pharmacological agents encapsulated with proteinoids. U.S. patent 4,925,673, May 15, 1990.

Stone, H. H., Kolb, L. D., Currie, C. A., Geheber, C. E., and Cuzzell, J. Z., 1974, Candida sepsis: Pathogenesis and principles of treatment, *Ann. Surg.* **179:**697–711.

Straus, W., 1969, Use of horseradish peroxidase as a marker protein for studies of phagolysosomes, permeability, and immunology, *Methods Achiev. Exp. Pathol.* **4:**54–91.

Swallen, L. C., 1941, Zein—a new industrial protein, *Ind. Eng. Chem.* **1941**(March)**:**394–398.

Tartakoff, A. M., 1987, *The Secretory and Endocytic Paths. Mechanism and Specificity of Vesicular Traffic in the Cell Cytoplasm*, John Wiley & Sons, New York.

Thomson, A. B. R., and Dietschy, J. M., 1984, The role of the unstirred water layer in intestinal permeation, in: *Pharmacology of Intestinal Permeation II* (T. Z. Csaky, ed.), Springer-Verlag, Berlin.

Tomita, Y., Katsura, T., Okano, T., Inui, K. I., and Hori, R., 1990, Transport mechanisms of bestatin in rabbit intestinal brush-border membrane: Role of H^+/dipeptide cotransport system. *J. Pharmacol. Exp. Ther.* **252:**859–862.

Tomizawa, H., Aramaki, Y., Fujii, Y., Hara, T., Suzuki, N., Yachi, K., Kikuchi, H., and Tsuchiya, S., 1993, Uptake of phosphatidyl-serine liposomes by rat Peyer's patches following intraluminal administration, *Pharm. Res.* **10:**549–552.

Tsuji, A., Hirooka, I., Tamai, I., and Terasaki, T., 1986, Evidence for a carrier mediated transport in the small intestine available for FK089, a new cephalosporin antibiotic without an amino group, *J. Antibiot.* **39:**1592–1597.

Tsuji, A., Terasaki, T., Tamai, O., and Hirooka, H., 1987, H^+-gradient-dependent and carrier-mediated transport of cefixime, a new cephalosporin antibiotic, across brush-border membrane vesicles from rat small intestine, *J. Pharmacol. Exp. Ther.* **241:**594–601.

Verdun, C., Couvreur, P., Vranckx, H., Lenaerts, V., and Roland, M., 1986, Development of a nanoparticle controlled-release formulation for human use, *J. Controlled Release* **3:**205–210.

Volkheimer, G., 1975, Hematogenous dissemination of ingested polyvinylchloride particles, *Ann. N.Y. Acad. Sci.* **246:**164–171.

Volkheimer, G., 1977, Persorption of particles: Physiology and pharmacology, *Adv. Pharmacol. Chemother.* **14:**163–187.

Volkheimer, G., and Schulz, F. H., 1968, The phenomenon of persorption, *Digestion* **1:**213–218.

Volkheimer, G., and Schulz, F. H., 1969, Effect of caffeine on the rate of persorption, *Nutr. Dieta* **11:**13–22.

Volkheimer, G., Schulz, F. H., Hofmann, I., Pioeser, J., Rack, O., Reichlt, G., Rothenbaecker, W., Schmelich, G., Schurig, B., Teicher, G., and Weiss, B., 1968, The effects of drugs on the rate of persorption, *Pharmacology* **1:**8–14.

Volkheimer, G., Schulz, F. H., John, H., Meier, Zu Eisen, J., and Niederkorn, K., 1969, Persorbed food particles in the blood of newborns, *Gynaecologia* **168:**86–92.

Walker, W. A., 1979, Gastrointestinal host defense: Importance of gut closure in control of macromolecular transport, in: *Development of Mammalian Absorptive Processes,* Ciba Foundation Symposium No. 70, Excerpta Medica, Amsterdam, pp. 201–216.

Walker, W. A., 1982, Mechanism of antigen handling by the gut, in: *Clinics in Immunology and Allergy*, Vol. w, No. 1, (J. Brostoff and S. J. Challacombe, eds.), W. B. Saunders, London, pp. 15–40.

Walker, W. A., and Isselbacher, K. J., 1974, Uptake and transport of macromolecules by the intestine: Possible role in clinical disorders, *Gastroenterology* **67:**531–550.

Walker, W. A., Cornell, R., Davenport, L. M., and Isselbacher, K. J., 1972, Macromolecular absorption mechanism of horesradish peroxidase uptake and transport in adult and neonatal rat intestine, *J. Cell. Biol.* **54:**195–205.

Warshaw, A. L., Walker, W. A., Cornell, R., and Isselbacher, K. J., 1971, Small intestinal permeability to macromolecules: Transmission of horseradish peroxidase into mesenteric lymph and portal blood, *Lab. Invest.* **25:**675–684.

Wells, C. L., Maddaus, M. A., Erlandsen, S. L., and Simmons, R. L., 1988, Evidence for the phagocytic transport of intestinal particles in dogs and rats, *Infect. Immun.* **56:**278–282.

Weltzin, R., Lucia-Jandris, P., Michetti, P., Fields, B. N., Kraehenbuhl, J. P., and Neutra, M. R., 1989, Binding and transepithelial transport of immunoglobulins by intestinal M cells: Demonstration using monoclonal IgA antibodies against enteric viral proteins, *J. Cell. Biol.* **108:**1673–1685.

Wessling-Resnick, M., and Braell, W. A., 1990, The sorting and segregation mechanism of the endocytic pathway is functional in a cell-free system, *J. Biol. Chem.* **265:**690–699.

Westergaard, H., 1987, The passive permeability properties of *in vivo* perfused rat jejunum, *Biochim. Biophys. Acta* **900:**129–138.

Wheeler, P. G., Menzies, I. S., and Creamer, B., 1978, Effect of hyperosmolar stimuli and coeliac disease on the permeability of the human gastrointestinal tract, *Clin. Sci. Mol. Med.* **54:**495–501.

Wilson, S. J., and Walzer, M., 1935, Absorption of undigested proteins in human beings, *Am. J. Dis. Child.* **50:**49–57.

Wolf, J. L., Rubin, D. H., Finberg, R., Kauffman, R. S., Sharpe, A. H., Trier, J. S., and Fields,

B. N., 1981, Intestinal M cells: A pathway for entry of reovirus into the host, *Science* **212**:471–472.

Wood, A. J., Maurer, G., Niederberger, W., and Reveridge, T., 1983, Cyclosporine—pharmacokinetics, metabolism, and drug interactions, *Transplant. Proc.* **15**:2409.

Woodward, S. C., Herrmann, J. B., Cameron, J. L., Brandes, G., Pulaski, E. J., and Leonard, F., 1965, Histotoxicity of cyanoacrylate tissue adhesives in the rat, *Ann. Surg.* **162**:113–122.

Wyvratt, M. J., and Patchett, A., 1985, Recent developments in the design of angiotensin-converting enzyme inhibitors, *Med. Res. Rev.* **5**:483–485.

Xi, M., Variano, B. F., Chaudary, K., Pastores, G. W., Lonardo, C. A., Milstein, S. J., Santiago, N., and Baughman, R. A., 1993, *In vitro* mechanistic investigation of the proteinoid microsphere oral delivery system, *Proc. Int. Symp. Control. Rel. Bioact. Mater.* **20**:334–335.

Yee, A., and Amidon, G. L., 1990, Intestinal absorption mechanism of three angiotensin-converting enzyme inhibitors: Quinapril, benazepril and CGS16617, *Pharm. Sci.* **7**:S-155.

Yokohama, S., Yoshioka, T., and Kitamori, N., 1984a, Absorption of γ-butyrolactone-γ-carbonyl-L-histidyl-L-prolinamide citrate (DN-1417), an analog of thyrotropin-releasing hormone, in rats and dogs. *J. Pharmacobio-Dyn.* **7**:527–535.

Yokohama, S., Yoshioka, T., Yamashita, K., and Kitamori, N., 1984b, Intestinal absorption mechanisms of thyrotropin releasing hormone, *J. Pharmacobio-Dyn.* **7**:445–451.

Chapter 11

Controlled Delivery of Somatotropins

Susan M. Cady and William D. Steber

1. INTRODUCTION

The somatotropins, also known as growth hormones, are protein hormones produced by the anterior pituitary that stimulate growth in virtually all vertebrate species. These effects are the result of direct receptor stimulation and stimulation of insulin-like growth factor-I (IGF-I), a related anabolic protein hormone (Hart and Johnson, 1986). Parenteral daily injections of porcine somatotropin (PST) into swine have shown an increase in the lean-to-fat ratio as well as increased feed efficiency (McLaren *et al.*, 1990; Campbell *et al.*, 1991). Parenteral administration of bovine somatotropin (BST) increases milk yield in dairy cows (Peel *et al.*, 1981, 1983) and increases the ratio of growth in steers (Moseley *et al.*, 1992). Somatotropin treatment accelerates growth, improves feed conversion efficiencies, and increases appetites in teleost fishes (McLean and Donaldson, 1993). Human growth hormone (HGH) has found uses in replacement therapy treatment for children with pediatric hypopituitary dwarfism and those suffering from low levels of HGH. HGH use has also been reported for wound healing (Jorgensen, 1991; Hoelgaard, 1991).

These molecules present a challenge to the formulators designing controlled delivery systems. They can be most efficiently delivered parenterally because they are extensively degraded in the gastrointestinal tract

---

Susan M. Cady • Hoechst Roussel Vet, Somerville, New Jersey 08876-1258. *William D. Steber* • Fort Dodge Animal Health, Cyanamid Agricultural Research Center, Princeton, New Jersey 08543-0400.

Protein Delivery: Physical Systems, Sanders and Hendren, eds., Plenum Press, New York, 1997.

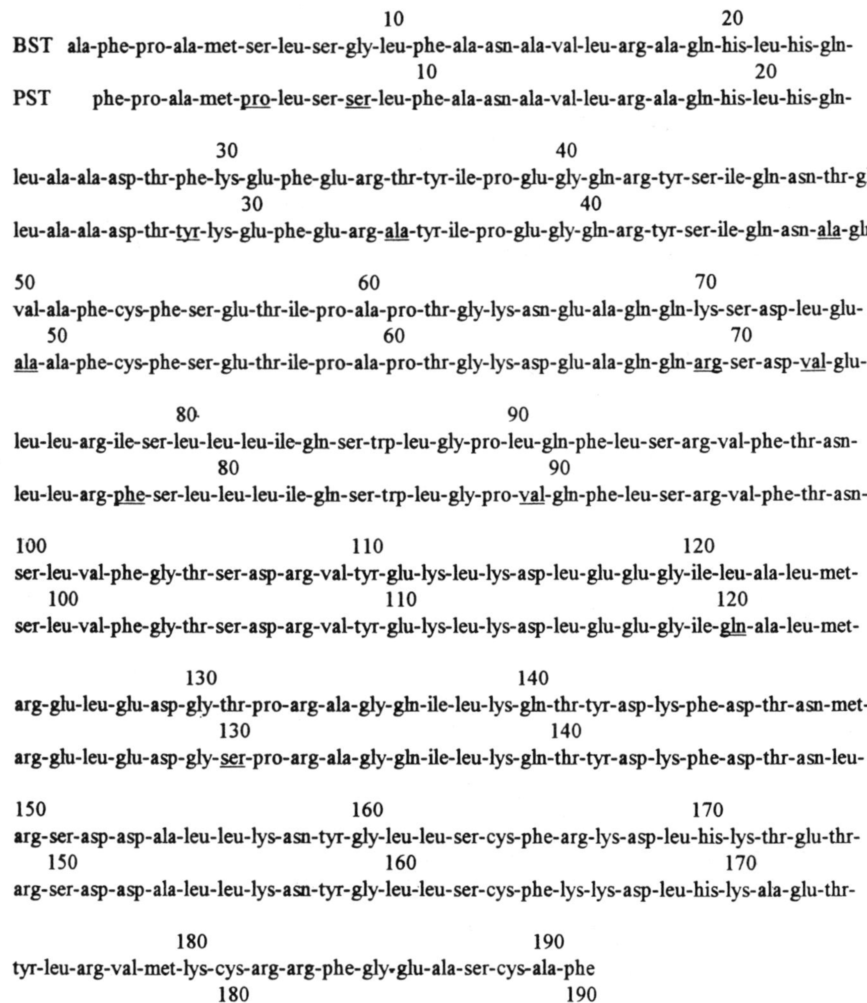

Figure 1. Amino acid sequences of naturally occurring porcine (PST) and bovine (BST) somatotropins. [Adapted from Souza (1993) for PST and from Lehrman *et al.* (1991) for BST.]

and are poorly absorbed by this route. These molecules have a tendency to form aggregates which are biologically inactive or poorly absorbed. Somatotropins are relatively large molecules having about 191 amino acid residues with a molecular weight of approximately 21,000, and they contain two disulfide linkages. The molecules have a high degree of species-to-species homology. The primary sequences of natural porcine and bovine somatotropins are shown in Fig. 1. The crystallographic data determined for PST indicate that the tertiary structure consists of four antiparallel α-helices (Abdel-Meguid *et al.*, 1987). Human growth hormone and bovine somatotropin characterization and preformulation studies have been recently reviewed (Pearlman and Bewley, 1993; Davio and Hageman, 1993). A description of the differences in primary structure between the natural and recombinant somatotropins was presented in these reviews. Most of the formulations work reported recently has been on recombinantly derived somatotropins. In the present chapter, it is generally to be understood that the formulation studies and patent literature involve use of the recombinantly derived somatotropins. Strategy for the parenteral delivery of proteins have been reviewed by Pitt (1990).

2. PREFORMULATION DEVELOPMENT

2.1. Solution Stability

The solution stability of somatotropin is an important consideration in the development of commercial long-acting formulations because practical dosage forms must be small to be easily administered (in the case of veterinary products, administered by the farmer) and contain a high loading of the protein. Upon hydration in a delivery system, which occurs shortly after insertion into animal tissues, the somatotropin will exist at high concentrations and will be exposed to elevated temperatures (37–39 °C) for the duration of the protein release. Typically, applications of extended release delivery systems for bovines and porcines call for duration of release of at least 1–2 weeks and up to about 6 weeks. Consequently, workers have investigated a variety of preformulation approaches in attempts to improve stability of the somatotropins, which would permit extending the duration of release.

Several groups have studied the solution aggregation problems of these molecules. Hageman *et al.* (1992a) illustrated that BST degraded in part by irreversible aggregation in solution (10 mg/ml) at pH 9.8 and 30 °C over 15 days as determined by size-exclusion and reversed-phase high-pressure liquid chromatography (HPLC). The extent and rate of degradation were far worse for lyophilized BST held at 96% relative humidity and 30 °C (Hageman *et al.*, 1992b). These authors showed that at high humidity BST adsorbed about its own weight in water. Clearly, these conditions of high humidity and significant water absorption mimic conditions of somatotopins in typical compact, long-acting drug delivery systems.

Brems and co-workers at Upjohn concluded in work done with BST that a significant portion of the aggregates form as a result of a complex multistep process initiated by hydrophobic intermolecular bonding at the interior surface of the third helix (Brems *et al.*, 1985, 1986, 1988, 1989; Havel *et al.*, 1986; Holzman *et al.*, 1986). There is continued discussion below of their work concerning molecular modifications of the somatotropins.

There is an extensive literature that discusses the stabilization of proteins in solution and for processing as a consequence of pH, temperature, and solutes (see, e.g., Lee and Timasheff, 1981; Carpenter *et al.*, 1986). The conditions that stabilize a given protein are usually unique to that particular protein. Following the development of large-scale production of somatotropins by recombinant technology, there was considerable development activity on the stabilization of the somatotropins, as revealed in the patent and journal literature.

Hamilton and Burleigh (1989), at International Minerals and Chemicals Corporation (IMC), demonstrated stabilization of highly concentrated somatotropins, especially BST, by incorporating into aqueous formulations certain amino acids, amino acid polymers, or choline derivatives. Their patent exemplifies the stabilizing effects of nonreducing sugars such as sucrose and sorbitol in isotonic physiological phosphate buffer solutions. Viswanathan and DePrince (1990), also of IMC, demonstrated the effectiveness *in vivo* of sucrose-stabilized PST implants placed inside silicone tubes. The release of somatotropin was designed to occur from the ends of the tubes via microporous polyethylene disks having 70-μm pores. Silicone-tube-covered implants were prepared containing either 40 mg of PST alone or 40 mg of PST with sucrose. When these devices were implanted subcutaneously behind the ears of pigs, blood level data showed that the implants with PST and sugar released considerably more PST over the 2-week test.

Azain *et al.* (1990) claimed stabilizing compositions for aqueous formulations for porcine somatotropin containing polyols that were effective for extended periods of time even at room temperature. A typical composition was about 30% porcine somatotropin, about 33.5% glycerol, and about 33.5% water, with buffers controlling pH to about 6–6.5. Compared to controls such as 30% porcine somatotropin in water with buffer or glycerol only, this type of formulation showed enhanced stability, remaining a clear viscous solution for 4 days at 37 °C. Analysis by size-exclusion HPLC showed little or no dimer increase or precipitate formed. Azain *et al.* showed this type of viscous aqueous formulation to be useful in Alzet (Alza Corporation) 14-day minipumps; when tested in pigs, these systems were shown to achieve average daily gain and feed efficiencies comparable to those obtained in positive control pigs receiving daily injections.

2.2. Molecular Modification

Molecular modification of somatotropins has been investigated for a basic understanding of protein structure and function. Practical application of directed modification of this molecule has been pursued vigorously by researchers in the pharmaceutical and biotechnology industry to improve production yields, to increase potency and selectivity of biological effects, and to design and optimize long-acting drug delivery systems. Harbour *et al.* (1987) claimed enhanced bioactivity of recombinant BST (rBST) in which the asparagine located between amino acid residues 96–101 is modified to isoaspartic acid, aspartic acid, or glutamic acid by either chemical or genetic processes. They found a potency increase of up to 2.4-fold relative to unmodified rBST for the selectively deamidated BST as determined in the hypophysectomized rat bioassay.

Gráf *et al.* (1975) reported retention of growth-promoting activity by reduction and alkylation of the disulfide linkage between cysteine-181 and cysteine-189 of BST. This disulfide linkage is frequently termed the small-loop disulfide. Subsequently, it was demonstrated that reduction and derivatization of the small-loop disulfide of porcine and human somatotropins could be accomplished without loss of biological activity (Schleyer *et al.*, 1982, 1983). Researchers have reported improved solution stability and formulation processability of somatotropins derivatized at the small-loop-disulfide (Randawa and Seely, 1990; Buckwalter *et al.*, 1992). Buckwalter and co-workers showed that the disulfide exchange reaction is involved in the rPST aggregation formation and that elimination of this process by chemical modification of the small-loop cysteines in rPST resulted in a marked improvement in solution thermostability at high concentration. These workers utilized selective reductive alkylation of the Cys-181–Cys-189 disulfide to prepare derivatives. Figure 2 shows the improved thermostability at concentrations of about 100 mg/ml in phosphate-buffered saline (pH 7.4) at 43 °C that are seen with the carboxymethylated PST (CM-PST) in comparison with rPST (Buckwalter *et al.*, 1992). Several groups have used site-directed mutagenesis to replace the small-loop cysteines with other amino acids (Cady *et al.*, 1990; Parcells and Mott, 1990). The improvements in solution stability of these analog molecules improve production processability and formulation stability. The cysteine modification or replacement extends the useful half-life of the protein in concentrated solutions with no deleterious effect on biological activity.

Lehrman, Brems, and co-workers have used site-directed mutagenesis to prepare somatotropin analogs with reduced α-helical stability or hydrophobicity in the primary structure between residues 96 and 133, the region called the third helix. These changes alter the hydrophobic intermolecular

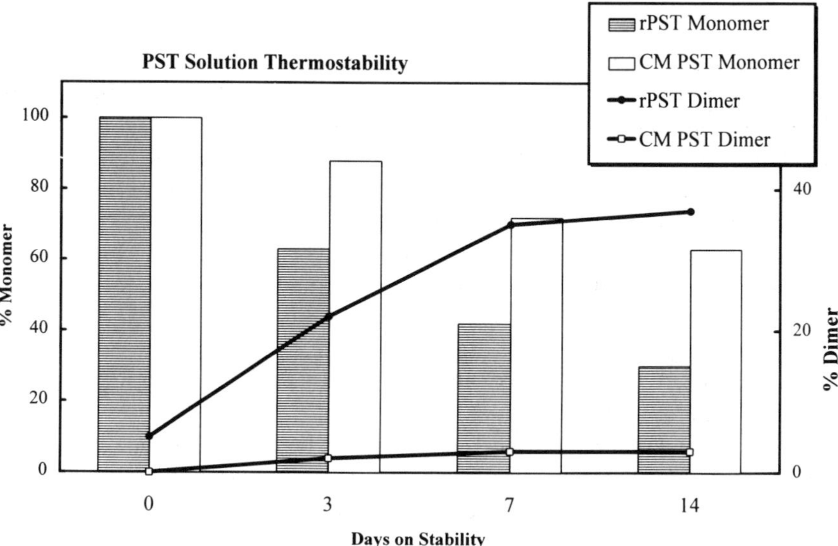

Figure 2. Solution thermostability of rPST and CM-PST. Samples were initially dissolved at 100 mg/ml in phosphate-buffered saline and incubated at 43 °C. The percentage of the original protein that remained in the soluble monomeric form is shown by bars for normal recombinant PST and chemically modified PST, which was reduced and alkylated at the small-loop disulfide. The amount of dimerization of these species as a percentage of the protein remaining in solutions is also shown. (Data from Buckwalter *et al.*, 1992.)

associations of the molecules, particularly in solution (Lehrman *et al.*, 1991). This work was an application of the research at Upjohn cited above. Lehrman and co-workers found that particular changes in the primary sequence of the third helix permitted marked improvements in the recovery yields of BST from *Escherichia coli.* According to their kinetic model for BST folding, a stable folding intermediate exists between the unfolded and the native conformation. Even at low concentrations, this folding intermediate can form an associated intermediate that has lower solubility than either the unfolded or the native conformation and therefore can precipitate more easily. The stability of the associated intermediate involves the hydrophobic intermolecular interactions of the third helix. In their patent application, Lehrman *et al.* described specific replacements of amino acid residues in this helix that stabilize or destabilize the associated intermediate. A less stable associated intermediate leads to less precipitation, which improves solution stability and production recovery yields; improvements up to 72% are exemplified. Forms with increased stability of the associated intermediate precipitate more easily and lead to poorer recovery. Lehrman and coworkers pointed out that, although the native conformation is relatively stable and soluble once achieved in production, downstream manufacturing may expose the BST to conditions that perturb the native configuration into the intermediate folding state and associated intermediate, which leads to precipitation. These researchers did not mention the consequences to stability of somatotropins in a hydrated delivery system. However, it may be expected that analogs of the variety they described, which have inherently better solution stability and recovery yields, would also have desirable properties to enhance *in vivo* formulation stability and thus increase the duration of release from a drug delivery system.

Despite the amount of preformulation and molecule stabilization work, most formulations that are reported in the patent literature have not yet been commercialized. It remains to be seen just how many of these delivery systems will be commercially viable. The advances in the understanding of protein structure and development of stable somatotropin compositions and analogs are seen to be essential to the development of sustained release systems for the somatotropins.

3. INJECTABLES

3.1. Oil-Based Gel Depots

Perhaps the most direct approach to achieving sustained release of these water-soluble somatotropins from a parenteral formulation is the use of nonaqueous gels. Typically, these systems are vegetable oils incorporating dispersions of the somatotropin and thickening agents that permit injection with a hypodermic syringe. Viscosity control is needed to achieve a compromise between injectability and physical stability of the depot after injection. A viscous gel upon injection lengthens the duration of release. These types of systems are compatible with relatively long-term shelf stability of the somatotropins, since they usually do not contain water. They may incorporate excipients with the somatotropin to aid in the sustained release properties or improve the stability of the somatotropins once injected into the animal. A variety of thickening agents and the range of hydrophobicity of the oils or other excipients permit optimization with respect to shelf stability, injectability, injection-site chemical and physical stability, drug loading, and duration of release. Oil-based gels are relatively easy to manufacture and are probably easier to manufacture aseptically than implants. Their principal disadvantage is the inconsistency of the depot

Table I
Milk Production Increases in Dairy Cows Treated with Zinc Salt of Bovine Somatotropin (ZnMBS)[a]

Treatment	Cumulative adjusted average daily milk production (kg milk/day)[b]			
	7 days	14 days	21 days	30 days
Control	23.3	22.4	22.0	21.1
ZnMBS	26.6(16.2)[c]	26.0(17.1)	24.9(13.7)	22.9(9.6)

[a]Adapted from Mitchell (1991).
[b]Data are from 5 or 6 animals per treatment.
[c]Numbers in parentheses represent percentage improvement relative to average control response.

geometry, which can be expected to influence the uniformity of the release profile from animal to animal. Finally, the release profiles from injection-site depots are not inherently zero-order but tend to be approximately first-order.

Mitchell (1991), at Monsanto, prepared methionine N-terminated bovine somatotropin zinc salt (ZnMBS) in an oil gel consisting of sesame oil thickened with 5% antihydration agent, aluminum monostearate (wt. ratio

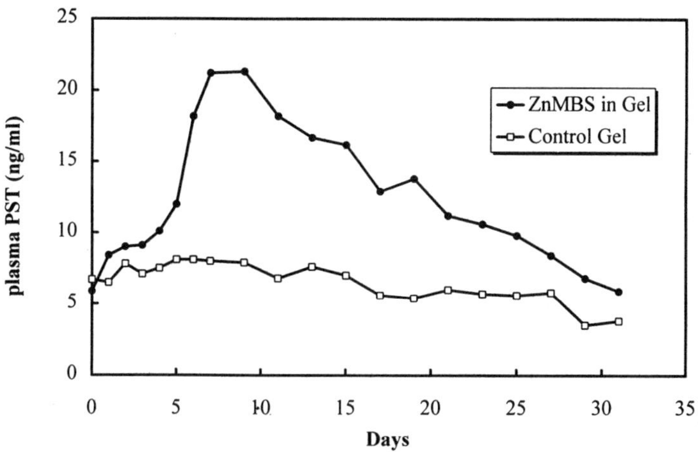

Figure 3. Bovine somatotropin (BST) plasma levels, measured by radioimmunoassay in nanograms per milliliter, in dairy cattle injected with a control gel formula or the same gel formula containing a zinc salt of BST (ZnMBS). (Data from Mitchell, 1991.)

of ZnMBS to aluminum monostearate = 9.4:1). This formulation was effective for increasing milk production, as exemplified in Holsteins which were given intramuscular injections of 2.54 g of formulation containing 805 mg of ZnMBS (the patent's nomenclature for zinc-complexed methionine BST). Compared to control cattle, given 2.4 g of placebo gel, milk production increased (Table I) in the treated group of cows, and sustained elevated blood levels of BST were seen (Fig. 3). Monsanto has been successfully marketing Posilac®, (Monsanto Chemical Co.) a single-dose injection of 500 mg of rBST lasting 14 days, for administration to dairy cows.

Ferguson *et al.* (1990), at Elanco, prepared various oil/wax depot injectables for sustained release of bovine somatotropin. To thicken the oils, waxes were melted into the oils, which were then cooled and homogenized until uniform in consistency. Solid BST was dispersed into the viscous oil and wax mixture. Three combinations of BST, oil, and wax were formulated and injected into dairy cattle, and these formulations gave sustained

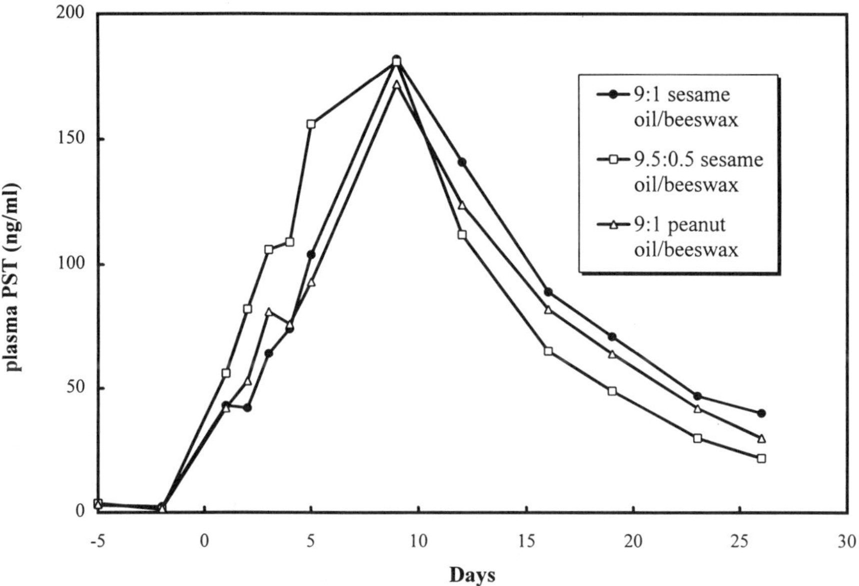

Figure 4. Bovine somatotropin (BST) plasma levels, measured by radioimmunoassay in nanograms per milliliter, in dairy cattle injected subcutaneously with formulations consisting of 9:1 sesame oil/white beeswax, 9.5:0.5 sesame oil/white beeswax, or 9:1 peanut oil/yellow beeswax containing about 12.5% suspended bovine somatotropin in a 4:9-g injection. (Data from Ferguson *et al.*, 1990.)

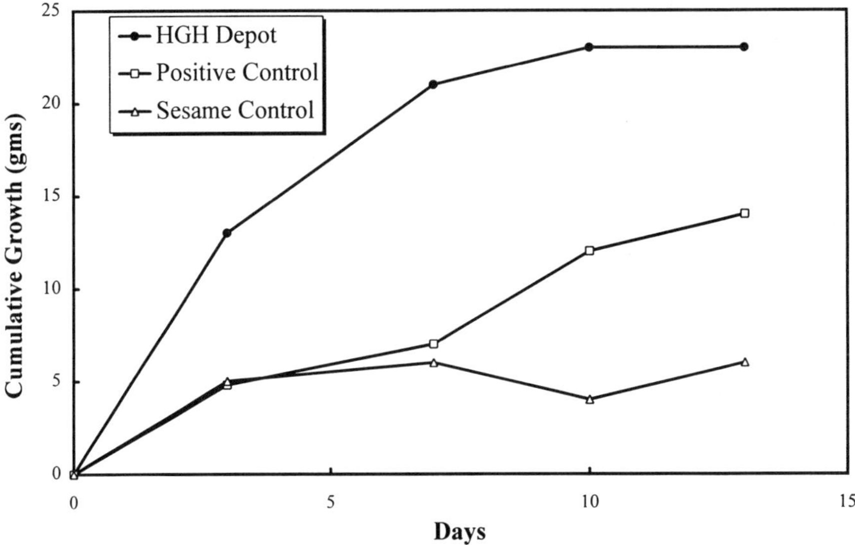

Figure 5. Hypophysectomized rat cumulative growth data from animals injected intramuscularly with negative control (sesame oil), positive control [0.01 mg of human growth hormone (HGH) in aqueous buffer daily], or 0.1 ml of HGH sesame oil and Gelucire® 64/02 mixtures as depots containing 1.004 mg of HGH.

elevated plasma levels which are indicative of sustained release of the somatotropin (Fig. 4).

Thakkar *et al.* (1988), also at Elanco, prepared depot matrices from oils and glyceride release modifying agents such as Gelucire® 64/02, which are esters of glycerol with one or more fatty acids having a melting point of ~64 °C and an HLB value of about 2. Sesame oil and Gelucire were stirred at about 65 °C until melted and homogeneous. The solution was cooled, and HGH (1.004 mg per 0.1 ml of formulation) was dispersed into the mixture. These formulations increased growth in hypophysectomized rats when injected intramuscularly, as shown in Fig. 5.

Other researchers prepared freeze-dried powders from L-α-phosphatidylcholine and a variety of somatotropins, and the powders were then homogenized in tocopheryl acetate. Sustained release was demonstrated by hypophysectomized rat growth, increased milk production in Holstein cattle, and increased feed efficiency in swine treated with the formulations (Kim *et al.*, 1991).

3.2. Microsphere Systems

Potentially, microspheres may achieve more consistent release patterns and alter the compromises among factors to achieve an effective sustained release formulation for somatotropins. Steber *et al.* (1989), at American Cyanamid Company, used glyceryl tristearate or glyceryl distearate to form microspheres for bovine somatotropin by spray prilling techniques. Microspheres loaded with 24.5% BST were suspended in soybean oil/Miglyol® 812 in a 1:4 ratio. Holstein dairy cattle were injected subcutaneously with 350 mg of BST in this formulation. Milk production increases are plotted in Fig. 6. Steber (1993) reported an improved fat-based microsphere formulation that gave better physical stability and shelf life for the microspheres when reconstituted in an oil vehicle. The formulation change in the microspheres was to incorporate a small amount of oil, semisoft fat, and/or fatty acid derivative into the microsphere composition prior to prilling; such

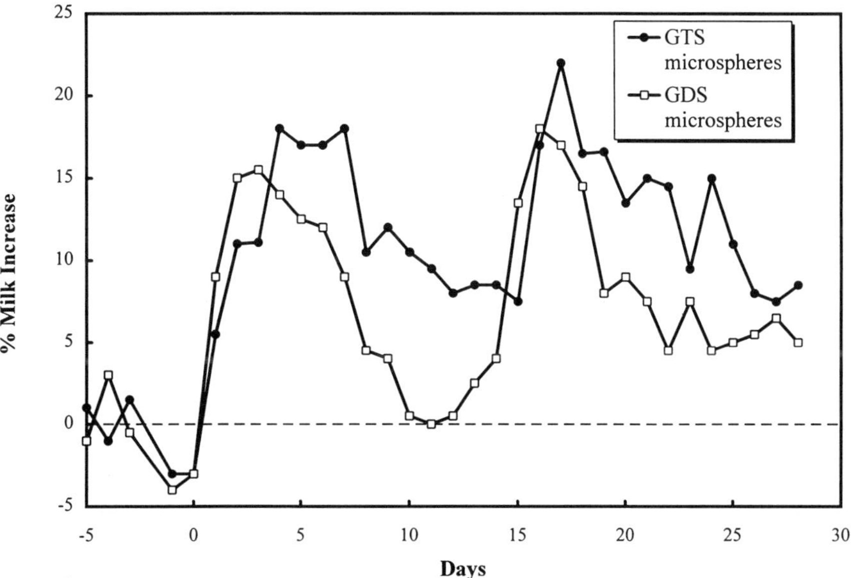

Figure 6. Percentage milk production increases in dairy cattle treated with BST-loaded microsphere formulations relative to untreated dairy cattle. Dairy cattle were injected at 0 and 14 days with either glyceryl tristearate (GTS) or glyceryl distearate (GDS) micropheres suspended in soybean oil/Miglyol® 812. Each injection contained approximately 350 mg of BST. (Data from Steber *et al.*, 1989.)

Table II
Physical Stabilization of Glyceryl Tristearate Microspheres[a]

	Viscosity at 25 °C (cP)[b]	
Time	Unstabilized[c]	Stabilized[d]
Initial	66	66
1 day	168	—
5 days	159	—
1 wk	159	70
2 wk	159	71
4 wk	—	79
8 wk	164	82
13 wk	270	74
26 wk	624	70
39 wk	741	71

[a]Adapted from Steber (1993).
[b]Viscosity was measured after storage at 4 °C. Samples were allowed to warm to room temperature before viscosity was measured with a Brookfield viscometer, type T, spindle C, @100 rpm.
[c]The unstabilized microsphere composition was, in weight percentage, 28% BST, 2% sodium benzoate, 0.14% block copolymer of ethylene oxide and propylene oxide, and 70% glyceryl tristearate.
[d]The stabilized microsphere composition was 28% BST, 2% sodium benzoate, 0.14% block copolymer of ethylene oxide and propylene oxide, 63% glyceryl tristearate, and 7% neutral triglyceride oil. Microspheres were suspended in neutral triglyceride oil, and the suspension was placed on stability.

addition was found to accelerate the transformation from the α-crystalline structure of the hard fats to the β-crystalline form. Because the β form is higher melting and more stable, the microspheres obtained by utilizing this approach were more resistant to physical alteration by the oil vehicles. Table II shows viscosity data for fat-based microsphere systems and illustrates the improved shelf-life stability achieved by incorporation of a neutral triglyceride oil.

Auer *et al.* (1994), at Alkermes Controlled Therapeutics, Inc., prepared polymer-based microspheres for human growth hormone. They used 2–5-μm-sized lyophilized HGH containing zinc ions in a 4:1 molar ratio to HGH and sodium bicarbonate (6:1 wt./wt.). The microsphere matrix was prepared from lactide/glycolide copolymers (D/L-lactide 50:50, inherent viscosity 0.16) using an atomization process in liquid nitrogen. The microspheres were about 10% HGH salts. The microspheres were evaluated for biological and controlled release properties using the hypophysectomized rat bioassay as shown in Fig. 7. A rapid growth response in rats given the HGH microspheres is shown during the first week followed by slower growth. This

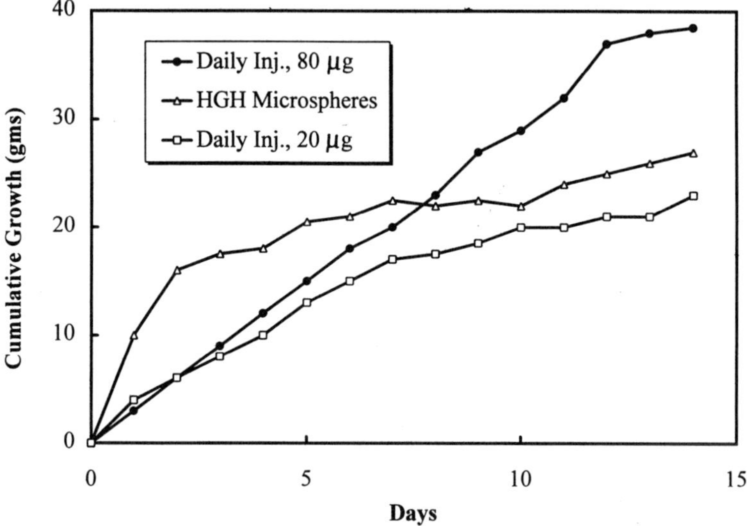

Figure 7. Cumulative growth data from hypophysectomized rats treated with either 20- or 80-μg daily injections of human growth hormone (HGH) compared to data from rats treated with a single injection of 18 mg of microspheres containing 1.8 mg of HGH salts. (Data from Auer *et al.*, 1994.)

growth pattern suggests a typical non-zero-order release pattern from monolithic microspheres.

3.3. Liposomes

Some sustained release of bovine somatotropin was demonstrated from liposome delivery systems. Egg phosphatidylcholine, ethanolamine and α-tocopheryl hemisuccinate, and Tris salt vesicles released bovine somatotropin, giving hypophysectomized rat growth for over a week (Janoff *et al.*, 1989). Hydrogenated soy phosphatidylcholine–cholesterol–BST liposomes injected into dairy cattle increased weekly milk production (Weiner *et al.*, 1989).

3.4 Emulsions

Tyle (1989) and Tyle and Cady (1990) reported sustained release of BST from water-in-oil-in-water multiple emulsions. In one example, the BST

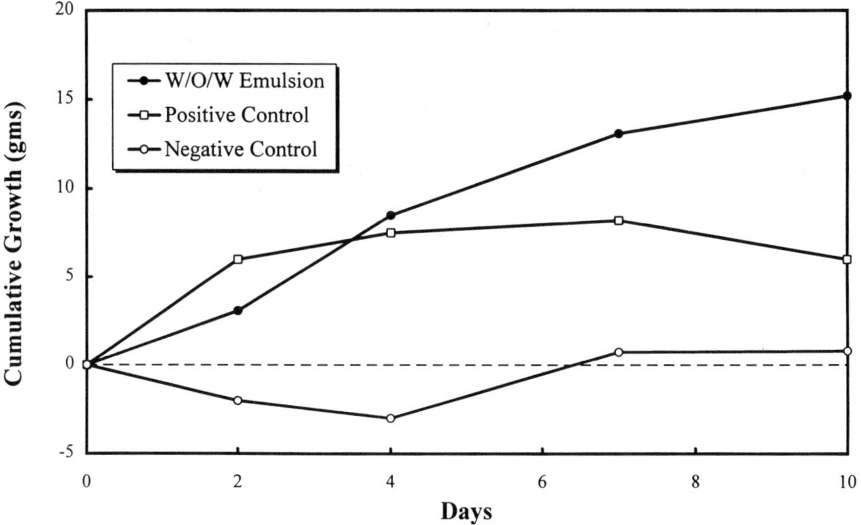

Figure 8. Cumulative growth data showing the efficacy of a W/O/W emulsion in hypophysectomized rats (rats with pituitaries surgically removed so they do not produce growth hormone). Each of 10 animals was injected with 2400 μg of BST in 0.2 ml of W/O/W emulsion. The positive control group of animals received a single bolus injection of 2400 μg of BST in carbonate-buffered saline. Negative control animals received no injections. (Data from Tyle and Cady, 1990.)

was dissolved in the primary carbonate-buffered saline aqueous phase (W1), and then mixed with the oil phase (O), consisting of light mineral oil and sorbitan trioleate. Addition of the resulting emulsion to a secondary aqueous phase (W_2) consisting of carbonate-buffered saline, sorbitol, and polysorbate 80 with mixing completed the multiple emulsion. Phase W_1 of the $W_1/O/W_2$ multiple emulsion contained the BST. The emulsion was injected subcutaneously into hypophysectomized rats and increased growth (Fig. 8).

3.5. Aqueous Gels and Complexes

Cady *et al.* (1993) described BST injectable formulations based on carbohydrate polymers such as dextrins, heteropolysaccharides (Biopolymer PS-87, supplied by Lever Brothers Company), and various gums that gave sustained release in dairy cattle.

Aston *et al.* (1991) prepared HGH and BST conjugates with bovine serum albumin (BSA) using glutaraldehyde or with immunoglobulin using carbodiimide. These conjugates had enhanced biological activity in Snell-dwarf strain mice, and the BST/BSA conjugate has improved solubility over BST in less than pH5 solutions.

4. IMPLANTS

In contrast to injectable gels, which form variably shaped depots, implants can achieve theoretically more consistent release rate patterns (injection to injection) because their geometry is fixed. Furthermore, depending on design, such as the use of rate-limiting coatings or release membranes, the implant offers the potential for a more constant or zero-order release. The disadvantages of implant systems are that the concentrations of the somatotropins in the implants are generally higher than in injectables, which may impact the *in vivo* stability problem. Relatively large amounts of somatotropin may be required to give formulations lasting for 1–6 weeks *in vivo*. Implantation usually requires a large-bore needle, with attendant undesirable features such as a more complex injection system. Finally, controlled release implants can require more processing steps than gels. The additional steps may complicate manufacturing, particularly if aseptic processing is required. The implant examples described below represent a range of approaches and considerable variation in the level of development toward commercially viable products. However, it is difficult to predict which approach could become commercially successful regardless of the apparent manufacturing complexity or the variable biological effectiveness revealed in the patents, because a given approach may incorporate unobvious advantages or features that eventually may be optimized easily and economically.

4.1. Uncoated Implants

Clark *et al.* (1993) prepared complexes of aromatic aldehydes and PST in an effort to produce stable compositions capable of prolonged release. Various compositions were prepared with 1% (w/w) 2-hydroxy-3-methoxybenzaldehyde, 4-hydroxy-3-methoxybenzaldehyde, or 4-methoxybenzaldehyde and 99% PST. The isolated complexes were pressed into implants that were injected subcutaneously behind the ear in swine. Table III reports the data obtained during the 21-day study. Pigs treated with a

Table III
Performance of Pigs Treated with Complexed PST-Containing Implants[a]

PST treatment[b]	Average daily gain (kg)	Percent increase over control	F/G ratio[c]	Percent change from control
Control	0.71	—	4.4	—
2H3MB-PST	0.79	11.9	3.52	21
Daily Pst	0.86	21	2.7	39
leu-PST	0.70	−1.4	4.12	6

[a]Adapted from Clark *et al.* (1993).
[b]Swine were treated with sham implant (control); 5 mg of PST per day (daily PST); 2-hydroxy-3-methoxybenzaldehyde—PST complex, 35-mg implant (2H3MB-PST); or leupeptin–PST complex, 35-mg implant (leu-PST).
[c]Feed-to-gain ratio.

35-mg 2-hydroxy-3-methoxybenzaldehyde (2H3MB)–PST complex implant showed a response over the sham implants control group. However, the pigs treated with a leupeptin–PST complex did not show a significant response. An increase over the sham control was observed in the 2H3MB–PST implant treatment group in terms of both average daily gain and feed-to-gain (F/G) ratio at the end of the 21-day study.

Azain *et al.* (1989) disclosed implant formulations for PST and BST that are essentially without binders and polymer matrices. One example was described which incorporated silicone rubber tube coverings to leave only the ends of cylindrical implants exposed for release. In numerous examples of implants with different configurations and amounts of somatotropins, there are demonstrated effective formulations in pigs and cows that release biologically active PST or BST, respectively, for at least 1 week. Example 2 in their patent shows an average increase in milk production of 30% compared to untreated controls, and their figure 2 indicates consistently higher blood levels of BST in treated cows over the 56-day study. The treated cows received 10 uncoated implants, each containing 50 mg of BST, at 0, 14, and 28 days of the study.

Raman and Gray (1994) layered PST, glycerin, and a wax and surfactant blend onto nonpareil seeds in three passes using a fluid-bed coater. These were then compressed into tablets. The *in vitro* dissolution results indicated some delayed release.

Sivaramakrishnan (1991) prepared wax and wax/water-insoluble surfactant slabs of the zinc salt of PST/L-arginine (1:1 ratio). The wax and surfactant were melt blended, PST was added, and the mixture was homogenized and quickly cooled. Slabs of bees wax (100%) or a 90:10

mixture of beeswax and Mazol® (Mazer Chemical Co.) 80 MG-K (ethoxylated mono-and diglycerides; MW 1400, HLB 13.5), containing about 10% w/w PST that measured 2 cm × 0.5 cm × 0.2 cm were tested *in vitro* in pH7.4 phosphate-buffered saline. The results indicated that there was more complete release from systems incorporating the surfactant; in 7 days, 21.1% of the PST was released from the beeswax and surfactant implants compared to 9.8% of the PST from the beeswax implants.

4.2. Coated Implants

Lindsey and Clark (1991) prepared peptide aldehyde complexes of PST in an effort to stabilize compositions for prolonged delivery in swine. PST and leupeptin, a tripeptide protease inhibitor, were complexed, and the dried mixture was pressed into pellets that were spray-coated with an ethyl cellulose and poly(ethylene glycol) coating dissolved in methylene chloride. The coated pellets were implanted subcutaneously behind the ear in swine. Table IV shows that a response was seen in all treated groups in average daily gain and feed-to-gain ratio compared to untreated controls. The total amount of PST utilized to achieve a response in implanted pigs (treatment 4) was about 65% higher than in pigs treated with daily injections, and this response was achieved with a relatively short dosing interval of 1 week. Nevertheless, these data are indicative of the potential performance improvements in pigs treated with PST implants.

Castillo *et al.* (1991) prepared compacted pellets from copper-associated porcine somatotropin that were spray-coated in a fluidized bed with poly(vinyl alcohol) [Evanol® (E.I. Dupont de Nemours) PVA, a fully hydrolyzed polymer having a molecular weight M_n of 50,000]. The core compositions are those exemplified in the patent of Azain *et al.* (1989); that is, they are compositions of Monsanto's PST and BST recombinant analogs incorporating metal salts to modulate the somatotropin solubility, and the cores are essentially binderless and without polymer matrices. Their examples show *in vivo* weight gain comparisons between adult female rats implanted with coated and uncoated pellets. Markedly better overall weight gain and longer duration of release were measured in rats implanted with certain coated implants compared to rats with uncoated implants. Thus, implants prepared by conventional pharmaceutical processes (tableting and aqueous coating operations) were shown to be potentially viable for commercial applications.

Steber *et al.* (1993) also used conventional pharmaceutical tableting and coating processes to prepare partially coated PST implants. The

Table IV
Performance of Pigs Treated with PST–Leupeptin-Containing Implants[a]

Treatment	Dose	ADG[b] (kg)	Percent increase over control	F/G ratio[c]	Percent change in efficiency
1. Sham implant	0	0.82[d]	–	3.77[d]	–
2. PST daily injection	60 μg/kg per day (5.2 mg/day)	0.9[d,e]	9.7	2.59[f]	+31.2
3. PST–leupeptin implant	1 pellet per 7-day interval; 30 mg PST (4.3 mg/day)	0.84[d]	2.4	3.10[d,e,f]	+15.3
4. PST–leupeptin implant	2 pellets per 7-day interval; 60 mg PST (8.6 mg/day)	1.02[c]	24.3	2.65[f]	+29.5
5. PST–leupeptin implant	1 pellet per 14-day interval; 30 mg PST (2.1 mg/day)	0.86[d]	4.9	3.45[d,e]	+8.4
6. PST–leupeptin implant	2 pellets per 14-day interval; 60 mg PST (4.3 mg/day)	0.87[d]	6.1	3.09[e,f]	+18.0

[a]Adapted from Lindsey and Clarke (1991).
[b]Average daily gain. The pigs were weighed (in kilograms) individually on day zero and at 7-day intervals throughout the study. The overall average daily gain was derived by subtracting the initial weight on day zero from the pigs' final weight and dividing by the number of days on the study.
[c]Feed-to-gain ratio per pen during the study was derived by adding all feed consumed per pen and dividing by total body weight gain per pen.
[d,e,f]Within columns, the means with different superscripts are significantly different ($p < 0.5$).

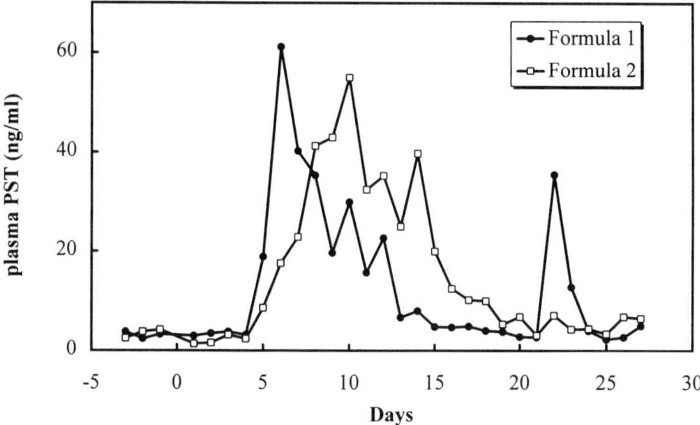

Figure 9. Porcine somatotropin (PST) levels in plasma, measured by radioimmunoassay in pigs implanted with PST implants. Formula 1 was a two-layer implant in which the layer closest to the indentation was 40 mg consisting of PST (45%), glyceryl trimyristate (43.8%), sucrose (8.4%), and monobasic/dibasic sodium phosphate (2.8%) and the second layer was 80 mg consisting of PST (60%), glyceryl trimyristate (25%), sucrose (11.3%), and monobasic/dibasic sodium phosphate (3.8%). The implant was coated with Eudragit® RL30D containing 15% (by weight) triethyl citrate followed by Eudragit NE30D containing 8% (by weight) talc. Formula 2 was a two-layer implant in which the layer closest to the indentation was 40 mg consisting of PST (50%), glyceryl trimyristate (37.5%), sucrose (9.4%), and monobasic/dibasic sodium phosphate (3.1%) and the second layer was 80 mg consisting of PST (60%), glyceryl trimyristate (25%), sucrose (11.3%), and monobasic/dibasic sodium phosphate (3.8%). The implant was coated with Eudragit RL30D containing 15% (by weight) triethyl citrate followed by Eudragit NE30D containing 15% (by weight) talc. (Data from Steber *et al.*, 1993.)

implant core compositions included a variety of fats, sugars, and salts, and the coating materials were poly(ethylacrylate methylmethacrylates) [Eudragit® (Rohm Pharma)]. Layered implant cores were prepared by sequentially filling a tablet press die with different granular compositions containing the PST. The top punch in the tablet press had a 3 mm × 1 mm projection on the center line of the punch, which projection left an indentation in the end of the cylindrical implant when it was compressed. When the implants were coated in a standard coating operation, the surface within the indentation remained essentially uncoated and became the only passageway for the active ingredient to exit. *In vitro* release was shown to vary depending on the implant and coating compositions and coating thickness. Figure 9 indicates sustained plasma levels in swine implanted subcutaneously in the ear with two implants of this invention.

Sivaramakrishnan *et al.* (1989) described BST matrix implants consisting of either BST/ethyl cellulose or BST/poly(lactic acid) (M_n 2700) contained within microporous polyethylene sleeves (average pore size; 30–70 μm). The *in vitro* dissolution studies showed significant initial release from uncoated tablets (70% cumulative release in the first 4 days to a total of 80% in 8 days); the coated tablets had sustained release over 14 days with 35% cumulative release over that time.

Sivaramakrishnan and Miller (1990) prepared pellets of the zinc complex of rPST, arginine, and sucrose (1:3:1 parts by weight). These were dip-coated with a molten blend of beeswax, carnauba wax, and Mazol® (80:20:2 parts by weight) kept at about 85 °C. These coated pellets were implanted in pigs. Elevated PST levels indicated sustained release from these compositions.

Pitt *et al.* (1992), prepared chitosan–porcine somatotropin blends (40:60 ratio), compressed into $\frac{1}{8}''$-diameter pellets. The pellets were then spray-coated in a fluidized bed with a 2.5% w/v solution of poly(vinyl alcohol) (PVA, 99.7 mol% hydrolyzed, M_w 78,000). The length of spray time controlled the thickness of the coating. Thicknesses of 25, 50, 70, and 75 μm were prepared. These were tested *in vitro* using a medium of phosphate-

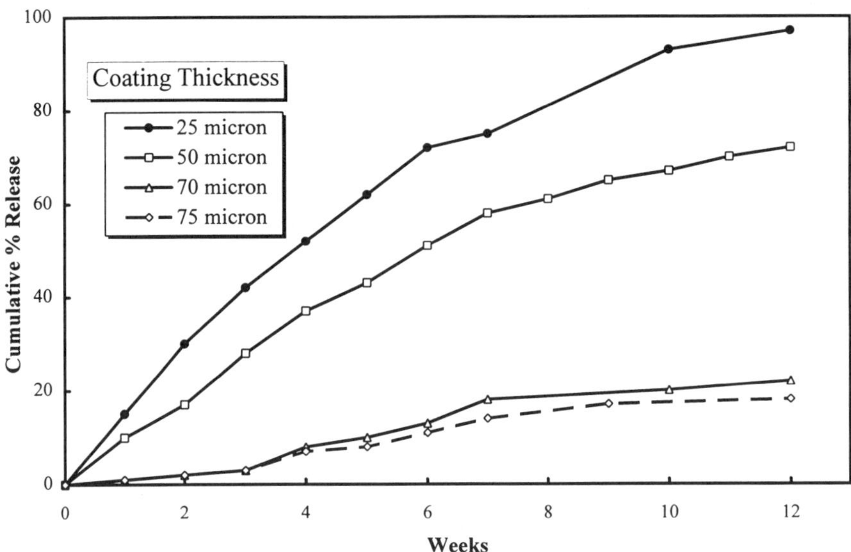

Figure 10. Release of PST from PST–chitosan PVA-coated implants. Implants were monitored for *in vitro* dissolution in phosphate-buffered saline. Coating thicknesses of 25 μm, 50 μm, 70 μm, and 75 μm were evaluated. (Data from Pitt *et al.*, 1992.)

buffered saline, pH 7.4, at 4 °C. Aliquots were analyzed to determine the PST release profile. Figure 10 shows the rate of release of PST from the various pellets. The release rate of PST from the coated pellets was found to be inversely proportional to the PVA coating thickness. Coho salmon were implanted with the pellets, and their weight gain and length were measured initially and once a week for 6 weeks and then every other week over a total of 20 weeks. Figure 11 indicates a positive response in the salmon; the fish with the pellets containing the thinnest PVA coating had the highest growth rates. McLean *et al.* (1994) recently reported 95-week data on similar implants placed in the peritoneal cavity of coho salmon. Salmon treated with the PST in PVA-coated pellets were longer than the control group over the first 77 weeks of the trial. The mean percentage body weight and length increases at 95 weeks were higher in the Fish with the 25- and 50-μm PVA-coated implants than in the other groups, indicating better sustained release.

Sanders and Domb (1990) described reservoir devices using hydrogel membranes to control the release of polypeptides and various growth hormones. The hydrogels used as the rate-limiting membrane barriers for the release of drug were copolymers of hydroxyethyl methacrylate (HEMA),

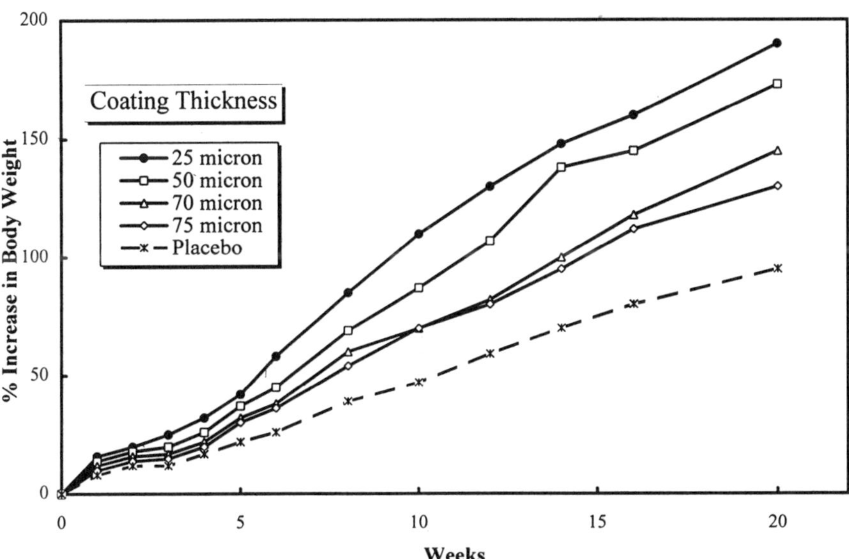

Figure 11. Increases in body weight of Coho salmon implanted with PST–chitosan PVA-coated pellets. Coating thicknesses of 25 μm, 50 μm, 70 μm, and 75 μm were evaluated compared to placebo pellets. (Data from Pitt *et al.*, 1992.)

glycerol methacrylate (GMA), and/or methyl methacrylate (MMA). The permeability and transport of active agents through these hydrogel membranes was controlled by the copolymer ratios of the hydrophilic constituents (e.g., HEMA) to the hydrophobic constituents (e.g., MMA) and the degree of cross-linking of the polymer, as well as such factors as the geometry of the rate-limiting membrane (e.g., surface area and thickness) and solubility of the polypeptide and protein drug. In their patent, Sanders and Domb give examples of membranes of various compositions and configurations that were capable of controlled release of the polypeptide nafarelin acetate. There are no data for the somatotropins.

5. OSMOTIC DEVICES

Researchers at Alza Corporation (Wong *et al.*, 1991, 1992) reported a patterned drug delivery device for porcine somatotropin. They prepared a dispenser designed to deliver several discrete, longitudinally aligned individual drug pellets by the linear expansion of a fluid-activated driving component. The dispenser housing was prepared with either a semipermeable membrane or one that was substantially impermeable. One of the porcine somatotropin device examples used a blend of polyvinylpyrrolidone, sodium salt of poly(acrylic acid) polymer, sodium chloride, and magnesium stearate that was compressed into osmotically active tablets as the expandable driving mechanism. The driving component consisted of a semipermeable wall that surrounded the compartment containing osmotically active tablets. Cellulose acetate butyrate and tributyl citrate were blended and injection-molded to form the wall that surrounded a compartment with an open end for receiving the osmotically active tablets and that mated with the dispensing component of the delivery system. The dispensing component was polycarbonate injection-molded into an impermeable wall that surrounded a compartment with one open end for receiving the drug pellets and for mating with the semipermeable wall. The second open end formed the exit port. The delivery device was assembled by putting three osmotic tablets into the semipermeable compartment, which was sealed with molten microcrystalline wax to form the driving component. The impermeable membrane compartment was filled with multiple compressed drug pellets (composition in weight percentage: 90% porcine somatotropin, 2% polyvinylpyrrolidone, 1% magnesium stearate, 3% hydrogenated vegetable oil, and 4% histidine HCl) containing about 112 mg of porcine somatotropin, which formed the dispensing component. The driving and dispensing components were joined at their open ends. Moisture-cured cyanoacrylic adhesive sealed the remaining open surface. No PST release data were disclosed; however, other examples gave limited data for gemfibrozil.

Eckenhoff *et al.* (1990) prepared osmotic implant devices containing PST in a glycerol, gelatin, and L-histidine gel. The device had an exit passageway for the gel and separate compartments for the osmotic driving excipients and the PST gel. The *in vitro* dissolution studies demonstrated controlled release for a 2-month period.

Magruder *et al.* (1992, 1993) in a series of patents showed release of PST from an implantable, nondegradable delivery device. The device had a semipermeable wall surrounding osmotically expandable tablets pushing a lubricated elastomeric piston compressing a compartment containing 340 mg of PST within an impermeable wall. The PST formulation consisted of porcine somatotropin, sodium phosphate monobasic, water, and polysorbate 80 (33.33/M4.53/M28.47/0.67 wt.%). The polycarbonate had been injection-molded with an orifice in place that was sealed with molten microcrystalline wax. In the improved version, a polycarbonate open-ended protective sleeve was placed over the semipermeable wall section to improve the *in vivo* robustness of the device. This protected the delivery device from transient mechanical forces such as those arising during the actual *in vivo* implanting procedures and animal-to-animal and animal-to-pen interactions. One example that was described has an implant body length of about 49 mm; the PST compartment contained therein is about 33 mm long and

Table V
Effects in Finishing Hogs Treated with PST-Containing Osmotic Devices[a]

Animal treatment	Cumulative average average daily gain	Cumulative feed consumption[b]	Cumulative feed efficiency[b,c]
Study I			
Control	0.71	2.77	3.91
Without sleeve	0.70	2.25 (17%)	3.55 (14%)
With sleeve	0.74	2.34 (16%)	3.10 (19%)
Study II			
Control	0.77	3.23	4.20
Without sleeve	0.81	2.87 (11%)	3.55 (15%)
With sleeve	0.81	2.73 (15%)	3.40 (19%)

[a]Adapted from Magruder *et al.* (1993).
[b]Percent improvement (%) is compared to the control group.
[c]The feed efficiency is determined by dividing the pounds of feed consumed by the number of pounds in body weight gained.

Table VI
Comparison of Delivery Rates at Six Weeks of Osmotic Devices Implanted in Finishing Hogs[a,b]

Implant type	Total pumps recovered	Number of pumps with a piston travel distance (mm) of 0	1–5	6–10	11–15	16–20	21–24	25–29	⩾30
Without sleeve	644	12	16	16	24	32	33	453	58
With sleeve	152	1	0	1	1	4	5	139	1

[a]Adapted from Magruder *et al.* (1993).
[b]The implanted devices were removed from the hogs and examined for piston travel, which correlates with the cumulative amount of PST released.

about 5 mm wide. Finishing hogs were implanted with devices with or without protective sleeves in the base of the ear with the orifice oriented up, using a modified trocar implanting device. Tables V and VI summarize the release and biological data from several studies. Moderately consistent piston travel, which correlates with PST delivery, was demonstrated for these prototype osmotic devices.

6. MISCELLANEOUS SYSTEMS

6.1. Wound Healing

Human growth hormone is capable of active stimulation of osteogenesis in cell culture. Downes *et al.* (1991) prepared ceramic disks loaded with HGH by placing the disks into solutions for absorption of the drug. The *in vitro* dissolution experiments indicated a two-phase release profile: a rapid release phase in which about 25% of the HGH was released within 10 days followed by a slower release phase in which another 10% was released through the next 10 days. Histology from the *in vivo* data from a rabbit model in which the HGH ceramic pins were implanted in the lateral cortex of each femur indicated an active remodeling process, with the HGH stimulating osteogenesis and matrix production at the ceramic–bone interface.

Hoelgaard (1991) prepared HGH in soft, porous, flexible sheets from hydroxyethyl cellulose and poly(ethylene glycol) 6000 using a lyophilization process. The HGH biological activity as measured by rat tibia test and HGH-receptor assay was reportedly maintained.

6.2. Nasal Delivery Systems

Illum (1991) demonstrated protein absorption from bioadhesive starch microspheres containing insulin or HGH delivered intranasally in experiments conducted in sheep.

Wearley (1991) described work by Osewein and Patton on an HGH aerosol formulation delivered by a nebulizer. A formulation tested in hypophysectomized rats demonstrated equivalent growth to that obtained with a subcutaneous injection.

7. CONCLUSIONS

With the advances made in recombinant DNA technology, fermentation and downstream refining of proteins is now a commercially viable process. These advances need to be combined with cost-effective controlled release technology to realize the commercial potential of the somatotropins. A preformulation effort to understand the complexities of the protein coupled with an evaluation of the feasibility of various approaches to sustained or controlled release of somatotropins is a complex undertaking involving potentially divergent considerations of the need for consistency and constancy of release profile, efficiency of dose utilization, the overall complexity and cost of the various approaches to formulation processing, end-user acceptance, and regulatory requirements. It is expected that the technologies studied for somatotropins will be applied to other applications of biotechnology that will, in turn, continue to assist farmers in maintaining production efficiency, resulting in increased return on investments for the food animal producers.

ACKNOWLEDGMENTS

The authors gratefully acknowledge the help of Harriet Maguire and Jackie White for typing the chapter and the editors for their encouragement and patience.

REFERENCES

Abdel-Meguid, S. S., Shieh, H. S., Smith, W. W., Dayringer, B. N., Violand, B. N., and Bentle, L. A. 1987, Three dimensional structure of genetically engineered variant of porcine growth hormone, *Proc. Natl. Acad. Sci. USA* **84:**6434–6437.

Aston, R., Bomford, R., and Holder, A. T., 1991, Physiologically active compositions of growth hormone and serum albumin or IgG, U.S. Patent 5,045,312 (Burroughs Wellcome Co., assignee).

Auer, H., Khan, M. A., and Bernstein, H., 1994, Controlled release growth hormone containing microspheres, WO 94/12158 (Alkermes Controlled Therapeutics, Inc. assignee).

Azain, M. J., Eigenberg, E. Kasser, T. R., and Sabacky, M. J., 1989, Somatotropin prolonged release, U.S. Patent 4,863,736 (Monsanto Company, assignee).

Azain, M. J., Kasser, T. R., and Sabacky, M. J., 1990, Compositions for controlled release of polypeptides, European Patent Application 374120A2 (Monsanto Company, assignee).

Brems, D. N., Plaisted, S. M., Havel, H. A., Kauffman, H. A., Stodola, J. D., Eaton, L. C., and White, R. D., 1985, Equilibrium denaturation of pituitary and recombinant-derived bovine growth hormone, *Biochemistry* **24:**7662–7668.

Brems, D. N., Plaisted, S. M., Kauffman, E. W., and Havel, H. A., 1986, Characterization of an associated equilibrium folding intermediate of bovine growth hormone, *Biochemistry* **25:**6539–6543.

Brems, D. N., Plaisted, S. M., Havel, H. A., and Tomich, C. S., 1988, Stabilization of an associated folded intermediate of bovine growth hormone by site-directed mutagenesis, *Proc. Natl. Acad. Sci. USA* **85:**3367–3371.

Brems, D. N., Havel, H. A., and Kauffman, H. A., 1989, Folding of bovine growth hormone is consistent with the molten globule hypothesis, *Proteins Struct. Funct. Genet.* **5:**93–95.

Buckwalter, B. L., Cady, S. M., Shieh, H. M., Chaudhuri, A. K., and Johnson, D. F., 1992, Improvement in the solution stability of porcine somatotropin by chemical modification of cysteine residues, *J. Agric. Food Chem.* **40:**358–362.

Cady, S. M., Steber, W. D., and Fishbein, R., 1989, Development of a sustained release delivery system for bovine somatotropin. *Proc. Int. Symp. Control. Rel. Bioact. Mater.* **16:**22–23.

Cady, S. M., Logan, J. S., Buckwalter, B. L., Stockton, G. W., and Chaleff, D. T., 1990, Stabilization of somatotropins by modification of cysteine residues utilizing site directed mutagenesis or chemical derivatization, European Patent Application 355,460. (American Cyanamid Co., assignee).

Cady, S. M., Fishbein, R., Schroder, U., Eriksson, H., and Probasco, B. L., 1993, Water dispersible and water soluble carbohydrate polymer compositions for parenteral administration of growth hormone, U.S. Patent 5,266,333. (American Cyanamid Co., assignee).

Campbell, R. G., Johnson, R. J., Tavener, M. R., and King, R. H., 1991, Interrelationships between exogenous porcine somatotropin (PST) administration and dietary protein and energy intake on protein deposition capacity and energy metabolism of pigs. *J. Anim. Sci.* **69:**1522–1531.

Carpenter, J. F., Hand, S. C., Crowe, L. M., and Crowe, J. H, 1986, Cryoprotection of phosphofructokinase with organic solutes: Characterization of enhanced protection in the presence of divalent cations, *Arch. Biochem. Biophys.* **250**(2)**:**505–512.

Castillo, E. J., Eigenberg, K. E., Patel, K. R., and Sabacky, M. J., 1991, Coated veterinary implants. European Patent Application 462,959A1 (Monsanto Company, assignee).

Clark, M. T., Gyurik, R. J., Lewis, S. K., Murray, M. C., and Raymond, M. J., 1993, Stabilized somatotropin for parenteral administration, U.S. Patent 5,198,422 (SmithKline Beecham Corporation, assignee).

Davio, S. R., and Hageman, M. J., 1993, Characterization and formulation considerations for recombantly derived bovine somatotropin, in: *Stability and Characterization of Protein and Peptide Drugs: Case Histories* (Y. J. Wang and R. Pearlman, eds.), Plenum Press, New York, pp. 59–89.

Downes, S., DiSilvio, L., Klein, C. P. A., and Kayser, M. V., 1991, Growth-hormone loaded bioactive ceramics, *J. Mater. Sci: Mater. Med.* **2:**176–180.

Eckenhoff, J. B., Magruder, J. A., Cortese, R., Perry, J. R., and Wright, J. C., 1990, Method for delivering somatotropin to an animal, U.S. Patent 4,959,218 (Alza Corp. assignee).

Ferguson, T. H., Harrison, R. G., and Moore, D. L., 1990, Injectable sustained release formulation, U.S. Patent 4,977,140 (Eli Lilly and Co., assignee).

Gráf, L., Li, C. H., and Bewley, T. W., 1975, Selective reduction and alkylation of the COOH-terminal disulfide bridge in bovine growth hormone, *Int. J. Pep. Protein Res.* **7:**467–473.

Hageman, M. J., Bauer, J. M., Possert, P. L., and Darrington, R. T., 1992a. Preformulation studies oriented toward sustained delivery of recombinant somatotropins, *J. Agric. Food Chem.* **40:**348–355.

Hageman, M. J., Possert, P. L., and Bauer, J. M., 1992b, Prediction and characterization of the water sorption isotherm for bovine somatotropin, *J. Agric. Food Chem.* **40:**342–347.

Hamilton, E. J., Jr., and Burleigh, B. D., 1989, Stabilization of growth hormones, U.S. Patent 4,816,568 (International Minerals and Chemical Corp., assignee).

Harbour, G. C., Hoogerheid, J. G., Garlick, R. L., 1987, Enhanced bioactivity of mammalian somatotropin through selective deamidation, International Patent Application PCT/US86/01860 (Upjohn Co., assignee).

Hart, I. C., and Johnson, I. D., 1986, Growth hormone and growth in meat producing animals, in: *Control and Manipulation of Animal Growth* (P. J. Buttery, D. B. Lindsay, and N. B. Haynes, eds.), Butterworths, London, pp 135–59.

Havel, H. A., Kauffman, E. W., Plaisted, S. M., and Brems, D. N., 1986, Reversible self-association of bovine during equilibrium unfolding, *Biochemistry* **25:**6533–6538.

Hoelgaard, A., 1991, Pharmaceutical preparation, WO91/19480 (Novo Nordisk A/S, assignee).

Holzman, T. F., Brems, D. N., and Dougherty, J. J., Jr., 1986, Reoxidation of reduced bovine growth hormone from a stable secondary structure, *Biochemistry* **25:**6907–6917.

Illum, L., 1991, The nasal delivery of peptides and proteins, *Trends Biotechnol.* **9**(8)**:**284–289.

Janoff, A. S., Bolcsak, L. E., Weiner, A. L., Trembley, P. A., Bergamini, M. V. W., and Suddith, R. L., 1989, Composition using salt form of organic acid derivative of alpha tocopherol, U.S. Patent 4,861,580 (The Liposome Co., Inc., assignee).

Jorgensen, J. O. L., 1991, Human growth hormone replacement therapy: pharmacological and clinical aspects, *Endrocr. Rev.* **12:**189–207.

Kim, N. J., Rhee, B. G., and Cho, H. S., 1991, A composition durably releasing bioactive polypeptides, Australian Patent 9,170,937 (Lucky Ltd., assignee).

Lee, J. L., and Timasheff, S. N., 1981, The stabilization of proteins by sucrose, *J. Biol. Chem.* **256:**7193–7201.

Lehrman, S. R., Havel, H. A., Tuls, J. L., Plaisted, S. M., and Brems, D. N., 1991, Somatotropin analogs, WO 91/00870 (The Upjohn Co., assignee).

Lindsey, T. O., and Clark, M. T., 1991, Stablized somatotropin for parenteral administration, U.S. Patent 5,015,627 (SmithKline Beecham Corp., assignee).

Magruder, J. A., Eckenhoff, J. B., Cortese, R., Wright, J. C., and Peery, J. R., 1992, Delivery system comprising means for delivering agent to livestock, U.S. Patent 5,110,596 (Alza Corp., assignee).

Magruder, J. A., Peery, J. R., and Eckenhoff, J. B., 1993, Delivery device with a protective sleeve, U.S. Patent 5,238,687 (Alza Corp., assignee).

McLaren, D. G., Bechtel, P. J., Grebner, G. L., Novakofski, J., McKeith, F. K., Jones, R. W., Darymple, R. H., and Easter, R. A., 1990, Dose response in growth of pigs injected daily with porcine somatotropin from 57 to 103 kilograms. *J. Anim. Sci.* **68:**640–651.

McLean, E., and Donaldson, E. M., 1993, The role of growth hormone in the growth of poikilotherms, in: *The Endocrinology of Growth, Development and Metabolism in Vertebrates* (M. P. Schriebmann, C. G. Scanes, and Pang, eds.), Academic Press, San Diego, pp 43–71.

McLean, E., Donaldson, E. M., Mayer, I., Teskeredzic, E., Teskeredzic, Z., Pitt, C., and Souza, L.M., 1994, Evaluation of a sustained-release polymer encapsulated form of recombinant porcine somatotropin upon long-term growth performance of Coho salmon, *Oncorhynchus kisutch, Aquaculture* **122:**359–368.

Mitchell, J. W., 1991, Prolonged release of biologically active somatotropin, U.S. Patent 5,013,713 (Monsanto Co., assignee).

Moseley, M., Paulissen, J. B., Goodwin, M. C., Alaniz, G. R., and Claflin, W.H., 1992, Recombinant bovine somatotropin improves growth performance in finishing beef steers, *J. Anim. Sci.* **70:**412–425.

Parcells, A. J., and Mott, J. E., 1990, Somatotropin analogs. WO90/08823 (The Upjohn Co., assignee).

Pearlman, R., and Bewley, T. A., 1993, Stability and characterization of human growth hormone in: *Stability and Characterization of Protein and Peptide Drugs: Case Histories* (Y. J. Wang and R. Pearlman, eds.), Plenum Press, New York, pp. 1–58.

Peel, C. J., Bauman, D. E., Gorewit, R. C., and Sniffen, C. J., 1981, Effect of exogenous growth hormone on lactational performance in high yielding dairy cows, *J. Nutr.* **111:**1662–1671.

Peel, C. J., Fronk, T. J., Bauman, D. E., and Gorewit, R. C., 1983, Effects of exogenous growth hormone in early and late lactation on lactational performance of dairy cows, *J. Dairy Sci.* **66:**776–782.

Pitt, C. G., 1990, The controlled parenteral delivery of polypeptides and proteins, *Int. J. Pharm.* **59:**173–196.

Pitt, C. G., Cha, Y, Donaldson, E. M, and Mclean, E., 1992, Polyvinyl alcohol coated pellet of growth hormone, WO92/07556 (Amgen Inc., assignee).

Raman, S. N., Gray, M. W., 1994, Compositions and processes for the sustained release of drugs, U.S. Patent 5,328,697 (Mallinckrodt Veterinary, Inc., assignee).

Randawa, Z. I., and Seely, J. E., 1990, Stable bioactive somatotropins, WO90/02758 (Pitman-Moore, Inc., assignee).

Sanders, L. M., and Domb, A., 1990, Delayed release of macromolecules, U.S. Patent 4,959,217 [Syntex (U.S.A.) Inc., assignee].

Schleyer, M., Etzrodt, H., Trah, T. H., and Voigt, H. H., 1982, Dependence of human somatotropin activity on interchain disulfide bridges, *Hoppe Seyler's Z. Physiol. Chem.* **363:**1111–1116.

Schleyer, M., Trah, T. H., Kornhuber, J., and Voigt, H. H., 1983, The importance of the intramolecular disulfide bridges in porcine somatotropin for its biological activity, *Hoppe Seyler's Z. Physiol. Chem.* **364:**291–294.

Sivaramakrishnan, K. N., 1991, Sustained release composition for macromolecular proteins, WO91/05548 (Pitman-Moore, Inc., assignee).

Sivaramakrishnan, K. N., and Miller, L. F., 1990, Controlled release delivery device for macromolecular proteins, WO90/11070 (Pitman-Moore, Inc., assignee).

Sivaramakrishnan, K. N., Rahn, B. M., Moore, B. M., and O'Neil, J., 1989, Sustained release of bovine somatotropin from implants, *Proc. Int. Symp. Control. Rel. Bioact. Mater.* **16:**14–15.

Souza, L. M., 1993, Porcine growth hormone analogs and compositions, U.S. Patent 5,244,882 (Amgen, Inc., SmithKline Beecham Corp., assignees).

Steber, W., 1993, Stable compositions for parenteral administration and method of making same, U.S. Patent 5,213,810 (American Cyanamid Co., assignee).

Steber, W. D., Cady, S. M., Johnson, D. F., and Rice, T., 1993, Implant compositions containing a biologically active protein, peptide or polypeptide, European Patent Application 523,330A1 (American Cyanamid Co., assignee).

Steber, W. D., Fishbein, R., and Cady, S. M., 1989, Compositions for parenteral administration and their use, U.S. Patent 4,837,381 (American Cyanamid Co., assignee).

Thakkar, A. L., Harrison, R. G., and Moore, D. L., 1988, Injectable sustained release formulation, U.S. Patent 4,977,659 (Eli Lilly and Co., assignee).

Tyle, P., 1989, Sustained release growth hormone compositions for parenteral administration and their use, U.S. Patent 4,857,506 (American Cyanamid Co., assignee).

Tyle, P., and Cady, S. M., 1990 Sustained release multiple emulsions for bovine somatotropin delivery, *Proc. Int. Symp. Control. Rel. Bioact. Mater.* **17**:49–50.

Viswanathan, R., and DePrince, R. B., 1990, Delivery device for the administration of stabilized growth promoting hormones, U.S. Patent 4,917,685 (International Minerals and Chemical Corp., assignee).

Wearley, L. L., 1991, Recent progress in protein and peptide delivery by noninvasive routes, *Crit. Rev. Ther. Drug Carrier Syst.* **8**(4):371–375.

Weiner, A. L., Estis, L. F., and Janoff, A. S., 1989, High integrity liposomes and method of preparation and use, WO89/05151 (The Liposome Co., Inc., assignee).

Wong, P. S. L., Theeuwes, F., Eckenhoff, J. B., Larson, S. D., and Huynh, H. T., 1991, Multi-unit delivery system, U.S. Patent 5,023,088 (Alza Corp., assignee).

Wong, P. S. L., Theeuwes, F., Eckenhoff, J. B., Larsen, S. D., and Huynh H. T., 1992, Multi-unit delivery system, U.S. Patent 5,110,597 (Alza Corp., assignee).

Chapter 12

Insulin Iontophoresis

Burton H. Sage, Jr.

1. INTRODUCTION

Before 1922, when Banting and Best (1922) first injected a pancreatic extract for treatment of juvenile diabetes, the disease was uniformly fatal, characterized by a wasting away of the body no matter how much food was consumed. In the years following this breakthrough, the active ingredient in the pancreatic extract, insulin, was identified and purified. Injection of insulin rapidly became the treatment of choice for juvenile diabetes, and the mortality rate rapidly declined. The treatment was so successful that the medical community thought a cure for the disease had been found.

As the first population of insulin-treated diabetics aged, it soon became clear that insulin did not cure the disease. While it added dramatically to life expectancy, insulin-treated diabetics began to exhibit a wide variety of seemingly unrelated disorders such as chronic leg ulcers, blindness, kidney disorders, and cardiovascular problems. Further, this population of patients frequently developed comas with either a very high or very low level of blood glucose.

As it became clear that these disorders were highly correlated with insulin-treated diabetics, research into the etiology of the disease began in earnest. In the early 1970s, reports began to appear (Fritz, 1972) showing that in healthy individuals insulin is responsible for the storage of glucose

---

Burton H. Sage, Jr. • Becton Dickinson Research Center, Research Triangle Park, North Carolina 27709.

Protein Delivery: Physical Systems, Sanders and Hendren, eds., Plenum Press, New York, 1997.

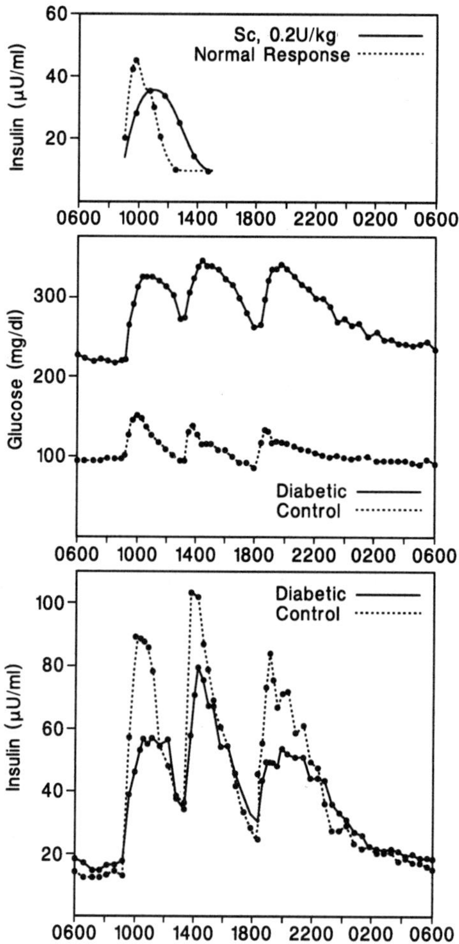

Figure 1. Insulin and glucose levels in healthy and diabetic individuals. In the upper panel, Normal insulin levels resulting from a meal are compared to insulin levels resulting from subcutaneous insulin, 0.2 U/kg. In the middle panel, Serum glucose levels in healthy individuals are compared to these in diabetic individuals. These glucose levels resulted from the insulin serum levels shown in the lower panel. (After Polonsky *et al.*, 1988.)

in certain cells of the body. In the absence of insulin, the body could not store glucose, and the body consumed itself in trying to stay alive. Injections of insulin allowed the body to store glucose, but the rate at which glucose was stored depended on the serum level of insulin. Too high a level of serum insulin, and the body tried to store all the blood glucose, resulting in hypoglycemic coma. Too low a serum level of insulin, and the body could not store the glucose fast enough, resulting in hyperglycemic coma.

The insulin injections were providing enough insulin to store enough glucose to keep the diabetic alive, but the serum level of insulin after a subcutaneous injection was dramatically different from the serum level of insulin in a healthy individual (Polonsky *et al.*, 1988). This is clearly shown in the upper panel of Fig. 1. Serum insulin from an injection rises too slowly and lasts too long. The impact of this improper administration of insulin is shown in the lower two panels of Fig. 1 (Polonsky *et al.*, 1988). In the healthy individual, after a meal the insulin rises rapidly and to higher levels than in the diabetic. The result is lower average glucose levels, and smaller changes in glucose in the healthy individual compared to the diabetic. It is important to note that the differences in insulin levels between a healthy individual and a diabetic are relatively small (Polonsky *et al.*, 1988), but the differences in serum glucose are much larger. While the relationship between insulin levels, serum glucose, levels, and diabetic sequelae is complex, it is clearly important to achieve tighter control of serum glucose in the diabetic patient.

While data showing the differences between serum glucose levels in healthy individuals and in diabetics have been available for several decades, the link between uncontrolled serum glucose and the increased morbidity seen in treated diabetics has been difficult to establish. Conventional wisdom has been that the better the control of serum glucose, the lower the prevalence and the later the onset of these secondary disorders. However, until recently, the proof was missing.

To establish the link between uncontrolled glucose levels and the secondary disorders, the National Institute of Health (NIH) began the 10-year Diabetes Control and Complications Trials (DCCT) in 1984. In 1993, the trial was ended one year early. The clear and conclusive results (Diabetes Control and Complications Trial Research Group, 1993) were that virtually all of the secondary disorders were either delayed or diminished, in some cases by as much as 70%, when glucose was tightly controlled. There remains little doubt that insulin administration regimens conventionally used prior to the DCCT were inadequate to prevent the secondary disorders of the disease and that intensive regimens were capable of a much better management of the disease. However, these intensive regimens are very user unfriendly. They require blood glucose monitoring, using finger sticks to get the blood up to eight times per day, and three to four injections of insulin daily—the actual amount of injected insulin depending on the measured blood glucose, the amount of food expected to be consumed, and qualitative factors such as levels of stress and physical activity.

To date, the only practical insulin delivery methods are the needle and syringe and pens. While these devices today are much more comfortable and convenient than those available in the past, they are fundamentally incapable of administering insulin in a pattern that could permit replication of the normal insulin secretion profile. The clear need is an artificial pancreas, that is, a system capable of continuous and variable administration of insulin at a rate depending on the serum level of glucose at the time. While such a mechanical pancreas has been developed for use during surgery of individuals with diabetes, portable systems for everyday use by diabetics remain unavailable. Improved insulin delivery systems are needed which can accomplish better control of serum glucose but are not as intensive and unfriendly as eight finger sticks and four injections daily. While this chapter will not review all the possibilities in detail, it will compare several to demonstrate that iontophoresis (use of low levels of electrical current to transport drugs through the skin) is among the most attractive.

2. SPECIFIC DRUG DELIVERY REQUIREMENTS FOR INSULIN

2.1. Duplicating the Function of the Pancreas

As mentioned above, an ideal insulin delivery system would supply insulin as a function of the concentration of glucose in the serum. This is what the pancreas does, among other things. Any candidate system for delivery of insulin should be evaluated in terms of the capabilities of the pancreas. The following performance parameters enter into this evaluation:

1. *Non-invasiveness.* The integrity of the body barrier functions should not have to be breached to administer the insulin.
2. *Delivery rate control.* The pancreas secretes insulin at a low "basal" rate that is specific to the individual. A replacement system should also be able to deliver insulin at a "basal" rate that is specific to the user. This "basal" rate is on the order of 1–2 units per hour.
3. *Bolus administration.* The pancreas is able to rapidly secrete insulin to respond to food intake. A replacement system should also be able to deliver a variable amount of insulin in a short amount of time to handle a meal. This bolus can range up to 20 units.
4. *Dose precision.* The pancreas delivers almost exactly what the body needs. A replacement system should, at a minimum, reproducibly deliver a given insulin dose.

5. *Portal delivery.* In order to supply the liver with all the insulin it needs, the pancreas delivers the insulin into the portal blood supply. Ideally, a replacement system would act similarly.
6. *Bioavailability.* The body utilizes virtually all of the insulin that the pancreas secretes. A replacement system should efficiently utilize the loaded dose.
7. *Compliance.* The healthy individual is completely unaware of what the pancreas does. While a diabetic must apply any replacement system, it should not be difficult for the diabetic to use.

2.2. Candidate Systems for Insulin Delivery

There are two basic concepts for the delivery of insulin—open- and closed-loop. A closed-loop system would be one that measured serum glucose and delivered an amount of insulin based on the level of serum glucose. In principle, this is the best way to approach the problem since this is what nature devised.* As mentioned above, a system that actually does this is routinely used in the surgical setting. However, it works much like an artificial heart/lung machine, in that blood is pumped out of the body, through a glucose measuring system, through an insulin infusion system, and back into the body—all the time remaining sterile. To date, a practical system for everyday use by ambulatory diabetics has not been developed.

All of the systems mentioned below are open-loop systems; that is, an amount of insulin is administered, or insulin is delivered at a given rate, and for several hours this insulin manages the serum glucose in an unknown way. After some time, the glucose may or may not be measured, and another dose of insulin is administered, or the delivery rate is adjusted.

Not all of the systems described below have been shown to be capable of administration of insulin. The intent is to compare the conceptual abilities of these systems in terms of the requirements listed above. Based on this comparison, the different systems can be ranked in terms of apparent attractiveness for this application. The following eight systems will be considered:

1. *Injection of insulin.* A syringe is filled with an insulin solution, and the solution is expressed into subcutaneous tissue using a needle.

*A biological solution to glucose control, encapsulated islet cells, has been under investigation for several years. Recently, some success has been reported in the dog. Because encapsulated islets are not physical systems, as the title of this volume requires, encapsulated islet cells will not be considered here.

2. *Oral insulin.* Insulin is formulated for enteral administration, and the dose is swallowed.
3. *Iontophoresis.* An aqueous formulation of insulin is placed on the skin, and delivery through the skin to the epidermal vasculature is accomplished using a weak electric field. Variable delivery is accomplished by adjusting the current.
4. *Nasal.* Insulin is inhaled through the nasal passages. It may be absorbed through the nasal mucosa, or in the lung through the bronchi.
5. *Bioerodible implant.* Insulin is formulated in a polymer that dissolves in tissue at a given rate. Once implanted in the tissue, insulin is released into the body at a rate determined by the rate at which the polymer dissolves.
6. *Infusion pump.* An aqueous formulation of insulin is placed in a pump connected to a catheter placed either in subcutaneous tissue or in a vein. Variable delivery of insulin is accomplished by adjusting the pumping rate.
7. *Intravenous microparticles.* Insulin is usually formulated either as microspheres or liposomes and injected into a vein. Insulin is released to the vasculature as the microparticles dissolve.
8. *Ultrasound.* An aqueous formulation of insulin is placed on the skin. An ultrasonic transducer is immersed in the formulation, and the ultrasonic vibrations cause aqueous pathways to form in the stratum corneum by a process called cavitation. Insulin reaches the epidermal vasculature by diffusion (see Tachibana and Tachibana, 1991).

The above eight alternative insulin delivery systems are compared on the basis of each of the insulin delivery requirements in Table I.

In the case of noninvasiveness, the method is either invasive or not. Hence, the only scores are $+2$ and -2. For delivery rate titration, a system is either capable of sustained release or not. Those not capable of sustained release are at a significant disadvantage with respect to the others. Those capable of sustained release may or may not be titratable during use. Those that are titratable during use have a significant advantage. For the other five criteria, the candidate systems can be scored in a similar way. However, the goal of this comparison is not to determine an absolute score nor to achieve a definitive ranking of the different systems. It is only to determine the relative attractiveness of the systems and to ascertain whether any of them are more attractive than the system now used. The reader is invited to make a similar comparison using personal algorithms. It is a very informative exercise. It is quite possible that a different ranking will result. The point of the exercise is to demonstrate that each of the candidate methods has

Table I
Comparison of Different Technologies for Insulin Administration[a]

Performance parameter	Injection	Oral	Iontophoresis	Nasal	Bioerodible implant	Infusion pump	IV microparticle	Ultrasound
Noninvasiveness	−2	2	2	2	−2	−2	−2	2
Delivery rate titration	−2	−2	2	−2	1	2	1	2
Bolus	2	1	1	2	−2	2	−2	−1
Dose precision	0	−1	0	−1	0	2	0	−1
Portal delivery	−2	2	−2	−2	−2	−2	−2	−2
Bioavailability	2	−2	0	−2	2	2	0	0
Compliance	−2	0	1	0	2	−2	2	1
Score	−4	0	4	−3	−1	2	−3	1

[a]Scoring: Strong advantage, +2; slight advantage, +1; no advantage, 0; slight disadvantage, −1; strong disadvantage, −2.

strengths and weaknesses and that none of the alternatives even approaches the maximum score of 14.

Interestingly, all of the candidate methods are ranked higher than injection, the system now routinely used. This is a good check; by virtually any standard, insulin delivery by syringe and needle is a poor substitute for the pancreas.

A second interesting result is that wearable infusion pumps for insulin delivery have been marketed for nearly a decade, and less than 1% of insulin- dependent diabetics now use the pump. Those that do, however, are able to demonstrate superior control of blood glucose. This fact is reflected in the relatively high score that the infusion pump concept received. However, the 99% who do not use pumps have determined that the improved glycemic control is just not worth the hassle of wearing the pump 24 hours a day, changing tube sets, and having a needle in their skin all the time.

By this analysis, iontophoresis is an attractive concept. It is basically a needleless miniature infusion pump, having many of the attractive features of the infusion pump without the needles and tubes. The remainder of this chapter will look in detail at iontophoretic delivery of insulin. The next section will describe briefly how iontophoresis works and will summarize the evidence which suggests utility for insulin. In section 4, insulin iontophoresis will be considered both from the theoretical point of view and in terms of the state of the art as reflected in the published literature. Section 5 will examine the insulin molecule itself from the point of view of deliverability by iontophoresis and discuss how the molecule might be changed to improve iontophoretic delivery. The final section will discuss prospects for future commercialization of an iontophoretic dosage form for treating diabetics.

3. CAPABILITIES OF IONTOPHORESIS RELATED TO INSULIN DELIVERY

This chapter is not intended to be a review of iontophoresis. Many excellent reviews have been published on both iontophoresis in general (Guy, 1992; Tyle, 1986; Sage, 1993) and iontophoresis of peptides and proteins (Chien *et al.*, 1989; Cullander and Guy, 1992). Based upon information already published for other molecules, iontophoresis can satisfy many of the delivery requirements for insulin as described below. The actual state of the art of insulin delivery by iontophoresis will be described in Section 4.

3.1. Noninvasive Delivery of Insulin

One of the inherent features of iontophoresis—in fact, one that is embodied in the definition of iontophoresis—is the ability to deliver a drug through the skin without prior preparation of the skin. An iontophoretic system for delivery of insulin would be a skin patch. Insulin passes through the skin and into the systemic circulation by action of a weak electric current. This current is barely perceptible as it passes through the skin and has been shown to not alter the barrier function of the skin (Ledger, 1992). The only apparent effect that iontophoresis has on the skin is a blushlike erythema of the skin under the patch, which resolves in a few hours.

3.2. Control of Delivery Rate of Insulin

A second major feature of drug delivery by iontophoresis is that the flux of the drug across the skin is proportional to the amount of electrical current that is flowing (Haak and Gupta, 1993; Sage and Riviere, 1992). Electrical current is very easily adjusted and programmed, and the level of current can be tightly controlled to provide the desired rate of administration of insulin. This capability can be used to titrate the amount of insulin a diabetic receives over a given period of time or to provide extra insulin during certain times of the day such as at morning wakeup (dawn phenomenon).

3.3. Bolus Administration

Because a change in current causes a change in the rate of delivery across the skin, bolus drug delivery can be obtained by increasing the current for a short period of time. For example, Heit *et al.* (1993) have been able to stimulate secretion of follicle-stimulating hormone (FSH) in swine by short-term iontophoresis of luteinizing hormone-releasing hormone (LHRH).

Such changes in the strength of the current can be preprogrammed to occur at selected times of the day or can be provided "on demand" by pressing a button on the patch.

3.4. Dose Precision

Ideally, a specific dose of insulin administered to an individual several times would provide the same amount of circulating insulin each time. As administered by syringe and needle, high precision might even be expected, given the accuracy to which a syringe may be loaded and its contents dispensed. When this ability is actually measured, the intraindividual variability (same person, many times) and the interindividual variability (different subjects, one time) in insulin absorption are as high as 25% and 50%, respectively (Kolendorf and Bojsen, 1982). This lack of precision has been attributed to many different factors, including metabolic state of the individual at time of dosing, actual site of injection (muscle, dermis, subcutaneous tissue, etc.), and body site used for the injection (leg, arm, trunk, etc.).

Iontophoresis of several drugs in different platforms (Haak and Gupta, 1993; Sage and Riviere, 1992) has exceeded this level of precision. It is within the reach of the technology, then, to exceed the precision of injection.

3.5. Portal Delivery

Since an iontophoresis system must be placed on the skin, it is beyond the capability of this technology to provide portal delivery of insulin. This may not be a fatal flaw, however, since the recent DCCT trials have shown that intensive management of diabetes via nonportal routes can provide dramatically improved outcomes. However, a delivery system that could provide portal delivery of insulin would be expected to provide even better results.

3.6. Bioavailability

There are two aspects to the bioavailability requirement—the fraction of the dose loaded into the patch that actually enters the vasculature and the percentage of this dose that is biologically active.

3.6.1. PERCENTAGE OF DOSE REACHING THE VASCULATURE

After leaving the iontophoresis system, a drug molecule may suffer several fates: It may be metabolized on its journey through the skin, it may partition to a nonaqueous skin moiety, it may bind irreversibly to cell receptors, or it may be absorbed by the skin vasculature. Only recently have data been published that have attempted to measure this aspect of bioavailability (Sage, 1995a). Using the porcine skin flap model (Riviere *et al.*, 1986) and iontophoresis conditions optimized to deliver the maximum amount of lidocaine hydrochloride from an iontophoresis system, an average of $82 \pm 11\%$ of the loaded dose was recovered from the venous side of the flap vasculature. This study shows that it is possible to deliver a large fraction of the loaded dose to the bloodstream.

3.6.2. PERCENTAGE OF DOSE THAT IS BIOACTIVE

Once a drug molecule enters the bloodstream, it can again suffer a variety of fates. It can again be metabolized, as it could have been in the skin. It can also be cleared out of the vasculature before it has a chance to interact with a receptor (the desired result).

On an absolute basis, bioactivity is an extremely difficult factor to measure. One usually resorts to comparison of plasma levels to those achieved with a reference method such as an intravenous (i.v.) injection. The ability of a drug to remain bioactive after iontophoresis was also reported recently (Sage, 1995a). The degree to which pyridostigmine bromide inactivates red blood cell (RBC) acetylcholinesterase in swine was determined by continuous i.v. infusions at different infusion rates. The porcine skin flap was then used to determine the rate of iontophoretic delivery of pyridostigmine as a function of current. Based on these data, it was determined that iontophoresis of pyridostigmine at 0.9 mA would inactivate 35% of the RBC acetylcholinesterase. Iontophoresis of pyridostigmine in swine was then performed, with the result that a steady 40% inhibition of RBC acetylcholinesterase was achieved. These results demonstrate that the process of iontophoresis is capable of delivering biologically active drug to the bloodstream.

3.7. Compliance

Recent results have shown that people like to wear their medicine as a skin patch. In a study comparing a seven-day clonidine patch with once-a-day oral verapamil (Burris *et al.*, 1991), a compliance of greater than 90% was achieved for the patch compared to about 50% for the oral dosage form. In a patient population that requires drug delivery by syringe and needle several times a day, a patch system should be expected to provide excellent compliance.

3.8. Summary of Capabilities Related to Insulin Delivery

Based on the above paragraphs, there is considerable evidence to suggest a strong fit between the characteristics of iontophoresis and the needs for insulin delivery. In the next section, the state of the art, in terms of theoretical predictions and laboratory results of insulin iontophoresis, will be presented.

4. THEORETICAL LIMITATIONS AND PUBLISHED RESULTS

In many ways, noninvasive delivery of insulin is the drug delivery holy grail. There is a compelling need, with over 2 million U.S. diabetics presently injecting themselves every day. Given that iontophoresis is over a hundred years old and that the need for a noninjectable delivery system has been recognized for at least 20 years, one might expect a large body of scientific literature related to iontophoresis of insulin. The opposite is in fact true—there are less than two dozen scientific articles that report the results of iontophoresis of insulin, and six of those are from the Rutgers group headed by Dr. Yie W. Chien. Perhaps the easiest way to review this literature is chronologically.

4.1. Published Results of Insulin Iontophoresis

The first report of insulin iontophoresis was not related to diabetes management. It was an attempt to see if insulin could affect the saline content of sweat in cystic fibrosis patients (Shapiro *et al.*, 1975). A decrease in sweat chloride was observed after insulin iontophoresis; however, the role of the insulin, and whether insulin reached the circulatory system for systemic effect, was not clear.

The first attempt at lowering blood glucose by iontophoresis of insulin was that of Stephen *et al.* (1984). In a first series of studies on humans with regular insulin, where passive delivery, cathodal iontophoresis, and anodal iontophoresis were compared to controls, no evidence of a change in glucose was discovered.

The authors then tried cathodal iontophoresis of a monomeric form of insulin, sulfated insulin, in pigs in the belief that the smaller molecular size of the monomeric insulin (MW 5800) versus hexameric regular insulin (MW 35,000) would iontophorese better. In one of the animals studied, a large rise in serum insulin which correlated with a large fall in serum glucose was observed. However, in each of the other animals studied, no similar result was obtained. The authors concluded that insulin iontophoresis is possible and that a smaller monomeric insulin is important. The reasons for failure in all the animals but one remain unknown.

In 1986, Kari (1986) reported the results of iontophoresis of regular insulin in rabbits that were made diabetic by injection of alloxan. Two series of studies were done—in one series, the rabbit fur was removed by electric clippers only, and in the other the fur was removed by the clippers and the skin abraded using a scalpel blade. The series using unabraded skin showed no significant glucose lowering compared to controls. The series where the skin had been abraded showed very large increases of serum insulin and large decreases in serum glucose—to the point where some of the animals nearly died due to hypoglycemia. This work clearly establishes the role of the stratum corneum in iontophoretic delivery of insulin, as is well known for passive transdermal therapy.

Meyer *et al.* (1989) also studied iontophoresis of regular insulin in alloxanized rabbits. Cathodal iontophoresis was used as in the work of Kari (1986), but the skin was not abraded. In order to reduce the average molecular weight of the insulin, urea was added to the formulation. While a significant increase in serum insulin and decrease in serum glucose were recorded on average versus controls (4 control animals, 15 treated animals), there were some animals that did not respond to the treatment.

By far the largest body of work is that reported by Chien and co-workers. The work was done *in vivo* using alloxanized rabbits (Siddiqui *et al.*, 1987a), and rats (Chien *et al.*, 1987; Siddiqui *et al.*, 1987b), and *in vitro* (Banga and Chien, 1993) using excised hairless rat skin, all with regular insulin. *In vivo*, evidence of insulin transport was obtained by measuring serum glucose in test and control animals. *In vitro*, evidence of insulin transport was obtained by counting radioactive decays in the receptor solution. Iontophoresis with direct current was compared to that with pulsed direct current as well.

In vivo, large changes in serum glucose were measured. The biggest changes occurred with pulsed direct current as opposed to direct current. *In vitro*, efforts were made to determine the nature of the barrier to iontophoretic transport of insulin. Skin that had been treated with delipidizing agents and skin with the stratum corneum removed using the tapestripping method showed rates of transport much higher than untreated skin. Again, it was clearly shown that the stratum corneum plays a significant role in limiting iontophoresis of insulin. More recently, we have studied cathodal iontophoresis of sulfated insulin in alloxanized rabbits (Sage, 1995a). We also observed large decreases in serum glucose compared to passive insulin and saline iontophoresis controls.

All of the above work, while similar in many respects and different in a few, demonstrates that iontophoresis is capable of moving insulin through skin and, most importantly, that the insulin that reaches the bloodstream retains some biological activity. However, this work also shows that the stratum corneum represents a major barrier to the transport of insulin. While some skin treatments have reduced that barrier, these treatments are not appropriate for routine use in the diabetic patient population.

It is further instructive to compare the results achieved by the above workers (only the cases of untreated skin will be considered). Using the data reported, the actual delivered dose can be estimated based on the area under the curve of the observed glucose drop and calibration results from subcutaneous injections of insulin in the rabbit (0.1 units of insulin per kilogram results in an area under the curve of about 1000 mg·hr/dl). These estimates, scaled to 24 hr, are shown in Table II. As can be seen, the rabbit data obtained with regular insulin by Chien *et al.* (1987) and Meyer *et al.* (1989) are in relatively good agreement (within a factor of about 2 when scaled to equal current). When sulfated insulin (one-fourth the potency of regular insulin) is used, higher transport is achieved. However, it is dangerous to try to scale these results to human diabetics. In the first place, the current required to deliver up to 100 units almost certainly will not be tolerated (see Section 4.2.1). Perhaps more importantly, though, rabbit

Table II
Estimates of Amount of Insulin Delivered to Rabbits by Iontophoresis

			Estimated insulin dose[b]	
Reference	Current (mA)	Area[a] (cm^2)	Units	mg
Chien *et al.* (1987)	4	60	5	0.2
Meyer *et al.* (1989)	0.4	40	1	0.04
Sage (1995a)	0.2	6	2	0.32

[a]Area of skin contact.
[b]Scaled to 24 hr for 3-kg rabbits.

stratum corneum is much thinner than human stratum corneum. It remains to be shown that insulin can be iontophoresed into humans.

4.2. Theoretical and Practical Limitations to Insulin Iontophoresis

An insulin delivery system, in order to be approved for sale, must be shown to be both safe and efficacious. Since an insulin iontophoresis product would be worn on the skin, it must not damage the skin in any way, and, since it would be worn all day, every day, it must also be comfortable. And, of course, it must be able to deliver the required insulin dose. The safety issue is not amenable to theory at this time; the limitations, however, can be estimated from recent studies. The delivery question is more amenable to theory—it does establish constraints on what may be considered possible.

4.2.1. TOLERABILITY OF AN IONTOPHORESIS DOSAGE FORM

An iontophoretic dosage form for the delivery of insulin would be of little benefit if diabetics could not wear it. Recently, the skin response to a 24-hr iontophoresis dosage form was measured in human volunteers (Maibach, 1994). In terms of comparisons of iontophoretic patches at 200 $\mu A/cm^2$ and similar patches without current, no significant changes were measured for transepithelial water loss (TEWL), skin capacitance, and skin temperature. The only effect was modest transient erythema. For a 24-hr application, this establishes a well-tolerated current density. In order to determine a well-tolerated total current, we need only know the skin contact area. Experience with passive patches has shown that a total area of $50\,cm^2$ for a system worn all day, every day, may be close to a maximum. Allowing $20\,cm^2$ for skin adhesion and $30\,cm^2$ for the anode and cathode, a 15-cm^2 electrode area is estimated. For a current density of $200\,\mu A/cm^2$, this yields an estimate of 3 mA as well tolerated for a system worn all day, every day.

4.2.2. THEORETICAL LIMITS TO IONTOPHORETIC DELIVERY OF INSULIN

In order to predict the rate at which iontophoresis is able to transport insulin across the skin, only two assumptions must be made. The first assumption is that the insulin molecule can be formulated as a charged molecule. This is sufficient to ensure that iontophoresis will deliver the molecule out of the reservoir. The second assumption is that this charge is retained at all anatomical locations along the route the molecule takes on its way into the skin. As will be seen in the next section, this last assumption is not satisfied for all insulins. For the present, though, we will assume that we have an insulin, such as sulfated insulin, which meets this requirement.

Using basic theory regarding the conductivity of ionic solutions and Faraday's law, the following relationship for the delivered dose of a drug can be derived (Sage, 1995b):

$$\text{Dose} = \frac{\mu_i C_i I T \times \text{MW}}{F \, \Sigma_{j \neq i} |Z_j| \mu_j C_j} \tag{1}$$

where μ_i is the mobility of drug molecule, μ_j is the mobility of nondrug molecules, C_i is the concentration (molar) of drug, C_j is the concentration (molar) of nondrug molecules, I is the current, T is the length of treatment, MW is the molecular weight of the drug molecule, $|Z_j|$ is the valence of nondrug molecules, and F is Faraday's constant.

To simplify this expression, we will take the case in which there is only insulin in the drug formulation (no other ions which may compete with the insulin for the current) and the contribution to the current from ions arising in the skin can be represented by the contribution of the sodium ion only, using its free solution mobility and its apparent tissue concentration of about 0.15 M. In studies of iontophoresis of lidocaine, the author has been able to obtain excellent agreement between predictions obtained by using this theoretical approach and experiments using the porcine skin flap (Sage, 1995b).

With these simplifications, the equation for the delivered dose becomes

$$\text{Dose} = 744 \, \mu_i C_i I T \tag{2}$$

To calculate the amount of insulin that could be delivered in 1 hr using the estimated current limit of 3 mA, Eq. (2) reduces to

$$\text{Dose} = 8.0 \, \mu_i C_i \times 10^3 \tag{3}$$

Using capillary zone electrophoresis (CZE), the free solution mobility of human insulin was measured to be 0.00027 $\text{cm}^2/(\text{V} \cdot \text{Sec})$ at pH 2.5. Since this mobility is the highest mobility measured for any pH, it can be used as

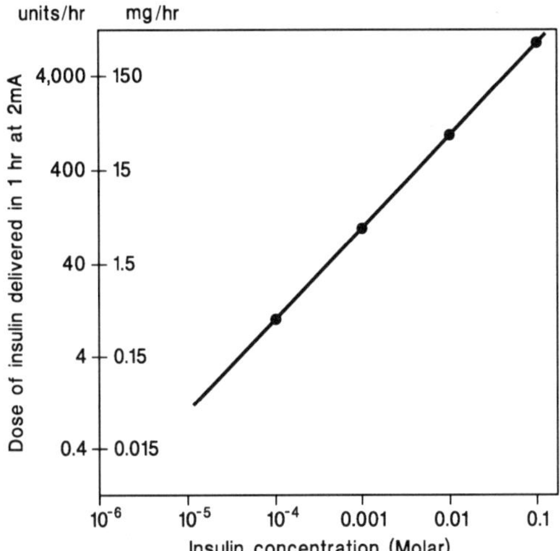

Figure 2. Theoretical insulin delivery by iontophoresis as a function of insulin concentration (molar).

an upper limit. The result is an expression for the delivered dose which is a function only of the concentration of the insulin. This expression is graphed in Fig. 2.

The graph in Fig. 2 can be used to address the most fundamental question regarding insulin iontophoresis—is it theoretically possible, under the most favorable conditions, to deliver the required dose? Based on the discussion in Section 2.1, the needs are a basal delivery rate of 1–2 units per hour coupled with a bolus of up to 20 units over about a half-hour. Figure 2 indicates that a delivery rate of 40 units per hour could be achieved with a 1 mM solution of insulin, which is equivalent to about 4 mg/ml or 100 units/ml. Regular (currently marketed human, pork, or beef) insulin has a water solubility that exceeds this value. Thus, it is theoretically possible to iontophorese insulin at the required rate. However, an idealized model has been used to reach this conclusion. Specifically, it was assumed that insulin exists as an ideal solution (with a MW of 5800), that the mobility of insulin is independent of pH and has a value close to its maximum value, and that the molecule is not degraded on its way through the skin. For regular insulins, these assumptions are not true. In the next section, the physicochemical properties of insulin that impact its deliverability by iontophoresis are described.

5. PHYSICOCHEMICAL PROPERTIES OF INSULIN RELATED TO IONTOPHORESIS

Several of the physicochemical properties of insulin play an important role in determining its deliverability by iontophoresis. In order of importance, they are charge titration, solubility, enzymatic susceptibility, and propensity for self-association.

5.1. Charge Titration

The basic assumption regarding iontophoresis of any molecule is that the molecule must be ionic and must remain ionic at all points on its journey into the skin. It is very easy to formulate insulin so that it is highly charged. It is virtually impossible to ensure that regular insulin will retain this charge, and hence its mobility, at all points in the skin. Two factors dictate this—the charge titration properties of regular insulin and the range of pHs encountered in the skin.

Figure 3 is a graph of the charge on regular insulin as a function of pH. As can be seen from this graph, it is dramatically different from the titration curve assumed in the theoretical calculation. Still, the difference between the assumed titration curve and the actual titration curve would not make any

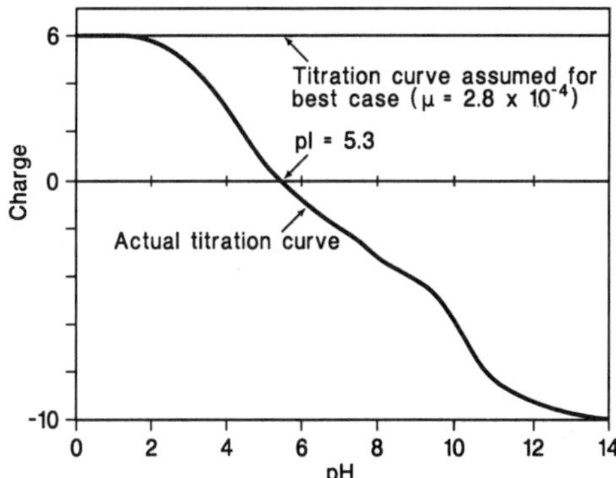

Figure 3. Titration curve of regular insulin. Also shown is the hypothetical titration curve used to show that insulin is theoretically deliverable by iontophoresis. Note that if the actual curve were shifted 5 pH units to the right, it would have properties close to ideal for iontophoretic delivery.

difference if the only pHs encountered were 2.5 or lower. The actual range of pHs encountered as the molecule progresses into human skin is from about pH 4.2 at the skin surface to pH 7.3 in the well-hydrated portions below the stratum corneum (Katz and Poulsen, 1971). For the assumed titration curve, which gives the molecule the same high mobility at all pHs, this would have little effect. For the actual titration curve, the result is that the mobility (roughly proportional to charge for a given molecule) of the molecule in the skin starts out lower than the assumed value, goes to zero at the isoelectric point, and actually changes sign before the molecule reaches the well-hydrated portions of the epidermis. In the presence of the iontophoretic electric field, the molecule would thus slow down, stop, and actually start to move back to the patch as it progressed into the skin.

It is these basic properties of human skin and the insulin molecule that best explain the lack of published data showing iontophoresis of regular insulin in humans. The above-cited data in rabbits are best interpreted in light of the difference between the stratum corneum of rabbits and that of humans. The rabbit stratum corneum is only 2–3 layers thick compared to the 15–20 layers in humans. It is reasonable to expect a quite different pH profile in rabbit skin, especially when occluded by an aqueous reservoir at a favorable pH.

Of the published data on *in vivo* iontophoresis of insulin, perhaps the most interesting are those obtained in pig studies with sulfated insulin (Stephen *et al.*, 1984) since pig skin is very similar to human skin. Sulfated insulin has an isoelectric point of about pH 2, which would give it a strong negative charge at all pHs encountered in skin. Such an insulin should be deliverable by cathodal iontophoresis. A large dose was delivered in one pig. It remains a mystery why a similar dose delivery was not seen in any of the other pigs.

5.2. Solubility

A second very important physicochemical property of insulin related to iontophoresis is its solubility. As is seen in Fig. 2, the rate of delivery is directly proportional to the molar concentration of the insulin in solution. Regular insulins that are marketed for the treatment of diabetes are available at a concentration of 100 units/ml, or 4 mg/ml. A higher molar concentration will, in principle (see Fig. 2), result in a higher flux, enabling the iontophoresis system to run at a lower current.

If the solubility of insulin were independent of the pH, the only constraint would be the concentration in the reservoir formulation, which could be set arbitrarily high. However, as the pH of the insulin environment approaches the isoelectric point, the solubility of insulin falls dramatically. Thus, the solubility at all pHs encountered must be considered.

5.3. Enzymatic Degradation

Another of the assumptions made in the theoretical section was that all of the insulin that left the reservoir eventually reached the bloodstream and an insulin receptor. It has been shown that hairless rat skin contains enzymes capable of metabolizing peptides and proteins such as insulin (Banga and Chien, 1993). These authors have shown that the addition of 0.25 mM aprotinin to an insulin formulation reduced insulin degradation during 3 hr of incubation in skin homogenate from 22% to 5%. They further showed that the aprotinin provided a few hours of protection from degradation during *in vitro* experiments using hairless rat skin. However, the aprotinin only slowed down the rate of degradation. These authors further reported that if the receptor solution was not assayed within a few hours, assay results were negative. The extent of insulin degradation in human skin during iontophoresis is not known.

5.4. Insulin Self-Association

The original hypothesis posed by Stephen *et al.* (1984) to account for lack of iontophoretic delivery of regular insulin in humans was that regular insulin formed polymers—mainly hexamers—at aqueous concentrations above about 100 μg/ml. His solution to this problem was to use sulfated insulin, which is monomeric at concentration up to about 100 mg/ml. His lack of consistent success with sulfated insulin suggests that average molecular size is not the only reason for lack of delivery.

The role of molecular size in iontophoretic delivery of drugs, beyond that of limiting the molar solubility of the drug, remains to be determined. Work by Ulashik (1976) with heparin and by Haak and Gupta (1993) with cytochrome *c* indicates that if there is a molecular weight limit for iontophoretic delivery, it is above 12–15 kDa. The fact that Chien *et al.* (1987) measured significant flux with regular insulin at 100 units/ml, where greater than 75% of the drug is nonmonomeric and mostly hexameric, suggests that molecules as large as 36 kDa can be iontophoresed.

6. FUTURE PROSPECTS FOR IONTOPHORETIC DELIVERY OF INSULIN

For regular insulin, such as beef, pork, or human insulin, the prospects for iontophoretic delivery seem quite poor. In spite of the attractiveness of the concept, there is not a single publication reporting success. Indeed, Stephen *et al.* (1984) reported failure.

The above paragraphs attempt to explain the lack of success in terms of the physicochemical properties of regular insulins and the nature of the path the insulin must take to become bioavailable. However, regular insulins are not the only insulins that can be made. Through chemical modification, site-directed mutagenesis, and other processes, insulin analogs and other compounds which possess insulinomimetic properties can be made. In the paragraphs above, the desirable properties of a new insulin that would be appropriate for iontophoresis have been identified. First, it should have an isoelectric point outside the pH range of the skin, and the farther outside this range, the better. Preferably, the isoelectric point would be pH 10 or higher, but a pI of 2 or lower may also be useful. Second, the new insulin should have a high charge. This would give it a relatively high mobility. A high charge combined with an isoelectric point remote from the pH range of skin would give the insulin a high mobility at all locations in the skin. This combination of features is considered necessary for iontophoretic delivery of insulin. Third, the new insulin should have a high water solubility. In general, the higher the solubility, the better, since this property, in theory, would reduce the current required for delivery at a specified rate. Fourth, the new insulin should resist degradation by skin enzymes. The degree of resistance needed is not known at this time. Finally, after all of this biological engineering, the new insulin must retain a significant degree of biological activity and be nonantigenic. If only 10% of the biological activity of regular insulin is retained, then 10 times the mass of the new insulin must be delivered, and 10 times the mass of a regular insulin must be manufactured. Further, the challenge of modifying insulin without introducing antigenicity will need to be addressed. Can such a new insulin be engineered? Probably yes, given the state of the art in the biotechnology industry. Will such a new insulin be engineered? Only if one of the major drug companies is willing to make the necessary investment. The incentive for a drug company to make such an investment has now appeared. The results from the DCCT trials show dramatically reduced complications when intensive therapy is used. Iontophoresis has the potential to be a userfriendly dosage form for intensive therapy.

REFERENCES

Banga, A. K., and Chien, Y. W., 1993, Characterization of *in vitro* transdermal iontophoretic delivery of insulin, *Drug Dev. Ind. Pharm.* **19:**2069–2087.

Banting, F. G., Best, C. H., Collip, J. B., Campbell, W. R., and Fletcher, A. A., 1922, Pancreatic extracts in the treatment of diabetes mellitus: Preliminary report, *J. Can. Med. Assoc.* **12:**141.

Burris, J. F., Papademetriou, V., Wallin, J. D., Cook, M. E., and Weidler, D. J., 1991, Therapeutic adherence in the elderly: Transdermal clonidine compared to oral verapamil for hypertension, *Am. J. Med.* **91:**225–285.

Chien, Y. W., Siddiqui, O., Sun, Y., Shi, W. M., and Liu, J. C., 1987, Transdermal iontophoretic delivery of therapeutic peptides/proteins I: Insulin, *Ann. N. Y. Acad. Sci.* **507:**32–51.

Chien, Y. W., Siddiqui, O., Shi, W. M., Lelawongs, P., and Liu, J. C., 1989, Direct current iontophoretic transdermal delivery of peptide and protein drugs, *J. Pharm. Sci.* **78:**376–383.

Cullander, C., and Guy, R. H., 1992, Transdermal delivery of peptides and proteins, *Adv. Drug Delivery Rev.* **8:**291–329.

Diabetes Control and Complications Trial Research Group, 1993, The effect of intensive treatment of diabetes on the development and progression of long term complications in insulindependent diabetes mellitus, *N. Engl. J. Med.* **329:**977–986.

Fritz, I. B., 1972, Insulin actions on carbohydrate and lipid metabolism, in: *Biochemical Actions of the Hormones,* Volume 2, pp. 166–214.

Guy, R. H. (Theme Editor), 1992, Iontophoresis, *Adv. Drug Delivery Rev.* **9:**119–307.

Haak, R., and Gupta, S. K., 1993, Pulsatile drug delivery from electrotransport therapeutic systems, in: *Pulsatile Drug Delivery, Current Applications and Future Trends* (R. Gurny, H. E. Juninger, and N. A. Peppas, eds.), Wissenschaftliche Verlagsgesellschaft, Stuttgart, pp. 99–112.

Heit, M. C., Williams, P. L., Jayes, F. L., Chang, S. K., and Riviere, J. E., 1993, Transdermal iontophoretic peptide delivery: *in vitro* and *in vivo* studies with luteinizing hormone releasing hormone, *J. Pharm. Sci.* **82:**240.

Kari, B., 1986, Control of blood glucose levels in alloxan-diabetic rabbits by iontophoresis of insulin, *Diabetes* **35:**217–221.

Katz, M., and Poulsen, B. J., 1971, Absorption of drugs through the skin, *Handb. Exp. Pharmacol.* **28:**103–174.

Kolendorf, K., and Bojsen, J., 1982, Kinetics of subcutaneous NPH insulin in diabetics, *Clin. Pharm. Ther.* **31:**494–500.

Ledger, P. W., 1992, Skin biological issues in electrically enhanced transdermal delivery, *Adv. Drug Delivery Rev.* **9:**289–307.

Maibach, H. I., 1994, Unpublished results.

Meyer, R. B., Katzeff, H. L., Eschbach, J. C., Trimmer, J., Zacharias, S. B., Rosen, S., and Sibalis, D., 1989, Transdermal delivery of human insulin to albino rabbits using electrical current, *Am. J. Med. Sci.* **297:**321–325.

Mitragotri, S., Edwards, D., Blankschtein, D., and Langer, R., 1995, A mechanistic study of ultrasonicallyenhanced transdermal drug delivery, *J. Pharm. Sci.* **84:**697.

Polonsky, K. S., Given, B. D., Hirsch, L. J., Tillil, H., Shapiro, E. T., Beebe, C., Frank, B. H., Galloway, J. A., and van Cauter, E., 1988, Abnormal patterns of insulin secretion in noninsulindependent diabetes mellitus, *N. Engl. J. Med.* **318:**1231–1239.

Riviere, J. E., Bowman, K. F., Monteiro-Riviere, N. A., Dix, L. P., and Carver, M. P., 1986, The isolated perfused porcine skin flap (IPPSF) 1. A novel *in-vitro* model for percutaneous absorption and cutaneous toxicology studies, *Fundam. Appl. Toxicol.* **7**:444–453.

Sage, B. H., 1993, Iontophoresis, in: *Encyclopedia of Pharmaceutical Technology,* Vol. 8 (J. Swarbrick and J. Boylan, eds.), Marcel Dekker, New York, pp. 217–247.

Sage, B. H., 1995a, Technological and developmental issues of iontophoretic transport of peptide and protein drugs, in: *Trends and Future Perspectives in Peptide and Protein Drug Delivery* (V. H. L. Lee, M. Hashida, and Y. Mizushima, eds.), Harwood Academic Publishers, Chur, Switzerland, pp. 111–134.

Sage, B. H., 1995b, Iontophoresis, in: *Percutaneous Penetration Enhancers* (E. Smith and H. Maibach, eds.), CRC Press, Boca Raton, Florida, pp. 351–368.

Sage, B. H., and Riviere, J. E., 1992, Model systems in iontophoresis—transport efficacy, *Adv. Drug Delivery Rev.* **9**:265–287.

Shapiro, B. L., Pence, T. V., Warwick, W. J., and Smith, Q. T., 1975, Insulin iontophoresis in cystic fibrosis, *Proc. Soc. Exp. Biol. Med.* **149**:592–593.

Siddiqui, O., Shi, W. M., and Chien, Y. W., 1987a, Transdermal iontophoretic delivery of insulin for blood glucose control in diabetic rabbits, *Proc. Int. Symp. Control. Rel. Bioact. Mater.* **14**:174.

Siddiqui, O., Sun, Y., Liu, J. C., and Chien, Y. W., 1987b, Facilitated transdermal transport of insulin, *J. Pharm. Sci.* **76**:341–345.

Stephen, R. L., Petelenz, T. J., and Jacobsen, S. C., 1984, Potential novel methods of insulin administration: 1. Iontophoresis, *Biomed. Biochim. Acta* **43**:553–558.

Tachibana, K., and Tachibana, S., 1991, Transdermal delivery of insulin by ultrasonic vibration, *J. Pharm. Pharmacol.* **43**:270.

Tyle, P., 1986, Iontophoretic devices for drug delivery, *Pharm. Res.* **3**:318–326.

Ulashik, V. S., 1976, *Theory and Practice of Medicinal Electrophoresis,* Minsk, Belarus.

Chapter 13

Insulin Formulation and Delivery

Jens Brange and Lotte Langkjær

1. INTRODUCTION

Ever since its introduction in 1922, insulin has provided a major stimulus for scientific research in numerous and diverse fields, including protein chemistry, structure, synthesis, and biosynthesis, polymer biochemistry, metabolism, endocrinology, cellular biology, immunogenicity, radioimmunoassay, receptor–ligand interactions, molecular genetics, and recombinant DNA technology. More importantly, in the context of this book, insulin has for many years served as a model compound for research into protein drug formulation and delivery.

The epoch-making discovery by Banting and Best (1922) prolonged the life expectancy for all insulin-dependent diabetic patients from two years to several decades. However, despite the major advances that have occurred relating to production, purification, and pharmaceutical formulation, insulin-replacement therapy is far from ideal (Zinman, 1989; Home *et al.*, 1989). The optimal method of insulin delivery must be safe, should provide insulin to diabetic patients in a way that will correct the metabolic abnormalities of diabetes mellitus, and must be psychologically and socially acceptable. Metabolic control should be maintained the closest possible to normal as this gives the best hope of preventing, delaying, arresting, or even reversing progression of long-term complications in diabetic patients. Sophisticated and sometimes complicated systems, such as continuous infusion pumps, have been developed to enable optimal regulation of

Jens Brange and Lotte Langkjær • Novo Nordisk A/S, DK-2880 Bagsvaerd, Denmark.
Protein Delivery: Physical Systems, Sanders and Hendren, eds., Plenum Press, New York, 1997.

diabetes (Reeves *et al.*, 1982). Current studies are directed toward optimizing the insulin molecule itself through protein engineering and determining the optimal means of its administration.

The objectives of the present chapter are to provide a brief overview of the complexity and diversity of insulin formulation, to give a historical review of the efforts to improve parenteral delivery of insulin, and, with an emphasis on the last 25 years' efforts, to survey the numerous studies performed in a search for acceptable insulin delivery through various noninvasive routes.

2. FORMULATION OF INSULIN

2.1. Introduction

During the first decade of the insulin era, only an acid solution of an impure form of the hormone was available for therapy. When it became possible to crystallize insulin, the purity, and hence the biological potency, of the hormone improved substantially (Table I). These improvements were mainly achieved by introduction of Zn^{2+}-crystallization (Scott, 1934) and recrystallization methods, the latter based on the observation that insulin recrystallized several times was better tolerated by patients suffering from allergic reactions (Jorpes, 1949). Until the late 1960s, recrystallized insulin was considered to be an essentially pure substance, but the introduction of new analytical methods made it possible to detect the presence of significant amounts of protein impurities by disc electrophoresis (Mirsky and

Table I
Development of the Purity of Insulin

		Content of			
International standard	Purification method	Desamido insulins	Proinsulin + derivatives	Di- and polymers of insulin	IU/mg
Animal, 1923	Amorphous precipitate	?	?	?	8
Animal, 1935	Crystallized	>15%	>5%	>5%	22
Animal, 1952	Recrystallized	5–15%	3–5%	2–3%	26.0
Animal, 1958	Recrystallized	5–15%	3–5%	2–3%	25.4
Human, 1987	Chromatographed + recrystallized	<1%	<1 ppm	0.2%	28.1

Kawamura, 1966) and gel filtration (Steiner, 1967; Steiner *et al.*, 1968). The purity of recrystallized insulin, as revealed by these methods, was only 80–90% (Table I; Brange *et al.*, 1987), and it was suggested that the contaminants served as adjuvants for immunization against insulin (Wehner *et al.*, 1973). The introduction of chromatographic purification in the 1970s reduced the content of impurities by at least one order of magnitude to less than 0.5%, and immunologic complications of insulin treatment (allergy, lipoatrophy, and insulin resistance) have essentially been eliminated. The major potential impurities in human insulin derived from recombinant DNA technology are those of host cell origin, i.e., *Escherichia coli* or yeast proteins, and those related to precursor protein.

Whereas extensive purification has resulted in a very pure and homogeneous insulin product, the protamine used for retardation in NPH type preparations (see Section 2.2.1) is highly inhomogeneous as it contains four major components, which are present in approximately equal amounts. All four components show close structural homology with one another, and consequently they generate nearly identical protamine–insulin complexes (Hoffman *et al.*, 1990).

As a result of derivatization of insulin during isolation and purification steps, and during storage of the pharmaceutical preparations, therapeutic insulin contains smaller amounts of desamido insulins, covalent insulin dimers, and other insulin derivatives. Physical and chemical stability of insulin in formulations for injection therapy has recently been reviewed in a previous volume of this series (Brange and Langkjær, 1993) and in a monograph (Brange, 1994).

2.2. Formulation for Parenteral Administration

2.2.1. CONTROLLED RELEASE

A shortcoming of the first therapeutic insulin, an acid solution of an impure form of the hormone, was that it required 4–6 injections daily, and nevertheless this regimen was fraught with the hazard of hyperglycemia during the night. The desirability of simplifying administration combined with the need to reduce the frequency of insulin injections and to ensure insulin action through the night led to a search for extended-acting formulations capable of prolonging the effects of insulin for periods of more than 24 hours. The first efforts were based upon formation of noncovalent complexes of insulin with other proteins or preparation of insulin crystals in a slow-release form. Thus, it was found that the addition of protamine, a highly basic protein (pI 13.8) isolated from fish sperm nuclei, notably slowed the absorption and hence sustained the action of the hormone (Hagedorn *et al.*, 1936; Scott and Fisher, 1936). Ten years later, the protamine–insulin concept was improved further through the development of NPH (Neutral Protamine Hagedorn), in which protamine and insulin are brought together in isophane (or stoichiometric) proportions (a molar ratio of ~1:9) at neutral pH, in the presence of zinc ions and phenol and/or cresol (Krayenbühl and Rosenberg, 1946). Under these conditions, the protamine–insulin complex forms oblong, tetragonal crystals without leaving any insulin or protamine in the supernatant. This type of intermediate-acting preparation is still—nearly 50 years after its invention—the most widely used insulin preparation. Recently, it has been demonstrated that phenol (or *m*-cresol), necessary for proper formation of protamine–insulin crystals, not only serves as a preservative but also is an essential part of the crystal structure, as six phenol molecules occupy cavities in the hexameric crystal unit (Balschmidt *et al.*, 1991).

Concern about the inclusion of protamine, being a foreign protein, in the preparation led to further investigations, and, a few years after the introduction of NPH, a sustained absorption of insulin was achieved without the use of protamine or other foreign additives by suspending carefully prepared zinc insulin crystals with a surplus of zinc ions in acetate rather than phosphate buffer. This retardation principle is based on the discovery that zinc ions in a molar ratio larger than one per insulin monomer dramatically reduce the solubility of insulin in neutral medium. This so-called Lente insulin principle, also invented in Denmark (Hallas-Møller, 1956; Hallas-Møller *et al.*, 1952), allowed gradation of the extent of the retardation. When the zinc ions are combined with the insulin in microcrystalline form only, the result is a long-acting form called Ultralente, and when the complex is made in such a way as to produce material that is entirely amorphous, the resulting suspension is shorter acting (Semilente). A 70:30 combination of Ultralente and Semilente is called Lente and is intermediate-acting.

NPH and Lente insulins differ in their ability to form mixtures with neutral insulin solutions (Berger *et al.*, 1982; Heine *et al.*, 1984; Forlani *et al.*, 1986; Brange *et al.*, 1987). Both can be mixed with regular insulin immediately before injection. Mixtures of NPH with regular insulin preserve their pharmacokinetic characteristics over time. This is not the case with Lente/regular mixtures because the surplus of zinc ions in the Lente part will combine with part of the insulin in solution and change it into a Semilente-like preparation.

Over the last two decades the increasing emphasis on an insulin therapy designed to produce near-normal glycemia has resulted in the development of premixed insulins for those patients who cannot accept a multiple-injection regime. These preparations are combinations of regular insulin and NPH in ratios from 10:90 to 50:50. Although a proportion of the added regular insulin is loosely bound to the NPH crystals (Nolte *et al.*, 1983; Brange *et al.*, 1987), this loss of dissolved insulin does not seem to be clinically significant. These mixtures are used in twice-daily injection regimens, which are regarded as the minimum desirable therapy for patients with insulin-dependent diabetes.

The biphasic-acting NPH/regular insulin mixtures were actually preceded by one of the more elegant and ingenious discoveries relating to insulin formulations. In a search for an intermediate-acting insulin preparation with a stronger initial effect, Schlichtkrull and co-workers (Schlichtkrull, 1959; Schlichtkrull *et al.*, 1965) found that bovine and porcine insulin crystals with a specific zinc-ion content differed sufficiently in their solubility between pH 7.0 and 7.1 to create a biphasic preparation. The preparation, named Rapitard®, contained 25% dissolved porcine insulin in a suspension of bovine insulin crystals with four zinc atoms per insulin hexamer. Later on, when bovine insulin was forsaken for immunogenicity reasons, the same principle was utilized to create a biphasic formulation of porcine insulin crystals in a solution of porcine monodesamido insulin (Brange, 1977), the latter insulin derivative showing sufficiently higher solubility than the parent insulin at neutral pH (Fig. 1; Brange *et al.*, 1987).

Based on the anisotropic arrangement of the insulin hexamers in rhombohedral insulin crystals (Schlichtkrull, 1957), it has also been possible to devise a formulation with a triphasic action profile. This was achieved by using crystals of porcine insulin to seed further crystallization of bovine

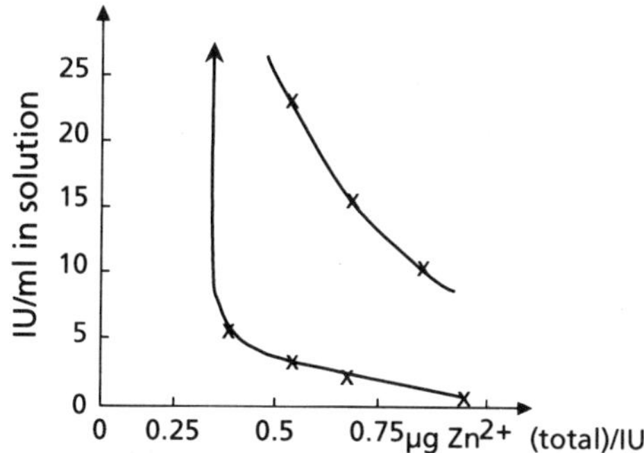

Figure 1. Solubility at pH 7.0 of monodesamido-(A21)-insulin (upper curve) and insulin of porcine origin (lower curve) as a function of zinc-ion content. (Adapted from Brange *et al.*, 1987.)

insulin followed by suspension of the mixed-species crystals in a solution of porcine insulin. When such double-layered crystals were studied in a rabbit prolongation test, a delayed burst release of the porcine insulin was apparent when the less soluble outer layer of bovine insulin had dissolved in a late phase of the course of absorption (Brange, unpublished results).

With the introduction, in the 1980s, of the homologous human insulin, it was no longer acceptable to utilize physicochemical differences between species of insulin in the formulation work (as in the examples mentioned above). All the present types of human insulin formulations are copied from their animal insulin counterparts, and only minor adjustment of the compositions has been necessary. Thus, human NPH seems to have a more pronounced tendency to form clumps of insulin which attach to the sides of the vial (termed flocculation or frosting) than NPH from animal insulins (Benson *et al.*, 1988). This propensity to flocculation, probably a macroscopic manifestation of insulin fibrillation (Brange and Langkjær, 1993), has been suppressed by increasing the zinc-ion content of the preparation, a recipe originally demonstrated to be effective when used in regular insulin preparations (Brange *et al.*, 1986).

The human zinc–insulin hexamer complexes in regular insulin dissociate more rapidly than the equivalent porcine insulin complexes (Brange *et al.*, 1990), resulting in a clinically advantageous faster onset of action (Kemmer *et al.*, 1982; Pramming *et al.*, 1984; Patrick *et al.*, 1988). On the other hand, a drawback of human NPH is the shorter duration of action of the injected suspension (Home *et al.*, 1989). Immunogenicity of human insulin with respect to formation of circulating antibodies to insulin is marginally lower than that of porcine insulin (Fineberg *et al.*, 1983; Marshall *et al.*, 1988), but insulin antibody responses to the highly purified porcine insulin preparations were already very low compared to those found with conventionally purified preparations. Delayed-acting human insulin formulations based on the Lente principle do not fit therapeutic requirements either, and current work is aimed at better simulating the natural pattern of insulin secretion, a constant baseline activity with a major increase following meals.

2.2.2. INSULIN PUMPS

In the late 1970s, the endeavors to obtain normoglycemia in diabetes treatment resulted in the introduction of continuous insulin infusion devices (insulin pumps) which were either worn externally (Pickup *et al.*, 1978; Irsigler and Kritz, 1979; Tamborlane *et al.*, 1979) or implanted (Buchwald *et al.*, 1980; Irsigler *et al.*, 1981; Schade *et al.*, 1982). Use of insulin solutions in mechanical systems involves the exposure of insulin to increased temperatures for extended periods of time, hydrophobic surfaces, and, because of the movement of the devices, shear forces, which together substantially increase the propensity of insulin to change its native conformation or "normal" pattern of assembly and to form precipitates due to insulin fibril formation. (For recent reviews, see Brange and Langkjær, 1993, and Brange, 1994.) In the 1980s it became evident that commercial insulin formulations were not sufficiently stable for long-term use in infusion pumps (Buchwald *et al.*, 1980; Lougheed *et al.*, 1983; Brennan *et al.*, 1985), and a comprehensive search for more stable formulations was started.

Numerous ways of stabilizing insulin solutions for use in delivery systems against aggregation and fibrillation were investigated in the following years (for a review, see Brange, 1994). These attempts included the addition of autologous serum (Albisser *et al.*, 1980), use of organic media such as glycerol (Blackshear *et al.*, 1983), introduction of organic (Lougheed *et al.*, 1980; Bringer *et al.*, 1981; Grau *et al.*, 1982; Brange and Havelund, 1983a; Quinn and Andrade, 1983; Thurow and Geisen, 1984; Chawla *et al.*, 1985) or inorganic additives (Brange *et al.*, 1982, 1986; Mecklenburg and Guinn, 1985; Ratner and Steiner, 1987), and derivatization of insulin (Thurow, 1980; Albisser *et al.*, 1982; Pongor *et al.*, 1983). Most of these methods are either impractical, inconvenient, or unsafe, in the sense that the additives used are not physiological or they have a deteriorating effect on the chemical and biological stability of the insulin. For example, while the addition of glycerol (Buchwald *et al.*, 1980) as well as certain polysaccharides (Brange and Havelund, 1983a) significantly increases the physical stability of neutral insulin solutions, the physical stabilization is accompanied by an unacceptable decrease in the chemical stability and biological activity of the preparations (Brange and Havelund, 1983b).

Addition of calcium and zinc ions, which probably also play an important role in the storage stability of insulin in the β cells of the pancreas (Howell *et al.*, 1978), significantly improves the physical stability of neutral insulin solutions (Fig. 2; Brange *et al.*, 1987), especially at higher concentrations of insulin (Brange *et al.*, 1982, 1986). These metal ions probably exert their fibrillation-inhibitory effect by neutralizing negative charges in the center of the insulin hexamer, whereby the hexameric assembly, and thereby the native structure, is stabilized. A marked fibrillation-inhibitory effect is obtained by the addition of low concentrations of lecithins (Hansen *et al.*, 1986) or of a synthetic detergent (Grau *et al.*, 1982). The positive influence of detergent additives is based on their ability to inhibit adsorption of insulin onto hydrophobic surfaces, which, subsequently, may lead to exposure of insulin's hydrophobic surfaces and structural rearrangement (Thurow and Geisen, 1984).

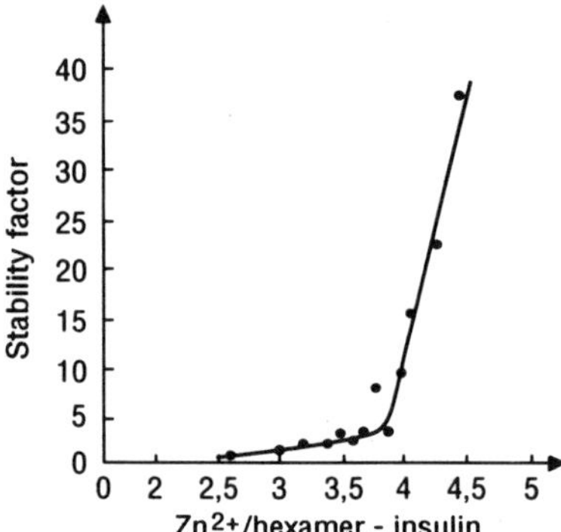

Figure 2. The effect of zinc-ion concentration on the relative physical stability of neutral porcine insulin solution in a shaking test at 37 °C (Brange *et al.*, 1986).

Recently, it has been demonstrated that an increase of the viscosity of insulin solutions by addition of 15% (w/w) of polygeline, a partially hydrolyzed and subsequently cross-linked gelatin product used as a plasma substitute, counteracts the promoting effect of shear forces on insulin fibrillation (Grau *et al.*, 1989). Most effective stabilization is seen when detergent additives are combined with elevation of the viscosity of the solution (Grau *et al.*, 1989) or with metal-ion stabilization (Brange and Langkjær, 1993).

The influence of materials used in delivery systems on the tendency of insulin to fibrillate has been studied by several groups (Brange *et al.*, 1982; Feingold *et al.*, 1984; Chawla *et al.*, 1985; Selam *et al.*, 1987; Grau *et al.*, 1989), and also the chemical stability of insulin in the environment of delivery systems or in contact with materials used in such systems has been reported (Blackshear *et al.*, 1983; Grau, 1985; Selam *et al.*, 1987; Melberg *et al.*, 1988; Grau *et al.*, 1989). These investigations have revealed that not only does the stability vary with the type and origin of the material in a very complex manner, but it is also dependent on the method by which the material has been sterilized (Melberg *et al.*, 1988). More systematic studies have recently demonstrated that the effect of contact materials on the tendency of insulin to fibrillate is correlated with the surface area and the hydrophobicity of the material (Sluzky *et al.*, 1991, 1992).

Loss of insulin by adsorption, due to its affinity for various compounds and materials, is a common phenomenon, especially at low concentrations of insulin, i.e., less than 1 IU/ml, and large surface areas. Insulin has binding affinity for hydrophobic (Hirsch *et al.*, 1981) as well as hydrophilic surfaces (Mitrano and Newton, 1982), although more insulin is adsorbed onto hydrophobic than onto hydrophilic materials (Sefton and Antonacci, 1984). Insulin adsorption can be counteracted, but normally not totally prevented, by addition of competing agents, such as albumin (Schildt *et al.*, 1978) or certain concentrations of urea (Sato *et al.*, 1983). A short review of the problems of binding of insulin to infusion sets can be found in a book by Brange *et al.* (1987).

2.3. Formulation for Alternative Routes

The stability and compatibility problems related to design of aqueous insulin solutions for use in mechanical pumps, mentioned above, are small compared with the myriad of challenges encountered by the formulation chemist working with insulin for delivery via alternative routes. These additional difficulties are related to the often harsh environment, including nonaqueous media and extremes of pH, during preparation of the dosage forms, and the effects of different excipients, absorption enhancers, and surrounding materials on the structural integrity of the insulin molecule. The few reports describing insulin preformulation considerations are mentioned in the following, but more systematic studies of stability and compatibility issues in relation to insulin formulation for alternative routes are lacking in the literature.

An aerosol dosage form developed by suspending zinc insulin crystals in a propellant together with oleyl alcohol to improve the wetting of the crystals was stated to be chemically stable (Lee and Sciarra, 1976), but as this conclusion was based on immunoassay results, it does not necessarily reflect reality. When insulin is formulated for pulmonary delivery by dry-powder generators (Byron, 1990), it is important to remember that insulin in the dry state, in an amorphous as well as a crystalline form, is hygroscopic, even when it contains 10–20% water, depending on the relative humidity.

Noninvasive delivery of insulin via most mucosal membranes requires the use of chemical enhancement for notable insulin absorption (see Section 3.3 and Table II). However, most permeation enhancers have, in addition to their effect on the mucosal membrane, an often pronounced influence on insulin three-dimensional structures. Thus, sodium salicylate (Touitou *et al.*, 1987) as well as bile salts (Gordon *et al.*, 1985) have been shown to dissociate insulin oligomers into monomers. This effect improves membrane permeability, but it may also reduce the physical stability and increase the susceptibility of insulin to enzymatic degradation. The exposure of new epitopes may also influence the immunological properties of the insulin formulation.

Table II
Insulin Bioavailability (Percent of Parenteral Availability) after Noninvasive Administration without Enhancer to Rats by Various Routes

Mode of administration	Insulin dose (U/kg)	Absorbed fraction (%)	Reference(s)
Ocular	3–4	1	Yamamoto *et al.* (1989)
Nasal	10	0.2	Irie *et al.* (1992)
	50	1.2	Aungst *et al.* 1988
	50	2.0	Aungst (1994)
Buccal	50	1.9	Aungst *et al.* (1988)
	50	0.8	Aungst (1994)
Intestinal (ligated segment)	20	0.2	Shao *et al.* (1994a)
Rectal	2–50	3–9	Aungst *et al.* (1988); Aungst (1994)
Pulmonary (by instillation)	1	9	Li *et al.* (1993)
	3	13	Okumura *et al.* (1992)
	6	15	Liu *et al.* (1993)
Pulmonary (by aerosol), at a depth of:			Okumura *et al.* (1992)
10 cm	7.5	37	
20 cm	7.5	98	

Surfactants are frequently used as enhancers, but such compounds interact with insulin, often in an unpredictable way. Thus, enthalpy measurements of the interaction of alkyl sulfates with insulin have shown a high degree of binding of these anionic surfactant to insulin, as compared to nonassociating globular proteins (Sarmiento *et al.*, 1992; Prieto *et al.*, 1993). Cationic surfactants also interact with insulin (Ushiwata *et al.*, 1975). In some cases insulin precipitation can be observed (Birdi, 1973; Shao *et al.*, 1992); in other situations, the solubility of insulin is markedly increased (Touitou *et al.*, 1987).

2.4. Insulin Analogs and Derivatives

Until recently, advances in insulin formulation were limited to improvements in insulin purity, insulin species, and adjustment of the vehicle composition. With the advent of recombinant DNA techniques, it has been possible to optimize also the insulin molecule itself for substitution therapy. The therapeutic need to improve the hormone's pharmacokinetic properties, storage stability, and suitability for less intrusive routes of administration is increasingly being addressed through the development of insulin analogs by protein engineering.

Native human insulin associates above physiologic concentrations into dimers and hexamers, an inherent property which limits the possibilities for the absorption of the molecule by various routes (Brange *et al.*, 1990). Through the use of molecular modeling and recombinant DNA technology, a series of human insulin analogs with reduced tendency to self-association have been developed (Fig. 3). Knowledge about the most important surfaces on the insulin monomer responsible for the subunit interactions and binding to the insulin receptor (Baker *et al.*, 1988) permitted the selection of mutations that would not compromise the potency of the hormone. The original principles for the selective replacement of amino acid residues have been to influence charge effects and hydrophobic interactions or to introduce steric hindrance at the interfaces between associating subunits in the insulin hexamer (Brange *et al.*, 1988).

Human insulin analogs with largely reduced self-association are absorbed substantially faster from subcutaneous tissue than current regular, hexameric insulin and are therefore better suited for bolus injection (see Section 3.2.1). The availability of the monomeric insulins has also contributed significantly to our understanding of the processes involved in subcutaneous insulin absorption (Brange *et al.*, 1990; Kang *et al.*, 1991b). Monomeric insulins also have the potential for improved absorption via some of the noninvasive routes, particularly those which do not utilize chemical enhancement resulting in dissociation of insulin hexamers, e.g., transdermal iontophoresis (see Section 3.3.9) and pulmonary inhalation.

More recently, manipulation of the structure of the B-chain, aiming at decreasing the potential for β-sheet interactions of the peptide backbone between insulin monomers, has also resulted in insulins with decreased tendency to self-associate. Examples of such analogs are the LysPro insulin (Brems *et al.*, 1992), in which the B28 and B29 residues are interchanged, as is the case in insulin-like growth factor-I, and analogs characterized by deletion of one of the amino acid residues in the B25–B28 sequence (Balschmidt and Brange, 1992).

With the aim of obtaining formulations that simulate basal insulin secretion, attention is also being focused on developing insulin analogs or derivatives with more constant and reproducible pharmacokinetics and which are extended-acting without an added retarding factor. Examples of such approaches are the introduction of extra positive charges, resulting in a less soluble insulin at physiological pH which crystallizes spontaneously

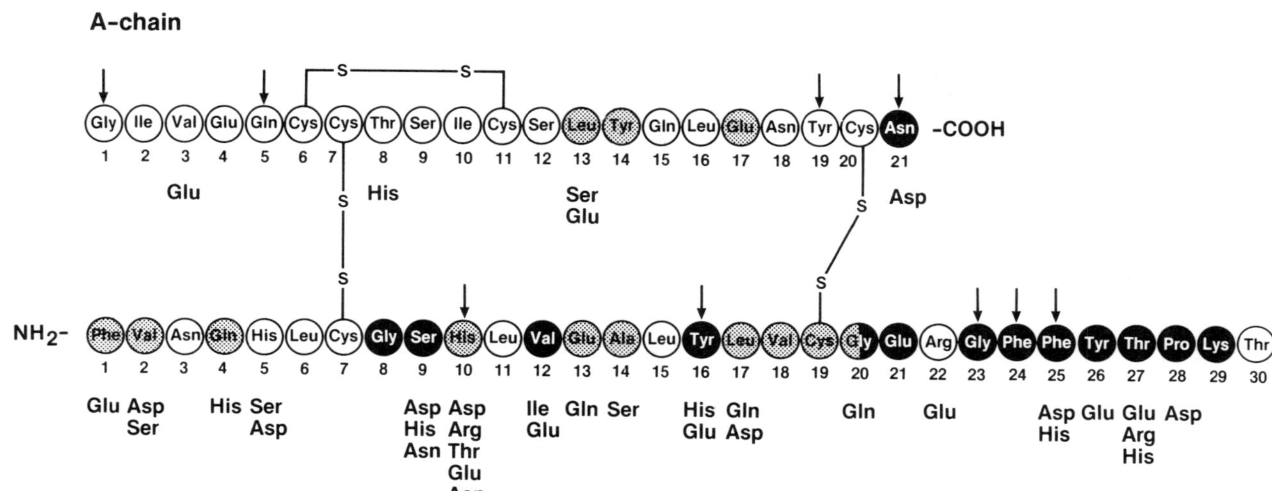

Figure 3. The primary structure of human insulin. The sites and types of mutation in the different analogs with reduced tendency to self-association are indicated. Also shown are the amino acid residues involved in the association of two insulin molecules into a dimer (black residues) and in the assembly of three dimers and two zinc ion into a Zn-insulin hexamer (shaded residues). The putative sites interacting with the receptor are indicated by arrows. (From Brange *et al.*, 1990, with permission.)

in subcutis after injection of a slightly acid solution of the modified insulin (Markussen *et al.*, 1988), and neutralization of charge in the center of the hexamer, which reduces its tendency to dissociate (Brange *et al.*, 1989; Bentley *et al.*, 1992). A more recent approach to obtaining a stable hexamer is to utilize the capability of Co^{3+} to form a very strong complex with the insulin hexamer, characterized by a reduced tendency to subunit dissociation (Sudmeier *et al.*, 1981; Storm and Dunn, 1985; Kurtzhals *et al.*, 1995a). After subcutaneous injection, a depot of the biologically inactive hexameric Co-insulin prodrug is established because the insulin prodrug is distributed and eliminated more slowly than native insulin. The gradual release and activation of the insulin monomer, probably occurring via reduction of Co^{3+} to Co^{2+} with complexing properties similar to those of Zn^{2+} (Cunningham *et al.*, 1955), provides a protracted action of the hormone (Kurtzhals *et al.*, 1995a; Kurtzhals and Ribel, 1995). Most recently, the search for soluble, prolonged-acting insulin for basal rate delivery has resulted in the development of insulin derivatives in which the ε-amino group of Lys^{B29} has been acylated with saturated fatty acids containing 10–16 carbon atoms (Kurtzhals *et al.*, 1995b). Like fatty acids, such fatty acid insulin derivatives have affinity for albumin, and neutral solutions of the derivatives showed, after subcutaneous injection in pigs, a protracted action strongly correlated to the albumin binding constant. Lys^{B29}-tetradecanoyl-des(B30)-insulin showed maximal albumin binding affinity and had a more retarded action after subcutaneous injection in pigs than NPH insulin (Markussen *et al.*, 1996).

The knowledge that has been obtained about the potential residues undergoing chemical changes during storage (Brange *et al.*, 1987, 1992; Brange, 1994) has also allowed the design of insulin analogs with improved properties with respect to chemical stability (Langkjær *et al.*, 1988; Markussen *et al.*, 1988; Brange and Havelund, 1991).

3. DELIVERY OF INSULIN

3.1. Introduction

There has long been a belief in the relationship between inadequate blood glucose control and the development of tissue complications such as blindness and renal failure. Thus, increasing attention has been given to optimizing insulin delivery. The Diabetes Control and Complications Trial (DCCT) has now definitively proved that tight glycemic control leads to a reduction in development and progression of these long-term diabetic complications (Nathan *et al.*, 1994). Recently, it has also been concluded that long-term normalization of blood glucose retards deterioration of nerve conduction velocity in the diabetic (Amthor *et al.*, 1994).

For the immediate future, parenteral administration remains the only viable route as well as the only means by which sustained, controlled release can be achieved. The challenge is to imitate, as closely as possible, the pattern of plasma insulin levels produced by a healthy pancreas (Fig. 4). Near-normal glycemia has been achieved in well-selected and motivated patients by using various intensive insulin-therapy regimens such as multiple injections and continuous subcutaneous insulin infusion (Hirsch *et al.*, 1990; Amiel, 1993). The attention toward meticulous glycemic control has brought renewed attention to the major risk of hypoglycemia that accompanies such control (Heine, 1993). The steady delivery of insulin that is needed overnight and between bolus doses at mealtimes can at present only be achieved with pump infusion of regular insulin (Home *et al.*, 1989).

In response to the discomfort, inconvenience, and variability of subcutaneous administration, a variety of improvements in syringe and needle technology have been introduced, such as pen systems and ultrafine needles or needle-free injection, with the aim of making insulin delivery easier, more accurate, and less painful. For the same reasons, several alternative nonparenteral routes for the administration of insulin have been investigated repeatedly over the years. The search for better ways of giving insulin reflects the physician's search for more physiological insulin delivery and the patient's desire to escape from the tyranny of the needle (Home *et al.*, 1985).

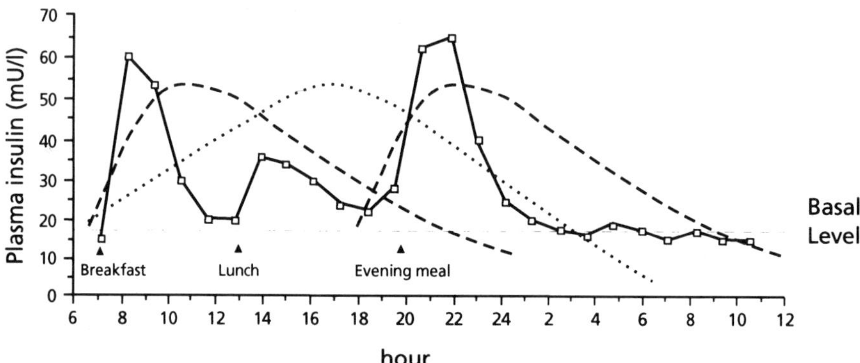

Figure 4. The diurnal plasma insulin levels in a nondiabetic subject (□—□), the approximate plasma insulin profile obtained by a twice-daily insulin injection regimen using premixed rapid- and intermediate-acting human insulins (———), and the approximate profile obtained by one injection of a long-acting human insulin (·······).

Controlled parenteral delivery of proteins has recently been reviewed by Pitt (1950) and Heller (1993a,b). Developments in insulin delivery systems have been reviewed by Saudek (1993), devices for insulin administration by Selam and Charles (1990), and subcutaneous insulin therapy by Houtzagers (1989), Home and Alberti (1992), Koivisto (1993), and Heine (1993).

3.2. Parenteral Insulin Delivery

3.2.1. INJECTION THERAPY

Modern insulin injection therapy is designed to mimic the two main rates of insulin secretion in a normal person, a low-level supply of the hormone between meals and during the night (basal insulin) and boosts at meals (bolus insulin). Bolus insulin is provided by injections of short-acting regular (unmodified) insulin, preferably given about 30 min before the start of the meal. Basal insulin is normally given in the form of twice-daily injections of the intermediate-acting Lente or NPH (isophane) insulin or once- to twice-daily injections of the slightly more long-acting Ultralente. From a double-blind crossover trial in 82 patients, it has been concluded that Lente- and NPH-based twice-daily human insulin regimens give indistinguishable metabolic control (Tunbridge *et al.*, 1989).

The absorption rate of both rapid- and delayed-acting insulins is, however, highly unsatisfactory (Fig. 4). Regular insulin peaks too late, and insulin concentration falls too slowly after the peak. None of the commercially available extended-acting insulin preparations are capable of delivering insulin at the low steady rates that characterize normal insulin secretion in the basal state. Ultralente formulated with bovine insulin is the most slowly absorbed extended-acting insulin and comes the closest of all the commercial insulin preparations to the ideal for basal delivery (Hildebrandt *et al.*, 1985a; Seigler *et al.*, 1991). However, the immunogenicity of this species of insulin reduces its clinical usefulness and restricts its general use.

Insulin absorption from subcutis is strongly affected by a multitude of factors (for recent reviews, see Brange *et al.*, 1990, and Christiansen and Lauritzen, 1991). These factors can be categorized as follows:

1. *Factors originating from the insulin preparation.* The rate of absorption evidently varies with the type of preparation, but also concentration (Binder, 1969; Hildebrandt *et al.*, 1983, 1984; Heinemann *et al.*, 1992), species of insulin (Owens *et al.*, 1984; Pramming *et al.*, 1984; Hildebrandt *et*

al., 1985a; Fernqvist *et al.*, 1986), and the volume (dose) injected (Binder, 1969; Galloway *et al.*, 1973) influence the absorption rate.

2. *Factors originating from the method of administration.* Administration conditions with influence on the rate of absorption include anatomical site and depth of injection (Frid *et al.*, 1988; Spraul *et al.*, 1988). The anatomical site of the injection has a major influence on the rate of absorption, with faster absorption from intramuscular versus subcutaneous sites (Frid *et al.*, 1988; Spraul *et al.*, 1988; Thow *et al.*, 1990) and from subcutis at the abdomen as compared to the limb regions (Binder *et al.*, 1967; Koivisto and Felig, 1980; Galloway *et al.*, 1981; Berger *et al.*, 1982; Süsstrunk *et al.*, 1982; Henriksen *et al.*, 1993). Absorption is also faster from the upper than the lower abdomen (Frid and Linde, 1993).

3. *Factors originating in the patient.* The patient-related factors with an influence on absorption kinetics include body posture (Hildebrandt *et al.*, 1985b), temperature of the skin (Koivisto *et al.*, 1981; Thow *et al.*, 1989; Sindelka *et al.*, 1994), subcutaneous fat thickness (Sindelka *et al.*, 1994), exercise of the muscle that underlies the site of insulin injection (Dandona *et al.*, 1978; Koivisto and Felig, 1978; Süsstrunk *et al.*, 1982; Fernqvist *et al.*, 1986), and massage of the injection site (Berger *et al.*, 1982; Dillon, 1983; Linde 1986; Linde and Philip, 1989). Massage may act by dissociating insulin hexamers rather than by expanding the insulin depot (Linde and Philip, 1989). Smoking has been reported to slow insulin absorption (Klemp *et al.*, 1982). Repeated insulin injections in the same area often result in marked disruption of subcutaneous tissue (Home *et al.*, 1989).

4. *External factors.* For quantified insulin delivery, a few external factors also have to be considered, including seasonal (Fahlén *et al.*, 1971) and circadian variations in insulin requirements, as well as the influence of sequence, composition, and preparation of meals (Schrezenmeir *et al.*, 1985).

A few diabetic subjects require large doses of insulin by subcutaneous injection, and increased insulin degradation by muscle or fat tissue has been demonstrated in rare cases (Paulsen *et al.*, 1979; Maberly *et al.*, 1982) and has been linked to the effect of proteolytic enzymes. The admixture of insulin with various oligopeptides (Hori *et al.*, 1983), collagen (Hori *et al.*, 1989), sodium cholate (Zhou and Po, 1991), or aprotinin (Berger *et al.*, 1980, 1982; Linde and Gunnarsson, 1985; Owens *et al.*, 1988) or pretreatment of the skin with other protease inhibitors (Takeyama *et al.*, 1991) has been demonstrated to inhibit degradation and/or increase insulin absorption from subcutis. Aprotinin probably augments absorption by increasing capillary permeability and/or blood flow rather than by inhibiting insulin proteolysis (Williams *et al.*, 1983). However, subcutaneous local degradation of insulin is of no clinical importance in the large majority of patients (Lauritzen *et al.*, 1983; Chisholm *et al.*, 1984).

Studies of the pharmacokinetics of insulin analogs with reduced tendency to self-association in animals (Brange *et al.*, 1990; Fischer *et al.*, 1993) and in normal human subjects (Kang *et al.*, 1991b) have shown that they are absorbed from subcutis at rates inversely proportional to their average state of self-association (Fig. 5) and lead to more rapid rises in plasma insulin concentrations and resulting hypoglycemic responses when compared with human insulin in the same formulation (Vora *et al.*, 1988; Brange *et al.*, 1990; Heinemann *et al.*, 1990, 1993a,b; Kang *et al.*, 1991b,c; Howey *et al.*, 1994). Thus, monomeric insulins were absorbed three times faster than human insulin, had no lag phase, and followed a monoexponential course

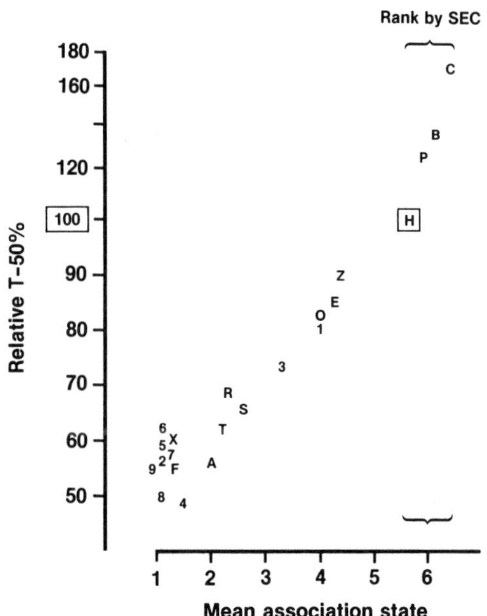

Figure 5. Correlation between time to 50% disappearance from the injection site in pig studies and the mean association state of various insulins at 1 mmol/1. The association state was deduced from osmometry, and, in the case of hexameric insulins, the tendency to dissociation upon dilution was assessed by size-exclusion chromatography (SEC). Note that T-50% > 100 is log scale. Each figure and letter in the diagram represents the mean results for one insulin or insulin analog H, Human 2Zn-, B, bovine 2Zn-, and P, porcine 2Zn-insulin; C, cobalt(III) human insulin; Z, human, zinc-free insulin. For other relevant codes, see Table VII. (From Brange *et al.*, 1990, with permission.)

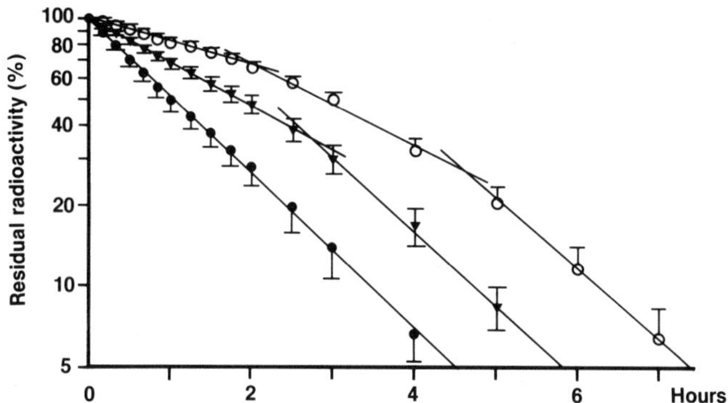

Figure 6. Disappearance curves of hexameric human insulin (○), the monomeric human insulin analog B9Asp, B27Glu (●), and the dimeric human insulin analog B10Asp (▼) in a logarithmic scale. Straight-line segments were calculated by linear regression analysis using the data from the relevant time intervals. (From Brange *et al.*, 1990, with permission.)

throughout the absorption process. In contrast, three phases in rate of absorption were identified for the normal hexameric human insulin (Fig. 6).

Studies in insulin-dependent diabetics have confirmed the beneficial pharmacokinetics and significant therapeutic value of monomeric insulin analogs (Kang *et al.*, 1990, 1991a; Brange *et al.*, 1990; Wiefels *et al.*, 1993; Luttermann *et al.*, 1993). They yield a degree of glycemic control, when injected just before a meal, at least comparable to that of human insulin administered 30 min before the meal. In addition, because of the more rapid decline in serum hormone concentrations, they are likely to reduce the frequency of between-meal hypoglycemia (Fig. 7).

3.2.2. PUMP INFUSION

3.2.2a. Externally Worn Pumps

When originally tested in the early 1980s, externally worn pumps were relatively large and cumbersome devices. Nevertheless, the early studies with continuous subcutaneous insulin infusion (CSII) were performed with great enthusiasm, and relatively good control was achieved in a number of studies (Pickup *et al.*, 1978; Tamborlane *et al.*, 1979; Champion *et al.*, 1980; Irsigler and Kritz, 1979) as later reviewed by Houtzagers (1989), Home *et al.* (1989), and Home and Alberti (1992). Because the technique uses only rapidly absorbed regular insulin, CSII results in less variation in insulin action when

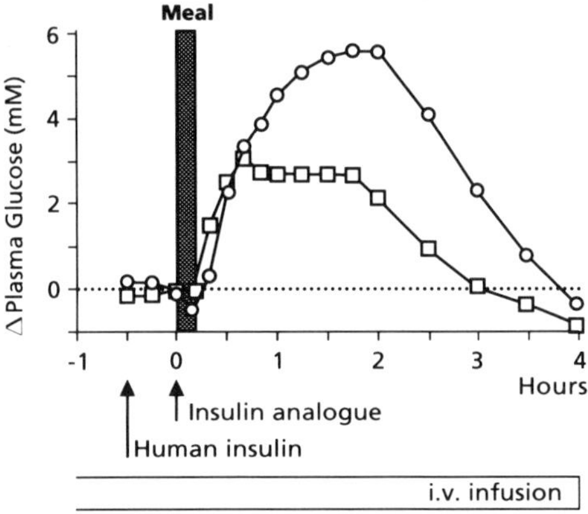

Figure 7. Mean incremental plasma glucose concentrations after 500-kcal test meal in insulin-dependent diabetic patients (n = 6) receiving constant insulin infusion (0.12 mU/min/kg regular insulin), and given subcutaneous injections of neutral solutions of 72 nmol insulin analog B28Asp (□) immediately following the meal, and 72 nmol human insulin (○) 30 min before the meal. (Adapted from Kang *et al.*, 1991a.)

compared with traditional injection of extended-acting insulins. It also removed some of the variables in absorption after injections, such as depth of injection and, because the infusion site normally is on the trunk, the accelerating effect of an exercising extremity on rate of absorption. CSII offers a more flexible lifestyle despite the discomfort of carrying an external device. However, CSII is an expensive therapy, the improvements relative to intensive insulin injection therapy have been small (Marshall *et al.*, 1987; Christiansen and Lauritzen, 1991), and only specific subgroups of highly motivated and appropriately selected patients can benefit from external pumps (Selam and Charles, 1990). In addition, localized skin reactions are a potential problem among users of insulin infusion pumps (Pietri and Raskin, 1981; Levandoski *et al.*, 1982).

3.2.2b. Implantable Pumps

Development of implantable pumps was a logical next step in the evolution of insulin infusion therapy, and different approaches have progressed gradually over the last decades. The original pump (Buchwald *et al.*, 1980) had only a single delivery rate, which was determined by constant vapor pressure above liquid freon. Programmable, variable-rate, implanted pumps with peristaltic roller-pump systems were later introduced and investigated in humans by Irsigler *et al.* (1981) and Schade *et al.* (1982). A more advanced design with a positive displacement piston and electronically controlled delivery rates which allows the patient to adjust bolus insulin doses, using an external radiotelemetry unit, has also been produced (Saudek *et al.*, 1990).

Pumps are surgically implanted subcutaneously, normally in the abdomen, and the catheter tip is usually placed in the peritoneal space, as intravenous delivery generally has been less successful. The catheter is undoubtedly the weakest point in implantable pumps, and the most significant complication with these pumps has been recurrent episodes of apparent pump slowdown related to catheter obstruction. In the early days, these events were most often attributed to insulin precipitation in the delivery cannulas. However, more careful inspection revealed that the problem in the majority of cases was, in fact, caused by small fibrin plugs at the end of the catheter or even macroscopic tissue encapsulation of the catheter tip.

In a recent study, it was found that treatment of diabetic patients by using an implantable pump to deliver insulin into the peritoneal cavity leads to a greater immunogenicity of insulin as compared to conventional therapy (Lassmann-Vague *et al.*, 1995).

Because an implanted pump contains thousands of units of insulin, safety considerations are of vital importance. Therefore, such a device must be extremely reliable and safe, and further long-term assessment of these devices is needed. At present, implanted pumps are considered research devices, which are only available as part of approved protocols (Saudek, 1993). The financial costs and need for recurrent surgery makes it unlikely that implantable insulin pumps ever will be a widely used treatment option.

3.2.3. INJECTION DEVICES

Jet injectors deliver insulin transcutaneously by an air-jet mechanism. The insulin solution or suspension is forced at high pressure through a fine nozzle, penetrates the skin without a needle, and creates a multitude of small depots. The dispersion of insulin deposited in the tissue explains the more rapid absorption of both rapid- and retarded-acting preparations (Taylor *et al.*, 1981; Malone *et al.*, 1986; Houtzagers *et al.*, 1988). Jet injection seems to affect the action profile of NPH insulin more markedly than that of the Lente type insulins (Houtzagers *et al.*, 1988). These devices are not painless and, in a European study, not well accepted by patients irrespective of the presence or absence of needle phobia (Houtzagers *et al.*, 1988). However, in a more recent American study, the majority of patients preferred to take insulin by jet injector compared to needle injection (Denne *et al.*, 1992). Jet injection has been found to be associated with a diminished antibody response compared with that to needle-injected insulin (Jovanovic-Peterson *et al.*, 1993), possibly because of the shortened stay in subcutis. There seem to be too many practical drawbacks for this type of device to play a significant role in insulin delivery, and no such device has ever achieved widespread acceptability. A so-called sprinkler needle may also be used to increase the area of insulin dispersion. Here, a number of small holes in the wall of a needle with a sealed tip are used to produce many smaller depots, resulting in faster absorption of short-acting insulin (Edsberg *et al.*, 1987).

Conceived as a means of combining the insulin vial and the syringe, together with a dose measuring device, in a single package (Paton *et al.*, 1981), design of an insulin-injection pen, a pocket-sized apparatus that resembles a fountain pen, was accomplished in the mid 1980s (Walters *et al.*, 1985). Pen injectors eliminate the inconvenience of carrying insulin and syringes to draw each dose, but they do not alter the characteristic absorption profile of the insulin preparation. However, because technical handling is simple and injection painless due to ultrafine and sharp needles, the use of pens improves the patient compliance, resulting in better metabolic control (Rosak *et al.*, 1993; Plevin and Sadur, 1993). The obvious convenience of insulin pens, allowing quick and safe injections, combined with their relatively modest price, should be attractive, especially for patients on multiple-injection regimens (Walters *et al.*, 1985; Jefferson *et al.*, 1985). Thus, from a long-term prospective study in a large group of diabetic patients receiving at least three daily injections, it was concluded that multiple insulin injections with a pen injector improved general well-being and diminished the risk of acute metabolic complications (Gall *et al.*, 1989). In a comparison of CSII with multiple daily injections using the NovoPen injector on a long-term basis, identical metabolic control was obtained. However, owing to physical discomfort during pump treatment, the insulin pen injector was preferred by a large majority of patients (Bak *et al.*, 1987).

Mechanical devices for insulin administration have been reviewed by Selam and Charles (1990).

3.2.4. IMPLANTABLE MATRICES

In sustained-release implants, the general approach is to provide a barrier between the body fluid and insulin. This barrier can be microporous, where the rate of insulin release is diffusion-controlled, or it can be an erodible material that dissolves gradually and releases the entrapped insulin. By using mechanical stress or transmitted energy, externally modulated bolus release of insulin can be obtained. Even more sophisticated self-regulating (closed-loop) systems are also under development.

Numerous synthetic or natural polymers have been used as matrices for micro- and nanoparticles or -capsules, most of which are biodegradable or bioerodible. Microparticles (or microspheres) are systems in which the drug is dispersed throughout the particle whereas capsules are vesicular systems in which the drug is contained in a cavity surrounded by the polymeric membrane (Couvreur and Puisieux, 1993).

Sustained-release implants for insulin delivery have been described fully in an excellent review by Wang (1991a). However, practical application of any of these systems in humans is not yet a reality.

3.2.4a. Nondegradable Systems

In an early study, using acrylamide polymerized in a solution of insulin, the release of insulin after implantation into rats was found to depend on the insulin content and the pore size of the hydrogel and to last for as long as 3 weeks (Davis, 1972). In a more advanced approach, insulin-loaded ethylene/vinyl acetate copolymer (Creque *et al.*, 1980) was formed into thin disks which were further coated with a layer of matrix to minimize an initial burst release of insulin (Siegel and Langer, 1984; Brown *et al.*, 1986a,b). An exit orifice drilled in the center of each disk ensured steady delivery from the device; *in vitro* as well as *in vivo*, the device was found to release insulin at a uniform rate for at least 5 months. Rod implants in which insulin was compressed with an inorganic, ceramic powder were found to lower blood glucose levels in diabetic rats for 7 days (Muzina *et al.*, 1989). The need to explant these devices and implant a new one is, however, a serious drawback of these systems.

3.2.4b. Biodegradable Polymers

The rate of *in vivo* degradation of most synthetic polymers is too slow for the sustained release of insulin, but several polyanhydrides and polyesters have been designed for use as biomaterial implants. In addition, natural materials such as serum albumin and several different lipids have been found useful for sustained release of insulin. Insulin-loaded polyanhydride microspheres implanted subcutaneously into rats provided blood glucose control for less than 4 days (Mathiowitz and Langer, 1987; Mathiowitz *et al.*, 1988). In addition, even with a stabilizer added, these types of implants seem to promote formation of covalent insulin oligomers (Ron *et al.*, 1993). Thus, despite their excellent chemical and biocompatibility characteristics, the polyanhydrides cannot be used readily for the sustained release of insulin (Wang, 1991a).

Release of insulin microencapsulated within poly(lactic acid) has been studied *in vitro* (Chang, 1976) and *in vivo* after administration as implantable pellets or injectable microspheres. After subcutaneous administration to rats, blood glucose levels were lowered for more than two weeks. However, the presence of insulin crystals on the surface of the microbeads and loosely bound insulin resulted in a burst effect, which increased with the insulin load (Kwong *et al.*, 1986). A reduced initial burst of insulin has been achieved by preparing the lactic acid polymer implant by using a W/O emulsion method (Tabata *et al.*, 1993). However, the emulsion method caused a 63% loss of insulin activity, probably due to the influence of the organic solution and ultrasound exposure on the stability of insulin. Using poly(lactic acid) with relatively low molecular weight and a double-layered implant, containing insulin only in the inner layer, sustained release of insulin has been obtained in rats for up to 19 days (Yamakawa *et al.*, 1990).

Although diffusion of proteins in lipophilic media is poor, several recent studies show that insulin can be released on a sustained basis *in vivo* through the tortuous channels present in compressed solid admixtures of insulin powder and various natural lipids such as cholesterol (Wang, 1987), palmitic acid (Wang, 1991b), or various other common fatty acids or their anhydrides or esters (Wang, 1989a). The prolongation of the sustained release of insulin from an implant made with palmitic acid seems to be inversely proportional to the insulin content. For implants containing 17% insulin and compressed to form 2 mm × 7 mm rods, readily inserted by a 12-gauge needle, the insulin release rate is about 2 IU/rod per day for 2 months (Wang, 1991a). For human therapy, a 10 times higher release rate would be necessary.

Serum albumin can be insolubilized by heat or by chemical cross-linking agents, and implantation of glutaraldehyde-cross-linked albumin microbeads containing insulin in diabetic rats has provided elevated blood insulin levels for periods of up to 3 weeks with biodegradation of the remnant implants within 1–2 months (Goosen *et al.*, 1982, 1983).

Nanoparticles with insulin adsorbed onto the surface (Douglas *et al.*, 1987) or liposomes loaded with insulin have also been utilized for sustained release after subcutaneous, intramuscular, or intraperitoneal administration (Patel and Ryman, 1976; Couvreur *et al.*, 1980; Stevenson *et al.*, 1982; Weiner *et al.*, 1985; Spangler, 1990). However, most injected liposomes and their content of insulin apparently remained at the subcutaneous injection site (Spangler, 1990).

Incorporation of insulin into polymeric matrices such as dextran or agar has been reported to prolong the action of the insulin after subcutaneous administration (Losse *et al.*, 1988). Another approach to prolonging the action of insulin is to condense the amino groups of insulin with the aldehyde groups of cellulose. The imino bonds hydrolyzed after intraperitoneal injection of the insulin derivative suspended in sesame oil and lowered the blood glucose in diabetic rabbits for 5 days (Singh *et al.*, 1981).

3.2.4c. Modulated Systems

Several research groups have been developing responsive implanted systems with the aim of being able to release drugs according to physiological needs (Kost and Langer, 1991). In these systems the release rates can be modulated externally by stimuli mechanisms including magnetism (Kost *et al.*, 1987), ultrasound (Kost, 1993), temperature changes (Bae *et al.*, 1989), and electrical effects (Kwon *et al.*, 1991).

When polymeric matrices containing insulin and magnetic beads were implanted in diabetic rats, glucose levels could be repeatedly decreased on demand by application of an oscillating magnetic field, and the blood glucose levels were additionally lowered by approximately 30% (Kost *et al.*, 1987). With ferrite microparticles dispersed together with insulin in alginate spheres, the rate of insulin release *in vitro* was reported to be 50 times higher in the presence than in the absence of a magnetic field (Saslawski *et al.*, 1988).

The release rate of insulin *in vitro* from a unit made of the ethylene/vinyl acetate copolymer described above (Section 3.2.4a) can be increased by a factor of 15 when exposed to ultrasound compared to the release rate during unexposed periods (Kost, 1993). Upon administration to rats of another type of copolymer implant, the release rates increased only by a factor of about 2 on exposure to ultrasound (Miyazaki *et al.*, 1988).

A silicone implant has been designed to provide a bolus dose when compressed externally over the skin with a finger. With an erodible palmitic acid implant to provide the basal insulin supply, as mentioned above, compression of a second implant of the silicone type has been used to deliver supplemental doses for better control of transient hyperglycemia episodes in diabetic rabbits. The dependable device is refillable percutaneously by injection of an insulin suspension (Wang, 1989b, 1993).

3.2.4d. Self-Regulating Systems

The ultimate goal of insulin replacement therapy is a closed-loop controlled system or device in which the insulin release rates are adjusted by the system in response to glucose levels in the blood. Several different approaches for glucose-responsive insulin delivery have been investigated, including competitive desorption, enzyme–substrate reactions, and glucose-sensitive polymer complexes (Heller, 1993b).

The principle of reversible, competitive binding of glycosylated insulin to carbohydrate-binding lectins isolated mainly from plant sources, originally introduced by Brownlee and Cerami (1979, 1983), has since been tested by several research groups. The glycosylated insulin derivative is complexed with the lectin concanavalin A, which binds specifically to carbohydrates with the D-glucose or D-mannose configuration. The dissociation of the complex depends on the free glucose concentration in the medium and on the binding constant of the particular glycosylated insulin derivative (Sato *et al.*, 1984a,b). The most comprehensive studies have concentrated on synthesis and of the different derivatives and their characterization with respect to release rate and biological properties (Sato *et al.*, 1984a,b; Seminoff *et al.*, 1989a,b; Kim *et al.*, 1990). Attempts have been made to compensate for the lag time for diffusion of glucose and the glycosylated insulin derivative across the necessary membrane barrier (Kim *et al.*, 1990; Pai *et al.*, 1992), but the kinetics of such systems remain the most serious obstacle to their implementation. Recently, a new system composed of insulin immobilized to an insoluble polymer membrane through one of its disulfide bonds and glucose dehydrogenase has been investigated *in vitro*. In response to the presence of glucose, the intermolecular disulfide bonds were cleaved upon oxidation of glucose with glucose dehydrogenase, resulting in release of insulin from the membrane (Chung *et al.*, 1992).

In approaches utilizing enzyme–substrate reactions, the rate-control mechanism is based on the oxidation of glucose to gluconic acid catalyzed by glucose oxidase, which, together with insulin or an insulin derivative, is incorporated in an immobilized form into a polymer system. One approach uses pH-induced changes in solubility of insulin. Thus, if trilysyl insulin, with an isoelectric point near physiological pH, is incorporated into an ethylene/vinyl acetate copolymer, a reduction of pH will increase its solubility and release rate from the implant (Fischel-Ghodsian *et al.*, 1988). Another approach uses acid-sensitive polymers whose erosion rate changes in response to pH changes (Heller *et al.*, 1990). However, insulin delivery requires a polymer that is capable of undergoing a significant increase in erosion rate with only a modest pH decrease, and it is essential that the hydrogel be macroporous (Heller, 1993b). Membranes that changes permeability with a decrease in pH have also been investigated as self-regulating insulin delivery systems (Horbett *et al.*, 1984; Ishihara *et al.*, 1984; Albin *et al.*, 1985, 1987; Ishihara and Matsui, 1986; Ito *et al.*, 1989).

In a somewhat different approach to a self-regulating insulin delivery system, Siegel and co-workers (Siegel *et al.*, 1988; Siegel and Firestone, 1990) are investigating the feasibility of a mechanochemical pump. The functioning of this pump is also based on enzymatic transformation of glucose to gluconic acid. The reaction controls the reversible swelling of a hydrogel which regulates expansion and contraction of the reservoir containing insulin.

In order to circumvent the use of glucose oxidase, direct glucose-responsive polymers composed of phenylboronic acid groups with affinity for hydroxyl-containing molecules have recently been synthesized. Gluconic acid, chemically attached to insulin, was bound to the system and shown to be released in response to varying glucose concentrations *in vitro* (Kitano *et al.*, 1992; Shiino *et al.*, 1994).

Common to all the self-regulating systems is the lack of evaluation of their *in vivo* performance. Critical issues in this respect include long-term stability of insulin in such hydrophobic environments and the stability and/or potential leakage and possible immunogenicity of the enzymes and other agents used in these systems. Also, the physical dimensions, response times, and reproducibility of release are pivotal questions with such systems.

3.3. Alternative Routes of Insulin Delivery

3.3.1. INTRODUCTION

The subcutaneous route for administration of insulin has many serious drawbacks, and alternative routes continue to attract considerable research interest. Nearly all available orifices of the human body seem to have gained attention as presenting possible noninvasive sites for insulin absorption. However, even by using modern enhancer techniques, only a small or minor fraction of the hormone becomes bioavailable when provided by most of these routes, except perhaps the pulmonary route. Key barriers to insulin absorption via the alternative routes are the resistance of those membranes to insulin penetration, the tendency of insulin to exist in associated form, and insulin proteolysis. Protection from proteolysis — through some sort of encapsulation, the use of complex emulsion systems, and/or the use of protease inhibitors — association of the hormone with polymeric particles, and addition of permeation enhancers have been utilized to overcome those barriers. The absorption and enzymatic barriers to nonparenterally administered protein drugs and the use of enhancers to modify absorption have been discussed in recent reviews (Lee, 1986; Lee *et al.*, 1991a; Zhou, 1994). The present review of alternative administration of insulin mainly covers investigations published since 1970.

3.3.2. ORAL DELIVERY

Oral administration of insulin seems to have been the ambitious dream ever since the hormone was introduced into medicine. Between 1922 and 1980, more than 125 reports on this subject were published (Berger, 1993), and the last 15 years have added more than 50 new failures in this area of unrealistic endeavors. The more recent attempts have explored the possibilities of:

1. protecting the insulin from enzymatic degradation (Fukushima and Toyoshima, 1983; Fuji *et al.*, 1985; Saffran *et al.*, 1986; Touitou and Rubinstein, 1986; Fukushima, 1987; Atchison *et al.*, 1989; Asada *et al.*, 1994, 1995; Bai and Chang, 1995);
2. promoting the uptake through the intestinal wall by adding a multitude of different enhancers (Touitou *et al.*, 1980; Mesiha and El-Bitar, 1981; Ziv *et al.*, 1981; Bar-on *et al.*, 1981; Mesiha, 1981; Nishihata *et al.*, 1981a; Schilling and Mitra, 1990; Liedtke *et al.*, 1990; Scott-Moncrieff *et al.*, 1994; Shao *et al.*, 1994a); or
3. increasing the gastrointestinal absorption of insulin by simultaneous use of enhancers and antiproteolytic precautions (Kidron *et al.*, 1982, 1989; Ziv *et al.*, 1984, 1987; Bendayan *et al.*, 1990; Gwinup *et al.*, 1991; Morishita *et al.*, 1992a,b, 1993a,b; Geary and Schlameus, 1993; Bendayan *et al.*, 1994; Ziv *et al.*, 1994).

The introduction of liposomes in the late 1960s as a drug delivery concept raised a considerable renewed interest in the oral administration of insulin, as reviewed by Woodley (1986) and Spangler (1990). The first reports (Dapergolas and Gregoriadis, 1976; Patel and Ryman, 1976) were, despite an extreme variability of the glycemic responses to peroral insulin-liposomes, quite optimistic. In the following years, many investigators tested the ability of liposomes to fulfill the dual role of preventing insulin degradation in the upper gastrointestinal (GI) tract (Patel and Ryman, 1977; Rowland and Woodley, 1981; Woodley and Prescott, 1988) and enhancing insulin absorption from various regions of the GI tract (Manosroi and Bauer, 1990) or after oral administration to rats (Weingarten *et al.*, 1981; Arrieta-Molero *et al.*, 1982; Shenfield and Hill, 1982; Dobre *et al.*, 1983, 1984; Das *et al.*, 1988) or dogs (Patel *et al.*, 1982). Also, the strategy of utilizing insulin-loaded microparticulate systems (nano- or microcapsules or -particles; for a review, see Couvreur and Puisieux, 1993) to circumvent the gastrointestinal, enzymatic barrier and to promote absorption by the intestinal mucosa has been tested a number of times (Couvreur *et al.*, 1980; Oppenheim *et al.*, 1982; Damgé *et al.*, 1988, 1990; Michel *et al.*, 1991; Morishita *et al.*, 1993a). Common to all these animal investigations of oral insulin delivery have been the use of exorbitant doses of insulin and extremely inconsistent and variable effects, if any at all, with respect to blood glucose lowering.

It is amazing that none of the above-mentioned studies has actually been performed in a way that allowed well-controlled evaluation of the bioavailability. However, the most carefully performed studies seem to indicate that the uptake of insulin via the oral route, despite all precautions, is less than 0.5%. Absolute bioavailability without enhancement from rat intestine has been found in a closed-loop *in situ* experiment to be on the order of 0.1% (Schilling and Mitra, 1992).

The few studies performed in human beings (Crane *et al.*, 1968; Earle, 1972; Galloway and Root, 1972; Patel *et al.*, 1978; Gwinup *et al.*, 1991; Cho and Flynn, 1993) have shown wide variation in responses, if any, to the massive doses, meaning that accurate dosing is not possible. It is therefore highly unlikely that any oral preparation will be forthcoming that can achieve consistent and economical insulin delivery.

Reviews on oral administration of insulin (Damgé, 1991) and of peptides and proteins (Lee *et al.*, 1991b; Smith *et al.*, 1992) have recently appeared.

3.3.3. BUCCAL DELIVERY

The human buccal mucosae of the oral cavity, i.e., the buccal and the sublingual epithelia, offer a robust and easily accessible area for systemic delivery and the advantage of low enzymatic activity (Lee *et al.*, 1987; Yamamoto *et al.*, 1990). Unfortunately, the multilayered buccal barrier is relatively thick and dense, and proteinaceous substances are not readily absorbed via this route. Thus, permeability of insulin via this route has been calculated to be 1–2 orders of magnitude less than via the nasal route (Harris and Robinson, 1990).

Insulin buccal absorption in animals has been studied in rats (Weingarten *et al.*, 1981; Aungst and Rogers, 1988, 1989; Aungst *et al.*, 1988; Aungst, 1994), rabbits (Ritschel *et al.*, 1989), and dogs (Ishida *et al.*, 1981; Nagai, 1985; Oh and Ritschel, 1990a,b). Without enhancer, the efficacy in rats relative to intramuscular injection was about 2% whereas addition of 5% of laureth-9 (a nonionic surfactant which reversibly removes membrane proteins or lipids) (Aungst and Rogers, 1988), of sodium glycocholate (Aungst *et al.*, 1988), or of dodecylmaltoside (Aungst, 1994) was able to increase the efficiency to 26–30%. The buccal membrane in rats differs from those in rabbits, dogs, and humans in being keratinized, which may render the epithelial membrane less permeable (Ho *et al.*, 1992). However, bioavailability in rabbits and dogs was not significantly higher than in rats, and addition of enhancer did not further increase availability.

One of the problems of buccal delivery is that the surface of the mucous membrane is constantly washed by a stream of saliva, which will exclude a major part of the drug from absorption and also reduce administration time. Increased residence time can be obtained by using a bioadhesive delivery system (Harris and Robinson, 1990), and such a system has been tested with insulin. A double-layered patch with an adhesive peripheral layer that stuck to the oral mucosa for 6 hr and a core consisting of insulin, sodium glycocholate, and cocoa butter was administered to dogs (Ishida *et al.*, 1981; Nagai, 1985). Bioavailability was calculated as 0.5% as compared with intramuscular administration.

A single study in humans has been performed by YoKosuka *et al.* (1977). An insulin solution containing 1% sodium glycocholate was sprayed on the sublingual mucosa of three healthy subjects, but no increase in serum insulin was observed.

Buccal delivery of peptide and protein drugs has recently been reviewed by Ho *et al.* (1992) and Merkle and Wolany (1992).

3.3.4. RECTAL DELIVERY

The rectal cavity has the potential of convenient access and easy administration of suppositories or gels, although patient acceptance of rectal delivery is low in some cultures. It has a limited surface area (de Boer *et al.*, 1992) and relatively high proteolytic activity (Lee *et al.*, 1987), but with respect to administration of insulin the rectal route offers the particular physiological advantage of potential delivery, via the upper rectal veins, into the portal system. This would mimic the natural secretion of insulin and result in reduced peripheral hyperinsulinemia. However, nearly two-thirds of the insulin absorbed from the rectum reaches the general circulation via the lymphatic pathway (Caldwell *et al.*, 1982).

The present review of rectal administration of insulin essentially covers studies performed after the review by Ritschel and Ritschel (1984). More recent reviews of this subject include those of Eppstein and Longenecker (1988) and Banga and Chien (1988a).

Because insulin absorption from the rectum in simple formulations is poor and erratic (Yagi *et al.*, 1983; Ritschel and Ritschel, 1984; Aungst *et al.*, 1988; Hoogdalem *et al.*, 1990; Yamamoto *et al.*, 1992), the use of promoters is imperative. An extensive array of enhancers, including nonionic, cationic, anionic, and amphoteric surfactants, bile salts, phospholipids, alkylglycosides, salicylates, enamines, chelating agents, and enzyme inhibitors have been experimentally tested in numerous animal studies during the last 15 years. In addition to the papers referenced in Table III, studies without proper comparative evaluation of the pharmacological availability of the administered dose of insulin have been performed in rats (Touitou *et al.*, 1978; Morimoto *et al.*, 1980; Bar-on *et al.*, 1981; Nishihata *et al.*, 1981b; Caldwell *et al.*, 1982; Nishihata *et al.*, 1983a; Touitou and Donbrow, 1983; Morimoto *et al.*, 1983; Kim *et al.*, 1984; Liversidge *et al.*, 1985, 1986; Hauss *et al.*, 1991), in rabbits (Bakth *et al.*, 1980; Morimoto *et al.*, 1980; Touitou and Donbrow, 1983; Nishihata *et al.*, 1983b; Ritschel and Ritschel, 1984; Losse *et al.*, 1988, 1989; Yamamoto *et al.*, 1992), in dogs (Yamasaki *et al.*, 1981b; Kim *et al.*, 1984; Liversidge *et al.*, 1985, 1986; Nishihata *et al.*, 1987), and in human normal or diabetic subjects (Fujita *et al.*, 1983; Nishihata *et al.*, 1986, 1989; Hosny *et al.*, 1994a,b; El-Shattawy *et al.*, 1994).

The hypoglycemic effect of rectally administered insulin is strongly influenced by numerous formulation factors such as nature of the suppository base, suppository size and shape, pH, type and concentration of the enhancer used, surfactant chain length, insulin dose, etc. (Touitou and Donbrow, 1983; Liversidge *et al.*, 1985, 1986; Eppstein and Longenecker, 1988). Consequently, the pharmacological availability, as judged by comparative blood glucose lowering effects, varies widely and ranges from 4 to 40% in rats, 30 to 50% in rabbits, and 10 to 45% in dogs (Table III). In

Table III
Pharmacological Availability of Insulin Administered Rectally

Model	Enhancer/formulation	Availability[a]	Reference
Rat	Taurodihydrofusidate	4–7	Hoogdalem *et al.* (1990)
	Alkylglycosides	6–8	Aungst (1994)
	Salicylates + EDTA	30–40	Aungst and Rogers (1988)
	Bile salt	40	Aungst *et al.* (1988)
Rabbit	Laureth-9	~30–50	Ichikawa *et al.* (1980)
	Brij 58	~50	Mesiha *et al.* (1981)
Dog	Laureth-9	~15	Shichiri *et al.* (1978)
	Laureth-9	~10	Yamasaki *et al.* (1978)
	Laureth-9	~10	Yagi *et al.* (1983)
	Salicylate + gelatine	~25	Nishihata *et al.* (1983c)
	Enamine	19	Nishihata *et al.* (1985)
	Enamine	19–45	Yagi *et al.* (1983)
Human	Laureth-9	~10	Yamasaki *et al.* (1981a)
	Brij 58	~4	Hildebrandt *et al.* (1984b)
	Bile salt	~7	Raz *et al.* (1984)

[a]Based on comparative serum insulin or hypoglycemic responses relative to parenteral administration.

humans the effectivity of rectal delivery of insulin, using a given type of enhancer, seems to be 5–10 times less than in rats or rabbits, whereas dogs apparently are better models for prediction of bioavailability in humans, at least with laureth-9 as the enhancer. The absorption of insulin from rectum, as compared to subcutis, gives more rapid initiation of action and shorter duration of the effect (Fujita *et al.*, 1983; R. Hildebrandt *et al.*, 1984b; Raz *et al.*, 1984; Nishihata *et al.*, 1986, 1989). Although rectal administration thus provides insulin profiles closer to those seen in normals following a meal, the overall bioavailability in humans is far too low for this route of administration to be an economical alternative to parenteral therapy. In addition, as observed in many studies (Ritschel and Ritschel, 1984), the pharmacological availability is not dose-dependent; i.e., increasing the dose does not result in a proportionally greater decrease in blood glucose.

3.3.5. VAGINAL AND INTRAUTERINE DELIVERY

As a rational systemic delivery route, application via the vagina or uterus, although technically viable, seems somewhat bizarre and exotic. Not only are these routes only applicable for one-half of the population, but the estrous cycle stages strongly influences thickness and porosity of the epithelia, and thereby their permeability to drug substances.

Nevertheless, during the last 15 years the vaginal route has been investigated for insulin delivery in a number of studies in rats (Touitou *et al.*, 1978; Morimoto *et al.*, 1982; Okada *et al.*, 1983a,b; Richardson *et al.*, 1992a), rabbits (Morimoto *et al.*, 1982), and sheep (Richardson *et al.*, 1992b). Also, a single study of intrauterine application of insulin to rats has recently appeared (Golomb *et al.*, 1993). These attempts, however, have not been very successful. Richardson *et al.* (1992a) found no effect on blood glucose in the absence of enhancer and also reported that the histological changes in rat vaginal epithelium after treatment with the enhancer systems were variable but often severe. With the use of enhancer, maximal bioavailability via the vaginal route has been found to be 18% (enhancement by organic acids) in rats (Okada *et al.*, 1983a) and 14% (lysophosphatidylcholine as enhancer) in sheep, with marked interanimal variations (Richardson *et al.*, 1992b). In a comparative study in rats, Touitou *et al.* (1978) showed that vaginal administration gave less blood glucose reduction than obtained with the same formulation and dose applied rectally. As expected, the vaginal absorption of insulin was markedly affected by the cyclical structural changes, which caused large fluctuations in absorption in rats (Okada *et al.*, 1983b); it also appeared to be influenced by the estrous cycle in sheep (Richardson *et al.*, 1992b).

The vaginal route of protein drug delivery has recently been reviewed by Richardson and Illum (1992).

3.3.6. OCULAR DELIVERY

As for all other alternative routes, the ocular delivery of proteins is impeded by a permeation barrier and enzymatic metabolism barriers (created by the existence of peptidases in the ocular tissue). In addition, major limitations of this route of delivery are the perceived sensitivity of the eye, the potential hypertonicity of the drug solutions, and the limited volumes that can be applied without spillage from the eye.

Ocular administration of insulin has been studied in rats (Pillion *et al.*, 1991; 1994a), in rabbits (Chiou *et al.*, 1988, 1989; Yamamoto *et al.*, 1989; Chiou and Chuang, 1989; Hayakawa *et al.*, 1992; Chiou and Li, 1993; Sasaki *et al.*, 1994), and in dogs (Nomura *et al.*, 1990, 1994). Absorption without enhancement is poor as bioavailability was observed to be only 1% in rabbits (Yamamoto *et al.*, 1989) and less than 0.5% in normal and 1–7% in diabetic dogs (Nomura *et al.*, 1990). Bartlett *et al.* (1994b) found that insulin eyedrops without surfactant agents were ineffective at lowering systemic glucose levels in human volunteers. Using laureth-9 or bile salt enhancers, Yamamoto *et al.* (1989) found only 4–13% bioavailability in rabbits.

In an investigation of the ability of different molecular substances to elicit ocular anaphylaxis when applied topically to the eye, Kahn *et al.* (1990) concluded that only substances having a molecular weight less than 3500 penetrate the conjunctival barrier. Therefore, systemic delivery via the ocular route seems to rely on overflow of the instilled drug to the nasal cavity. Thus, the nasal mucosa contributed about 4 times more than the conjunctival mucosa to the systemic absorption of ocularly applied insulin (Yamamoto *et al.*, 1989). Topically applied insulin administered chronically without surfactant seems to be nontoxic to the external human eye (Bartlett *et al.*, 1994a).

Ocular delivery of other peptide and protein drugs has been reviewed by Harris *et al.* (1992).

3.3.7. NASAL DELIVERY

From the early 1920s, attempts have been made to administer insulin by the nasal route (Jensen, 1938; Drejer *et al.*, 1991). Delivery of peptide and protein drugs via the nasal mucous membrane has, especially during the last decades, attracted considerable interest from academia as well as from the pharmaceutical industry. It is probably the most extensively investigated of al the mucosal routes, and, as with rectal delivery, the protein most often studied has been insulin.

Intranasal delivery is potentially an attractive alternative site for insulin administration on the basis of the relatively large surface area available for absorption (150–170 cm^2 in humans) and its highly vascularized subepithelial layer. The nasal mucosa is considerably more permeable than that of the buccal region but is, on the other hand, also enzymatically more active (Lee *et al.*, 1987; Yamamoto *et al.*, 1990). From *in vitro* studies it has, however, been concluded that enzymatic degradation in the nasal cavity is not a limiting factor for the intranasal application of insulin (Gizurarson and Bechgaard, 1991a).

Mucociliary clearance, with a half-life of 10–20 min in humans, is one of the major physical barriers for the nasal absorption of insulin. This rapid clearance substantially limits the residence time of insulin at the nasal mucosa, and it is probably the major contributory factor in the loss of insulin from potential absorption and the resulting low bioavailability (Vora and Owens, 1991). The pH of 5.5–6.5 in the mucous layer of the nasal cavity further contributes to reduce the absorption of insulin, which possesses limited solubility in this pH range.

Nasal insulin delivery has been studied in several animal models, *in vitro* (Bechgaard *et al.*, 1992; Maitani *et al.*, 1992; Carstens *et al.*, 1993) and *in vivo* using a plethora of different types of promoters (Table IV). Intranasal insulin is better absorbed when administered as a spray than as drops (Pontiroli *et al.*, 1987), and lyophilized insulin in an aerosol is more efficient than spray of an insulin solution (Nagai *et al.*, 1984). As would be expected, large interspecies differences in the nasal absorption appear to exist, and enhancers differ substantially in efficacy and safety between species (Merkus *et al.*, 1993). It is therefore extremely difficult to extrapolate absorption results obtained from a particular animal study to humans.

A systematic study in rats, in which the absorption-enhancing effects of a number of surfactants were compared, was reported by Hirai *et al.* (1981a,b). The promoting effect of several nonionic surfactants was paralleled by their ability to lyse red blood cells and to release protein from the nasal mucosa of rats. Bile salts, however, were found to be less lytic and damaging to the rat nasal mucosa than nonionic ether-type surfactants. Generally, a positive correlation was observed between the damaging effect on the biomembrane and the absorption-promoting effect of the surfactants.

The bile salt promotion of insulin absorption is probably a combined effect of several factors, including dissociation of insulin oligomers (Li *et al.*, 1992; Shao *et al.*, 1992), inhibition of proteolysis (Hirai *et al.*, 1981b), and a direct effect on the permeability of the mucosa. However, monomers of insulin are not absorbed in the absence of a promoter (Pontiroli and Pozza, 1990).

Table IV
Animal Models in Studies of Nasal Absorption of Insulin

Model	Type of enhancement	Reference(s)
Rat	Fatty acids	Mishima *et al.* (1987)
	Surfactants	Aungst and Rogers (1988); Hirai *et al.* (1981a, b)
	Bile salts	Shao and Mitra (1992); Aungst *et al.* (1988)
	Chitosan	Illum *et al.* (1994a)
	Cyclodextrins	Irie *et al.* (1992); Merkus *et al.* (1991a, b); Schipper *et al.* (1992); Shao *et al.* (1992); Gill *et al.* (1994); Verhoef *et al.* (1994)
	Phospholipids	Illum *et al.* (1989); Chandler *et al.* (1994)
	Dihydrofusidates	Merkus *et al.* (1991b); Shao and Mitra (1992); Deurloo *et al.* (1989); Lee *et al.* (1992)
	Alkylglycosides	Aungst (1994); Pillion *et al.* (1994b)
	Glycyrrhetinic acid	Mishima *et al.* (1989)
	Microsperes	Edman *et al.* (1992)
Rabbit	Dihydrofusidates	Deurloo *et al.* (1989); Merkus *et al.* (1991b)
	Sterol, sterol glycosides	Maitani *et al.* (1995)
	Cyclodextrin	Merkus *et al.* (1991b); Schipper *et al.* (1993)
	Bioadhesive polymers	Dondeti *et al.* (1995)
Dog	Surfactant	Hirai *et al.* (1978)
	Bile salt	Hirai *et al.* (1978)
Sheep	Phosphatidylcholines	Farraj *et al.* (1990)
	Fusidate and bile salt	Longenecker *et al.* (1987)
	Cationic polysaccharide	Illum *et al.* (1994a)
	Hyaluronic acid ester microspheres	Illum *et al.* (1994b)

As mentioned above, mucociliary clearance is an important limiting factor for insulin absorption, and, in addition to modifying membrane permeability, another approach to increase absorption has been to prolong the time of contact between insulin and the nasal mucosa. Methods include development of powder formulations (Nagai *et al.*, 1984; Nagai, 1985; Pontiroli *et al.*, 1986; Schipper *et al.*, 1993), bioadhesive polymers or microspheres (Björk and Edman, 1988; Farraj *et al.*, 1990; Edman *et al.*, 1992; Rydén and Edman, 1992; Dondeti *et al.*, 1995), viscous polymer solutions (Rydén and Edman, 1992), or gel dosage forms (Morimoto *et al.*, 1985). The promoting effects achieved by the use of such techniques are generally comparable to those obtained by surfactant enhancement although combination with chemical enhancement has been reported to further improve the extent of absorption (Farraj *et al.*, 1990; Dondeti *et al.*, 1995).

A great number of clinical studies of nasal insulin administration have been performed during the last 15 years (Table V; Gizurarson and Bechgaard, 1991b). Negligible effect on blood glucose was observed without addition of enhancer, and the effect differed in type I and type II diabetics compared to normal subjects (Moses *et al.*, 1983). Pharmacokinetics of intranasal insulin are close to those of intravenous insulin, as delivery to

Table V
Pharmacokinetic Data for Nasal Insulin Administration to Humans

Enhancer	Subjects	Peak insulin (min)	BG nadir[a] (min)	Bioavailability (%)	Reference
Laureth-9	4 normals	15–30	45	NS[b]	Salzman *et al.* (1985)
	4 normals	20	30–40	NS	Pontiroli *et al.* (1987)
	8 normals	15	45	⩽10	Paquot *et al.* (1988)
	5 type I	30	NS	NS	Kimmerle *et al.* (1991)
Bile salt	5 normals	15	NS	NS	YoKosuka *et al.* (1977)
	6 normals	14	50–60	12	Pontiroli *et al.* (1982)
	24 normals	5–15	30	10	Moses *et al.* (1983)
	7 normals	5–15	30–45	NS	Hirata *et al.* (1983)
	40 normals	10–15	NS	10–20	Gordon *et al.* (1985)
	9 normals	10	30–40	NS	Pontiroli *et al.* (1987)
	8 type I	NS	20–40	11	Pontiroli *et al.* (1982)
	4 type I	10–20	NS	NS	Moses *et al.* (1983)
	9 type I	NS	NS	5	Frauman *et al.* (1987a)
	5 type I	NS	NS	<10	Lassmann-Vague *et al.* (1988)
	8 type I	15	NS	NS	Sinay *et al.* (1990)
	2 type II	20–40	NS	NS	Moses *et al.* (1983)
	4 type II	30	60	NS	Hirata *et al.* (1983)
	9 type II	10	NS	12	Frauman *et al.* (1987b)
	8 type II	10	20–35	NS	El-Etr *et al.* (1987)
	6 type II	5–10	NS	NS	Bruce *et al.* (1991)
Fusidate	8 normals	8	21–24	7–9	Nolte *et al.* (1990)
Phosphatidylcholine					
	11 normals	23	44	8	Drejer *et al.* (1992)
	12 normals	23–28	40–50	9–15	Jacobs *et al.* (1993)
	10 type I	36	NS	NS	Holman (1993)
	17 type II	80	NS	NS	Coates *et al.* (1995)
	31 type I	NS	NS	~5	Hilsted *et al.* (1995)

[a]BG, Blood glucose.
[b]NS, Not stated.

normal and diabetic subjects results in peak plasma insulin levels within 5–30 min and in a nadir in plasma glucose between 20 and 60 min post administration. Bioavailability in these studies varied, depending on type and concentration of the enhancer, dosage form, and the dose and concentration of insulin administered, but optimally no more than about 10% seems to be absorbed in diabetic patients, independent of the enhancer used (Table V). Intranasal insulin was observed to be less effective during the afternoon (Pontiroli *et al.*, 1982). In a study assessing the efficacy in type II diabetic patients (Bruce *et al.*, 1991), it was concluded that the nasal insulin preparation failed to influence postprandial blood glucose rise despite peak insulin levels that were at least equal to those seen in nondiabetic subjects, although at an earlier time. An examination of the initial distribution and subsequent clearance from the nasal cavity of healthy volunteers revealed that the distribution of insulin varied according to the bolus volume but was not significantly affected by the deposition techniques, and none of the insulin reached the lung (Newman *et al.*, 1994). Some degree of mild local nasal irritation, such as stinging, congestion, and rhinorrhea, has been common in almost all clinical trials. The irritation produced by bile salts did not correlate with the absorption-enhancing effect (Gordon *et al.*, 1985).

Three long-term studies have been conducted to evaluate reliability of intranasal insulin treatment (Salzman *et al.*, 1985; Frauman *et al.*, 1987a; Lassmann-Vague *et al.*, 1988). Drop-out rate in two of the studies was 30% due to inadequate glycemic control, and one patient developed chronic rhinitis which disappeared when subcutaneous therapy was resumed (Lassmann-Vague *et al.*, 1988).

Nasal delivery of peptide and proteins has recently been reviewed by Pontiroli and Pozza (1990), Su (1991), Lassmann-Vague (1991), and Edman and Björk (1992).

3.3.8. PULMONARY DELIVERY

The lungs are an attractive site for systemic delivery and offer a vast surface area (140 m^2 in humans) capable of rapid absorption but have to date received much less attention for systemic delivery than most other routes. However, the growing awareness that the lung, due to the very thin epithelium of the alveoli, is permeable to proteins, combined with the fact that inhalation is well accepted in most Western societies, makes the investigation of this route of administration a rapidly growing new field in drug delivery (Byron, 1990; Patton and Platz, 1992).

The types of devices for medication of the lung include metered dose inhalers, nebulizers, and powder inhalers or insufflators. It is important to design the device so that the dose predominantly reaches the alveoli, where mucociliary clearance is absent. Deposition in the alveoli is optimal for aerosols with aerodynamic diameters of around 2–3 μm (Byron, 1990). In general, pulmonary absorption rates are inversely related to molecular size (Hastings *et al.*, 1992). However, some poorly absorbed proteins are much smaller than well-absorbed ones, and size is therefore not the only limiting factor (Wall, 1995). Additional determinants of bioavailability for macromolecules may be their charge and site of deposition (Byron and Patton, 1994).

Dosing presents difficulties in animal investigations, and most studies of pulmonary insulin administration have been performed by instillation of an insulin solution into the lung of rats or rabbits (Table VI). However, drug is absorbed to a greater extent when inhaled as an aerosol than when

Table VI
Absorption of Insulin Administered to Animals via the Respiratory Tract

Animal	Enhancer	Mode of administration	Bioavailability (%)	Reference
Rat	None	Instillation, pH 3	42	Okumura *et al.* (1992)
	None	Instillation, pH 7	7	Okumura *et al.* (1992)
	Glycocholate (50 mM)	Instillation, pH 7	25	Okumura *et al.* (1992)
	None	Powder aerosol	98	Okumura *et al.* (1992)
Rat	None	Instillation	15	Liu *et al.* (1993)
	with liposomes	Instillation	26–30	Liu *et al.* (1993)
Rat	None	Installation	9	Li *et al.* (1993)
	Glycocholate (20 mM)	Instillation	79	Li *et al.* (1993)
Rat	None	Instillation, pH 3	~100	Jones *et al.* (1987)
Rat	None	Instillation, pH 7	9.1	Shao *et al.* (1994b)
	Hydroxypropyl β-cyclodextrin (1%)	Instillation, pH 7	22	Shao *et al.* (1994b)
	Dimethyl β-cyclodextrin (1%)	Instillation, pH 7	103	Shao *et al.* (1994b)
Rabbit	None	Liquid aerosol	36	Sakr (1992)
Rabbit	None	Instillation	6	Colthorpe *et al.* (1992)
	None	Liquid aerosol	57	Colthorpe *et al.* (1992)
Rabbit	Acetylglycerin monostearate	Powder aerosol	40	Yoshida *et al.* (1979)

administered by instillation (Brown and Schanker, 1983; Schanker *et al.*, 1986; Colthorpe *et al.*, 1992). The depth of deposition into the lung has a significant influence on the absorption of insulin. Thus, the apparent relative bioavailability of insulin increased more than twofold after the administration of aerosol delivered via a tube placed in the trachea of rats to a depth of 20 mm as compared to a depth of 10 mm (Okumura *et al.*, 1992; Table II). An increase in the viscosity of an aqueous formulation facilitated insulin absorption from rat lung, probably because of reduced mucocilliary clearance (Li and Mitra, 1994). Hypotonicity as well as low pH of the administered insulin solution significantly improved hypoglycemic responses, presumably as a result of membrane damage (Table VI; Okumura *et al.*, 1992; Li and Mitra, 1994; Wall, 1995).

Lung homogenates metabolize insulin (Liu *et al.*, 1992), although the proteolytic enzyme activity of the lung in degrading insulin is weaker than that of subcutaneous tissue (Hori *et al.*, 1983). In accordance with this finding, addition of protease inhibitors has been reported to increase the pulmonary absorption of insulin (Okumura *et al.*, 1992; Yamamoto *et al.*, 1994).

Only a few studies of pulmonary insulin delivery in humans have been reported. Shortly after the discovery of insulin, a fall in blood glucose was demonstrated in diabetic patients after insulin inhalation (Gänsslen, 1925), but there seem to have been no further investigations of this approach at that time (Jensen, 1938). Many years later, work by Wigley *et al.* (1971) demonstrated the feasibility of aerosol insulin delivery to normal and diabetic subjects, as an increase in plasma insulin was observed within 15–45 min after aerosol administration. Later, using administration by inhalation of an aerosolized "mist" of concentrated (U 500) neutral regular insulin, Elliott *et al.* (1987) demonstrated blood glucose control in diabetic children, at least as good as that on a control day with administration of their usual subcutaneous dose. The pulmonary dose was 4–5 times higher than the subcutaneous dose. The bioavailability in normal subjects, as judged by C-peptide measurements, was 20–25%. The efficiency of pulmonary absorption seems to be three times higher for smokers as compared to nonsmokers (Köhler *et al.*, 1987). In a recent study, using aerosolized insulin generated by a raindrop nebulizer, Laube *et al.* (1993) were able to maximize deposition of the dose of insulin within the lungs of diabetic patients. They found 50–93% of the inhaled dose to be deposited below the larynx, and average time to peak insulin level was 40 min.

The lung and the respiratory tract are immunologically competent, and elicitation of an immune response is a general concern for insulin therapy. However, the healthy lung may be designed to be particularly unresponsive to self-proteins, as suggested by studies with growth hormone (Patton and

Platz, 1992). The potential effects of exposing patients to inhaled insulin on a chronic basis are unexplored and will need to be investigated in clinical studies. Abnormal pulmonary function has been documented in 60% of a cross section of a diabetic population. Impaired diffusion due to a reduced pulmonary capillary blood volume and reduced lung volumes were the most consistent abnormalities (Sandler *et al.*, 1986; Sandler, 1990). Therefore, it may be necessary to exclude certain categories of risk patients from pulmonary insulin administration (Byron, 1990).

Pulmonary systemic delivery of peptides and proteins has recently been reviewed by Patton and Platz (1992), Adjei and Gupta (1994), Byron and Patton (1994), and Wall (1995).

3.3.9. TRANSDERMAL DELIVERY

The skin covers a surface area of more than 1 m^2 and is the most readily accessible organ in terms of drug delivery. However, the skin is also the most impermeable of all the epithelial membranes, and, owing to the multiple layers of horny cells in the stratum corneum, it is an excellent barrier to the penetration of hydrophilic substances. Therefore, the rate of passive permeation of proteins is far too slow to be therapeutically significant.

Enhancement of the skin permeation of peptides and proteins by electrical or ultrasonic means has recently been receiving a lot of attention from both academia and industry, as recently reviewed by Chien (1991) and Cullander and Guy (1992). The concurrent development of technology relating to patch materials, microprocessors, and miniaturized batteries (Singh and Maibach, 1993) makes iontophoretic delivery of certain drugs an attractive concept (Banga and Chien, 1988b).

The principle of iontophoresis is that ionized drug molecules can be driven into the skin when an electrical potential is applied across the skin by placing two electrodes on the skin in electrolyte solution, one of which contains the drug. The ionic current flow in the skin seems to take place at discrete pores, which have been identified mainly to be the sweat glands and hair follicles (Cullander, 1992; Scott *et al.*, 1993). Human skin contains 40 to 70 hair follicles and more than 200 sweat ducts per square centimeter, but these appendages occupy only 0.1% of the total skin surface (Chien, 1991). Additional pores may, however, be created by the applied current (Amsden and Goosen, 1995).

During the last decade, transdermal iontophoretic delivery of insulin has been investigated in a number of studies (Stephen *et al.*, 1984; Kari, 1986; Chien *et al.*, 1987; Siddiqui *et al.*, 1987; Liu *et al.*, 1988; Meyer *et al.*,

1989). These studies are comprehensively reviewed and discussed by Sage in the preceding chapter of this volume.

The transdermal iontophoretic transport of normal, hexameric insulin through rodent skin *in vitro* has been reported to be very low (Banga and Chien, 1993; Langkjær *et al.*, 1994) and far from clinically relevant. In contrast, it has been demonstrated *in vitro* that specific monomeric insulin analogs (Table VII) can be iontophoretically transported across hairless mouse skin, some at rates sufficient to cover the basal need for a typical diabetic, i.e., approximately one unit per hour (Fig. 8; Langkjær *et al.*, 1994). In these experiments hexameric human insulin was found to be virtually unable to penetrate the skin. For the monomeric insulins, in which 1–5 amino acid residues in human insulin were substituted by aspartate or glutamate residues, it was found that increase in charge and/or hydrophilicity generally correlated with insulin delivery (Langkjær *et al.*, 1994). To obtain clinically relevant doses, it was necessary to wipe the skin gently with absolute ethanol prior to iontophoresis, a procedure that is expected to solubilize and remove some of the lipids (sebum) blocking the hair follicles (Cullander and Guy, 1992). In contrast, wiping with 70% alcohol had only a slight effect (Table VIII).

Extracts of skin homogenates exhibit proteolytic activity, but the proteases are probably intracellular and may therefore not interact with insulin permeating via the extracellular routes (Cullander and Guy, 1992). One recent paper reports substantial degradation of ^{125}I-insulin during *in vitro* iontophoretic studies using hairless rat skin (Banga and Chien, 1993). However, these results are controversial because the iodinated insulin presumably is unstable when exposed to the Pt electrode used in the studies owing to the electrochemical reactions taking place at this type of inert electrode material (Sage and Riviere, 1992).

Table VII
Monomeric Human Insulin Analogs

Code in Fig. 5	Human insulin analog (substitutions)	Negative charge added (pH 7)
	B16Gln	0
F	B28Asp	1
5	B9Asp	1
2	B9Asp, B27Glu	2
8	B16Glu, B27Glu	2
6	A21Asp, B9Asp, B27Glu	3
	B1Asp, B4Asp, B10Asp, B16Asp, B27Glu	5

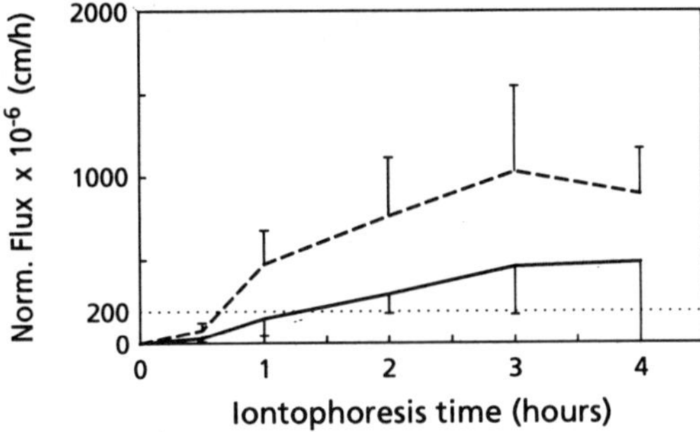

Figure 8. Transdermal flux of monomeric insulins during *in vitro* iontophoresis (cathodal delivery) of neutral solutions of the insulin analog B9Asp, B27Glu (———) and the insulin analog B16Glu, B27Glu (— — —) through hairless mouse skin (mean ± SEM, $N = 3$), wiped gently with ethanol prior to iontophoresis. Current, $0.5\,mA/cm^2$; donor area, $0.55\,cm^2$. The dotted line represents the rate of delivery necessary for basal treatment (~1 IU/hr) of diabetic subjects, assuming a donor area of $10\,cm^2$. (Data from Langkjær *et al.*, 1994.)

Another method being investigated to enhance the transport of peptides and proteins through the skin is the use of ultrasound (phonophoresis). Ultrasonic energy can perturb mammalian tissue via its heating, radiation pressure, cavitation, and acoustic microstreaming effects (Meidan *et al.*, 1995). Ultrasound-mediated transdermal delivery of insulin has been studied in mice (Tachibana and Tachibana, 1991), in rats (Mitragotri *et al.*, 1995), and in rabbits (Tachibana, 1992). A significant fall in blood glucose was observed in both mice and rabbits, and in the rabbits the plasma insulin concentration increased by a factor of 4 to 46 μU/ml during 90 min of phonophoresis. After ultrasound exposure was discontinued, the insulin concentration returned to the initial level within 2 hr. Mitragotri *et al.* (1995) reported that low-frequency ultrasound induced significant transdermal

Table VIII
Influence of Skin Pretreatment on Normalized Flux ($\times 10^6$ cm/hr ± SD) of Different Insulins through Mouse Skin *in Vitro* during Iontophoresis

	Insulin		
Skin pretreatment	Human	Asp^{B9}, Glu^{B27}	Glu^{B16}, Glu^{B27}
None	<0.2	<0.2	0.5 ± 0.05
2-Propanol (70%)	<0.2	4.2 ± 5.0	4.5 ± 4.9
Ethanol (100%)	7.1 ± 5.7	303 ± 168	757 ± 343

transport of insulin. Ultrasound (20 kHz) applied for 1 hr at an intensity of 225 mW/cm^2 was shown to induce a decrease in the blood glucose concentration in hairless rats similar to that induced by a subcutaneous injection of 1 IU of insulin.

Transdermal delivery of peptide and protein drugs has recently been reviewed by Amsden and Goosen (1995).

3.3.10. CONCLUSIONS

Administration of insulin via noninvasive routes is, even with the use of modern enhancement techniques, characterized by poor bioavailability and unpredictable delivery of insulin to the blood. Most recent attention has been directed at the nasal and the oral mucosa, but neither of these alternative routes looks like being a clinically useful proposition. Insulin can be absorbed from the gastrointestinal trace in measurable amounts but only when very large doses are administered. Furthermore, the fraction absorbed is low, variable, and inconsistent. The dream of an oral insulin substitution therapy appears highly unrealistic. Even if efficient enzymatic protection would be possible, and even if enhancers are eventually identified that are able to sufficiently increase the absorption of intact insulin through the intestinal wall, the dosage and timing problems would be of such magnitude that oral insulin would remain "a dim star in a distant part of the galaxy" (Home *et al.*, 1989).

The quest for acceptable insulin delivery via the nasal route has not been very successful as the problems of low bioavailability, local irritation, and probable toxicity in long-term use are unresolved despite the huge efforts that have been made. The major problem, which seems very hard to solve, is to reach an adequate biological availability. A common concern is also that local disease, such as the common cold or chronic rhinitis, might impede the efficiency of intranasal absorption and alter the systemic availability.

The ocular, buccal, rectal, and vaginal routes of administration all have inherent limitations and serious drawbacks for delivery of insulin and therefore are highly unlikely to be of any use for chronic insulin therapy.

The pulmonary and transdermal routes are the least investigated of the alternative, nonparenteral delivery routes for insulin, and perhaps therefore they still hold some promise. The potential of pulmonary delivery for bolus therapy has been demonstrated as this route is feasible from a bioavailability standpoint, even without the addition of enhancers. However, long-term safety remains to be established. Inhalation of insulin may offer fairly reproducible absorption kinetics, but it is a major challenge from a device point of view to ensure reproducibility of the administered dose.

Monomeric insulin analogs can, in contrast to human, hexameric insulin, be iontophoretically delivered across hairless mouse skin. Although the insulin transport rates were too low to satisfy meal-related insulin requirements, clinically relevant delivery rates for basal insulin supply were obtained with such insulin after wiping the skin with absolute alcohol prior to iontophoresis. The potential clinical advantages of iontophoretic transdermal delivery of insulin for basal therapy include that controlled delivery, e.g., during the night, will be possible, that the insulin depot is outside the body, reducing the risk of hypoglycemia, and that variation in delivery rate is expected to be low.

4. SUMMARY AND FUTURE PERSPECTIVES

Insulin formulation and delivery is presently at a crossroads, and over the next decade dramatic changes in diabetes therapy are likely to take place, many of which will be technology-driven. Advances in protein chemistry and molecular biology are being utilized to produce new insulins to optimize therapy and more effectively meet the clinical needs of diabetes mellitus. Analogs or derivatives of insulin have been and are being designed to meet formulation and delivery challenges. Conventional delivery clearly has deficiencies that must be addressed in order to bring blood glucose into as normal a range as possible, with a minimum of hypoglycemia and hyperinsulinism. A more convenient insulin therapy is expected to result in improved patient compliance, and, together with more physiologic insulin levels, this could lead to better metabolic control, and thereby a reduced risk of late-diabetic complications. Injection devices, e.g., jet injectors or pen systems, may improve the patient's comfort although they cannot be regarded as a major breakthrough in the quest for improved control. Implantable polymer matrices for sustained release can possibly be developed for basal rate delivery, but it is hard to imagine that they can be improved sufficiently to provide efficient bolus delivery. Provided a reliable long-duration glucose sensor becomes available, the implantable pump may represent the ultimate solution to optimal insulin therapy although the cost will presumably prevent its widespread use.

The quest for complete restoration of physiology, possibly by finding the optimal route of administration of the hormone, has been continuously and strenuously pursued. Nonetheless, the dream of noninvasive insulin delivery by alternative routes of administration is far from realization, and the fantasy of oral insulin treatment "has been one of great ambition and continuous failure" (Berger, 1993). The only routes that still hold some promise are transdermal delivery by iontophoresis and pulmonary inhalation. Iontophoretic delivery offers the possibility of a zero-order delivery, adjustable according to varying requirements, and pulmonary delivery has the potential to provide bolus therapy very close to physiological needs. However, much work remains to be done in these areas of delivery research before the clinical utility of these routes for insulin substitution therapy can be fully evaluated.

Insulin has always been in the forefront of research into therapeutic technology, and such research is ongoing to discover more efficient insulins and insulin formulations and to find less intrusive methods or routes of insulin administration.

REFERENCES

Adjei, A., and Gupta, P., 1994, Pulmonary delivery of therapeutic peptides and proteins, *J. Controlled Release* **29**:361–373.

Albin, G., Horbett, T. A., and Ratner, B. D., 1985, Glucose sensitive membranes for controlled delivery of insulin: Insulin transport studies, *J. Controlled Release* **2**:153–164.

Albin, G. W., Horbett, T. A., Miller, S. R., and Ricker, N. L., 1987, Theoretical and experimental studies of glucose sensitive membranes, *J. Controlled Release* **6**:267–291.

Albisser, A. M., Lougheed, W. D., Perlman, K., and Bahoric, A., 1980, Nonaggregating insulin solutions for long-term glucose control in experimental and human diabetes, *Diabetes* **29**:241–243.

Albisser, A. M., Williamson, J. R., and Lougheed, W. D., 1982, Desired characteristics of insulins to be used in infusion pumps. *in: Hormone Drugs* (J. L. Gueriguian, E. D. Bransome, and A. S. Outschoorn, eds.), U.S. Pharmacopeial Convention, Rockville, Maryland, pp. 84–95.

Amiel, S. A., 1993, Intensified insulin therapy, *Diabetes/Metab. Rev.* **9**:3–24.

Amsden, B. G., and Goosen, M. F. A., 1995, Transdermal delivery of peptide and protein drugs: An overview, *AIChE J.* **41**:1972–1997.

Amthor, K.-F., Dahl-Jørgensen, K., Berg, T. J., Heier, M. S., Sandvik, L., Aagenæs, Ø, and Hanssen, K. F., 1994, The effect of 8 years of strict glycaemic control on peripheral nerve function in IDDM patients: The Olso study, *Diabetologia* **37**:579–584.

Arrieta-Molero, J. F., Aleck, K., Sinha, M. K., Brownscheidle, C. M., Shapiro, L. J., and Sperling, M. A., 1982, Orally administered liposome-entrapped insulin in diabetic animals, *Hormone Res.* **16**:249–256.

Asada, H., Douen, T., Mizokoshi, Y., Fujita, T., Murakami, M., Yamamoto, A., and Muranishi, S., 1994, Stability of acyl derivatives of insulin in the small intestine: Relative importance of insulin association characteristics in aqueous solution, *Pharm. Res.* **11**:1115–1120.

Asada, H., Douen, T., Waki, M., Adachi, S., Fujita, T., Yamamoto, A., and Muranishi, S., 1995, Absorption characteristics of chemically modified-insulin derivatives with various fatty acids in the small and large intestine, *J. Pharm. Sci.* **84**:682–687.

Atchison, J. A., Grizzle, W. E., and Pillion, D. J., 1989, Colonic absorption of insulin: An *in vitro* and *in vivo* evaluation, *J. Pharmacol. Exp. Ther.* **248**:567–572.

Aungst, B. J., 1994, Site-dependence and structure–effect relationships for alkylglycosides as transmucosal absorption promoters for insulin, *Int. J. Pharm.* **105:**219–225.

Aungst, B. J., and Rogers, N. J., 1988, Site dependence of absorption-promoting actions of laureth-9, Na salicylate, Na2EDTA, and aprotinin on rectal, nasal, and buccal insulin delivery, *Pharm. Res.* **5:**305–308.

Aungst, B. J., and Rogers, N. J., 1989, Comparison of the effects of various transmucosal absorption promoters on buccal insulin delivery, *Int. J. Pharm.* **53:**227–235.

Aungst, B. J., Rogers, N. J., and Shefter, E., 1988, Comparison of nasal, rectal, buccal, sublingual and intramuscular insulin efficacy and the effects of a bile salt absorption promoter, *J. Pharmacol. Exp. Ther.* **244:**23–27.

Bae, Y. H., Okano, T. O., and Kim, S. W., 1989, Insulin permeation through thermo-sensitive hydrogels, *J. Controlled Release* **9:**271–279.

Bai, J. P. F., and Chang, L. L., 1995, Transepithelial transport of insulin: I. Insulin degradation by insulin-degrading enzyme in small intestinal epithelium, *Pharm. Res.* **12:**1171–1175.

Bak, J. F., Nielson, O. H., Pederson, O., and Beck-Nielson, H., 1987, Multiple insulin injections using a pen injector versus pump treatment in young diabetic patients, *Diabetes Res.* **6:**155–158.

Baker, E. N., Blundel, T. L., Cutfield, J. F., Cutfield, S. M., Dodson, E. J., Dodson, G. G., Hodgkin, D. M. C., Hubbard, R. E., Isaacs, N. W., Reynolds, C. D., Sakabe, K., Sakabe, N., and Vijayan, N. M., 1988, The structure of 2Zn pig insulin crystals at 1.5 Å resolution, *Phil. Trans. R. Soc.* **319:**369–456.

Bakth, S., Ertel, H. H., May, P. B., Jr., and Akgun, S., 1980, Rectal insulin: A potent hypoglycemic preparation, *Horm. Metab. Res. Suppl. Ser.* **12:**86–87.

Balschmidt, P., and Brange, J., 1992, Fast acting human insulin analogues with a single amino acid deletion in the B-chain, *Diabetologia* **35**(Suppl. 1)**:**A4(Abstract).

Balschmidt, P., Hansen, F. B., Dodson, E. J., Dodson, G. G., and Korber, F., 1991, Structure of porcine insulin cocrystallized with clupeine Z *Acta Crystallogr.* **B47:**975–986.

Banga, A. K., and Chien, Y. W., 1988a, Systemic delivery of therapeutic peptides and proteins, *Int. J. Pharm.* **48:**15–50.

Banga, A. K., and Chien, Y. W., 1988b, Iontophoretic delivery of drugs: Fundamentals, developments and biomedical applications, *J. Controlled Release* **7:**1–24.

Banga, A. K., and Chien, Y. W., 1993, Characterisation of *in vitro* transdermal iontophoretic delivery of insulin, *Drug. Dev. Ind. Pharm.* **19:**2069–2087.

Banting, F. G., and Best, C. H., 1922, Pancreatic extracts, *J. Lab. Clin. Med.* **7:**464–472.

Bar-on, H., Berry, E. M., Eldor, A., Kidron, M., Lichtenberg, D., and Ziv, E., 1981, Enteral administration of insulin in the rat, *Br. J. Pharmacol.* **73:**21–24.

Bartlett, J. D., Slusser, T. G., Turner-Henson, A., Singh, K. P., Atchison, J. A., and Pillion, D. J., 1994a, Toxicity of insulin administered chronically to human eye *in vivo*, *J. Ocul. Pharmacol.* **10:**101–107.

Bartlett, J. D., Turner-Henson, A., Atchison, J. A., Wooley, T. W., and Pillion, D. J., 1994b, Insulin administration to the eyes of normoglycemic human volunteers, *J. Ocul. Pharmacol.* **10:**683–690.

Bechgaard, E., Gizurarson, S., Jørgensen, L., and Larsen R., 1992, The viability of isolated rabbit nasal mucosa in the Ussing chamber, and the permeability of insulin across the membrane, *Int. J. Pharm.* **87:**125–132.

Bendayan, M., Ziv, E., Ben-Sasson, R., Bar-on, H., and Kidron, M., 1990, Morpho-cytochemical evidence for insulin absorption by the rat ileal epithelium, *Diabetologia* **33:**197–204.

Bendayan, M., Ziv, E., Gingras, D., Ben-Sasson, R., Bar-on, H., and Kidron, M., 1994, Biochemical and morpho-cytochemical evidence for the intestinal absorption of insulin in control and diabetic rats. Comparison between the effectiveness of duodenal and colon mucosa, *Diabetologia* **37:**119–126.

Benson, E. A., Benson, J. W., Jr., Fredlund, P. N., Mecklenburg, R. S., and Metz, R., 1988, Flocculation and loss of potency of human NPH insulin, *Diabetes Care* **11**:563–566.

Bentley, G. A., Brange, J., Derewenda, Z., Dodson, E. J., Dodson, G. G., Markussen, J., Wilkinson, A. J., Wollmer, A., and Xiao, B., 1992, Role of B13 Glu in insulin assembly. The hexamer structure of recombinant mutant (B13 Glu-Gln) insulin, *J. Mol. Biol.* **228**:1163–1176.

Berger, M., 1993, Oral insulin 1922–1992: The history of continuous ambition and failure, in; *Frontiers in Insulin Pharmacology* (M. Berger and F. A. Gries, eds.), Thieme Verlag, Stuttgart, pp. 144–148.

Berger, M., Cüppers, H. J., Halban, P. A., and Offord, R. E., 1980, The effect of aprotinin on the absorption of subcutaneously injected regular insulin in normal subjects, *Diabetes* **29**:81–83.

Berger, M., Cüppers, H. J., Hegner, J., Jörgens, V., and Berchtold, P., 1982, Absorption kinetics and biological effect of subcutaneously injected insulin preparations, *Diabetes Care* **5**:77–91.

Binder, C., 1969, Absorption of injected insulin, *Acta Pharmacol. Toxicol. Suppl.* **2**:1–87.

Binder, C., Nielsen, A., and Jørgensen, K., 1967, The absorption of an acid and a neutral insulin solution after subcutaneous injection into different regions in diabetic patients, *Scand. J. Clin. Lab. Invest.* **19**:156–163.

Birdi, K. S., 1973, The binding of cationic detergent to insulin, zinc-insulin and iso-insulin, *Biochem. J.* **135**:253–255.

Björk, E., and Edman, P., 1988, Degradable starch microspheres as a nasal delivery system for insulin, *Int. J. Pharm.* **47**:233–238.

Blackshear, P. J., Rohde, T. D., Palmer, J. L., Wigness, B. D., Rupp, W. M., and Buchwald, H., 1983, Glycerol prevents insulin precipitation and interruption of flow in an implantable insulin infusion pump, *Diabetes Care* **6**:387–392.

Brange, J., 1977, Improvements in or relating to biphasic insulin preparations comprising crystalline insulin and monodesamido insulin, United Kingdom Patent 1560232.

Brange, J., 1994, Stability of Insulin. Studies on the Physical and Chemical Stability of Insulin in Pharmaceutical Formulation, Thesis, Royal Danish School of Pharmacy; Kluwer Academic Publishers, Boston.

Brange, J., and Havelund, S., 1983a, Insulin pumps and insulin quality—requirements and problems, *Acta Med. Scand.* **671** (Suppl.):135–138.

Brange, J., and Havelund, S., 1983b, Properties of insulin in solution, in: *Artificial Systems for Insulin Delivery* (P. Brunetti, K. G. M. M. Alberti, A. M. Albisser, K. D. Hepp, and M. M. Benedetti, eds.), Raven Press, New York, pp. 83–88.

Brange, J., and Havelund, S., 1991, Novel insulin analogues stabilized against chemical modifications, European Patent 0419504.

Brange, J., and Langkjær, L., 1993, Insulin structure and stability, in: *Stability and Characterization of Protein and Peptide Drugs: Case Histories* (Y. J. Wang and R. Pearlman, eds.), Plenum Press, New York, pp. 315–350.

Brange, J. Havelund, S., Hansen, P., Langkjaer, L., Sørensen, E., and Hildebrandt, P., 1982, Formulation of physically stable neutral solutions for continuous infusion by delivery systems, in: *Hormone Drugs* (J. L. Gueriguian, E. D. Bransome, and A. S. Outschoorn, eds.), U.S. Pharmacopeial Convention, Rockville, Maryland, pp. 96–105.

Brange, J., Havelund, S., Hommel, E., Sørensen, E., and Kühl, C., 1986, Neutral insulin solutions physically stabilized by addition of Zn^{2+}, *Diabetic Med.* **3**:532–536.

Brange, J., Skelbaek-Pedersen, B., Langkjaer, L., Damgaard, U., Ege, H., Havelund, S., Heding, L. G., Jørgensen, K. H., Lykkeberg, J., Markussen, J., Pingel, M., and Rasmussen, E., 1987, *Galenics of Insulin: The Physico-chemical and Pharmaceutical Aspects of Insulin and Insulin Preparations*, Springer-Verlag, Berlin.

Brange, J., Ribel, U., Hansen, J. F., Dodson, G., Hansen, M. T., Havelund, S., Melberg, S. G., Norris, F., Norris, K., Snel, L., Sørensen, A. R., and Voigt, H. O., 1988, Monomeric insulins obtained by protein engineering and their medical implications, *Nature* **333:**679–682.

Brange, J., Drejer, K., Hansen, J. F., Havelund, S., Kaarsholm, N. C., Melberg, S. G., and Sørensen, A. R., 1989, Design of novel insulins with changed self-association and ligand binding properties, in: *Advances in Protein Design, International Workshop 1988* (H. Blocker, J. Collins, R. D. Schmid, and D. Schomburg, eds.), VCH Publishers, Weinheim, pp. 139–144.

Brange, J., Owens, D. R., Kang, S., and Vølund, A., 1990, Monomeric insulins, and their experimental and clinical implications, *Diabetes Care* **13:**923–954.

Brange, J., Hallund, O., and Sørensen, E., 1992, Chemical stability of insulin: 5. Isolation, characterization and identification of insulin transformation products, *Acta Pharm. Nord.* **4:**223–232.

Brems, D. N., Alter, L. A., Beckage, M. J., Chance, R. E., DiMarchi, R. D., Green, L. K., Long, H. B., Pekar, A. H., Shields, J. E., and Frank, B. H., 1992, Altering the association properties of insulin by amino acid replacement, *Protein Eng.* **5:**527–533.

Brennan, J. R., Gebhart, S. P., and Blackard, W. G., 1985, Pump-induced insulin aggregation. A problem with the Biostator, *Diabetes* **34:**353–359.

Bringer, J., Heldt, A., and Grodsky, G. M., 1981, Prevention of insulin aggregation by dicarboxylic amino acids during prolonged infusion, *Diabetes* **30:**83–85.

Brown, L., Munoz, C., Siemer, L., Edelman, E., and Langer, R., 1986a, Controlled release of insulin from polymer matrices. Control of diabetes in rats, *Diabetes* **35:**692–697.

Brown, L., Siemer, L., and Munoz, C., 1986b, Controlled release of insulin from polymer matrices. *In vitro* kinetics, *Diabetes* **35:**684–691.

Brown, R. A., Jr., and Schanker, L. S., 1983, Absorption of aerosolized drugs from the rat lung, *Drug Metab. Dispos.* **11:**355–360.

Brownlee, M., and Cerami, A., 1970, A glucose-controlled insulin-delivery system: Semisynthetic insulin bound to lectin, *Science* **206:**1190–1191.

Brownlee, M., and Cerami, A., 1983, Glycosylated insulin complexed to concanavalin A. Biochemical basis for a closed-loop insulin delivery system. *Diabetes* **32:**499–504.

Bruce, D. G., Chisholm, D. J., Storlien, L. H., Borkman, M., and Kraegen, E. W., 1991, Meal-time intranasal insulin delivery in type 2 diabetes, *Diabetic Med.* **8:**366–370.

Buchwald, H., Rohde, T. D., Dorman, F. D., Skakoon, J. G., Wigness, B. D., Prosl, F. R., Tucker, E. M., Rublein, T. G., Blackshear, P. J., and Varco, R. L., 1980, A totally implantable drug infusion device: Laboratory and clinical experiences using a model with single flow rate and new design for modulated insulin infusion, *Diabetes Care* **3:**351–358.

Byron, P. R., 1990, Determinants of drug and polypeptide bioavailability from aerosols delivered to the lung, *Adv. Drug Delivery Rev.* **5:**107–132.

Byron, P. R., and Patton, J. S., 1994, Drug delivery via the respiratory tract, *J. Aerosol Med. Deposition Clear. Eff. Lung* **7:**49–75.

Caldwell, L., Nishihata, T., Rytting, J. H., and Higuchi, T., 1982, Lymphatic uptake of water-soluble drugs after rectal administration, *J. Pharm. Pharmacol.* **34:**520–522.

Carstens, S., Danielsen, G., Guldhammer, B., and Frederiksen, O., 1993, Transport of insulin across rabbit nasal mucosa *in vitro* induced by didecanoyl-L-α-phosphatidylcholine, *Diabetes* **42:**1032–1040.

Champion, M. C., Shepherd, G. A. A., Rodger, N. W., and Dupre, J., 1980, Continuous subcutaneous infusion of insulin in the management of diabetes mellitus, *Diabetes* **29:**206–212.

Chandler, S. G., Thomas, N. W., and Illum, L., 1994, Nasal absorption in the rat. III. Effect of lysophospholipids on insulin absorption and nasal histology, *Pharm. Res.* **11:**1623–1630.

Chang, T. M. S., 1976, Biodegradable semipermeable microcapsules containing enzymes, hormones, vaccines, and other biologicals, *J. Bioeng.* **1**:25–32.

Chawla, A. S., Hinberg, I., Blais, P., and Johnson, D., 1985, Aggregation of insulin, containing surfactants, in contacts with different materials, *Diabetes* **34**:420–424.

Chien, Y. W., 1991, Transdermal route of peptide and protein drug delivery, in: *Peptide and Protein Drug Delivery* (V. H. L. Lee, ed.), Marcel Dekker, New York, pp. 667–689.

Chien, Y. W., Siddiqui, O., Sun, Y., Shi, W. M., and Liu, J. C., 1987, Transdermal iontophoretic delivery of therapeutic peptides/proteins. I: Insulin, *Ann. N.Y. Acad. Sci.* **507**:32–50.

Chiou, G. C. Y., and Chuang, C. Y., 1989, Improvement of systemic absorption of insulin through eyes with absorption enhancers, *J. Pharm. Sci.* **78**:815–818.

Chiou, G. C. Y., and Li, B. H. P., 1993, Chronic systemic delivery of insulin through the ocular route, *J. Ocul. Pharmacol.* **9**:85–90.

Chiou, G. C. Y., Chuang, C. Y., and Chang, M. S., 1988, Reduction of blood glucose concentration with insulin eye drops, *Diabetes Care* **11**:750–751.

Chiou, G. C. Y., Chuang, C. Y., and Chang, M. S., 1989, Systemic delivery of insulin through eyes to lower the glucose concentration, *J. Ocul. Pharmacol.* **5**:81–91.

Chisholm, D. J., Kraegen, E. W., Hewett, M. J., and Furler, S., 1984, Low subcutaneous degradation and slow absorption of insulin in insulin-dependent diabetic patients during continuous subcutaneous insulin infusion at basal rate, *Diabetologia* **27**:238–241.

Cho, Y. W., and Flynn, M., 1993, Oral delivery of insulin, *Lancet* **ii**:1518–1519.

Christiansen, J. S., and Lauritzen, T., 1991, Pharmacokinetic aspects of insulin therapy—with special reference to pumps and pens versus conventional insulin treatment, in: *Pharmacology of Diabetes. Present Practice and Future Perspectives* (C. E. Mogensen and E. Standl, eds.), de Gruyter, Berlin, pp. 23–38.

Chung, D.-J., Ito, Y., and Imanishi, Y., 1992, An insulin-releasing membrane system on the basis of oxidation reaction of glucose, *J. Controlled Release* **18**:45–54.

Coates, P. A., Ismail, I. S., Luzio, S. D., Griffiths, I., Ollerton, R. L., Vølund, A., and Owens, D. R., 1995, Intranasal insulin: The effect of three dose regimens on postprandial glycaemic profiles in type II diabetic subjects, *Diabetic Med.* **12**:235–239.

Colthorpe, P., Farr, S. J., Taylor, G., Smith, I. J., and Wyatt, D., 1992, The pharmacokinetics of pulmonary-delivered insulin: A comparison of intratracheal and aerosol administration to the rabbit, *Pharm. Res.* **9**:764–768.

Couvreur, P., and Puisieux, F., 1993, Nano- and microparticles for the delivery of polypeptides and proteins, *Adv. Drug Delivery Rev.* **10**:141–162.

Couvreur, P., Lenaerts, V., Kante, B., Roland, M., and Speiser, P. P., 1980, Oral and parenteral administration of insulin associated to hydrolysable nanoparticles, *Acta Pharm. Technol.* **26**:220–222.

Crane, C. W., Path, M. C., and Luntz, G. R. W. N., 1968, Absorption of insulin from the human small intestine, *Diabetes* **17**:625–627.

Creque, H. M., Langer, R., and Folkman, J., 1980, One month of sustained release of insulin from a polymer implant, *Diabetes* **29**:37–40.

Cullander, C., 1992, What are the pathways of iontophoretic current flow through mammalian skin?, *Adv. Drug Delivery Rev.* **9**:119–135.

Cullander, C., and Guy, R. H., 1992, Transdermal delivery of peptides and proteins, *Adv. Drug Delivery Rev.* **8**:291–329.

Cunningham, L. W., Fischer, R. L., and Vestling, C. S., 1955, A study of the binding of zinc and cobalt by insulin, *J. Am. Chem. Soc.* **77**:5703–5707.

Damgé, C., 1991, Oral insulin, in: *Biotechnology of Insulin Therapy* (J. C. Pickup, ed.), Blackwell Scientific Oxford, pp. 97–112.

Damgé, C., Michel, C., Aprahamian, M., and Couvreur, P., 1988, New approach for oral administration of insulin with polyalkylcyanoacrylate nanocapsules as drug carrier, *Diabetes* **37:**246–251.

Damgé, C., Michel, C., Aprahamian, M., Couvreur, P., and Devissaguet, J. P., 1990, Nanocapsules as carriers for oral peptide delivery, *J. Controlled Release* **13:**233–239.

Dandona, P., Hooke, D., and Bell, J., 1978, Exercise and insulin absorption from subcutaneous tissue, *Br. Med. J.* **1:**479–480.

Dapergolas, G., and Gregoriadis, G., 1976, Hypoglycaemic effect of liposome-entrapped insulin administered intragastrically into rats, *Lancet* **ii:**824–827.

Das, N., Basu, M. K., and Das, M. K, 1988, Oral application of insulin encapsulated liposomes, *Biochem. Int.* **16:**983–989.

Davis, B. K., 1972, Control of diabetes with polyacrylamide implants containing insulin, *Experientia* **28:**348.

de Boer, A. G., Hoogdalem, E. J., and Breimer, D. D., 1992, Rate-controlled rectal peptide drug absorption enhancement, *Adv. Drug. Delivery Rev.* **8:**237–251.

Denne, J. R., Andrews, K. L., Lees, D. V., and Mook, W., 1992, A survey of patient preference for insulin jet injectors versus needle and syringe, *Diabetes Educ.* **18:**223–227.

Deurloo, M. J. M., Hermens, W. A. J. J., Romeyn, S. G., Verhoef, J. C., and Merkus, F. W. H. M., 1989, Absorption enhancement of itranasally administered insulin by sodium taurodihydrofusidate (STDHF) in rabbits and rats, *Pharm. Res.* **6:**853–856.

Dillon, R. S., 1983, Improved serum insulin profiles in diabetic individuals who massaged their insulin injection sites, *Diabetes Care* **6:**399–401.

Dobre, V., Georgescu, D., Simionescu, L., Aman, E., Stroescu, V., and Motas, C., 1983, The entrapment of biologically active substances into liposomes. I. Effects of the oral administration of liposomally entrapped insulin in normal rats. *Rev. Roum. Biochim.* **20:**15–20.

Dobre, V., Simionescu, L., Stroescu, V., Georgescu, D., Aman, E., Trandaburu, T., and Motas, C., 1984, The entrapment of biological active substances into liposomes. II. Effects of oral administration of liposomally entrapped insulin in normal and alloxanized rats, *Rev. Roum. Med. Endocrinol.* **22:**253–260.

Dondeti, P., Zia, H., and Needham, T. E., 1995, *In vivo* evaluation of spray formulations of human insulin for nasal delivery, *Int. J. Pharm.* **122:**91–105.

Douglas, S. J., Davis, S. S., and Illum, L., 1987, Poly(alkyl 2-cyanoacrylate) microspheres as drug carrier systems, in: *Polymers in Controlled Drug Delivery* (L. Illum and S. S. Davis, eds.), IOP Publishing, Bristol, U.K., pp. 66–72.

Drejer, K., Vaag, A., Bech, K., Hansen, P., Sørensen, A. R., and Mygind, N., 1991, Intranasal insulin administration, in: *Diabetes 1991* (H. Rifkin, J. A. Colwell, and S. Taylor, eds.), Elsevier, Amsterdam, pp. 1261–1264.

Drejer, K., Vaag, A., Bech, K., Hansen, P., Sørensen, A. R., and Mygind, N., 1992, Intranasal administration of insulin with phospholipid as absorption enhancer: Pharmacokinetics in normal subjects, *Diabetic Med.* **9:**335–340.

Earle, M. P., 1972, Experimental use of oral insulin, *Isr. J. Med. Sci.* **8:**899–900.

Edman, P., and Björk, E., 1992, Nasal delivery of peptide drugs, *Adv. Drug Delivery Rev.* **8:**165–177.

Edman, P., Björk, E., and Rydén, L., 1992, Microspheres as a nasal delivery system for peptide drugs, *J. Controlled Release* **21:**165–172.

Edsberg, B., Herly, D., Hildebrandt, P., and Kühl, C., 1987, Insulin bolus given by sprinkler needle: Effect on absorption and glycaemic response to a meal, *Br. Med. J.* **294:**1373–1376.

El-Etr, M., Slama, G., and Desplanque, N., 1987, Preprandial intranasal insulin as adjuvant therapy in type II diabetes, *Lancet* **ii:**1085–1086.

Elliott, R. B., Edgar, B. W., Pilcher, C. C., Quested, C., and McMaster, J., 1987, Parenteral absorption of insulin from the lung in diabetic children, *Aust. Paediatr. J.* **23:**293–297.

El-Shattawy, H. H., El-Ahmady, O., Hosny, E. A., Nabib, A. E., El-Damasy, H., El-Deen, S. G., and El-Kabbany, N., 1994, Effectiveness of rectal insulin suppositories containing sodium cholate in normal and insulin dependent diabetic subjects, *Drug Dev. Ind. Pharm.* **20:**357–366.

Eppstein, D. A., and Longenecker, J. P., 1988, Alternative delivery systems for peptides and proteins as drugs, *Crit. Rev. Ther. Drug Carrier Syst.* **5:**99–139.

Fahlén, M., Odén, A., Björntorp, P., and Tibblin, G., 1971, Seasonal influence on insulin secretion in man, *Clin. Sci.* **41:**453–458.

Farraj, N. F., Johansen, B. R., Davis, S. S., and Illum, L., 1990, Nasal administration of insulin using bioadhesive microspheres as a delivery system, *J. Controlled Release* **13:**253–261.

Feingold, V., Jenkins, A. B., and Kraegen, E. W., 1984, Effect of contact material on vibration-induced insulin aggregation, *Diabetologia* **27:**373–378.

Fernqvist, E., Linde, B., Östman, J., and Gunnarsson, R., 1986, Effects of physical exercise on insulin absorption in insulin-dependent diabetics: A comparison between human and porcine insulin, *Clin. Physiol.* **6:**489–498.

Fineberg, S. E., Galloway, J. A., Fineberg, N. S., Rathbun, M. J., and Hufferd, S., 1983, Immunogenicity of recombinant DNA human insulin, *Diabetologia* **25:**465–469.

Fischel-Ghodsian, F., Brown, L., Mathiowitz, E., Brandenburg, D., and Langer, R., 1988, Enzymatically controlled drug delivery, *Proc. Natl. Acad. Sci. USA* **85:**2403–2406.

Fischer, U., Rebrin, K., Salzsieder, E., and Volund, A., 1993, Pharmacodynamics of fast-absorbed human insulin (HM) analogues, *Diabetologia* **36**(Suppl. 1)**:**A30 (Abstract).

Forlani, G., Santacroce, G., Ciavarella, A., Capelli, M., Mattioli, L., and Vannini, P., 1986, Effects of mixing short- and intermediate-acting insulins on absorption course and biologic effect of short-acting preparation, *Diabetes Care* **9:**587–590.

Frauman, A. G., Cooper, M. E., Parsons, B. J., Jerums, G., and Louis, W. J., 1987a, Long-term use of intranasal insulin in insulin-dependent diabetic patients, *Diabetes Care* **10:**573–578.

Frauman, A. G., Jerums, G., and Louis, W. J., 1987b, Effects of intranasal insulin in non-obese type II diabetics, *Diabetes Res. Clin. Pract.* **3:**197–202.

Frid, A., and Linde, B., 1993, Clinically important differences in insulin absorption from abdomen in IDDM, *Diabetes Res. Clin. Pract.* **21:**137–141.

Frid, A., Gunnarsson, R., Güntner, P., and Linde, B., 1988, Effects of accidental intramuscular injection on insulin absorption in IDDM, *Diabetes Care* **11:**41–45.

Fujii, S., Yokoyama, T., Ikegaya, K., Sato, F., and Yokoo, N., 1985, Promoting effect of the new chymotrypsin inhibitor FK-448 on the intestinal absorption of insulin in rats and dogs, *J. Pharm. Pharmacol.* **37:**545–549.

Fujita, S., Kamado, K., Takimoto, T., Yoshioka, H., Yamasaki, Y., Shichiri, M., and Hoshi, M., 1983, Insulin suppositories to stabilize brittle diabetes, in: *Current and Future Therapies with Insulin* (N. Sakamoto and K. G. M. M. Alberti, eds.), Excerpta Medica, Amsterdam, pp. 268–271.

Fukushima, K., 1987, Experimental approaches to oral administration of insulin by cinnamoyl phenylalanine derivatives as adjuvants in streptozotocin-diabetic animals, in: *Best Approach to the Ideal Therapy of Diabetes Mellitus* (Y. Shigeta, H. E. Lebovitz, J. E. Gerich, and W. J. Malaisse, eds.), Elsevier, Amsterdam, pp. 367–370.

Fukushima, K., and Toyoshima, S., 1983, Oral administration of insulin in combination with amino acid derivatives in animal experiments, in: *Current and Future Therapies with Insulin,* (N. Sakamoto and K. G. M. M. Alberti, eds.), Excerpa Medica, Amsterdam, pp. 257–262.

Gall, M.-A., Mathiesen, E. R., Skøtt, P., Musaeus, L., Damm, S., Beck-Nielsen, H., and Parving, H.-H., 1989, Effect of multiple insulin injections with a pen injector on metabolic control and general well-being in insulin-dependent diabetes mellitus, *Diabetes Res.* **11:**97–101.

Galloway, J. A., and Root, M. A., 1972, New forms of insulin, *Diabetes* **21**(Suppl. 2)**:**637–648.

Galloway, J. A., Root, M. A., Rathmacher, R. P., and Carmichael, R. H., 1973, A comparison of acid regular and neutral regular insulin: Responses of normal fasted subjects to varying doses of regular insulin, *Diabetes* **22:**471–479.

Galloway, J. A., Spradlin, C. T., Nelson, R. L., Wentworth, S. M., Davidson, J. A., and Swarner, J. L., 1981, Factors influencing the absorption, and blood glucose responses after injections of regular insulin and various insulin mixtures, *Diabetes Care* **4:**366–376.

Gänsslen, M., 1925, Über Inhalation von Insulin, *Klin. Wochenschr.* **4:**71.

Geary, R. S., and Schlameus, H. W., 1993, Vancomycin and insulin used as models for oral delivery of peptides, *J. Controlled Release* **23:**65–74.

Gill, I. J., Fisher, A. N., Hinchcliffe, M., Whetstone, J., Farraj, N., Ponti, R. D., and Illum, L., 1994, Cyclodextrins as protection agents against enhancer damage in nasal delivery systems II. Effect on *in vivo* absorption of insulin and histopathology of nasal membrane, *Eur. J. Pharm. Sci.* **1:**237–248.

Gizurarson, S., and Bechgaard, E., 1991a, Study of nasal enzyme activity towards insulin. *In vitro*, *Chem. Pharm. Bull.* **39:**2155–2157.

Gizurarson, S., and Bechgaard, E., 1991b, Intranasal administration of insulin to humans, *Diabetes Res. Clin. Pract.* **12:**71–84.

Golomb, G., Avramoff, A., and Hoffman, A., 1993, A new route of drug administration: Intrauterine delivery of insulin and calcitonin, *Pharm. Res.* **10:**828–833.

Goosen, M. F. A., Leung, Y. F., Chou, S., and Sun, A. M., 1982, Insulin-albumin microbeads: An implantable, biodegradable system, *Biomater. Med. Devices Artif. Organs* **10:**205–218.

Goosen, M. F. A., Leung, Y. F., O'Shea, G. M., Chou, S., and Sun, A. M., 1983, Long-acting insulin. Slow release of insulin from a biodegradable matrix implanted in diabetic rats, *Diabetes* **32:**478–481.

Gordon, G. S., Moses, A. C., Silver, R. D., Flier, J. S., and Carey, M. C., 1985, Nasal absorption of insulin: Enhancement by hydrophobic bile salts, *Proc. Natl. Acad. Sci. USA* **82:**7419–7423.

Grau, U., 1985, Chemical stability of insulin in a delivery system environment, *Diabetologia* **28:**458–463.

Grau, U., Seipke, G., Obermeier, R., and Thurow, H., 1982, Stabile Insulinlösungen für automatische Dosiergeräte, in: *Neue Insuline. 1. Internationales Symposium* (K.-G. Petersen, K. J. Schlüter, and L. Kerp, eds.), Freiburg Graphischer Betriebe, Freiburg, pp. 411–419.

Grau, U., Geisen, K., and Jährling, P., 1989, Preclinical evaluation of a remote-controlled implantable pump with a compatible insulin preparation: Studies on long-term stability of the insulin, *Diabetes Nutr. Metab.* **2:**43–52.

Gwinup, G., Elias, A. N., and Domurat, E. S., 1991, Insulin and c-peptide levels following oral administration of insulin in intestinal-enzyme protected capsules, *Gen. Pharmacol.* **22:**243–246.

Hagedorn, H. C., Jensen, B. N., Krarup, N. B., and Wodstrup, I., 1936, Protamine insulinate, *J. Am. Med. Assoc.* **106:**177–180.

Hallas-Møller, K., 1956, The 'Lente' insulins, *Diabetes* **5:**7–14.

Hallas-Møller, K., Petersen, K., and Schlichtkrull, J., 1952, Crystalline and amorphous insulin–zinc compounds with prolonged action, *Science* **116:**394–399.

Hansen, P. E., Brange, J., and Havelund, S., 1986, Stabilized insulin preparations and a process for preparation thereof, U.S. Patent 4,614,730.

Harris, D., and Robinson, J. R., 1990, Bioadhesive polymers in peptide drug delivery, *Biomaterials* **11:**652–658.

Harris, D., Liaw, J.-H., and Robinson, J. R., 1992, Ocular delivery of peptide and protein drugs, *Adv. Drug Delivery Rev.* **8:**331–339.

Hastings, R. H., Grady, M., Sakuma, T., and Matthay, M. A., 1992, Clearance of different-sized proteins from the alveolar space in humans and rabbits, *J. Appl. Physiol.* **73:**1310–1316.

Hauss, D. J., Ando, H. Y., Grasela, D. M., and Sugita, E. T., 1991, The relationship of salicylate lipophilicity to rectal insulin absorption enhancement and relative lymphatic uptake, *J. Pharmacobio-Dyn.* **14:**139–151.

Hayakawa, E., Chien, D.-S., Inagaki, K., Yamamoto, A., Wang, W., and Lee, V. H. L., 1992, Conjunctival penetration of insulin and peptide drugs in the albino rabbit, *Pharm. Res.* **9:**769–775.

Heine, R. J., 1993, Insulin therapy, in: *Diabetes Annual/7* (S. Marshall, P. D. Home, K. G. M. M. Alberti, and L. P. Krall, eds.), Elsevier, Amsterdam, pp. 284–302.

Heine, R. J., Bilo, H. J. G., Fonk, E. A., Veen, E. R., and Meer, J., 1984, Absorption kinetics and action profiles of mixtures of short- and intermediate-acting insulins, *Diabetologia* **27:**558–562.

Heinemann, L., Starke, A. A. R., Heding, L., Jensen, I., and Berger, M., 1990, Action profile of fast onset insulin analogues, *Diabetologia* **33:**384–386.

Heinemann, L., Chantelau, E. A., and Starke, A. A. R., 1992, Pharmacokinetics and pharmacodynamics of subcutaneously administered U40 and U100 formulations of regular human insulin, *Diabetes Metab.* **18:**21–24.

Heinemann, L., Heise, T., Ampudia, J., and Starke, A. A. R., 1993a, Action profiles of rapid-acting human insulin analogues, in: *Frontiers in Insulin Pharmacology* (M. Berger and F. A. Gries, eds.), Thieme Verlag, Stuttgart, pp. 87–96.

Heinemann, L., Heise, T., Jorgensen, L. N, and Starke, A. A. R., 1993b, Action profile of the rapid acting insulin analogue: Human insulin B28Asp, *Diabetic Med.* **10:**535–539.

Heller, J., 1993a, Modulated release from drug delivery devices, *Crit. Rev. Ther. Drug Carrier Syst.* **10:**253–305.

Heller, J., 1993b, Polymers for controlled parenteral delivery of peptides and proteins, *Adv. Drug. Delivery Rev.* **10:**163–204.

Heller, J., Chang, A. C., Rodd, G., and Grodsky, G. M., 1990, Release of insulin from pH-sensitive poly(orthoesters), *J. Controlled Release* **13:**295–302.

Henriksen, J. E., Djurhuus, M. S., Vaag, A., Thye-Rønn, P., Knudsen, D., Hother-Nielsen, O., and Beck-Nielsen, H., 1993, Impact of injection sites of soluble insulin on glycaemic control in Type 1 (insulin dependent) diabetic patients treated with a multiple insulin injection regimen, *Diabetologia* **36:**752–758.

Hildebrandt, P., Sestoft, L., and Nielson, S. L., 1983, The absorption of subcutaneously injected short-acting soluble insulin: Influence of injection technique and concentration, *Diabetes Care* **6:**459–462.

Hildebrandt, P., Birch, K., Sestoft, L., and Vølund, A., 1984a, Dose-dependent subcutaneous absorption of porcine, bovine and human NPH insulins, *Acta Med. Scand.* **215:**69–73.

Hildebrandt, P., Ilius, A., Lotz, U., and Schiliack, V., 1984b, Effect of insulin suppositories in type I diabetic patients, *Exp. Clin. Endocrinol.* **83:**168–172.

Hildebrandt, P., Berger, A., Vølund, A., and Kühl, C., 1985a, The subcutaneous absorption of human and bovine ultralente insulin formulations, *Diabetic Med.* **2:**355–359.

Hildebrandt, P., Birch, K., Sestoft, L., and Nielsen, S. L., 1985b, Orthostatic changes in subcutaneous blood flow and insulin absorption, *Diabetes Res.* **2:**187–190.

Hilsted, J., Madsbad, S., Hvidberg, A., Rasmussen, M. H., Krarup, T., Ipsen, H., Hansen, B., Pedersen, M., Djurup, R., and Oxenbøll, B., 1995, Intranasal insulin therapy: The clinical realities, *Diabetologia* **38:**680–684.

Hirai, S., Ikenaga, T., and Matzuzawa, T., 1978, Nasal absorption of insulin in dogs, *Diabetes* **27:**296–299.

Hirai, S., Yashiki, T., and Mima, H., 1981a, Effect of surfactants on the nasal absorption of insulin in rats, *Int. J. Pharm.* **9:**165–172.

Hirai, S., Yashiki, T., and Mima, H., 1981b, Mechanism for the enhancement of the nasal absorption of insulin by surfactants, *Int. J. Pharm.* **9:**173–184.

Hirata, Y., Kohama, T., and Ooi, K., 1983, Nasal administration of insulin in healthy subjects and diabetic patients, in: *Current and Future Therapies with Insulin* (N. Sakamoto and K. G. M. M. Alberti, eds.), Excerpa Medica, Amsterdam, pp. 263–267.

Hirsch, I. B., Farkas-Hirsch, R., and Skyler, J. S., 1990, Intensive insulin therapy for treatment of type I diabetes, *Diabetes Care* **13:**1265–1283.

Hirsch, J. I., Wood, J. H., and Thomas, R. B., 1981, Insulin adsorption to polyolefin infusion bottles and polyvinyl chloride administration sets, *Am. J. Hosp. Pharm.* **38:**995–997.

Ho, N. F. H., Barsuhn, C. L., Burton, P. S., and Merkle, H. P., 1992, Mechanistic insight to buccal delivery of proteinaceous substances, *Adv. Drug Delivery Rev.* **8:**197–235.

Hoffmann, J. A., Chance, R. E., and Johnson, M. G., 1990, Purification and analysis of the major components of chum salmon protamine contained in insulin formulations using high-performance liquid chromatography, *Protein Expr. Purif.* **1:**127–133.

Holman, R. R., 1993, Intranasal insulin in type I diabetes, in: *Frontiers in Insulin Pharmacology* (M. Berger and F. A. Gries, eds.), Thieme Verlag, Stuttgart, pp. 138–143.

Home, P. D., and Alberti, K. G. M. M., 1992, Insulin therapy, in: *International Textbook of Diabetes Mellitus* (K. G. M. M. Alberti, R. A. DeFronzo, H. Keen, and P. Zimmet, eds.), John Wiley & Sons, Chichester, U.K., pp. 831–863.

Home, P. D., Gill, G. V., Husband, D., Kruszynska, Y. T., Marshall, S., and Alberti, K. G. M. M., 1985, Alternative routes and methods of insulin delivery, *Neth. J. Med.* **28:**(Suppl. 1)**:**32–36.

Home, P. D., Thow, J. C., and Tunbridge, F. K. E., 1989, Insulin treatment: A Decade of change, *Br. Med. Bull.* **45:**92–110.

Hoogdalem, E. J., Heijligers-Feijen, C. D., Verhoef, J. C., de Boer, A. G., and Breimer, D. D., 1990, Absorption enhancement of rectally infused insulin by sodium tauro-24,25-dihydrofusidate (STDHF) in rats, *Pharm. Res.* **7:**180–183.

Horbett, T. A., Ratner, B. D., Kost, J., and Singh, M., 1984, A bioresponsive membrane for insulin delivery, in: *Recent Advances in Drug Delivery Systems* (J. M. Andersen and S. W. Kim, eds.), Plenum Press, New York, pp. 209–220.

Hori, R., Komadea, F., and Okumura, K., 1983, Pharmaceutical approach to subcutaneous dosage forms of insulin, *J. Pharm. Sci.* **72:**435–439.

Hori, R., Komada, F., Iwakawa, S., Seino, Y., and Okumura, K., 1989, Enhanced bioavailability of subcutaneously injected insulin coadministered with collagen in rats and humans, *Pharm. Res.* **6:**813–816.

Hosny, E. A., Bayomi, M., El-Shattawy, H. H., El-Kabany, N., El-Deen, S. G. El-Damasy, H., and Nabih, A. E., 1994a, Absorption promoters for delivery of insulin from suppositories in humans, *Pharm. Ind.* **56:**576–578.

Hosny, E., El-Ahmady, O., El-Shattawy, H., Nabih, A. El-M., El-Damacy, H., Gamal-El-Deen, S., and El-Kabbany, N., 1994b, Effect of sodium salicylate on insulin rectal absorption in humans, *Arzneim.-Forsch./Drug Res.* **44**(I)**:**611–613.

Houtzagers, C. M. G. J., 1989, Subcutaneous insulin delivery: Present status, *Diabetic Med.* **6:**754–761.

Houtzagers, C. M. G. J., Berntzen, P. A., Stap, H., Heine, R. J., and Veen, E. A., 1988, Absorption kinetics of short- and intermediate-acting insulins after jet injection with Medijector II, *Diabetes Care* **11:**739–742.

Howell, S. L., Tyhurst, M., Duvefelt, H., Anderson, A., and Hellerström, C., 1978, Role of zinc and calcium in the formation and storage of insulin in the pancreatic B-cell, *Cell Tissue Res.* **188:**107–118.

Howey, D. C., Bowsher, R. R., Brunelle, R. L., and Woodworth, J. R., 1994, [Lys(B28), Pro(B29)]-human insulin. A rapidly absorbed analogue of human insulin, *Diabetes* **43:**396–402.

Ichikawa, K., Ohata, I., Mitomi, M., Kawamura, S., Maeno, H., and Kawata, H., 1980, Rectal absorption of insulin suppositories in rabbits, *J. Pharm. Pharmacol.* **32:**314–318.

Illum, L., Farraj, N. F., Critchley, H, Johansen, B. R., and Davis, S. S., 1989, Enhanced nasal absorption of insulin in rats using lysophosphatidylcholine, *Int. J. Pharm.* **57:**49–54.

Illum, L., Farraj, N. F., and Davies, S. S., 1994a, Chitosan as a novel nasal delivery system for peptide drugs, *Pharm. Res.* **11:**1186–1189.

Illum, L., Farraj, N. F., Fisher, A. M., Gill, I, Miglietta, M., and Benedetti, L. M., 1994b, Hyaluronic acid ester microspheres as a nasal delivery system for insulin, *J. Controlled Release* **29:**133–141.

Irie, T., Wakamatsu, K., Arima, H., Aritomi, H., and Uekama, K., 1992, Enhancing effects of cyclodextrins on nasal absorption of insulin in rats, *Int. J. Pharm.* **84:**129–139.

Irsigler, K, and Kritz, H., 1979, Long-term continuous intravenous insulin therapy with a portable insulin dosage-regulating apparatus, *Diabetes* **28:**196–203.

Irsigler, K., Kritz, H., Hagmüller, G., Franetzki, M, Prestele, K., Thurow, H., and Geisen, K., 1981, Long-term continuous intraperitoneal insulin infusion with an implanted remote-controlled insulin infusion device, *Diabetes* **30:**1072–1075.

Ishida, M., Machida, Y., Nambu, N., and Nagai, T., 1981, New mucosal dosage form of insulin, *Chem. Pharm. Bull.* **29:**810–816.

Ishihara, K., and Matsui, K., 1986, Glucose-responsive insulin release from polymer capsule, *J. Polym. Sci.* **24:**413–417.

Ishihara, K., Kobayashi, M., Ishimaru, N., and Shinohara, I., 1984, Glucose induced permeation control of insulin through a complex membrane consisting of immobilized glucose oxidase and a poly(amine), *Polym. J.* **16:**625–631.

Ito, Y., Casolaro, M., Kono, K., and Imanishi, Y., 1989, An insulin system that is responsive to glucose, *J. Controlled Release* **10:**195–203.

Jacobs, M. A. J. M., Schreuder, R. H., Jap-a-joe, K., Nauta, J. J., Andersen, P. M., and Heine, R. J., 1993, The pharmacodynamics and activity of intranasally administered insulin in healthy male volunteers, *Diabetes* **42:**1649–1655.

Jefferson, I. G., Marteau, T. M., Smith, M. A., and Baum, J. D., 1985, A multiple injection regimen using an insulin injection pen and pre-filled cartridged soluble human insulin in adolescents with diabetes, *Diabetic Med.* **2:**493–497.

Jensen, H. F., 1938, *Insulin. Its Chemistry and Physiology*, Commonwealth Foundation, New York, pp. 93–112.

Jones, A. L., Taylor, G., and Kellaway, I. W., 1987, An investigation of the pulmonary absorption of insulin in the rat, in: *Proceedings of the Third European Congress of Biopharmaceutics and Pharmacokinetics in Freiburg—Volume II. Experimental Pharmacokinetics* (J. M. Aiache and J. Hirtz, eds.), pp. 143–149.

Jorpes, J. E., 1949, Recrystallized insulin for diabetic patients with insulin allergy, *Arch. Intern. Med.* **833:**363–371.

Jovanovic-Peterson, L., Sparks, S., Palmer, J. P., and Peterson, C. M., 1993, Jet-injected insulin is associated with decreased antibody production and postprandial glucose variability when compared with needle-injected insulin in gestational diabetic women, *Diabetes Care* **16:**1479–1484.

Kahn, M., Barney, N. P., Briggs, R. M., Bloch, K. J., and Allansmith, M. R., 1990, Penetrating the conjunctival barrier. The role of molecular weight, *Invest. Ophthalmol. Visual Sci.* **31:**258–261.

Kang, S., Owens, D. R., Vora, J. P., and Brange, J., 1990, Recombinant DNA derived insulin analogue B9AspB27Glu: Comparisons with soluble human insulin in insulin treated diabetics, *Lancet* **335:**303–306.

Kang, S., Creag, F. M., Peters, J. R., Brange, J., Vølund, A., and Owens, D. R., 1991a,

Comparison of subcutaneous soluble human insulin and insulin analogues (AspB9, GluB27; AspB10; AspB28) on meal-related plasma glucose excursions in type I diabetic subjects, *Diabetes Care* **14**:571–577.

Kang, S., Brange, J., Burch, A., Vølund, A., and Owens. D. R., 1991b, Subcutaneous insulin absorption explained by insulin's physicochemical properties. Evidence from absorption studies of soluble human insulin and insulin analogues in humans, *Diabetes Care* **14**:942–948.

Kang, S., Brange, J., Burch, A., Vølund, A., and Owens, D. R., 1991c, Absorption kinetics and action profiles of subcutaneously administered insulin analogue (AspB9GluB27, AspB10, AspB28) in healthy subjects, *Diabetes Care* **14**:1057–1965.

Kari, B., 1986, Control of blood glucose levels in alloxan diabetic rabbits by iontophoresis of insulin, *Diabetes* **35**:217–221.

Kemmer, F. W., Sonnenberg, G., Cüppers, H. J., and Berger, M., 1982, Absorption kinetics of semisynthetic human insulin and biosynthetic (recombinant DNA) human insulin, *Diabetes Care* **5**:23–28.

Kidron, M., Bar-on, H., Berry, E. M., and Ziv, E., 1982, The absorption of insulin from various regions of the rat intestine, *Life Sci.* **31**:2837–2841.

Kidron, M., Krausz, M. M., Raz, I., Bar-on, H., and Ziv, E., 1989, The absorption of insulin. The absorption of insulin from the intestine in dogs, *Tenside Surf. Deterg.* **26**:352–354.

Kim, S., Nishihata, T., Kawabe, S., Okamura, Y., Kamada, A., Yagi, T., Kawamori, R., and Shichiri, M., 1984, Effect of gelatin in the formulation on rectal insulin absorption in the presence of enamine in normal rats and depancreatized dogs, *Int. J. Pharm.* **21**:179–186.

Kim, S. W., Pai, C. M., Makino, K., Seminoff, L. A. Holmberg, D. L., Gleeson, J. M., Wilson, D. E., and Mack, E. J., 1990, Self-regulated glycosylated insulin delivery, *J. Controlled Release* **11**:193–201.

Kimmerle, R., Griffing, G., McCall, A., Ruderman, N. B., Stoltz, E., and Melby, J. C., 1991, Could intranasal insulin be useful in the treatment of non-insulin-dependent diabetes mellitus? *Diabetes Res. Clin. Pract.* **13**:69–76.

Kitano, S., Koyama, Y., Kataoka, K., Okano, T., and Sakurai, Y., 1992, A novel drug delivery system utilizing a glucose responsive polymer complex between poly(vinyl alcohol) and poly(*N*-vinyl-2-pyrrolidone) with a phenylboronic acid moiety, *J. Controlled Release* **19**:161–170.

Klemp, P., Staberg, B., Madsbad, S., and Kølendorf, K., 1982, Smoking reduces insulin absorption from subcutaneous tissue, *Br. Med. J.* **284**:237.

Köhler, D., Schlüter, K. J., Kerp, L., and Matthys, H., 1987, Nicht radioaktives Verfahren zur Messung der Lungenpermeabilität: Inhalation von Insulin, *Atemwegs.-Lungenkrankh.* **13**:230–232.

Koivisto, V. A., 1993, Insulin therapy in type II diabetes, *Diabetes Care* **16**(Suppl. 3):29–39.

Koivisto, V. A., and Felig, P., 1978, Effect of leg exercise on insulin absorption in diabetic patients, *N. Engl. J. Med.* **298**:79–83.

Koivisto, V. A., and Felig, P., 1980, Alterations in insulin absorption and in blood glucose control associated with varying insulin injection sites in diabetic patients, *Ann. Intern. Med.* **92**:59–61.

Koivisto, V. A., Fortney, S., Hendler, R., and Felig, P., 1981, A rise in ambient temperature augments insulin absorption in diabetic patients, *Metabolism* **30**:402–405.

Kost, J., 1993, Ultrasound induced delivery of peptides, *J. Controlled Release* **24**:247–255.

Kost, J., and Langer, R., 1991, Responsive polymeric delivery systems, *Adv. Drug Delivery Rev.* **6**:19–50.

Kost, J., Wolfrum, J., and Langer, R., 1987, Magnetically enhanced insulin release in diabetic rats, *J. Biomed. Mater. Res.* **21**:1367–1373.

Krayenbühl, C., and Rosenberg, T., 1946, Crystalline protamine insulin, *Rep. Steno Mem. Hosp. Nord. Insulinlab.* **1**:60–73.

Kurtzhals, P., and Ribel, U., 1995, Action profile of cobalt(III)-insulin. A novel principle of protraction of potential use for basal insulin delivery, *Diabetes* **44:**1381–1385.

Kurtzhals, P., Kiehr, B., and Sørensen, A. R., 1995a, The cobalt(III)-insulin hexamer is a prolonged-acting insulin prodrug, *J. Pharm. Sci.* **84:**1164–1168.

Kurtzhals, P., Havelund, S., Jonassen, I., Kiehr, B., Larsen, U. D., Ribel, U., and Markussen, J., 1995b, Albumin binding of insulins acylated with fatty acids: Characterization of the ligand–protein interaction and correlation between affinity and timing of the insulin effect *in vivo*, *Biochem. J.* **312:**725–731.

Kwon, I. C., Bae, Y. H., and Kim, S. W., 1991, Electrically erodible polymer gel for controlled release of drugs, *Nature* **354:**291–293.

Kwong, A. K., Chou, S., Sun, A. M., Sefton, M. V., and Goosen, M. F. A., 1986, *In vitro* and *in vivo* release of insulin from poly(lactic acid) microbeads and pellets, *J. Controlled Release* **4:**47–62.

Langkjær, L., Kornfeldt, T., Markussen, J, Norris, K., Plum, A., Sørensen, A. R., and Sørensen, E., 1988, Stable insulin solutions with prolonged action, *Diabetologia* **31:**512A.

Langkjær, L., Brange, J., Grodsky, G. M., and Guy, R. H., 1994, Transdermal delivery of monomeric insulin analogues by iontophoresis, *Proc. Int. Symp. Control. Rel. Bioact. Mater.* **21:**172–173.

Lassmann-Vague, V., 1991, The intranasal route for insulin administration, in: *Biotechnology of Insulin Therapy* (J. C. Pickup, ed.), Blackwell Scientific, Oxford, pp. 113–125.

Lassmann-Vague, V., Thiers, D., Vialettes, B., and Vague, P., 1988, Preprandial intranasal insulin, *Lancet* **i:**367–368.

Lassmann-Vague, V., Belicar, P., Raccah, D., Vialettes, B., Dodoyez, J. C., and Vague, P., 1995, Immunogenicity of long-term intraperitoneal insulin administration with implantable programmable pumps, *Diabetes Care* **18:**498–503.

Laube, B. L., Georgopoulos, A., and Adams, G. K., 1993, Preliminary study of the efficacy of insulin aerosol delivered by oral inhalation in diabetic patients, *J. Am. Med. Assoc.* **269:**2106–2109.

Lauritzen, T., Pramming, S., Deckert, T., and Binder, C., 1983, Pharmacokinetics of continuous subcutaneous insulin infusion, *Diabetologia* **24:**326–329.

Lee, S., and Sciarra, J. J., 1976, Development of an aerosol dosage form containing insulin, *J. Pharm. Sci.* **65:**567–572.

Lee, V. H. L., 1986, Enzymatic barriers to peptide and protein absorption and the use of penetration enhancers to modify absorption, in: *Delivery Systems for Peptide Drugs* (S. S. Davis, L. Illum, and E. Tomlinson, eds.), Plenum Press, New York, pp. 87–104.

Lee, V. H. L., Kashi, S. D., Patel, R. M., Kayakawa, E., and Inagaki, K., 1987, Mucosal peptide and protein delivery: Proteolytic activities in mucosal homogenates, *Proc. Int. Symp. Control. Rel. Bioact. Mater.* **14:**23–24.

Lee, V. H. L., Traver, R. D., and Taub, M. E., 1991a, Enzymatic barriers to peptide and protein drug delivery, in: *Peptide and Protein Drug Delivery* (V. H. L. Lee, ed.), Marcel Dekker, New York, pp. 303–358.

Lee, V. H. L., Dodda-Kashi, S., Grass, G. M., and Rubas, W., 1991b, Oral route of peptide and protein delivery, in: *Peptide and Protein Drug Delivery* (V. H. L. Lee, ed.), Marcel Dekker, New York, pp. 691–738.

Lee, W. A., Lu, H. F.-L., Maffuid, P. W., Botet, M. T., Baldwin, P. A., Benkert, T. A., and Klingbeil, C. K., 1992, The synthesis, characterization and biological testing of a novel class of mucosal permeation enhancers, *J. Controlled Release* **22:**223–237.

Levandoski, L. A., White, N. H., and Santiago, J. V., 1982, Localized skin reactions to insulin: Insulin lipodystrophies and skin reactions to pumped subcutaneous insulin therapy, *Diabetes Care* **5**(Suppl. 1)**:**6–10.

Li, Y., and Mitra, A. K., 1994, Effects of formulation variables on the pulmonary delivery on insulin, *Drug Dev. Ind. Pharm.* **20:**2017–2024.

Li, Y., Shao, Z., and Mitra, A. K., 1992, Dissociation of insulin oligomers by bile salt micelles and its effects on α-chymotrypsin-mediated proteolytic degradation, *Pharm. Res.* **9:**864–869.

Li, Y., Shao, Z., DeNicola, D. B., and Mitra, A. K., 1993, Effects of a conjugated bile salt on the pulmonary absorption of insulin in rats, *Eur. J. Pharm. Biopharm.* **39:**216–221.

Liedtke, R. K., Suwelack, K., and Karzel, K., 1990, Wirkung peroraler und transdermaler Insulin-Präparationen auf die Blutglukose-Konzentration bei Mäusen, *Arzneim.-Forsch./Drug Res.* **40:**880–883.

Linde, B., 1986, Dissociation of insulin absorption and blood flow during massage of a subcutaneous injection site, *Diabetes Care* **9:**570–574.

Linde, B., and Gunnarsson, R., 1985, Influence of aprotinin on insulin absorption and subcutaneous blood flow in type 1 (insulin-dependent) diabetes, *Diabetologia* **28:**645–648.

Linde, B., and Philip, A., 1989, Massage-enhanced insulin absorption—increased distribution or dissociation of insulin? *Diabetes Res.* **11:**191–194.

Liu, F., Kildsig, D. O., and Mitra, A. K., 1992, Pulmonary biotransformation of insulin in rat and rabbit, *Life Sci.* **51:**1683–1689.

Liu, F., Shao, Z., Kildsig, D. O., and Mitra, A. K., 1993, Pulmonary delivery of free and liposomal insulin, *Pharm. Res.* **10:**228–232.

Liu, J.-C., Sun, Y., Siddiqui, O., Chien, Y. W., Shi, W. M., and Li, J., 1988, Blood glucose control in diabetic rats by transdermal iontophoretic delivery of insulin, *Int. J. Pharm.* **44:**197–204.

Liversidge, G. G., Nishihata, T., Engle, K. K., and Higuchi, T., 1985, Effect of rectal suppository formulation on the release of insulin and on the glucose plasma levels in dogs, *Int. J. Pharm.* **23:**87–95.

Liversidge, G. G., Nishihata, T. Engle, K. K., and Higuchi, T., 1986, Effect of suppository shape on the systemic availability of rectally administered insulin and sodium salicylate, *Int. J. Pharm.* **30:**247–250.

Longenecker, J. P., Moses, A. C., Flier, J. S., Silver, R. D., Carey, M. C., and Dubovi, E. J., 1987, Effects of sodium taurodihydrofusidate on nasal absorption of insulin in sheep, *J. Pharm. Sci.* **76:**351–355.

Losse, G., Müller, F., Raddatz, H., Naumann, W., and Kossowicz, J., 1988, Verzögerungsformen des Insulins durch Matrixeinschluss oder Komplexbildung, *Pharmazie* **43:**355–357.

Losse, G., Müller, F., Fischer, S., Wunderlich, G., and Hacker, E., 1989, Szintigraphische Verfolgung der retardierten Insulinliberation aus Agar-Einbettungen in Suppositorien, *Pharmazie* **44:**331–332.

Lougheed, W. D., Woulfe-Flanagan, H., Clement, J. R., and Albisser, A. M., 1980, Insulin aggregation in artificial delivery systems, *Diabetologia* **19:**1–9.

Lougheed, W. D., Albisser, A. M., Martindale, H. M., Chow, J. C., and Clement, J. R., 1983, Physical stability of insulin formulations, *Diabetes* **32:**424–432.

Luttermann, J. A., Pijpers, E., Netten, P. M., and Jørgensen, L. N., 1993, Glycaemic control in IDDM patients during one day with injection of human insulin or the insulin analogues insulin X14 and insulin X14(+Zn), in: *Frontiers in Insulin Pharmacology* (M. Berger and F. A. Gries, eds.), Thieme Verlag, Stuttgart, pp. 102–109.

Maberly, G. F., Wait, G. A., Kilpatric, J. A., Loten, E. G., Gain, K. R., Stewart, R. D. H., and Eastman, C. J., 1982, Evidence for insulin degradation by muscle and fat issue in an insulin resistant diabetic patient, *Diabetologia* **23:**333–336.

Maitani, Y., Asano, S., Takahashi, S., Nakagaki, M., and Nagai, T., 1992, Permeability of insulin entrapped in liposomes through the nasal mucosa of rabbits, *Chem. Pharm. Bull.* **40:**1569–1572.

Maitani, Y., Yamamoto, T., Takayama, K., and Nagai, T., 1995, The effect of soybeam-derived sterol and its glycoside as an enhancer of nasal absorption of insulin in rabbits *in vitro* and *in vivo*, *Int. J. Pharm.* **117:**129–137.

Malone, J. I., Lowitt, S., Grove, N. P., and Shah, S. C., 1986, Comparison of insulin levels after injection by jet stream and disposable insulin syringe, *Diabetes Care* **9:**637–640.

Manosroi, A., and Bauer, K. H., 1990, Effects of gastrointestinal administration of human insulin and human insulin–DEAE–dextran complex entrapped in different compound liposomes on blood glucose in rats, *Drug Dev. Ind. Pharm.* **16:**1521–1538.

Markussen, J., Diers, I., Hougaard, P., Langkjaer, L., Norris, K, Snel, L., Sørensen, A. R., Sørensen, E., and Voigt, H. O., 1988, Soluble, prolonged-acting insulin derivatives. III. Degree of protraction, crystallizability and chemical stability of insulins substituted in position A21, B13, B23, B27 and B30, *Protein Eng.,* **2:**157–166.

Markussen, J., Havelund, S., Kurtzhals, P., Andersen, A. S., Halstrøm, J., Hasselager, E., Larsen, U. D., Ribel, U., Schäffer, L., Vad, K., and Jonassen, I., 1996, Soluble, fatty acid acylated insulins bind to albumin and show protracted action in pigs, *Diabetologia* **39:**281–288.

Marshall, M. O., Heding, L. G., Villumsen, J., Åkerblom, H. K., Baevre, H., Dahlquist, G., Kjaergaard, J.-J., Knip, M., Lindgren, F., Ludvigsson, J., Persson, B., Rilva, A., Stenhammar, L., Strömberg, L., Søvik, O., Thalme, B., Vidnes, J., and Wefring, K., 1988, Development of insulin antibodies, metabolic control and B-cell function in newly diagnosed insulin dependent diabetic children treated with monocomponent human or monocomponent porcine insulin, *Diabetes Res.* **9:**169–175.

Marshall, S. M., Home, P. D., Taylor, R., and Alberti, K. G. M. M., 1987, Continuous subcutaneous insulin infusion versus injection therapy: A randomised cross-over trial under usual diabetic clinic conditions, *Diabetic Med.* **4:**521–525.

Mathiowitz, E., and Langer, R., 1987, Polyanhydride microspheres as drug carriers. I. Hot-melt microencapsulation, *J. Controlled Release* **5:**13–22.

Mathiowitz, E., Saltzman, W. M., Domb, A., Dor, P., and Langer, R. 1988, Polyanhydride microspheres as drug carriers. II. Microencapsulation by solvent removal, *J. Appl. Polym. Sci.* **35:**755–774.

Mecklenburg, R. S., and Guinn, T. S., 1985, Complications of insulin pump therapy: The effect of insulin preparation, *Diabetes Care* **8:**367–370.

Medan, V. M., Walmsley, A. D., and Irwin, W. J., 1995, Phonophoresis—is it a reality? *Int. J. Pharm.* **118:**129–149.

Melberg, S. G., Havelund, S., Villumsen, J., and Brange, J., 1988, Insulin compatibility with polymer materials used in external pump infusion systems, *Diabetic Med.* **5:**243–247.

Merkle, H. P., and Wolany, G., 1992, Buccal delivery for peptide drugs, *J. Controlled Release* **21:**155–164.

Merkus, F. W. H. M., Verhoef, J. C., Romeijn, S. G., and Schipper, N. G. M., 1991a, Absorption enhancing effect of cyclodextrins on intranasally administered insulin in rats, *Pharm. Res.* **8:**588–592.

Merkus, F. W. H. M., Verhoef, J. C., Romeijn, S. G., and Schipper, N. G. M., 1991b, Interspecies differences in the nasal absorption of insulin, *Pharm. Res.* **8:**1343.

Merkus, F. W. H. M., Schipper, N. G. M., Hermens, W. A. J. J., Romeijn, S. G., and Verhoef, J. C., 1993, Absorption enhancers in nasal drug delivery: Efficacy and safety, *J. Controlled Release* **24:**201–208.

Mesiha, M. S., 1981, Bioavailability of insulin from oral solid dispersion systems, *Arch. Pharm. Chemi., Sci. Ed.* **9:**137–142.

Mesiha, M. S., and El-Bitar, H. I., 1981, Hypoglycaemic effect of oral insulin preparations containing Brij 35, 52, 58, or 92 and stearic acid, *J. Pharm. Pharmacol,* **33:**733–734.

Mesiha, M. S., Lobel, S., Salo, D. P., Khaleeva, L. D., and Zekova, N. Y., 1981, Biopharmaceutical study of insulin suppositories, *Pharmazie* **36:**29–32.

Meyer, B. R., Katzeff, H. L., Eschbach, J. C., Trimmer, J., Zacharias, S. B., Rosen, S., and Sibalis, D., 1989, Transdermal delivery of human insulin to albino rabbits using electrical current, *Am. J. Med. Sci.* **297:**321–325.

Michel, C., Aprahamian, M., Defontaine, L., Couvreur, P., and Damgé, C., 1991, The effect of site of administration in the gastrointestinal tract on the absorption of insulin from nanocapsules in diabetic rats, *J. Pharm. Pharmacol.* **43:**1–5.

Mirsky, I. A., and Kawamura, K., 1966, Heterogeneity of crystalline insulin, *Endocrinology* **78:**1115–1119.

Mishima, M., Wakita, Y., and Nakano, M., 1987, Studies on the promoting effects of medium chain fatty acid salts on the nasal absorption of insulin in rats, *J. Pharmacobio-Dyn.* **10:**624–631.

Mishima, M., Okada, S., Wakita, Y., and Nakano, M., 1989, Promotion of nasal absorption of insulin by glycyrrhetinic acid derivatives, *J. Pharmacobio-Dyn.* **12:**31–36.

Mitragotri, S., Blankschtein, D., and Langer, R., 1995, Ultrasound-mediated transdermal protein delivery, *Science* **269:**850–853.

Mitrano, F. P., and Newton, D. W., 1982, Factors affecting insulin adherence to type I glass bottles, *Am. J. Hosp. Pharm.* **39:**1491–1495.

Miyazaki, S., Yokouchi, C., and Takada, M., 1988, External control of drug release: Controlled release of insulin from a hydrophilic polymer implant by ultrasound irradiation in diabetic rats, *J. Pharm. Pharmacol.* **40:**716–717.

Morimoto, K., Hama, I., Nakamoto, Y., Takeeda, T., Hirano, E., and Morisaka, K., 1980, Pharmaceutical studies of polyacrylic acid aqueous gel bases: Absorption of insulin from polyacrylic acid aqueous gel bases following rectal administration in alloxan diabetic rats and rabbits, *J. Pharmacobio-Dyn.* **3:**24–32.

Morimoto, K., Takeeda, T., Nakamoto, Y., and Morisaka, K., 1982, Effective vaginal absorption of insulin in diabetic rats and rabbits using polyacrylic acid aqueous gel bases, *Int. J. Pharm.* **12:**107–111.

Morimoto, K., Kamiya, E., Takeeda, T., Nakamoto, Y., and Morisaka, K., 1983, Enhancement of rectal absorption of insulin in polyacrylic acid aqueous gel bases containing long chain fatty acid in rats, *Int. J. Pharm.* **14:**149–157.

Morimoto, K., Morisaka, K., and Kamada, A., 1985, Enhancement of nasal absorption of insulin and calcitonin using polyacrylic acid gel, *J. Pharm. Pharmacol.* **37:**134–136.

Morishita, M., Morishita, I., Takayama, K., Machida, Y., and Nagai, T., 1992a, Novel oral microspheres of insulin with protease inhibitor protecting from enzymatic degradation, *Int. J. Pharm.* **78:**1–7.

Morishita, I., Morishita, M., Takayama, K., Machida, Y., and Nagai, T., 1992b, Hypoglycemic effect of novel oral microspheres of insulin with protease inhibitor in normal and diabetic rats, *Int. J. Pharm.* **78:**9–16.

Morishita, I., Morishita, M., Takayama, K., Machida, Y., and Nagai, T., 1993a, Enteral insulin delivery by microspheres in 3 different formulations using Eudragit L100 and S100, *Int. J. Pharm.* **91:**29–37.

Morishita, M., Morishita, I., Takayama, K., Machida, Y., and Nagai, T., 1993b, Site-dependent effect of aprotinin, sodium caprate, Na2EDTA and sodium glycocholate on intestinal absorption of insulin, *Biol. Pharm. Bull.* **16:**68–72.

Moses, A. C., Gordon, G. S., Carey, M. C., and Flier, J. S., 1983, Insulin administered intranasally as an insulin-bile salt aerosol. Effectiveness and reproducibility in normal and diabetic subjects, *Diabetes* **32:**1040–1047.

Muzina, D. J., Snow, K. R., and Bajpai, P. K., 1989, Regulation of blood glucose by sustained delivery of insulin via ALCAP ceramics in rats, in: *Biomedical Sciences Instrumentation*, Vol. 25 (D. Carlson, ed.), Instrument Society of America, Research Triangle Park, North Carolina, pp. 191–202.

Nagai, T., 1985, Adhesive topical drug delivery system, *J. Controlled Release* **2:**121–134.

Nagai, T., Nishimoto, Y., Nambu, N., Suzuki, Y., and Sekine, K., 1984, Powder dosage form of insulin for nasal administration, *J. Controlled Release* **1:**15–22.

Nathan, D. M., Siebert, C., and Genuth, S., 1994, DCCT: Design, outcome and implications, *IDF Bull.* **39:**(Suppl.):5–10.

Newman, S. P., Steed, K. P., Hardy, J. G., Wilding, I. R., Hooper, G., and Sparrow, R. A., 1994, The distribution of an intranasal insulin formulation in healthy volunteers: Effect of different administration techniques, *J. Pharm. Pharmacol.* **46:**657–660.

Nishihata, T., Rytting, J. H., Kamada, A., and Higuchi, T., 1981a, Enhanced intestinal absorption of insulin in rats in the presence of sodium 5-methoxysalicylate, *Diabetes* **30:**1065–1067.

Nishihata, T., Rytting, J. H., Higuchi, T., and Caldwell, L., 1981b, Enhanced rectal absorption of insulin and heparin in rats in the presence of non-surfactant adjuvants, *J. Pharm. Pharmacol.* **33:**334–335.

Nishihata, T., Liversidge, G., and Higuchi, T., 1983a, Effect of aprotinin on the rectal delivery of insulin, *J. Pharm. Pharmacol.* **35:**616–617.

Nishihata, T., Kim, S., Morishita, S., Kamada, A., Yata, N., and Higuchi, T., 1983b, Adjuvant effects of glyceryl esters of acetoacetic acid on rectal absorption of insulin and insulin in rabbits, *J. Pharm. Sci.* **72:**280–285.

Nishihata, T., Rytting, J. H., Kamada, A., Higuchi, T., Routh, M., and Caldwell, L., 1983c, Enhancement of rectal absorption of insulin using salicylates in dogs, *J. Pharm. Pharmacol.* **35:**148–151.

Nishihata, T., Okamura, Y., Kamada, A., Higuchi, T., Yagi, T., Kawamori, R., and Shichiri, M., 1985, Enhanced bioavailability of insulin after rectal administration with enamine as adjuvant in depancreatized dogs, *J. Pharm. Pharmacol.* **37:**22–26.

Nishihata, T., Okamura, Y., Inagaki, H., Sudho, M., Kamada, A., Yagi, T., Kawamori, R., and Shichiri, M., 1986, Trials of rectal insulin suppositories in healthy humans, *Int. J. Pharm.* **34:**157–161.

Nishihata, T., Sudoh, M., Inagaki, H., Kamada, A., Yagi, T., Kawamori, R., and Shichiri, M., 1987, An effective formulation for an insulin suppository; examination in normal dogs, *Int. J. Pharm.* **38:**83–90.

Nishihata, T., Kamada, A., Sakai, K., Yagi, T., Kawamori, R., and Shichiri, M., 1989, Effectiveness of insulin suppositories in diabetic patients, *J. Pharm. Pharmacol.* **4:**799–801.

Nolte, M. S., Poon, V., Grodsky, G. M., Forsham, P. H., and Karam, J. H., 1983, Reduced solubility of short-acting soluble insulins when mixed with longer-acting insulins, *Diabetes* **32:**1177–1181.

Nolte, M. S., Taboga, C., Salamon, E., Moses, A., Longenecker, J., Flier, J., and Karam, J. H., 1990, Biological activity of nasally administered insulin in normal subjects, *Horm. Metab. Res. Suppl. Ser.* **22:**170–174.

Nomura, M., Kuboto, M. A., Sekiya, M., Hoshiyama, S., Imano, E., Matushima, Y., Ishimoto, I., Kawamori, R., and Kamada, T., 1990, Insulin absorption from conjunctiva studied in normal and diabetic dogs, *J. Pharm. Pharmacol.* **42:**292–294.

Nomura, M., Kubota, M. A., Kawamori, R., Yamasaki, Y., Kamada, T., and Abe, H., 1994, Effect of addition of hyaluronic acid to highly concentrated insulin on absorption from the conjunctiva in conscious diabetic dogs, *J. Pharm. Pharmacol.* **46:**768–770.

Oh, C. K., and Ritschel, W. A., 1990a, Biopharmaceutic aspects of buccal absorption of insulin, *Methods Find. Exp. Clin. Pharmacol.* **12:**205–212.

Oh, C. K., and Ritschel, W. A., 1990b, Absorption characteristics of insulin through the buccal mucosa, *Methods Find. Exp. Clin. Pharmacol.* **12:**275–279.

Okada, H., Yamazaki, I., Yashiki, T., and Mima, H., 1983a, Vaginal absorption of a potent luteinizing hormone-releasing hormone analogue (leuprolide) in rats II: Mechanism of absorption enhancement with organic acids, *J. Pharm. Sci.* **72:**75–78.

Okada, H., Yashiki, T., and Mima, H., 1983b, Vaginal absorption of a potent luteinizing hormone-releasing analogue (leuprolide) in rats. III. Effect of estrous cycle on vaginal absorption of hydrophilic model compounds, *J. Pharm. Sci.* **72:**173–176.

Okumura, K., Iwakawa, S., Yoshida, T., Seki, T., and Komada, F., 1992, Intratracheal delivery of insulin. Absorption from solution and aerosol by rat lung, *Int. J. Pharm.* **88:**63–73.

Oppenheim, R. C., Stewart, N. F., Gordon, L., and Patel, H. M., 1982, The production and evaluation of orally administered insulin nanoparticles, *Drug Dev. Ind. Pharm.* **8:**531–546.

Owens, D. R., Jones, M. K., Birtwell, A. J., Burge, C. T. R., Jones, I. R., Heyburn, P. J., Hayes, T. M., and Heding, L. G., 1984, Pharmacokinetics of subcutaneously administered human, porcine and bovine neutral soluble insulin to normal man, *Horm. Metab. Res.* **16:**195–199.

Owens, D. R., Vora, J. P., Luzio, S., and Hayes, T. M., 1988. The influence of aprotinin on regional absorption of soluble human insulin, *Br. J. Clin. Pharmacol.* **25:**453–456.

Pai, C. M., Bae, Y. H., Mack, E. J., Wilson, D. E., and Kim, S. W., 1992, Concanavalin A microspheres for a self-regulating insulin delivery system, *J. Pharm. Sci.* **81:**532–536.

Paquot, N., Scheen, A. J., Franchimont, P., and Lefebvre, P. J., 1988, The intra-nasal administration of insulin induces significant hypoglycaemia and classical counterregulatory hormonal responses in normal man, *Diabetes Metab.* **14:**31–36.

Patel, H. M., and Ryman, B. E., 1976, Oral administration of insulin by encapsulation within liposomes, *FEBS Lett.* **62:**60–63.

Patel, H. M., and Ryman, B. E., 1977. Orally administered liposomally entrapped insulin, *Biochem. Soc. Trans.* **5:**1739–1741.

Patel, H. M., Harding, N. G. L., Logue, F., Kesson, C., MacCuish, A. C., MacKenzie, J. C., Ryman, B. E., and Scobie, I., 1978, Intrajejunal absorption of liposomally entrapped insulin in normal man *Biochem. Soc. Trans.* **6:**784–785.

Patel, H. M., Stevenson, R. W., Parsons, J. A., and Ryman, B. E., 1982, Use of liposomes to aid intestinal absorption of entrapped insulin in normal and diabetic dogs, *Biochim. Biophys. Acta* **716:**188–193.

Paton, J. S., Wilson, M., Ireland, J. T., and Reith, S. B. M., 1981, Convenient pocket insulin syringe, *Lancet* **i:**189–190.

Patrick, A. W., Collier, A., Matthews, D. M., Macintyre, C. C. A., and Clarke, B. F., 1988, The importance of the time interval between insulin injection and breakfast in determining postprandial glycaemic control: A comparison between human and porcine insulin, *Diabetic Med.* **5:**32–35.

Patton, J. S., and Platz, R. M., 1992, Pulmonary delivery of peptides and proteins, *Adv. Drug. Delivery Rev.* **8:**179–196.

Paulsen, E. P., Courtney, J. W., III, and Duckworth, W. C., 1979, Insulin resistance caused by massive degradation of subcutaneous insulin, *Diabetes* **28:**640–645.

Pickup, J. C., Keen, H., Parsons, A. J., and Alberti, K. G. M. M., 1978, Continuous subcutaneous insulin infusion: An approach to achieving normoglycaemia, *Br. Med. J.* **1:**204–207.

Pietri, A., and Raskin, P., 1981, Cutaneous complications of chronic continuous subcutaneous infusion therapy, *Diabetes Care* **4:**624–626.

Pillion, D. J., Bartlett, J. D., Meezan, E. Yang, M. Crain, R. J., and Grizzle, W. E., 1991, Systemic absorption of insulin delivered topically to the rat eye, *Invest. Ophthalmol. Visual Sci.* **32:**3021–3027.

Pillion, D. J., Atchison, J. A., Wang, R.-X., and Meezan, E., 1994a, Alkylglycosides enhance systemic absorption of insulin applied topically to the rat eye, *J. Pharmacol. Exp. Ther.* **271:**1274–1280.

Pillion, D. J., Atchison, J. A., Gargiulo, C., Wang, R.-X., Wang, P., and Meezan, E., 1994b,

Insulin delivery in nosedrops: New formulations containing alkylglycosides, *Endocrinology* **135:**2386–2391.

Pitt, C. G., 1990, The controlled parenteral delivery of polypeptides and proteins, *Int. J. Pharm.* **59:**173–196.

Plevin, S., and Sadur, C., 1993, Use of a prefilled insulin syringe (Novolin Prefilled™) by patients with diabetes, *Clin. Ther.* **15:**423–431.

Pongor, S., Brownlee, M., and Cerami, A., 1983, Preparation of high-potency, non-aggregating insulins using a novel sulfation procedure, *Diabetes* **32:**1087–1901.

Pontiroli, A. E., and Pozza, G., 1990, Intranasal administration of peptide hormones: Factors affecting transmucosal absorption, *Diabetic Med.* **7:**770–774.

Pontiroli, A. E., Alberetto, M., Secchi, A., Dossi, G., Bosi, I., and Pozza, G., 1982, Insulin given intranasally induces hypoglycaemia in normal and diabetic subjects, *Br. Med. J.* **284:**303–306.

Pontiroli, A. E., Alberetto, M., Calderara, A., Pajetta, E., and Pozza, G., 1986, Metabolic effects of intranasally administered insulin and glucagon in man, in: *Delivery Systems for Peptide drugs* (S. S. Davis, L. Illum, and E. Tomlinson, eds.), Plenum Press, New York, pp. 243–248.

Pontiroli, A. E., Alberetto, M., Pajetta, E., Calderara, A., and Pozza, G., 1987, Human insulin plus sodium glycocholate in a nasal spray formulation: Improved bioavailability and effectiveness in normal subjects, *Diabetes Metab.* **13:**441–443.

Pramming, S., Lauritzen, T., Thorsteinsson, B., Johansen, K., and Binder, C., 1984, Absorption of soluble and isophane semi-synthetic human and porcine insulin in insulin-dependent diabetic subjects, *Acta Endocrinol.* **105:**215–220.

Prieto, G., Rio, J. M., Andrade, M. I. P., Sarmiento, F., and James, M. N., 1993, Interaction between sodium *n*-undecyl sulfate and insulin, *Int. J. Biol. Macromol.* **15:**343–345.

Quinn, R., and Andrade, J. D., 1983, Minimizing the aggregation of neutral insulin solutions, *J. Pharm. Sci.* **73:**1472–1473.

Ratner, R. E., and Steiner, M.L., 1987, Insulin-pump therapy: Effect of phosphate-buffered human insulin, *Diabetes Care* **10:**787–788.

Raz, I., Bar-on, H., Kidron, M., and Ziv, E., 1984, Rectal administration of insulin, *Isr. J. Med. Sci.* **20:**173–175.

Reeves, M. L., Seigler, D. E., Ryan, E. A., and Skyler, J. S., 1982, Glycemic control in insulin-dependent diabetes mellitus: Comparison of outpatient intensified conventional therapy with continuous subcutaneous insulin infusion, *Am. J. Med.* **72:**637–680.

Richardson, J. L., and Illum, L., 1992, The vaginal route of peptide and protein drug delivery, *Adv. Drug Delivery Rev.* **8:**341–366.

Richardson, J. L., Illum, L., and Thomas, N. W., 1992a, Vaginal absorption of insulin in the rat: Effect of penetration enhancers on insulin uptake and mucosal histology, *Pharm. Res.* **9:**878–883.

Richardson, J. L., Farraj, N. F., and Illum, L., 1992b, Enhanced vaginal absorption of insulin in sheep using lysophosphatidylcholine and a bioadhesive microsphere delivery system, *Int. J. Pharm.* **88:**319–325.

Ritschel, W. A., and Ritschel, G. B., 1984, Rectal administration of insulin, *Methods Find. Exp. Clin. Pharmacol.* **6:**513–529.

Ritschel, W. A., Ritschel, G. B., Forusz, H., and Kraeling, M., 1989, Buccal absorption of insulin in the dog, *Res. Commun. Chem. Pathol. Pharmacol.* **63:**53–67.

Ron, E., Turek, T., Mathiowitz, E., Chasin, M., Hageman, M., and Langer, R., 1993, Controlled release of polypeptides from polyanhydrides, *Proc. Natl. Acad. Sci. USA* **90:**4176–4180.

Rosak, C., Burkard, G., Hoffman, J. A., Humburg, E., Look, D., and Landgraf, R., 1993, Metabolic effect and acceptance of an insulin pen treatment in 20,262 diabetic patients, *Diabetes Nutr. Metab.* **6:**139–145.

Rowland, R. N., and Woodley, J. F., 1981, Uptake of free and liposome-entrapped insulin by rat intestinal sacs *in vitro*, *Biosci. Rep.* **1:**345–352.

Rydén, L., and Edman, P., 1992, Effect of polymers and microspheres on the nasal absorption of insulin in rats, *Int. J. Pharm.* **83:**1–10.

Saffran, M., Kumar, G. S., Savariar, C., Burnham, J. C., Williams, F., and Neckers, D. C., 1986, A new approach to the oral administration of insulin and other peptide drugs, *Science* **233:**1081–1084.

Sage, B. H., Jr., and Riviere, J. E., 1992, Model systems in iontophoresis—transport efficacy, *Adv. Drug Delivery Rev.* **9:**265–287.

Sakr, F. M., 1992, A new approach for insulin delivery via the pulmonary route: Design and pharmacokinetics in non-diabetic rabbits, *Int. J. Pharm.* **86:**1–7.

Salzman, R., Manson, J. E., Griffing, G. T., Kimmerle, R., Ruderman, N., McCall, A., Stolz, E. I., Mullin, C., Small, D., Armstrong, J., and Melby, J. C., 1985, Intranasal aerosolized insulin. Mixed-meal studies and long-term use in type I diabetes, *N. Engl. J. Med.* **312:**1078–1084.

Sandler, M., 1990, Is the lung a 'target organ' in diabetes mellitus? *Arch. Intern. Med.* **150:**1385–1388.

Sandler, M., Bunn, A. E., and Stewart, R. I., 1986, Pulmonary function in young insulin-dependent diabetic subjects, *Chest* **90:**671–675.

Sarmiento, F., Prieto, G., and Jones, M. N., 1992, Thermodynamic studies on the interaction of *n*-alkyl sulfates with insulin in aqueous solution, *J. Chem. Soc., Faraday Trans.* **88:**1003–1007.

Sasaki, H., Tei, C., Yamamura, K., Nishida, K., and Nakamura, J., 1994, Effect of preservatives on systemic delivery of insulin by ocular instillation in rabbits, *J. Pharm. Pharmacol.* **46:**871–875.

Saslawski, O., Weingarten, C., Benoit, J. P., and Couvreur, P., 1988, Magnetically responsive microspheres for the pulsed delivery of insulin, *Life Sci.* **42:**1521–1528.

Sato, S., Jeong, S. Y., McRea, J. C., and Kim, S. W., 1984a, Glucose stimulated insulin delivery systems, *Pure Appl. Chem.* **56:**1323–1328.

Sato, S., Jeong, S. Y., McRea, J. C., and Kim, S. W., 1984b, Self-regulating insulin delivery systems. II. *In vitro* studies, *J. Controlled Release* **1:**67–77.

Sato, T., Ebert, C. D., and Kim, S. W., 1983, Prevention of insulin self-association and surface adsorption, *J. Pharm. Sci.* **72:**228–232.

Saudek, C. D., 1990, Implantable insulin pumps: A current look, *Diabetes Res. Clin. Pract.* **10:**109–114.

Saudek, C. D., 1993, Future developments in insulin delivery systems, *Diabetes Care* **16:**(Suppl. 3)**:**122–132.

Saudek, C. D., Fishell, R. E., and Swindle, M. M., 1990, The programmable implantable medication systems (PIMS): Design features and pre-clinical trials, *Horm. Metab. Res.* **22:**201–206.

Schade, D. S., Eaton, R. P., Edwards, S., Doberneck, R. C., Spencer, W. J., Carlson, G. A. B., Bair, R. E., Love, J. T., Urenda, R. S., and Gaona, J. I., Jr., 1982, A remotely programmable insulin delivery system. Successful short-term implantation in man, *J. Am. Med. Assoc.* **247:**1848–1853.

Schanker, L. S., Mitchell, E. W., and Brown, R. A., Jr., 1986, Species comparison of drug absorption from the lung after aerosol inhalation or intratracheal injection, *Drug Metab. Dispos.* **14:**79–88.

Schildt, B., Ahlgren, T., Berghem, L., and Wendet, Y., 1978, Adsorption of insulin by infusion materials, *Acta Anaesthesiol. Scand.* **22:**556–562.

Schilling, R. J., and Mitra, A. K., 1990, Intestinal mucosal transport of insulin, *Int. J. Pharm.* **62:**53–64.

Schilling, R. J., and Mitra, A. K., 1992, Pharmacodynamics of insulin following intravenous and enteral administrations of porcine-zinc insulin to rats, *Pharm. Res.* **9:**1003–1009.

Schipper, N. G. M., Verhoef, J., Romeijn, S. G., and Merkus, F. W. H. M., 1992, Absorption enhancers in nasal insulin delivery and their influence on nasal ciliary functioning, *J. Controlled Release* **21:**173–186.

Schipper, N. G. M., Romeijn, S. G., Verhoef, J. C., and Merkus, F. W. H. M., 1993, Nasal insulin delivery with dimethyl-β-cyclodextrin as an absorption enhancer in rabbits: Powder more effective than liquid formulations, *Pharm. Res.* **10:**682–686.

Schlichtkrull, J., 1957, The anisotropic growth of insulin crystals, *Acta Chem. Scand.* **11:**484–486.

Schlichtkrull, J., 1959, New insulin crystal suspensions with various timings of action and containing no added zinc, in: *Diabetes Mellitus III. Kongress der International Diabetes Federation 1958* (K. Oberdisse and K. Jahnke, eds.), Georg Thieme, Stuttgart, pp. 773–777.

Schlichtkrull, J., Munck, O., and Jersild, M., 1965, Insulin Rapitard and insulin Actrapid, *Acta. Med. Scand.* **177:**103–113.

Schrezenmeir, J., Tato, F., Tato, S., Achterberg, H., Schulz, G., Strack, T., Aerssen, M., Kasper, H., and Beyer, J., 1985, Food-dependent insulin consumption, in: *Computer Systems for Insulin Adjustment in Diabetes Mellitus* (J. Beyer, M. Albisser, J. Schrezenmeir, and L. Lehmann, eds.), Panscienta-Verlag, Hedingen, Switzerland, pp. 43–53.

Scott, D. A., 1934, Crystalline insulin, *Biochem. J.* **28:**1592–1602.

Scott, D. A., and Fisher, A. M., 1936, Studies on insulin with protamine, *J. Pharmacol. Exp. Ther.* **58:**78–92.

Scott, E. R., Laplaza, A. L., White, H. S., and Phipps, J. B., 1993, Transport of ionic species in skin: Contribution of pores to the overall skin conductance, *Pharm. Res.* **10:**1699–1709.

Scott-Moncrieff, J. C., Shao, Z., and Mitra, A. K., 1994, Enhancement of intestinal insulin absorption by bile salt–fatty acid mixed micelles in dogs, *J. Pharm. Sci.* **83:**1465–1469.

Sefton, M. V., and Antonacci, G. M., 1984, Adsorption isotherms of insulin onto various materials, *Diabetes* **33:**674–680.

Seigler, D. E., Olsson, G. M., Agramonte, R. F., Lohman, V. L., Ashby, M. H., Reeves, M. L., and Skyler, J. S., 1991, Pharmacokinetics of long-acting (ultralente) insulin preparations, *Diabetes Nutr. Metab.* **4:**267–273.

Selam, J.-L., and Charles, M. A., 1990, Devices for insulin administration, *Diabetes Care* **13:**955–979.

Selam, J.-L., Zirinis, P., Mellet, M., and Mirouze, J., 1987, Stable insulin for implantable delivery systems: *In vitro* studies with different containers and solvents, *Diabetes Care* **10:**343–347.

Seminoff, L. A., Olsen, G. B., and Kim, S. W., 1989a, A self-regulating insulin delivery system. I. Characterization of a synthetic glycosylated insulin derivative, *Int. J. Pharm.* **54:**241–249.

Seminoff, L. A., Gleeson, J. M., Zheng, J. Olsen, G. B., Holmberg, D., Mohammad, S. F., Wilson, D., and Kim, S. W., 1989b, A self-regulating insulin delivery system. II. *In vivo* characteristics of a synthetic glycosylated insulin, *Int. J. Pharm.* **54:**251–257.

Shao, Z., and Mitra, A. K., 1992, Nasal membrane and intracellular protein and enzyme release by bile salts and bile salt-fatty acid mixed micelles: Correlation with facilitated drug transport, *Pharm. Res.* **9:**1184–1189.

Shao, Z., Krishnamoorthy, R., and Mitra, A. K., 1992, Cyclodextrins as nasal absorption promotors of insulin: Mechanistic evaluations, *Pharm. Res.* **9:**1157–1163.

Shao, Z., Li, Y., Chermak, T., and Mitra, A. K., 1994a, Cyclodextrins as mucosal absorption promotors of insulin. II. Effects of β-cyclodextrin derivatives on α-chymotryptic degradation and enteral absorption of insulin in rats, *Pharm. Res.* **11:**1174–1179.

Shao, Z., Li, Y., and Mitra, A. K., 1994b, Cyclodextrins as mucosal absorption promotors of insulin. III. Pulmonary route of delivery, *Eur. J. Pharm. Biopharm.* **40:**283–288.

Shenfield, G. M., and Hill, J. C., 1982, Infrequent response by diabetic rats to insulin-liposomes, *Clin. Exp. Pharmacol. Physiol.* **9:**355–361.

Shichiri, M., Yamasaki, Y., Kawamori, R., Kikuchi, M., Hakui, N., and Abe, H., 1978, Increased absorption of insulin: An insulin suppository, *J. Pharm. Pharmacol.* **30:**806–808.

Shiino, D., Murata, Y., Kataoka, K., Koyama, Y., Yokoyama, M., Okana, T., and Sakurai, Y., 1994, Preparation and characterization of a glucose-responsive insulin-releasing polymer device, *Biomaterials* **15:**121–128.

Siddiqui, O., Sun, Y., Liu, J.-C., and Chien, Y. W., 1987, Facilitated transdermal transport of insulin, *J. Pharm. Sci.* **76:**341–345.

Siegel, R. A., and Firestone, B. A., 1990, Mechanochemical approaches to self-regulating insulin pump design, *J. Controlled Release* **11:**181–192.

Siegel, R. A., and Langer, R., 1984, Controlled release of polypeptides and other macromolecules, *Pharm. Res.* **1:**2–10.

Siegel, R. A., Falamarzian, M., Firestone, B. A., and Moxley, B. C., 1988, pH-controlled release from hydrophobic/polyelectrolyte copolymer hydrogels, *J. Controlled Release* **8:**179–182.

Sinay, I. R., Schlimovich, S., Damilano, S., Cagide, A. L., Faingold, M. C., Facco, E. B., Gurfinkiel, M. S., and Arias, P., 1990, Intranasal insulin administration in insulin dependent diabetes: Reproducibility of its absorption and effects, *Horm. Metab. Res. Suppl. Ser.* **22:**307–308.

Sindelka, G., Heinemann, L., Berger, M., Frenck, W., and Chantelau, E., 1994, Effect of insulin concentration, subcutaneous fat thickness and skin temperature on subcutaneous insulin absorption in healthy subjects, *Diabetologia* **37:**377–380.

Singh, J., and Maibach, H. I., 1993, Topical iontophoretic drug delivery *in vivo*: Historical development, devices and future perspectives, *Dermatology* **187:**235–238.

Singh, M, Vasudevan, P., Sinha, T. J. M., Ray, A. R., Misro, M. M., and Guha, K., 1981, An insulin delivery system from oxidized cellulose, *J. Biomed. Mater. Res.* **15:**655–661.

Sluzky, V., Tamada, J. A., Klibanov, A. M., and Langer, R., 1991, Kinetics of insulin aggregation in aqueous solutions upon agitation in the presence of hydrophobic surfaces, *Proc. Natl. Acad. Sci. USA* **88:**9377–9381.

Sluzky, V., Klibanov, A. M., and Langer, R., 1992, Mechanism of insulin aggregation and stabiization in agitated aqueous solutions, *Biotechnol. Bioeng.* **40:**895–903.

Smith, P. L., Wall, D. A., Gochoco, C. H., and Wilson, G., 1992. Oral absorption of peptides and proteins, *Adv. Drug Delivery Rev.* **8:**253–290.

Spangler, R. S., 1990, Insulin administration via liposomes, *Diabetes Care* **13:**911–922.

Spraul, M., Chanteleau, E., Koumoulidou, J., and Berger, M., 1988, Subcutaneous or nonsubcutaneous injection of insulin, *Diabetes Care* **11:**733–736.

Steiner, D. F., 1967, Evidence for a precursor in the biosynthesis of insulin, *Trans. N.Y. Acad. Sci.* **30**(Ser. II):60–68.

Steiner, D. F., Hallund, O., Rubenstein, A., Cho, S., and Bayliss, C., 1968, Isolation and properties of proinsulin, intermediate forms, and other minor components from crystalline bovine insulin, *Diabetes* **17:**725–736.

Stephen, R. L., Petelenz, T. J., and Jacobsen, S. C., 1984, Potential novel methods for insulin administration: I. Iontophoresis, *Biomed. Biochim. Acta* **43:**553–558.

Stevenson, R. W., Patel, H. M., Parsons, J. A., and Ryman, B. E., 1982, Prolonged hypoglycemic effect in diabetic dogs due to subcutaneous administration of insulin in liposomes, *Diabetes* **31:**506–511.

Storm, M. C., and Dunn, M. F., 1985, The Glu(B13) carboxylates of the insulin hexamer form a cage for Cd^{2+} and Ca^{2+} ions, *Biochemistry* **24:**1749–1756.

Su, K. S. E., 1991, Nasal route of peptide and protein drug delivery, in: *Peptide and Protein Drug Delivery* (V. H. L. Lee, ed.), Marcel Dekker, New York, pp. 595–631.

Sudmeier, J. L., Bell, S. J., Storm, M. C., and Dunn, M. F., 1981, Cadmium-113 nuclear magnetic resonance studies of bovine insulin: Two-zinc hexamer specifically binds calcium, *Science* **212**:560–562.

Süsstrunk, H., Morell, B., Ziegler, W. H., and Froesch, E. R., 1982, Insulin absorption from the abdomen and the thigh in healthy subjects during rest and exercise: Blood glucose, plasma insulin, growth hormone, adrenaline and noradrenaline levels, *Diabetologia* **22**:171–174.

Tabata, Y., Takebayashi, Y., Ueda, T., and Ikada, Y., 1993, A formulation method using D,L-lactic acid oligomer for protein release with reduced initial burst, *J. Controlled Release* **23**:55–64.

Tachibana, K., 1992, Transdermal delivery of insulin to alloxan-diabetic rabbits by ultrasound exposure, *Pharm. Res.* **9**:952–954.

Tachibana, K., and Tachibana, S., 1991, Transdermal delivery of insulin by ultrasonic vibration, *J. Pharm. Pharmacol.* **43**:270–271.

Takeyama, M., Ishida, T., Kokubu, N., Komada, F., Iwakawa, S., Okumura, K., and Hori, R., 1991, Enhanced bioavailability of subcutaneously injected insulin by pretreatment with ointment containing protease inhibitors, *Pharm. Res.* **8**:60–64.

Tamborlane, W. V., Sherwin, R. S., Genel, M., and Felig, P., 1979, Reduction to normal of plasma glucose in juvenile diabetes by subcutaneous administration of insulin with a portable infusion pump, *N. Engl. J. Med.* **300**:573–578.

Taylor, R., Home, P. D., and Alberti, K. G. M. M., 1981, Plasma free insulin profiles after administration of insulin by jet and conventional syringe injection, *Diabetes Care* **4**:377–379.

Thow, J. C., Johnson, A. B., Antsiferov, M., and Home, P. D., 1989, Effect of raising injection-site skin temperature on isophane (NPH) insulin crystal dissociation, *Diabetes Care* **12**:432–434.

Thow, J. C., Johnson, A. B., Fulcher, G., and Home, P. D., 1990, Different absorption of isophane (NPH) insulin from subcutaneous and intramuscular sites suggests a need to reassess recommended insulin injection technique, *Diabetic Med.* **7**:600–602.

Thurow, H., 1980, Studies on the denaturation of dissolved insulin, in: *Insulin. Chemistry Structure and Function of Insulin and Related Hormones* (D. Brandenburg and A. Wollmer, eds.), de Gruyter, Berlin, pp. 215–221.

Thurow, H., and Geisen, K, 1984, Stabilisation of dissolved proteins against denaturation at hydrophobic interfaces, *Diabetologia* **27**:212–218.

Touitou, E., and Donbrow, M., 1983, Promoted rectal absorption of insulin: Formulative parameters involved in the absorption from hydrophilic bases, *Int. J. Pharm.* **15**:13–24.

Touitou, E., and Rubinstein, A., 1986, Targeted enteral delivery of insulin to rats, *Int. J. Pharm.* **30**:95–99.

Touitou, E., Donbrow, M., and Azaz, E., 1978, New hydrophilic vehicle enabling rectal and vaginal absorption of insulin, heparin, phenol red and gentamicin, *J. Pharm. Pharmacol.* **30**:662–663.

Touitou, E., Donbrow, M., and Rubinstein, A., 1980, Effective intestinal absorption of insulin in diabetic rats using a new formulation approach, *J. Pharm. Pharmacol.* **32**:108–110.

Touitou, E., Alhaique, F., Fisher, P., Memoli, A., Riccieri, F. M., and Santucci, E., 1987, Prevention of molecular self-association by sodium salicylate: Effect on insulin and 6-carboxyfluorescein, *J. Pharm. Sci.* **76**:791–793.

Tunbridge, F. K. E., Newens, A., Home, P. D., Davis, S. N., Murphy, M., Burrin, J. M., Alberti, K. G. M. M., and Jensen, I., 1989, Double-blind crossover trial of isophane (NPH)- and lente-based insulin regimens, *Diabetes Care* **12**:115–119.

Ushiwata, A., Nakaya, K., and Nakamura, Y., 1975, Interaction between proteins and detergents which contains a hydrocarbon chain longer than 16 carbon atoms. II. Difference spectra of various proteins in cetyldimethyl-benzylammonium chloride, *Biochim. Biophys. Acta* **393:**215–224.

Verhoef, J. C., Schipper, N. G. M., Romeijn, S. G., and Merkus, F. W. H. M., 1994, The potential of cyclodextrins as absorption enhancers in nasal delivery of peptide drugs, *J. Controlled Release* **29:**351–360.

Vora, J. P., and Owens, D. R., 1991, Future trends in insulin therapy: Clinical implications of novel insulin analogues and nasal administration of insulin, in: *Pharmacology of Diabetes. Present Practice and Future Perspectives* (C. E. Mogensen and E. Standl, eds.), de Gruyter, Berlin, pp. 39–56.

Vora, J. P., Owens, D. R., Dolben, J., Atiea, J. A., Dean, J. D., Kang, S., Burch, A., and Brange, J., 1988, Recombinant DNA derived monomeric insulin analogue: Comparison with soluble human insulin in normal subjects, *Br. Med. J.* **297:**1236–1239.

Wall, D. A., 1995, Pulmonary absorption of peptides and proteins, *Drug Delivery* **2:**1–20.

Walters, D. P., Smith, P. A., Marteau, T. M., Brimble, A., and Borthwick, L. J., 1985, Experience with NovoPen, an injection device using cartridged insulin, for diabetic patients, *Diabetic Med.* **2:**496–497.

Wang, P. Y., 1987, Prolonged release of insulin by cholesterol-matrix implant, *Diabetes* **36:**1068–1072.

Wang, P. Y., 1989a, Lipids as excipient in sustained release insulin implants, *Int. J. Pharm.* **54:**223–230.

Wang, P. Y., 1989b, Implantable reservoir for supplemental insulin delivery on demand by external compression, *Biomaterials* **10:**197–201.

Wang, P. Y., 1991a, Sustained-release implants for insulin delivery, in: *Biotechnology of Insulin Therapy* (J. C. Pickup, ed.), Blackwell Scientific, Oxford, pp. 42–74.

Wang, P. Y., 1991b, Palmitic acid as an excipient in implants for sustained release of insulin, *Biomaterials* **12:**57–62.

Wang, P. Y., 1993, Control of hyperglycaemia in diabetic rabbits by a combination of implants, *J. Biomed. Eng.* **15:**106–112.

Wehner, H., Huber, H., and Kronenberg, K. H., 1973, The glomerular basement membrane of the rabbit kidney on long-term treatment with heterologous insulin preparations of different purity, *Diabetologia* **9:**255–263.

Weiner, A. L., Carpenter-Green, S. S., Soehngen, E. C., Lenk, R. P., and Popescu, M. C., 1985, Liposome-collagen gel matrix: A novel sustained drug delivery system, *J. Pharm. Sci.* **74:**922–925.

Weingarten, C., Moufti, A., Desjeux, J. F., Luong, T. T., Durand, G., Devissaguet, J. P., and Puisieux, F., 1981, Oral ingestion of insulin liposomes: Effect of the administration route, *Life Sci.* **28:**2747–2752.

Wiefels, K., Kuglin, B., Hübinger, A., Jorgensen, L. N., and Gries, F. A., 1993, Insulin kinetics and dynamics in insulin-dependent diabetic patients after injections of human insulin or the insulin analogues X14 and X14 + Zn, in: *Frontiers in Insulin Pharmacology* (M. Berger and F. A. Gries, eds.), Thieme Verlag, Stuttgart, pp. 97–101.

Wigley, F. M., Londono, J. H., Wood, S. H., Shipp, J. C., and Waldman, R. H., 1971, Insulin across respiratory mucosae by aerosol delivery, *Diabetes* **20:**552–556.

Williams, G., Pickup, J. C., Bowcock, S., Cooke, E., and Keen, H., 1983, Subcutaneous aprotinin causes local hyperaemia. A possible mechanism by which aprotinin improves control in some diabetic patients, *Diabetologia* **24:**91–94.

Woodley, J. F., 1986, Liposomes for oral administration of drugs, *Crit. Rev. Ther. Drug Carrier Syst.* **2:**1–18.

Woodley, J. F., and Prescott, A. R., 1988, Body distribution of liposome-entrapped ^{125}I-labelled insulin after oral administration to rats, *Biochem. Soc. Trans.* **16:**343–344.

Yagi, T., Hakui, N., Yamasaki, Y., Kawamori, R., Shichiri, M., Abe, H., Kim, S., Miyake, M., Kamikawa, K., Nishihata, T., and Kamada, A., 1983, Insulin suppository: Enhanced rectal absorption of insulin using an enamine derivative as a new promotor, *J. Pharm. Pharmacol.* **35:**177–178.

Yamakawa, I., Kawahara, M., Watanabe, S., and Miyake, Y., 1990, Sustained release of insulin by double-layered implant using poly(d,l-lactic acid), *J. Pharm. Sci.* **79:**505–509.

Yamamoto, A., Luo, A. M., Dodda-Kashi, S., and Lee, V. H. L., 1989, The ocular route for systemic insulin delivery in the albino rabbit, *J. Pharmacol. Exp. Ther.* **249:**249–255.

Yamamoto, A., Hayakawa, E., and Lee, V. H. L., 1990, Insulin and proinsulin proteolysis in mucosal homogenates of the albino rabbit: Implications in peptide delivery from nonoral routes, *Life Sci.* **47:**2465–2474.

Yamamoto, A., Hayakawa, E., Kato, Y., Nishiura, A., and Lee, V. H. L., 1992, A mechanistic study on enhancement of rectal permeability to insulin in the albino rabbit, *J. Pharmacol. Exp. Ther.* **263:**25–31.

Yamamoto, A., Umemori, S., and Muranishi, S., 1994, Absorption enhancement of intrapulmonary administered insulin by various absorption enhancers and protease inhibitors in rats, *J. Pharm. Pharmacol.* **46:**14–18.

Yamasaki, Y., Shichiri, M., Kawamori, R., Shigeta, Y., and Abe, H., 1978, Efficacy of rectal administration of insulin in diabetic dogs, *J. Jpn. Diabetes Soc.* **21:**249–252.

Yamasaki, Y., Shichiri, M., Kawamori, R., Kikuchi, M. Yagi, T., Arai, S., Tohdo, R., Hakui, N., Oji, N., and Abe, H., 1981a, The effectiveness of rectal administration of insulin suppository on normal and diabetic subjects, *Diabetes Care* **4:**454–458.

Yamasaki, Y., Shichiri, M., Kawamori, R., Morishima, T., Hakui, N., Yagi, T., and Abe, H., 1981b, The effect of rectal administration of insulin on the short-term treatment of alloxan-diabetic dogs, *Can. J. Physiol. Pharmacol.* **59:**1–6.

YoKosuka, T., Omori, Y., Hirata, Y., and Hirai, S., 1977, Nasal and sublingual administration of insulin in man, *J. Jpn. Diabetes Soc.* **20:**146–152.

Yoshida, H., Okumura, K., Hori, R., Anmo, T., and Yamaguchi, H., 1979, Absorption of insulin delivered to rabbit trachea using aerosol dosage form, *J. Pharm. Sci.* **68:**670–671.

Zhou, X. H., 1994, Overcoming enzymatic and absorption barriers to non-parenterally administered protein and peptide drugs, *J. Controlled Release* **29:**239–252.

Zhou, X. H. and Po, A. L. W., 1991, Effects of cholic acid and other enhancers on the bioavailability of insulin from a subcutaneous site, *Int. J. Pharm.* **69:**29–41.

Zinman, B., 1989, The physiologic replacement of insulin: An elusive goal, *N. Engl. J. Med.* **321:**363–370.

Ziv, E., Kidron, M., Berry, E. M., and Bar-on, H., 1981, Bile salts promote the absorption of insulin from the rat colon, *Life Sci.* **29:**803–809.

Ziv, E., Kleinman, Y., Bar-on, H., and Kidron, M., 1984, Treatment of diabetic rats with enteral insulin, in: *Lessons from Animal Diabetes* (E. Shafrir and A. E. Renold, eds.), Libbey, London, pp. 642–647.

Ziv, E., Lior, O., and Kidron, M., 1987, Absorption of protein via the intestinal wall. A quantitative model, *Biochem. Pharmacol* **36:**1035–1039.

Ziv, E., Kidron, M., Raz, I., Krausz, M., Blatt, Y., Rotman, A., and Bar-on, H., 1994, Oral administration of insulin in solid form to nondiabetic and diabetic dogs, *J. Pharm. Sci.* **83:**792–794.

Index

Abbott Laboratories, 208, 247, 250
Absorption/desorption kinetics, 155
Absorption/desorption method, 154, 157–160
ACE: *see* Angiotensin-converting enzyme
Acetone, 176
Acrylamide, 364
Actimmune, pharmacokinetic parameters, 240–241
Adam, M., 162
Additives, effects on ATRIGEL formulation, 101*f*, 102
Adjei, A., 380
Administration routes, 240, 242–243; *see also* Oral administration; Parenteral administration
Adriamycin, 172
Affinity Biotech, Inc., 275
AFM: *see* Atomic force microscopy
Ag/AgCl electrodes, 232
Agar, 365
Alafosfalin, 260
Alberti, K.G.M.M., 357
Albumin, 3–4, 71–72, 149, 152, 168, 351, 364
  hydrogels and, 143
  molecular weight or matrix and release of, 129
Alcaligenes eutrophus, 56
Alcohol dehydrogenase, 78
Alkaline phosphatase, 78
Alkermes Controlled Therapeutics, Inc., 300
Alkermes Inc., 271
Alkylcyanoacrylates, 273–275
Alkylglycosides, 371, 376
Alum, 30–31
Aluminum monostearate, 296–297
Alza Corporation, 57, 119, 310
Alzamer, 57
Alzet minipump, 31, 35, 37, 141, 292
American Cyanamid Co., 299
Amino acid sequence, bovine and porcine somatotropins, 290*f*
Amis, E.J., 162
Amsden, B.G., 384
Analogs, of insulin, 352–355
Analytical diffusion models, 131
Anatomical site of injection, 358
Ando, H., 72
Andrade, J.D., 171
Angiogenesis inhibitors, poly(ethylene-*co*-vinyl acetate) matrix and, 123
Angiotensin-converting enzyme, 260
Angiotensin-converting enzyme inhibitors, 256
Anhydrides, 365
Animal models, preclinical, 37–38
Anionic peptides, 224
Antigen delivery systems, 66
Antigens
  cholesterol and, 79
  poly(ethylene-*co*-vinyl acetate) matrix and, 123
Applications
  hydrophobic polymers, 147
  matrix systems, 149
  multiple emulsions, 203–205
  protein/polymer matrix systems, 132–134

Aprotinin, 338, 358
Aqueous gels and complexes, somatotropins and, 302–303
Aqueous two-phase partitioning, 182
Arbutamine, 227
Arginine, 308
Artursson, P., 78
Asano, M., 54
Aseptic operation, 20
Aspenberg, P., 78
Aston, R., 303
Atomic force microscopy, 177–179
ATR–FTIR spectroscopy: *see* Attenuated total reflectance–FTIR spectroscopy
ATRIGEL drug delivery system, 54, 93–95
  release kinetics, 95–102
ATRIGEL formulation
  in vitro characterization, 102–110
  in vivo evaluations, 110–115
Atriopeptin III, 10
Atrix Laboratories, 54, 103
Atropine, 268
Attenuated total reflectance-FTIR spectroscopy, 161
Auer, H., 300
Azain, M.J., 292, 304–305
Azocoll assay, 108

B. Braun, 248
BALB/c 3T3 cells, [^{3}H]thymidine incorporation in, 109
Balloon pumps, 250, 253
Banga, A.K., 371
Banting, F.G., 319, 343
Barbituric acid, 268
Basal insulin, 357
Baxter, 247, 250
Baxter Travenol, 245, 248
BCA: *see* Bicinchoninic acid assay
BCNU: *see N,N*-bis(2-Chloroethyl)-*N*-nitrosourea
Becton Dickinson, 248
Beeswax, 304–305, 308
Berens, A.R., 158
Bernstein, H., 79
Best, C.H., 319, 343
bFGF: *see* Fibroblast growth factor
Bhargava, K., 72
Bicinchoninic acid assay, 97–99, 101
Bile salts, 351, 371, 374–377
Bioabsorption, defined, 48
Bioactivity
  in ATRIGEL formulation, 108–110, 113–115
  of insulin dose, 329
Bioadhesive polymers, 376
Bioavailability of insulin, 323, 325, 328–329, 370
  intranasal, 378
  pulmonary, 379
Biocompatibility
  in ATRIGEL formulation, 110–111
  of poly(ethylene-*co*-vinyl acetate) in rabbits, 121
  of polyurethanes, 122
  of silicone elastomers, 121–122
  of zein microspheres, 272
Biocompatible polymers, 120–122
Biodegradable hydrogels, 149–151
Biodegradable lipids, defined, 70
Biodegradable materials for controlled protein delivery, 4–7
Biodegradable microspheres: *see* Microspheres
Biodegradable polymeric biomaterials, defined, 70
Biodegradable polymers containing PEG blocks, 174–176
Biodegradable polymers, implantable matrices, 364–366
Biodegradable systems, 141
Biodegradation, 110
  defined, 48
Bioerodible implant, insulin delivery and, 324–325
Bioerosion, defined, 48
BioLure, 147
Biopol, 56
Biopolymer PS-87, 302
Bioresorption, defined, 48
Biostator, 248
Björk, E., 378
Blood coagulation factor VIII, 276
Blood half-life, of PEG-coated nanospheres, 189*f*, 190*f*
BMPs: *see* Osteoinductive bone morphogenetic proteins
Bolus administration, insulin administration and, 322, 325, 327–328, 357, 362
Bolus injections, 252
Bone morphogenic protein, 74
Bovine growth hormone, 52, 78, 80

Bovine insulin
association state, 359*f*
FPLC chromatograms from ATRIGEL formulation, 105*f*
release kinetics in ATRIGEL formulation, 104*f*
Bovine serum albumin, 11, 51, 63, 65, 67, 71–72, 78, 80–81, 95, 106, 144, 152, 258, 303
effect of protein load on release of, 100*f*, 101
tortuosity value, 127–130
Bovine somatotropin, 63–64, 289–291, 294–295, 299, 301–304
administration in cows, 296–297
amino acid sequence, 290*f*
levels in cattle, 296*f*, 297*f*
solution stability, 292
Bovine somatotropin/ethyl cellulose implants, 308
Bovine somatotropin/poly(lactic acid) implants, 308
Brain, targeted delivery of proteins using poly(ethylene-*co*-vinyl acetate) matrices, 133–134
Brange, J., 351
Breast implants, 122
Brems, D.N., 292–293
Brij, 93, 200, 203
Bronsted, H., 150
Brooks, S., 76
Brownlee, M., 367
BSA: *see* Bovine serum albumin
BST: *see* Bovine somatotropin
D-Ser(Bu^t)^6, Azgly^{10}-LHRH: *see* Goserelin
[D-Ser(Bu^t)^6, Pro^9 Net]-LHRH: *see* Buserelin
Buccal administration, of insulin, 352, 370–371
Buchner, J.E., 61
Buckwalter, B.L., 293
Bulk diffusion, 150; *see also* Diffusion
Bulk erosion, 47, 50; *see also* Erosion
of PEG-coated nanospheres, 174
of polyanhydrides, 61–62
Burleigh, B.D., 292
Buserelin, 56
Byron, P.R., 380

C.R. Bard, 248
C3, 168
Cady, S., 78
Cady, S.M., 301–302
Caffeine, 268
Calcein, 226
Calcitonin, 51, 76, 216, 222, 226, 271, 273, 275
Calcium, 349
Calcium phosphate, 4, 107
Caliceti, P., 152
Canal, T., 142
Candida albicans, 267
Capillary zone electrophoresis, 334
Captopril, 260
Carbon, 169
Carbonic anhydrase, 77–78
Carboxylic acids, 224
Carboxymethylated porcine somatotropin, 293
Carboxymethylcellulose, 33–34
bis(*p*-Carboxyphenoxy)propane, 61, 63
Cardinal, J., 78
Carelli, V., 144
Carmustine, 183, 192
Carothers, W.H., 61
Carrier-mediated transport, intestinal absorption of proteins and peptides, 259–260
Castillo, E.J., 305
Catalase, 77, 203
Cationic polysaccharides, 376
Cats, ethyl acetate toxicity in, 36
Cattle, milk production increase following bovine somatotropin administration, 299*f*
CD4 binding, 22
Cefixime, 260
Ceftibuten, 260
Cellular bioactivity, ATRIGEL formulation and, 108–110
Cellulose, 146, 366
Cellulose acetate butyrate, 310
Center for Devices and Radiological Health, 110
Cerami, A., 367
Charge
degradation rate and, 274
effect on transcellular pathway, 265
endocytosis and, 262
surface hydrophobicity and determination of, 181–183
Charge titration, insulin and, 336*f*, 337
Charles, M.A., 357, 363
Chelating agents, 371
Chen, C., 75
Chickens, proteinoid microspheres in, 273
Chien, G.C.Y., 381
Chien, Yie W., 330–332, 338, 371
Chitosan, 376

Chitosan–porcine somatotropin, release of porcine somatotropin, 308*f*, 309*f*
N,N-bis(2-Chloroethyl)-*N*-nitrosourea, 7
Cholesterol, 79–80, 171, 204, 275–276, 365
Cholesterol liposomes, 268
Chondrogenic stimulating proteins, 63
Chymotrypsinogen, 73
Ciba-Geigy Corporation, 119
Circular dichroism, 22
Citric acid, 176
Clark, M.T., 303–306
Clinical experiments, 37–38
Clisis, 244
Clonidine patch, 329–330
Closed-loop system, insulin delivery via, 323
CMC: *see* Carboxymethylcellulose
Coacervation, 8, 12, 13*f*, 16, 51–52, 76
Coated implants, somatotropins and, 305–310
Collagen, 4, 72–74, 80, 358
Colonizing factor antigen, 10
Competitive desorption, 367
Complement, 168
Compliance
 of diabetic patient, 323, 325
 insulin administration and, 329–330
Compression molding, 63, 120
Concanavalin A, 367
Conformation-sensitive immunoassay, 65
Consensus Conference of the European Society for Biomaterials, 48
Continuous release, 14*f*
Continuous subcutaneous insulin infusion, 356, 360–361, 363
Controlled release, 45
 insulin formulation and, 345–348
Controlled release injectable implant: *see* ATRIGEL drug delivery system
Copolymers
 LA, 176
 lactide/glycolide, 50, 68
 PEG–PLA, 177, 182
 PEG–PPO–PEG block, 200
 PLGA, 55
 poly(fatty acid dimer–sebacic acid), 63
Cortecs International Limited, 275
Cows
 milk production and somatotropin administration in, 296
 use of uncoated implants in, 304
 zinc somatotropin release in, 80
Critical flocculation temperature method, 182
Critical porosity, 132
Cross-linking agents, degradable, 150
Crystalline-rubbery PEG, 146
Crystallinity, 192
CSII: *see* Continuous subcutaneous insulin infusion
Cullander, C., 381
Cyclodextrins, 376
Cyclosporin, 256, 258, 270
Cyclosporin A, 10
Cytochrome *c*, 54, 78, 95
 hydrogels and, 143
 molecular size and iontophoretic delivery, 338
Cytokines, 37
CZE: *see* Capillary zone electrophoresis

D_A^*: *see* Self diffusion coefficient
D_A: *see* Intrinsic diffusion coefficient
Dacron, 73
Daymate Infusion Device 2ml/hr, 246*f*
DCCT: *see* Diabetes Control and Complications Trial
dDAVP: *see* Desmopressin
Deatherage, J., 74
Decapeptyl, 7
de Gennes, P.G., 142, 162
Degradable cross-linking agents, 150
Degradable delivery systems, 150
Degradable pendant chains, 150
Degradation
 controlled release, 45–48
 defined, 48
 of encapsulated proteins, 23
 of insulin, 338
 of polylactides in rats, 33
 of polymer backbone, 149–150
Degradation kinetics, of PEG-coated nanospheres, 174
Degradation profile
 of poly[bis(*p*-carboxyphenoxy)propane anhydride], 62*f*
 for polylactides in rats, 32*f*
Degradation rate, 27, 50
Degradation test, 180
Dehydrated delivery device, 140*f*
Delivery rate, of insulin, 322, 327
Denaturing gel electrophoresis, 21
DePrince, R.B., 292
Derivatives of insulin, 352–355
Desmopressin, 271–272

Dextran, 33, 78, 365
Dextran sulfate, 169
Dextrin, 78, 302
Diabetes Control and Complications Trial, 321
Dicetyl phosphate, 276
Dichloromethane, 188
Diethylenetriaminepentaacetic acid stearyl amide, 188
Differential scanning calorimetry, 153, 186
Diffusion, 12; *see also* Bulk diffusion
  defined, 153
  macroscopic models in porous polymer matrices, 125–130
  mechanisms, 140–142
  microscopic models in porous polymer matrices, 131–132
  protein, 151–153
Diffusion coefficient, 122–123
  measuring techniques, 153–155, 162
    absorption/desorption method, 154, 157–160
    FTIR spectroscopy, 154, 160–161
    membrane permeation method, 154–157
    quasi-elastic light scattering method, 154, 161–162
    scanning electron microscopy, 154, 160
Diffusional release, 70–71
  collagen and, 73
Diffusion-controlled delivery systems, 145–150
Diffusion-controlled reservoir systems, 47
Diffusion-controlled systems, 139, 141–142; *see also* Hydrogels; Hydrophilic polymers
Dihydrofusidates, 376
Diketene acetals, 57
Dilatancy effects, 33
Dimethyl β-cyclodextrin, 379
Dimethyl sulfoxide, 94, 99–100, 103, 111–112
Dimethylsiloxane, 121
Dipalmitoyl phosphatidylcholine, 276
Dipeptides, 259
Diphtheria toxoid, 11, 68
Disc electrophoresis, 344
Dispersive energy X-ray fluorescence, 155
Distearoyl phosphatidylcholine, 268, 276
DMSO: *see* Dimethyl sulfoxide
DNA transfection, 214
Dodecylmaltoside, 370
Dogs
  insulin administration in
    buccal, 370–371
    nasal, 376
    ocular, 374
    oral, 369
    rectal, 372
  liposomes in, 276
  Lupron Depot in, 26
  proteinoid microspheres in, 273
  somatostatin infusion via syringe and osmotic pumps in, 250*f*, 251*f*
Domb, A., 143, 309–310
Domb, A.J., 61
Dong, L.C., 151
Donnan effect, 152
Dose precision, insulin and, 322, 325, 328
Dose response, MN rgp120 controlled release vaccine in guinea pigs, 28*f*
Double-emulsion method, 12, 63, 144, 199, 200*f*; *see also* Emulsions; Multiple emulsions
  process variables, 19
Dow Chemical Corporation, 121
Downes, S., 312
Draize scoring method, 110
Drug encapsulation, of PEG-coated nanospheres, 183–187
Drug Master Files, 8
Drug release, parameters affecting, 184–187
DSC: *see* Differential scanning calorimetry
DTPA-SA: *see* Diethylenetriaminepentaacetic acid stearyl amide
Du Pont, 120
D^V: *see* Mutual diffusion coefficient

E.I. Dupont de Nemours, 305
Eckenhoff, J.B., 311
Edema, 233–234
Edman, P., 77, 378
Effective diffusion coefficient, 127, 130
EGF: *see* Epidermal growth factor
Egg lecithin, 200
Egg phosphatidylcholine, 301
Elanco, 297–298
Electrical effects, modulated systems and, 366
Electron microscopy, 203
Electron spectroscopy for chemical analysis, 181
Electropermeabilization: *see* Electroporation
Electrophoresis, 144

Electrophoretic loading, 144
Electroporation, 213–214
changes in lipid bilayer structure, 215*f*
vs. iontophoresis, 217–227
in isolated perfused porcine skin flap, 227–232
reversible changes in LHRH flux, 220*f*
skin toxicology following, 232–235
ELISA: *see* Enzyme-linked immunosorbent assay
Elliott, R.B., 380
ELVAX-40, 120
Emisphere Technologies Inc., 272–273
Emulsion parameters, 17, 19–20
Emulsion-evaporation method, 176, 180
Emulsions, 168; *see also* Double emulsion method; Multiple emulsions
oral drug delivery of proteins and, 275
somatotropins and, 301–302
Enalapril, 260
Enalaprilat, 260
Enamines, 371
Encapsulated islet cells, 323
Encapsulation: *see* Microsphere formulation
Encapsulation efficiency, 17
Encapsulation properties of PEG–PLGA, PEG_3–PLA, and PEG–PSA polymers, 185
Endocytosis, 268
receptor-mediated and non-receptor-mediated, 261–264
Engstrom, S., 80
Enterocytes, 262
Entrapment, 144
Environmental conditions affecting protein diffusion, 151–152
Enzyme activity, ATRIGEL formulation and, 107–108
Enzyme immobilization, 208
Enzyme inhibitors, 371
Enzyme-linked immunosorbent assay, 21–22
Enzymes
degradation and, 50–51, 369
insulin degradation and, 338
release kinetics in ATRIGEL formulations, 107*f*
Enzyme-substrate reactions, 367
EO: *see* Ethylene oxide
Epidermal growth factor, 95, 113–115
[^{125}I]-EGF release kinetics, 111*f*, 112*f*
release profile in rats, 103-104
Eppstein, D.A., 371
Equipment, microencapsulation, 17
Erosion, 12, 47–48; *see also* Bulk erosion
Erythema, 233–235, 333
Erythropoietin, 79, 271
ESCA: *see* Electron spectroscopy for chemical analysis
Escherichia coli, 10, 270, 294, 345
Estraderm, 119
Estradiol, 216
Ethanolamine, 301
Ethyl acetate, 13, 50, 176
toxicity data, 35–36
Ethylene oxide, 51
Ethylene vinyl acetate, 147
Eudragit, 307
EVAc: *see* Poly(ethylene-*co*-vinyl acetate)
Evanol, 305
Experiments, clinical 37–38; *see also names of individual animals*
External infusion pumps, 250–252, 360–361
EZ-Flow 80, 246*f*

Faraday's constant, 221, 334
Faraday's law, 334
Fatty acids, 376
FDA: *see* U.S. Food and Drug Administration
Feijen, J., 72
Ferguson, T.H., 297
Ferritin
release from poly(ethylene-*co*-vinyl acetate) matrix, 123*f*
tortuosity value, 127–128
FGF: *see* Fibroblast growth factor
Fibrin, 74–75
Fibrinogen, 73–75, 152, 168, 170
Fibroblast growth factor, 78, 95, 109
Fibronectin, 95, 109, 168
Fickian diffusion, 158
Fick's law, 125, 142, 151, 157, 161
Final processing parameters, 17, 19–20
Fluorescence polarization immunoassay, 208
Flux, 220–223
insulin, 327
during iontophoresis, 383*f*
Flux values, 225–226
FN: *see* Fibronectin
Folkman, J., 147
Follicle-stimulating hormone, 95, 327
effect of polymer concentration on release of, 97*f*
Formalinized staphylococcal enterotoxin B, 68

Formulation: *see* ATRIGEL formulation; Insulin formulation; Microsphere formulation
Formulation variables
effect on protein release
in hydrogels, 146–147
in matrix systems, 148–149
Fourier transform infrared spectroscopy: *see* FTIR spectroscopy
Freeze-fracture, 177
fSEB: *see* Formalinized staphylococcal enterotoxin B
FSH: *see* Follicle-stimulating hormone
FTIR spectroscopy, 160–161
Fujioka, K., 74
Fusidate, 376–377

β-Galactosidase, 81
GALT: *see* Gut-associated lymphoid tissue
Gamma irradiation, 51, 110
Gamma-scintigraphy studies, 189
Ganirelix, 98
Ganirelix acetate, 54
Gastrointestinal tract; *see also* Intestinal absorption
absorption of insulin in, 369
oral delivery of microencapsulated proteins and, 255–257
somatotropins and, 289–290
Gay, S., 72
GDS: *see* Glyceryl distearate
Gel filtration, 345
Gelatin, 75–76, 311
Gelucire, 298
Genetic engineering, 82
Gentamicin, 203
Geometric descriptions, poly(ethylene-*co*-vinyl acetate) matrices, 131
Gerbils, insulin multiple emulsions in, 205
GHRF: *see* Growth hormone-releasing factor
Gilbert, D., 73
γ-Globulin, 152
tortuosity value, 127–130
GLP: *see* Good Laboratory Practice
Glucagon, 248
Gluconic acid, 367
Glucose dehydrogenase, 367
Glucose levels, 319–322, 331–332
Glucose oxidase, 367
Glucose-sensitive polymer complexes, 367
des-Gly^{10}-(D-Leu^{6})-LHRH ethylamide acetate: *see* Leuprolide acetate
Glycerides, 80
Glycerin, 304
Glycerol, 311, 349
Glycerol methacrylate, 310
Glyceryl distearate, 299
Glyceryl monocaprate, 25
Glyceryl monooleate, 25
Glyceryl tristearate, 299
Glyceryl tristearate microspheres, physical stabilization of, 300
Glycocholate, 379
Glycolic acid, 54, 94, 150
Glycoproteins, 168, 273
Glycyrrhetinic acid, 376
GMA: *see* Glycerol methacrylate
GMP: *see* Good Manufacturing Practices
Golumbek, P., 76
Gonadotropins, 242
Good Laboratory Practice, 34, 37
Good Manufacturing Practices, 3
Goosen, M., 71
Goosen, M.F.A., 384
Goserelin, 53
Goserelin acetate, 7
Gráf, L., 293
Graham, N.B., 153
Granulocyte-macrophage colony-stimulating factor, 74, 76
Gravity controller, 253
Gray, M.W., 304
Greenley, R.Z., 151
Growth factors, poly(ethylene-*co*-vinyl acetate) matrix and, 123
Growth hormone-releasing factor, 74, 95
Growth hormones, 37; *see also individual growth hormones;* Somatotropins
GTS: *see* Glyceryl tristearate
Guinea pigs
bioactivity in ATRIGEL formulation in, 113, 114*f*, 115*f*
intestinal absorption of proteins in, 256
liposome adjuvants of tetanus toxoid, 276
MN rgp120 controlled release vaccine in, 28–30
proteinoid microspheres in, 273
Gupta, P., 380
Gupta, S.K., 338
Gut-associated lymphoid tissue, 265–266
Guy, R.H., 381

Haak, R., 338
Halogenated hydrocarbons, 50
Hamilton, E.J., 292
Harbour, G.C., 293
Harris, D., 374
Harvard syringe pump, 251
Hayashi, Y., 146
hCG: *see* Human chorionic gonadatropin
Heine, R.J., 357
Heller, J., 48, 69, 357
HEMA: *see* Hydroxyethyl methacrylate
Hemoglobin, 204
Heparin, 72, 149, 273
  molecular size and iontophoretic delivery, 338
  poly(ethylene-*co*-vinyl acetate) matrix and, 123
  tortuosity value, 128
Heparinase, 63
Hepatic avoidance, 169; *see also* Mononuclear phagocyte system
Hepatic blockade, 169; *see also* Mononuclear phagocyte system
Heteropolysaccharides, 302
Heyman, M., 263
HGH: *see* Human growth hormone
HIC: *see* Hydrophobic interaction chromatography
High-pressure liquid chromatography, 65, 219, 291–292
High-temperature vacuum drying, 35
Higuchi equation, 159
Hill, J., 61
L-Histidine gel, 311
Histidine HCl, 310
HLB: *see* Hydrophilic–lipophilic balance
Ho, H., 75
Ho, N.F.H., 371
Hoechst, 56
Hoelgaard, A., 312
Hofenberg, H.B., 159
Hoffman, A.S., 151
Hogs, porcine somatotropin-containing osmotic devices in, 311–312
Home, P.D., 357
Homogeneous erosion: *see* Bulk erosion
Homogenization system, 16–17
Hori, R., 73
Horisaka, Y., 74
Horseradish peroxidase, 11, 52, 95, 256, 258
  relative activity in ATRIGEL formulation, 108*f*
  release kinetics in ATRIGEL formulation, 107*f*
Hot melt microencapsulation, 63
Houtzagers, C.M.G.J., 357
HPLC: *see* High-pressure liquid chromatography
HRP, 261–263
Human chorionic gonadatropin, poly(ethylene-*co*-vinyl acetate) matrix and, 123
Human growth hormone, 73, 79, 80, 273, 275, 289, 291, 300, 303
  administration in rats, 298
    growth data, 301*f*
  nasal administration of, 313
  wound healing and, 312
Human insulin
  association state, 359*f*
  disappearance curve, 360*f*
  primary structure, 354*f*
Human insulin analogs, 353
Human serum albumin, 51, 71, 77–78
Human skin
  charge titration of insulin and, 337
  iontophoretic transport through, 217–227
  LHRH delivery after application of iontophoresis, 217*f*, 218*f*, 220*f*, 222*f*, 226*f*
  transport numbers for iontophoretic flux of various peptides, 223
Humans
  biocompatibility of poly(ethylene-*co*-vinyl acetate) in, 121
  immunogenicity in, 31
  insulin administration in
    buccal, 371
    pulmonary, 380
    rectal, 372–373
  insulin and glucose levels in, 320*f*, 356*f*
  intranasal insulin pharmacokinetics, 377–378
  methylene chloride and ethyl acetate toxicity in, 36
  oral uptake of calcitonin in, 275
  OraLease formulation of desmopressin in, 272
  plasma glucose concentrations in, 361*f*
  polycyanoacrylate microspheres in, 274
  use of osmotic pumps in, 251
  use of polymers in, 7
Hyaluronic acid, 76–77
Hyaluronic acid ester microspheres, 376
Hydrocortisone, 146
Hydrogel delivery device, 140*f*

Hydrogels, 4–5, 139–140, 364, 367–368; *see also* Hydrophilic polymers
biodegradable, 149–151
diffusion coefficient measuring techniques, 154
diffusion-controlled release delivery systems, 145–150
effect of environmental conditions, 151–152
formulation variable effect on protein release, 146–147
methods for loading proteins into, 144–145
microporous, 142–143
structure, 142, 143*f*, 152–153
synthetic, 81
Hydrophilic polymer membranes, 146
Hydrophilic polymer slab geometry, protein transport mechanism, 159
Hydrophilic polymeric biomaterials, 77–79
albumin, 71–72
collagen, 72–74
fibrinogen/fibrin, 74–75
gelatin, 75–76
hyaluronic acid, 76–77
zein, 79
Hydrophilic polymers, 139; *see also* Hydrogels
structure, 142–143
used in matrix systems, 148
Hydrophilic–lipophilic balance, 201–202
Hydrophobic interaction chromatography, 182
Hydrophobic matrices, 120–122
Hydrophobic nonpolymeric biomaterials, 70–71, 79–81
Hydrophobic polymers, applications, 147
Hydrophobicity, 46, 58
effect on transcellular pathway, 265
mononuclear phagocyte system and, 169
zein microspheres and, 272
Hydroxybutyric acid, 56
Hydroxycaproic acid, 94
α-Hydroxy-carboxylic acid, 176
Hydroxyethyl cellulose, 148
Hydroxyethyl methacrylate, 147, 152, 309, 310
Hydroxypropyl β-cyclodextrin, 379
Hydroxypropyl cellulose, 148
Hydroxypropyl methyl cellulose, 148
2-Hydroxy-3-methoxybenzaldehyde, 303–304
4-Hydroxy-3-methoxybenzaldehyde, 303
Hydroxyvaleric acid, 56
Hyperglycemic coma, 321
Hyperinsulinism, 385
Hypoglycemia, 331, 360, 372, 380, 385
Hypoglycemic coma, 321

i.v. Controllers, 244
IBP: *see* Ibuprofen
Ibuprofen, 183–184, 192
ICI Biological Products, 56
IF: *see* Intrinsic factor
I-Flow, 250
I-Flow Model Paragon, 246*f*
I-Flow Vivus 4000, 246*f*
IFN-β: *see* Interferon-β
IgA: *see* Immunoglobulin A
IGF-I: *see* Insulin-like growth factor-I
IgG: *see* Immunoglobulin G
Ikada, Y., 76
IL-2: *see* Interleukin-2
Illum, L., 313, 374
IMC: *see* International Minerals and Chemicals Corporation
IMED, 247
Immunizations: *see* Vaccines
Immunogen, 65–66
Immunogenicity, 30–34, 70
Immunoglobulin, 63–64, 261, 264
Immunoglobulin A, 264
Immunoglobulin G, 77–78, 168, 264
molecular weight of matrix and release of, 129
Implantable matrices, insulin and, 363–368
Implantable pumps, 249, 361–362
Implants, 2, 56; *see also* ATRIGEL drug delivery system
bioerodible, 324–325
somatotropins and, 303–310
In vitro characterization, ATRIGEL drug delivery system, 102–110
In vitro transport, electroporation vs. iontophoresis, 217–227
In vivo evaluation, ATRIGEL drug delivery system, 110–115
^{111}In-labeled nanospheres, 189
IND: *see* Investigational New Drug application
Influenza vaccine, 273
Infusaid Corporation, 249
Infusion devices, comparison of, 253
Infusion pumps; *see also* Insulin pumps
advantages, 239–242
disadvantages, 242–243
history, 243–245
insulin administration via, 324–326, 343–344
portable, 245, 249–252
stationary, 245–249

Infusion therapy
  limitations, 242–243
  rationale, 239–242
Inhalers, 378
Initial burst, 17
Injectable implant, 54–55; *see also* ATRIGEL drug delivery system
Injectable somatotropins
  aqueous gels and complexes, 302–303
  emulsions, 301–302
  liposomes, 301
  microsphere systems, 299–301
  oil-based gel depots, 295–298
Injection devices, insulin, 362–363
Injection molding, 51, 63
Injection therapy, insulin, 357–360
Injection-site considerations, 31–34
Instillation, 380
Insufflators, 378
Insulin, 1, 37, 52, 63, 71, 73, 78, 80, 95, 149, 223, 256, 271, 273, 275
  association states, 359*f*
  cholesterol and, 79
  disappearance curves, 360*f*
  effect of environment on release of, 151
  hydrogels and, 143
  poly(ethylene-*co*-vinyl acetate) matrix and, 123, 134
  solubility as function of zinc-ion content, 347*f*
  structure of human, 354*f*
  titration curve, 336*f*
  tortuosity value, 128
  treatment of diabetes and, 319–322
  zein and, 79
Insulin administration, 343–344, 355–357
  bolus dose, 322, 325, 327–328, 357, 362
  buccal, 352, 370–371
  comparison of different technologies, 325
  continuous subcutaneous, 244–245
  implantable matrices, 363–368
  injection, 357–360
    devices, 362–363
  iontophoresis, 214, 216, 330–333
    capabilities of, 326–330
    as function of insulin concentration, 335*f*
    future prospects of, 339
    limitations, 333–335
    physicochemical properties of, 336–338
  nasal, 313, 374–378
  ocular, 374
Insulin administration (*cont.*)
  oral, 204, 369–370
  pulmonary, 378–381, 384
  pump infusion, 360–362
    stationary infusion pumps, 248–249
    syringe pumps, 247
  rectal, 371–373
  requirements of, 322–326
  transdermal, 381–384
  vaginal and intrauterine, 373–374
Insulin analogs and derivatives, 352–355
Insulin bioavailability, 323, 325, 328–329, 370, 378–379
  after noninvasive administration to rats, 352
Insulin formulation, 344–345
  alternative routes, 351–352
  analogs and derivatives, 352–355
  parenteral, 345–351
Insulin levels, humans, 356*f*
Insulin polybutylcyanoacrylate nanocapsules, 274
Insulin pumps, 252, 348–351, 360–362
Insulin purity, development of, 344
Insulin self-association, 338
Insulin-injection pen, 363
Insulin-like growth factor-I, 95, 109, 141, 289
  [^{125}I]-IGF-I release kinetics, 111*f*
Interference microscopy, 162
Interferon-α, 74, 76, 80
  gelatin and, 76
Interferon-β, 52, 95
Interleukin-2, 52, 95, 109
International Minerals and Chemicals Corporation, 292
Intestinal absorption, *see also* Gastrointestinal tract
  lipid-based systems and, 275–277
  of microparticulates
    mechanisms, 264–268
  polycyanoacrylate microspheres and, 273–275
  polyester microspheres and, 270–271
  proteinoid microspheres and, 272–273
  of proteins
    carrier-mediated transport, 259–260
    passive diffusion, 257–258
    receptor-mediated and non-receptor-mediated endocytosis, 261–264
  zein microspheres and, 271–272
Intestinal administration, of insulin, 352
Intralipid, 202

Intramuscular administration, 240, 365
Intraperitoneal administration, 365
Intrauterine administration, of insulin, 373–374
Intravenous administration, 240, 242–243
Intravenous microparticles, insulin delivery and, 324–325
Intrinsic diffusion coefficient, 154
Intrinsic factor, 264
Investigational New Drug application, 34, 37, 38
Ionic mobility, vs. molecular weight for iontophoretic delivery of peptides, 224*f*
Ionic strength, effect on protein diffusion, 151–152
Iontophoresis, 213–214; *see also* Electroporation
  vs. electroporation, 217–227
    in isolated perfused porcine skin flap, 227–232
    reversible changes in LHRH flux, 220*f*
  insulin and, 324–333, 381–384
    future prospects for, 339
    limitations, 333–335
    physicochemical properties, 336–338
    theoretical delivery as function of insulin concentration, 335*f*
Iontophoretic flux, transport numbers of various peptides, 223
IPPSF: *see* Isolated perfused porcine skin flap
Iron uptake, 264
Irsigler, K., 362
Isolated perfused porcine skin flap
  iontophoresis vs. electroporation, 227–232
  LHRH concentration in
    following iontophoresis, 228*f*, 230*f*, 231*f*
    following iontophoresis and electroporation, 229*f*
Isopropyl myristate, 201
IVAC, 247

Jabbari, E., 161
Jamas, S., 78
Janoff, A., 80
Jeffery, H., 144
Jeon, S.I., 171
Jet injectors, 362

Kahn, M., 374
Kari, B., 331
Kawamura, M., 75
Kendall McGraw Intelligent Pump, 246*f*
Kent, J., 79
Khan, M., 80
Kim, C.J., 152
Kim, S., 73
Koivisto, F.A., 357
Kopecek, J., 150
Kost, J., 78
Kwon, G., 72

LA, 177
LA copolymers, 176
α-Lactalbumin, 75
β-Lactam antibiotics, 256, 260
Lactic acid, 50, 94, 150
D-Lactic acid, 54
DL-Lactic acid, 54
L-Lactic acid, 54
D-Lactide, 25
Lactide/glycolide copolymers, 35, 50, 68
β-Lactoglobulin, tortuosity value, 127–129
Langer, R., 61, 140, 147, 149
Lassmann-Vague, V., 378
Lattice-walk simulations, 131
Laube, B.L., 380
Laureth-9, 370, 374, 377
Lecithin, 80, 349
Lectins, 263–264, 367
Lee, P.I., 152, 159
Lehrman, S.R., 293–295
Lente insulin, 346, 348, 357, 362
Leupeptin, 304–305
Leuprolide acetate, 7, 25, 54, 219; *see also* Lupron Depot
Leuprolide acetate–PLGA microspheres, 32, 34
Lever Brothers Co., 302
Levonorgestrel, 119, 147
Lewis, D.H., 51
LHRH, 7, 216, 231–214, 327
  delivery through human skin after application of iontophoresis, 217*f*, 218*f*, 220*f*, 226*f*
  in isolated perfused porcine skin flap, 227–232
    after iontophoresis, 228*f*, 230*f*, 231*f*
    after iontophoresis and electroporation, 229*f*
LHRH agonist formulations in PLGA microspheres, 7
LHRH agonist market, 39
LHRH agonists, 11, 24, 33
LHRH analogs, 52, 143, 225

LHRH antagonists, 11
LHRH flux, 232
  electroporation vs. iontophoresis, 220*f*, 221, 222*f*
Lidocaine, 183–184, 192
Lidocaine hydrochloride, 329
Lidocaine release kinetics, 186*f*, 187*f*
Lindsey, T.O., 305–306
Linear polyesters, structure, 49*f*
Lipid matrices, 47
Lipid-based systems, 275–277
Lipids, biodegradable, defined, 70
Liposomes, 168–169, 193, 204, 365, 369
  absorption of microparticulates and, 268
  oral delivery of proteins and, 276–277
  PEG-coated, 171
  somatotropins and, 301
Lisinopril, 260
Liver uptake, of ^{111}In-labeled nanospheres in mice, 190
Lohmander, S., 78
Longenecker, J.P., 371
Low-shear mixing methods, 203
Lucas, P., 74
Lupron Depot, 7, 24–26, 33, 35, 53–54; *see also* Leuprolide acetate
  release rate in rats, 26*f*
Lustig, S.R., 158
Luteinizing hormone-releasing hormone: *see* LHRH
Lymphatic pathway, 371
Lyophilization, 12, 15
Lypressin, 52
Lys^{B29}-tetra-decanoyl-des(b30)-insulin, 355
Lysozyme, 52, 63–64, 72–73, 75, 95
  relative activity in ATRIGEL formulation, 108*f*
  release from poly(ortho esters), 69*f*
  release kinetics in ATRIGEL formulation, 107*f*
  tortuosity value, 128
LysPro insulin, 353

M cells, 262–267, 275–276
mAb, poly(ethylene-*co*-vinyl acetate) matrices and, 133
Macromol, 275
Macroscopic models of diffusion in porous polymer matrices, 125–130
Magnesium stearate, 310
Magnetism, modulated systems and, 366
Magruder, J.A., 311–312
Malarial antigen, 68
Manufacturing issues, microspheres, 36–37
Manufacturing process, microencapsulation, 16–21
Massage, injection therapy and, 358
Mathiowitz, E., 79
Matrix, 2
  poly(ethylene-*co*-vinyl acetate), 123–134
  retardation of release from, 126*f*
Matrix fabrication parameters, dependence of tortuosity values on, 130
Matrix macroscopic geometry, 126–127
Matrix systems, 132–134, 147, 148*f*, 149
Mazer Chemical Co., 305
Mazol, 305, 308
McGee, J.P., 144
McLean, E., 309
McNeil, M.E., 153
Mechanochemical pump, insulin administration via, 367–368
Medisorb Technologies International, 50
MedMate 1100, 246*f*
Medtronic, 249
Membrane permeation method, 154–157
Merkle, H.P., 371
Mesh size, 162
  collagen and, 73–74
  defined, 70
  fibrinogen and, 75
  gelatin and, 76
  hyaluronic acid and, 77
  hydrogels and, 142
Methionine N-terminated bovine somatotropin zinc salt, 296–297
4-Methoxybenzaldehyde, 303
Methyl methacrylate, 310
Methyl *p*-hydroxybenzoate, 25
N-Methyl-2-pyrrolidone, 54, 94, 96–102, 104–105, 107, 110–112
Methylene chloride, 13, 63, 176, 183, 305
  toxicity data, 35–36
Methylvinylsiloxane, 121
Meyer, R.B., 331–332
Mice
  biocompatibility of ATRIGEL system in, 110
  bovine serum albumin delivery via poly(CTTH-iminocarbonate) polymers, 67
  bovine serum albumin release in, 80
  ethyl acetate toxicity in, 36

Mice (*cont.*)
fibrin use in, 75
interferon and granulocyte-macrophage colony-stimulating factor in, 76
interferon release in, 74
iontophoretic transport through mouse skin, 225
organ distribution of PEG-coated nanospheres in, 188*f*, 189*f*, 190*f*
proteinoid microspheres in, 273
transcellular pathway in, 265
ultrasound-mediated insulin administration in, 383
use of PLGA microspheres in, 270
Micelles, 168, 174
PEG-containing, 172
Microcapsules, 199
Micrococcus lysodeikticus, 107–108
Microencapsulation: *see* Microspheres
Microfluidizer, 203
Microfold cells: *see* M cells
Microgeometric models, protein diffusion in polymer matrices and, 131
Micromachining, 252
Microparticles, 51
mechanisms of intestinal absorption, 264–268
Microporous hydrogels, 142–143
Microscopic models of diffusion in porous polymer matrices, 131–132
Microsphere formulation
encapsulation methods, 8–14
manufacturing process design, 16–21
novel methods, 15–16
polylactides and proteins, 8–14
polymer chemistry, 3–8
protein stability, 21–24
Microspheres, 2, 51–53, 63–64, 93, 300, 364, 376
albumin and, 71–72
case studies, 24–30
gelatin and, 76
PLGA, 21, 31
polycyanoacrylate, 273–275
polyester, 270–271
polylactide, 10–11, 18, 146
proteinoid, 272–273
regulatory requirements for development of, 34–38
somatotropins, 299–301
zein, 271–272
Miglyol, 299
Miles Laboratories, 248
Miller, E., 72, 74
Miller, L., 80
Miller, L.F., 308
Mineral oil, 201
Mini-Pump, 250
Model 524, 246*f*
Mitchell, J., 80
Mitchell, J.W., 296
MMA: *see* Methyl methacrylate
MN rgp120: *see* Recombinant glycoprotein 120
MOA3, 200
Modulated systems, implantable matrices, 366
Molecular biology, ix
Molecular modification, somatotropins, 293–295
Molecular weight, 27, 32, 46, 50, 59, 110, 192, 365, 374
collagen and, 73–74
degradation rate and, 274
drug release and, 184
effect on ATRIGEL formulation, 98*f*, 99*f*
effect on drug absorption, 258
effect on release rate, 25
fibrin and, 75
vs. ionic mobility for iontophoretic delivery of peptides, 224*f*
iontophoresis and, 214
lysozyme release and, 69
PEG-coated nanospheres and, 180–181
polysaccharides and, 77–78
somatotropins and, 290
tortuosity values of proteins and, 129–130
transport numbers for iontophoretic flux of various peptides, 223
Monkeys, proteinoid microspheres in, 273
Monoclonal antibody IgG 2a, 273
Monolinolein, 80
Monolithic systems, 46–47
Monomeric insulins, 353, 359–360, 382, 385
Mononuclear phagocyte system, 168–169, 173, 188–192
Monsanto Chemical Co., 296–297, 305
Morawiecki, A., 74
Morphology studies, nanospheres, 177–179
Mouse: *see* Mice
MPEG: *see* Poly(ethylene glycol) methyl ether
MPS: *see* Mononuclear phagocyte system
MTT dye assay, 109

Mucic acid, 176
Multiday Infusor 0.5ml/hr, 246*f*
Multiphase emulsions: *see* Multiple emulsions
Multiphase encapsulation system, 15–16
Multiphase release, 8
Multiple emulsions, *see also* Double emulsion method; Emulsions
  applications, 203–305
  defined, 199
  enzyme immobilization and, 208
  preparation, 200–201
  solid-state emulsions, 205–208
  stability issues, 201–203
  vaccine adjuvants, 208
Mutual diffusion coefficient, 154–155
Myoglobin, 54, 78, 95
  effect of molecular weight on release of, 98*f*, 99

Nafarelin, 59
  release from cross-linked poly(ortho ester), 60*f*
Nafarelin acetate, 310
  release from PGLA microspheres, 52*f*
Nanoparticles, 365
Nanosphere characterization
  detection and stability of PEG coating, 180–181
  morphology studies, 177–179
  size distribution measurement, 179–180
  surface hydrophobicity and charge determination, 181–183
Nanospheres
  PEG-coated, 173–177, 180–181, 183–190
  PEG–PGLA, 189–191
  PEG–PLA-coated PLGA, 182–184
  PEG–R, 185–186
  PGLA, 178*f*, 182*f*, 189*f*, 190*f*
  preparation procedure, 175*f*
Nasal administration
  of insulin, 324–325, 352, 374–378
  of somatotropins, 313
National Institute of Health, 321
Nebulizers, 378
Neostigmine, 268
Nerve growth factor, 52, 77, 80
  poly(ethylene-*co*-vinyl acetate) matrix and, 123
Neupogen, pharmacokinetic parameters, 240–241
Neurotensin, 216, 223
Neurotensin analog, 10
Neutral Protamine Hagedorn: *see* NPH
Nicotine, 216
NIH: *see* National Institute of Health
p-Nitroaniline, 67–68
Nitroglycerin, 216
NMP: *see* *N*-Methyl-2-pyrrolidone
NMR contrast agents, 193
NMR spectroscopy, 153, 162
Nondegradable, defined, 119
Nondegradable hydrophobic polymers, 119
Nondegradable systems, implantable matrices, 364
Noninvasive administration, *see also* Iontophoresis
  of insulin, 322, 325, 327
    bioavailability after, in rats, 352
Nonporous hydrogels, 142
Non-receptor-mediated endocytosis, 261–264
Non-stealth PLGA nanospheres, 183–184
Norplant, 119, 121–122, 147
NovoPen, 363
NPH, 346–348, 355, 357, 362
N205, 200

Octane, 201
Ocular administration, of insulin, 352, 374
Ocusert, 119
Ogawa, Y., 24, 26, 53
Oil-based gel depots, somatotropins, 295–298
Oil-in-water-in-oil systems, 199, 200*f*; *see also* Multiple emulsions
Oil phase, 8
Okada, H., 53
Oleic acids, structure, 62*f*
Opdebeeck, J., 79
Open-loop system, insulin delivery via, 323
Opsonization, 168, 170
Optical densities, 180–181
Oral administration
  of insulin, 324–325, 369–370, 384
  multiple emulsions and, 204–205
  of proteins and gastrointestinal tract, 255–257
  vancomycin solid-state emulsion and, 207–208
Oral vaccines, 270, 272
OraLease, 271–272
Organ distribution, of PEG-coated nanospheres, 188, 189*f*, 190*f*, 191*f*, 192
Osmotic delivery systems, 141

Osmotic devices, somatotropins and, 310–312
Osmotic gradients, 202
Osmotic pumps, 250–253
Osteoinductive bone morphogenetic proteins, 55
Ovalbumin, 11, 22, 63–64, 68, 72, 75, 78, 95, 144
  effect of additives on release of, 101*f*, 102

PAA: *see* Poly(acrylic acid)
PAAm: *see* Polyacrylamide
PAGE: *see* Polyacrylamide gel electrophoresis
Palmitic acid, 201, 365
Pancreas, duplicating the function of, 322–323
Pancretec Provider One, 246*f*
Paolisse, G., 248
Papini, D., 77
Paracellular transport, microparticulate absorption and, 267–268
Parenteral administration, *see also* Injectable somatotropins
  multiple emulsions and, 203–204
  insulin delivery and, 356–357
    formulation for, 345–351
    implantable matrices, 363–368
    injection, 357–360
devices, 362–363
    pump infusion, 360–362
Park, K., 81
Particle blood circulation time, 169
Particle size, effect of absorption of PEG–PLA copolymers and poloxamine 908 on, 182
Partition coefficient, 257
Passive diffusion, intestinal absorption of proteins and peptides and, 257–258
Patton, J.S., 380
PBS: *see* Phosphate buffer solution
PBSA: *see* Phosphate-buffered saline containing azide
PCL: *see* Polycaprolactone
PCPP: *see* bis(*p*-Carboxyphenoxy)propane *or see* Poly[bis(carboxylatophenoxy)-phosphazene]
PCPP:SA: *see* Poly[bis(*p*-carboxyphenoxy)propane] anhydride and sebacic acid
PCS: *see* Photon correlation spectroscopy
PDGF-BB: *see* Platelet-derived growth factor
PDL: *see* Periodontal ligament cells
PEG, 5, 143, 149, 168
PEG blocks, in biodegradable polymers, 174–176
PEG coating, detection and stability of, 180–181
PEG hydrophilic coatings, 170–171
PEG methyl ether, 152
PEG-coated long-circulating drug carriers, 171–173
PEG-coated nanospheres, 173–177
  drug encapsulation, 183–187
  preparation, 176–177
PEG–PGLA nanospheres, 189–191
  lidocaine loading and, 186
PEG–PLA, 180
PEG–PLA copolymers, 182
PEG–PLA nanospheres, 181, 187
$(PEG20K)_3$–PLA nanospheres, 178*f*
PEG–PLA-coated PLGA nanospheres, 182–184
PEG–PLGA, 8
  encapsulation properties, 185
PEG5K-PLGA5K nanospheres, 178*f*
PEG–PSA, encapsulation properties, 185
PEG–R nanospheres, 185–186
  lidocaine loading and, 186
Pendant chains, degradable, 150
Penicillin, 203
PEO: *see* Poly(ethylene oxide)
Peppas, N., 140
Peppas, N.A., 142, 158
Peptide hormone, effect of solvent on release of, 99*f*, 100
Peracchia, M.T., 184
Percolation descriptions, poly(ethylene-*co*-vinyl acetate) matrices, 131–132
Percolation theory, protein diffusion in polymer matrices, 131
Periodontal ligament cells, 109
Peristaltic pumps, 253
Persorption, 267
Peschier, L.J.C., 153
Petechiae, 233–234
Peyer's patch, 262, 265–266, 268, 270, 275–276
PFAD: *see* Poly(fatty acid dimer)
P(FAD-SA): *see* Poly(fatty acid dimer: sebacic acid)
[P(FAD-SA)] 25:75 microspheres, protein release from, 64*f*
PGA: *see* Poly(glycolic acid)
PGLA, 38, 52, 141, 148
PGLA microspheres, release of nafarelin, 52*f*

pH, 22, 272, 367, 372
charge titration of insulin and, 336–339
effect on protein diffusion, 151–152
insulin administration and, 351
nasal, 375
insulin solubility and, 338
PHA, 263–264
Phagocytosis, 191
Phagocytosis assay, PEG-coated nanospheres and, 187*f*, 188*f*
Pharmacia Deltec, 250
Pharmacodynamics, 1
Pharmacokinetic parameters, Neupogen, Actimmune, Proleukin, 240–241
Pharmacokinetics, 1
Pharmetrix, 250–251
Phase I clinical experiment, 37
Phaseolus vulgaris, 263
PHB: *see* Poly(hydroxybutyrate)
PHEMA: *see* Poly(2-hydroxyethyl methacrylate)
Phonophoresis: *see* Ultrasound
Phosphate buffer solution, 96–99, 188
Phosphate-buffered saline containing azide, 102–104
Phosphatidylcholine, 268, 276, 376–377
L-α-Phosphatidylcholine, 298
Phosphatidylserine, 268
Phospholipids, 371, 376
Photomicroscopy, 203
Photon correlation spectroscopy: *see* Quasi-elastic light scattering
PHV: *see* Poly(hydroxyvalerate)
Physical properties, solid-state emulsions, 206
Physical stability, methods to determine in multiple emulsions, 203
Physicochemical properties of insulin related to iontophoresis, 336–338
Pierce Chemical Co., 97
Piez, K., 72
Pigs
charge titration of insulin and, 337
insulin delivery via iontophoresis in, 331
mean association state of insulins in, 359*f*
porcine somatotropin in, 292
porcine somatotropin-leupeptin-containing implants in, 305–306
skin toxicology following electroporation in, 232
use of uncoated implants in, 304
Piston pumps, 253
Pitt, C.G., 291, 308, 357
PLA, 9–11, 13, 25, 38, 49, 52, 68, 96–102, 107, 110, 168, 173–174, 184, 192, 365
structure, 6*f*
PLA homopolymers, 55
PLA microspheres, 31
PLA nanospheres, 187
PLA–PEG copolymers, 177
Plasma glucose concentrations, constant insulin infusion and, 361*f*
Plasma levels, after infusion of insulin and glycagon with Biostator, 248*f*
Plasmin, 78
Platelet-derived growth factor, 95
Platelet-derived growth factor-BB, 109
[^{125}I]-PDGF-BB release kinetics, 111*f*
saline vs. polymer formation, 113–115
Platz, R.M., 380
PLC: *see* Poly(DL-lactide-*co*-caprolactone)
PLG: *see* Poly(DL-lactide-*co*-glycolide)
PLGA, 9–11, 13, 25, 50, 68, 173–175, 177, 184, 192, 270
structure, 6*f*
PLGA copolymers, 55
PLGA microspheres, 21, 31
release of MN rgp120, 27*f*, 28*f*
PLGA nanospheres, 178*f*, 182, 189–190
PLL: *see* Poly(L-lysine)
Pluronic, 16, 177
PMAA: *see* Poly(methacrylic acid)
PMMA: *see* Poly(methyl methacrylate)
P(MMA-*co*-MAA), 152
pNA: *see p*-Nitroaniline
POE: *see* Poly(ortho esters)
Polio vaccine 273
Poloxamer, 172, 174
Poloxamine, 172, 174
Polyacryl dextrans, 77
Polyacryl starch, 168
Polyacrylamide, 143–144, 146
Polyacrylamide gel electrophoresis, 103
Poly(acrylic acid), 143, 146, 148, 150
Polyamine, 200
Poly[(amino acid ester) phosphazenes], 65
Polyamino acids, 4–5
Polyanhydrides, 4–6, 61–65, 173, 184, 192, 364
structure, 61*f*, 62*f*
Poly(aspartic acid), 172
Poly[bis(carboxylatophenoxy)-phosphazene], structure, 81*f*

Poly[bis(*p*-carboxyphenoxy)propane anhydride], degradation profiles, 62*f*
Poly[bis(*p*-carboxyphenoxy)propane anhydride] and sebacic acid, 6–7
Poly(bispherol A-iminocarbonate), 67–68
Polycaprolactone, 4–5, 55–56, 173, 175, 184, 192
  structure, 55*f*
Polyclonal antibodies, poly(ethylene-*co*-vinyl acetate) matrix and, 123
Poly(CTTH-iminocarbonate), structure, 67*f*
Polycyanoacrylate, 168
Polycyanoacrylate microspheres, 273–275
Polydimethylsiloxane, 121
Polyester microspheres, 270–271
Polyesters, 4, 364
  linear, 49*f*
Poly(ethyl acrylate methyl methacrylates), 307
Poly(ethylene-*co*-vinyl acetate), 119–121
  matrices
    applications, 132–134
    geometric descriptions, 131
    macroscopic models of diffusion in, 125–130
    microscopic models of diffusion in, 131–132
    protein release and, 124
    tortuosity values and, 127–130
  release of ferritin and, 123*f*
  structure, 120*f*
Poly(ethylene glycol): *see* PEG
Poly(ethylene oxide), 143, 148
Poly(ethylene oxide)–poly(propylene oxide)–poly(ethylene oxide) block copolymer, 200
Polyethyleneimine, 16
Poly(fatty acid dimer-sebacic acid) copolymers, 6, 63
Polygeline, 350
Poly(glycolic acid), 38, 49, 52
  structure, 6*f*
Polyglycolide microspheres, 32
Poly(hydroxybutyrate), 56
Poly(hydroxyvalerate), 56
Poly(imidazole methylphenoxy) phosphazene, 65
Polyiminocarbonates, 4–5, 66–67
  structure, 67*f*
Polyisobutylcyanoacrylate, 274
Poly(lactic acid): *see* PLA
Poly(DL-lactic acid), 175
Poly(L-lactic acid), 175
Poly(lactic-*co*-glycolic acid): *see* PGLA; PLGA
Poly(D-lactide), 50
Poly(DL-lactide), 50, 54, 94
Poly(L-lactide), 50
Polylactide microspheres
  formulation examples, 10–11
  production of, 18*f*
  residual solvent concerns, 35
Poly-L-lactide microspheres, 146
Polylactide vaccine, process variables in production of, 19
Poly(DL-lactide-*co*-caprolactone), 54, 94, 110
Poly(DL-lactide-*co*-glycolide), 54, 94, 96, 102–103, 110, 144
Polylactides, 4–6, 8, 22–23, 50
  degradation profile in rats, 32*f*
  degradation times in rats, 33
  use in humans, 7
Poly(L-lysine), 5, 81, 262
Polymer adsorption, 172
Polymer backbone, degradable, 149–150
Polymer chemistry, microsphere formulation and, 3–8
Polymer concentration, effects on ATRIGEL formulation, 97*f*, 98
Polymer erosion, 47
Polymer grafting, 171–172
Polymer matrices
  macroscopic models of diffusion in, 125–130
  microscopic models of diffusion in, 131–132
  protein release, 122–123
Polymer molecular weight, effects on ATRIGEL formulation, 98*f*, 99
Polymer type, effects on ATRIGEL formulation, 96*f*, 97
Polymeric biomaterials, biodegradable, defined, 70
Polymers, hydrophilic, 139; *see also* Hydrogels
Poly(methacrylic acid), 143, 151
Poly(methyl methacrylate), 143, 152
Polymixin, 203
Poly-(*N*-vinylpyrrolidone), 107, 143
Poly-*N*-vinylpyrrolidone biodegradable hydrogels, 150
Poly[1,3-bis(*p*-carboxyphenoxy)hexane], 64
Poly(ortho esters), 4–5, 57–61
  lysozyme release and, 69*f*
  release of nafarelin and, 60*f*
  structure, 57*f*, 58*f*

Polyoxamers, 200
Polyoxyethylene aliphatic alcohol ether, 200
Polyoxyethylene alkylphenol ether, 200
Polyphosphazene, 4–5, 65, 81
  structure, 65*f*
Poly(propylene glycol), 172
Poly(propylene oxide), 174
Polysaccharides, 77–79, 143
Polysiloxanes: *see* Silicone elastomer
Polysorbate 80, 311
Poly(2-hydroxyethyl methacrylate), 120, 143, 146, 148, 152
Polyurethanes, 120, 122
Poly(vinyl alcohol), 12, 63, 143–144, 147, 177, 305, 308–309
Poly(vinyl chloride)–poly(ε-caprolactone), 160
Polyvinylpyrrolidone, 310
Pontiroli, A.G., 378
Porcine insulin, association state, 359*f*
Porcine skin flap model
  insulin percentage reaching vasculature in, 329
  iontophoresis using, 334
Porcine somatotropin, 289–291, 303–305, 310–312
  amino acid sequence, 290*f*
  levels in pigs with porcine somatotropin implants, 307*f*
Porex electrodes, 228, 232
Porous chitosan, 78
Portable infusion pumps, 245, 246*f*, 249–252
  defined, 245
Portal administration, insulin and, 323, 325, 328
Posilac, 297
Pozza, E., 378
PPG: *see* Poly(propylene glycol)
PPO: *see* Poly(propylene oxide)
Precipitation–solvent evaporation method, 180
Preclinical animal models, 37–38
Prednisolone, 183, 192
Preformulation developments, somatotropins, 291–295
Preparation
  of multiple emulsion systems, 200–201
  of solid-state emulsions, 206
Price, F.P., 160
Process denaturation of protein, 202–203
Process variables, 17, 19–20
Progestasert, 119, 147
Progesterone, 146–147, 152
Prolamine, 271
Proleukin, pharmacokinetic parameters, 240–241
Protamine, 1
Protamine–insulin complex, 345–346
Protein delivery, regulatory requirements for encapsulation development, 34–38
Protein diffusion
  factors affecting, 151–153
  mechanisms of, 140–142
Protein encapsulation methods, 8–14
Protein load, effects on ATRIGEL formulation, 100*f*, 101
Protein parenterals, 1–2
Protein–polymer interactions, 22–23
Protein quantitation in different release media, 102–105
Protein rejection, PEG hydrophilic coatings as mechanism of, 170–171
Protein release
  effect of formulation variables on, 146–147
    in matrix systems, 148–149
  from matrices, 124
    macroscopic models of diffusion, 125–130
    microscopic models of diffusion, 131–132
  from [P(FAD-SA)] 25:75 microspheres, 64*f*
  from polymer matrices, 122–123
Protein release kinetics, 95
  additive effects, 101–102
  in ATRIGEL formulation, 111–113
  polymer concentration effects, 97–98
  polymer molecular weight effects, 98–99
  polymer type effects, 96–97
  protein load effects, 100–101
  solvent effects, 99–100
Protein release rates, 125–130
  percolation theory and, 132
Protein reservoir system, 140*f*
Protein structure, ATRIGEL formulation and, 105–106
Protein/hydrophobic polymer matrix system, 124*f*
Protein/polymer matrix systems, applications, 132–134
Proteinoid microspheres, 272–273
Proteins
  cellular bioactivity in ATRIGEL formulation, 109
  mechanisms of intestinal absorption of, 255–264
  methods for loading into hydrogels, 144–145
  process denaturation of, 202–203
  stability, 21–24
  studied in ATRIGEL drug delivery system, 95

PSA: *see* Polyanhydrides
PST: *see* Porcine somatotropin
Pulmonary administration, of insulin, 351–352, 378–381, 384
Pulsatile release, 14, 27, 45
  of formalinized staphylococcal enterotoxin B, 68–69
Pump infusion, of insulin, 360–362
PVA: *see* Poly(vinyl alcohol)
PVP: *see* Poly-(*N*-vinylpyrrolidone)
Pyridostigmine bromide, 329
Pyrogallol activity assay, 107

QELS: *see* Quasi-elastic light scattering
QS-21, 30
Quasi-elastic light scattering, 142, 155, 161–162, 179–180, 185, 187

Rabbits
  bioactivity in ATRIGEL formulation in, 113, 114*f*
  biocompatibility
    of ATRIGEL system in, 110
    of poly(ethylene-*co*-vinyl acetate) in, 121
  charge titration of insulin in, 337
  endocytosis in, 263
  ethyl acetate toxicity in, 36
  human growth hormone retention in, 80
  implantable matrices in, 366
  insulin administration in
    buccal, 370–371
    iontophoresis, 331–332
    nasal, 376
    ocular, 374
    pulmonary, 379
    rectal, 372–373
    ultrasound-mediated, 383
    vaginal, 373
  insulin release in, 71
  interferon release in, 74
  polyanhydride microspheres in, 63
Radioimmunoassay, 65, 296–297, 307
Raman, S.N., 304
Random-walk simulations, 131
Rapitard, 347
Rational drug design, ix
Rats
  Alzet minipump use in, 141
  biocompatibility
    of ATRIGEL system in, 110
    of poly(ethylene-*co*-vinyl acetate) in, 121
Rats (*cont.*)
  buserelin release in, 56
  coated implants in, 305
  degradation profile for polylactides in, 32*f*
  degradation times of polylactides in, 33
  efficacy of water-in-oil-in-water emulsion in, 302*f*
  human growth hormone in
    growth data following administration of, 298*f*, 301*f*
  ibuprofen release in PLGA nanospheres in, 183–184
  implantable matrices in, 364–365
  insulin administration in
    buccal, 370–371
    iontophoresis, 331
    nasal, 375–376
    ocular, 374
    oral, 369
    pulmonary, 379–380
    rectal, 372–373
    ultrasound-mediated, 383
    vaginal, 373
  insulin bioavailability after noninvasive administration, 352
  insulin multiple emulsions in, 204–205
  insulin polybutylcyanoacrylate nanocapsules in, 274
  insulin release in, 71, 73
  intestinal absorption of proteins in, 256
  LHRH release in, 53
  liposome absorption in, 268, 276
  Lupron Depot in, 25, 26*f*
  methylene chloride and ethyl acetate toxicity in, 36
  nafarelin release in, 60*f*
  nasal administration systems in, 313
  non-receptor-mediated endocytosis in, 262
  passive diffusion in, 258
  PEG-coated nanospheres and the mononuclear phagocyte system in, 191
  Peyer's patch uptake studies in, 276
  polyanhydride microspheres in, 63
  proteinoid microspheres in, 273
  release profile of [^{125}I]-epidermal growth factor in, 103*f*
  serum concentrations of vancomycin solid-state emulsions in, 207*f*
  Zoladex release in, 53
Rayleigh line width: *see* Quasi-elastic light scattering

rBST: *see* Recombinant bovine somatotropin
Receptor-mediated endocytosis, 261–264
Recombinant bovine somatotropin, 293
  solution stability with CM-PST, 294*f*
Recombinant glycoprotein 120, 22
Recombinant glycoprotein 120 controlled release vaccine, 26–30
  release rate, 27*f*
Recombinant human bone morphogenetic protein-2, 55
Recombinant human fibroblast growth factor, release profile, 102*f*, 103*f*
Recombinant human transforming growth factor-β, release profile, 102*f*, 103*f*, 104*f*
Recombinant human tumor necrosis factor-β, release profile, 102*f*, 103*f*
Recombinant porcine somatotropin, 308
Recrystallized insulin, 344–345
Rectal administration, of insulin, 352, 371–373
Red blood cell acetylcholinesterase, 329
Regulatory requirements, microsphere development, 34–38
Relaxin, 78
Release kinetics,
  effects of formulation variables, 95–102
  in vitro, 102–105
  of lidocaine, 186*f*, 187*f*
  of proteins and polymer matrices, 122–123
Release patterns, polylactide encapsulation, 14*f*
Release properties, of PEG-coated nanospheres, 183–184
Release rates, 17, 25
  effect of formulation variables on, 146–147
  matrix systems and, 148–149
  percolation theory and, 132
  of proteins from matrices, 125–130
Reproducibility, 20
Reservoir systems, 145–147
  controlled-release devices, 145*f*
Residual solvent concerns, 35–36
Reversed-phase chromatography, 22
Reversed-phase high-pressure liquid chromatography, 105
rhbFGF: *see* Recombinant human fibroblast growth factor
rhBMP-2: *see* Recombinant human bone morphogenetic protein-2
rhTGF-β: *see* Recombinant human transforming growth factor-β
rhTNF-β: *see* Recombinant human tumor necrosis factor-β
Richardson, J.L., 374
Ricin toxoid, 68
Ritger, P.L., 158
Ritschel, G.B., 371
Ritschel, W.A., 371
Rodents, endocytosis in, 261–262
Rohm Pharma, 307
Roller clamps, 243–244
Roorda, W., 153
Rose Bengal binding methods, 182
Rosenblatt, J., 73
Rotor stator, 16–17
Routes of administration: *see* Administration routes; Oral administration; Parenteral administration
Royer, G., 71
RPHPLC: *see* Reversed-phase high-pressure liquid chromatography

SA: *see* Sebacic acid
Sabratek, 246*f*
Sage, B.H., 332
Sahlin, J.J., 161
Salicylates, 371
Salmon, porcine somatotropin in PVA-coated pellets in, 309*f*
Salmonella enteritidis, 22
Sanders, L.M., 52, 309–310
Sandwich ELISA, 103
Saudek, C.D., 357
Scalability, 20
Scanning electron microscopy, 152, 155, 160, 177
Schade, D.S., 362
Schlichtkrull, J., 347
Schroder, U., 77
Screw extrusion, 51
SDS: *see* Sodium dodecyl sulfate
SDS-PAGE: *see* SDS-polyacrylamide gel electrophoresis; *see also* Denaturing gel electrophoresis
SDS-polyacrylamide gel electrophoresis, 103, 106
SEB: *see* Staphylococcal enterotoxin B
Sebacic acid, 6, 61–63
  structure, 62*f*
Selam, J.-L., 357, 363
Self diffusion coefficient, 155
Self-regulating systems, implantable matrices, 366–368
SEM: *see* Scanning electron microscopy

Semilente, 346
Serum albumin, 365
Serum glucose levels, insulin delivery via iontophoresis and, 331–332
Serum levels, of insulin, 320–321
Shah, N.H., 54
Shalaby, W.S., 144
Sheep
  insulin administration in
    nasal, 376
    vaginal, 373
Shefer, S., 78
Siegel, R.A., 149, 367
Sigma Chemical Co., 107
Silastic, 121
Silicone elastomer, 119–122
  structure, 121*f*
Silicone implant, 366
Silicone prosthetic devices, 122
Sivaramakrishnan, K.N., 80, 304, 308
Size distribution measurement, nanospheres and, 179–180
Skin patch, insulin administration and, 327, 329–330, 333
Skin toxicology following electroporation, 232–235
Slade, W.C., 61
Smoking, injection therapy and, 358
Sodium chloride, 310
Sodium cholate, 358
Sodium dodecyl sulfate, 102
Sodium glycocholate, 370–371
Sodium hydroxide, 181
Sodium phosphate monobasic, 311
Sodium salicylate, 351
Solid-state emulsions, 205–208
Solubility, of insulin, 337–338
  as function of zinc-ion content, 347*f*
Solution loading, 144–145
Solution stability, somatotropins, 291–295
Solvent
  effect on protein diffusion, 151–152
  effects on ATRIGEL formulation, 99*f*, 100
Solvent diffusion method, 176–177, 180
Solvent evaporation, 8, 9*f*, 12, 15–16, 51, 120–121, 146
  production of polylactide microspheres, 18*f*
Solvent extraction: *See* Coacervation
Solvent removal microencapsulation, 63
Somatostatin analog, 10
Somatotropins, 80, 289–291; *see also* individual growth hormones
  aqueous gels and complexes and, 302–303
  coated implants and, 305–310
  emulsions and, 301–302
  liposomes and, 301
  microsphere systems, 299–301
  nasal administration systems and, 313
  oil-based gel depots, 295–298
  osmotic devices and, 310–312
  preformulation developments, 291–295
  uncoated implants and, 303–305
  wound healing and, 312
Song, S., 74
Sonication, 16, 203
Sorbitan oleate, 200
Sorbitol, 33
Span 80, 200, 203
Spangler, R.S., 369
Spray drying, 15
Spray freeze drying, 15
Spray prilling techniques, 299
Spring pumps: *see* Balloon pumps
Sprinkler needle, 363
Square lattice representation, 132*f*
S-shaped release: *see* Triphasic release
Stability
  effect of zinc-ion concentration on insulin solution, 350*f*
  of multiple emulsions, 201–203
Stannous octanoate, 176
Staphylococcal enterotoxin B, 270
Starch, 4
Stationary infusion pumps, 245–249
  defined, 245
Stealth PLGA nanospheres, 183–184
Steber, W., 80
Steber, W.D., 299–300, 305
Stephen, R.L., 330, 338–339
Sterilization, 51
Sterol, 376
Stratum corneum structure, schematic representation, 216*f*
Streptococcus mutans, 276
Streptomycin, 203
Su, K.S.E., 378
Subcutaneous administration, 240, 242, 365; *see also* Parenteral administration
  of insulin, 356–357, 368

Subcutaneous injection, 1
  insulin administration and, 321, 323, 325, 358, 384
Sucrose, 64, 206, 308
Supercritical fluid extraction, 15
Surface analysis techniques, 177, 186
Surface erodible systems, 141
Surface erosion, 47
  poly(ortho esters) and, 58
Surface hydrophobicity and charge determination, 181–183
Surfactant migration, in multiple emulsions, 202
Surfactants, 15, 101, 371, 376
  as enhancers with insulin, 352
Sustained release, 45, 70–71
Sustained-release implants, 364
Swelling-controlled delivery systems, 141
Swine
  prolonged delivery of porcine somatotropin using coated implants in, 305
  uncoated porcine somatotropin implants in, 393–394
Syntex, 52
Synthetic biodegradable polymers, 5–7
Synthetic human chorionic gonadotropin subunit, 68
Synthetic hydrogels, 81
Synthetic hydrophobic degradable polymers
  polyanhydrides, 61–65
  polycaprolactone, 55–56
  poly(hydroxybutyrate) and poly(hydroxyvalerate), 56
  poly(lactic acid) and poly(glycolic acid), 49–55
  poly(ortho esters), 57–61
  polyphosphazenes, 65
  vaccine delivery and, 65–69
Syringe pumps, 244–245, 247–248, 253
Systemic delivery, using poly(ethylene-*co*-vinyl acetate) matrix, 134

Tabata, Y., 76
Takaoka, K., 74
Takeda-Abbott, 53
Takeda Chemical Industries, 24
Targeted delivery, using poly(ethylene-*co*-vinyl acetate) matrices, 133–134
Tartaric acid, 176
Temperature
  effect on protein diffusion, 151–152
  modulated systems and, 366
Terminal sterilization, 21
Tetanus toxoid, 11, 22, 68, 276
Tetrahydrofuran, 50
TGF-β: *see* Transforming growth factor-β
Thakkar, A.L., 298
Therapeutic window, 24
Thermostability of recombinant porcine somatotropin and CM-PST, 294*f*
Thin-layer chromatography, 219
Thyroid releasing hormone, 260
Thyrotropin, 10
Thyrotropin-releasing hormone, 214
TLC: *see* Thin-layer chromatography
TNF-β: *see* Tumor necrosis factor-β
α-Tocopherol, 203–204
α-Tocopheryl hemisuccinate, 301
Tolerability, of iontophoresis dosage form, 333
Topical delivery, using poly(ethylene-*co*-vinyl acetate) matrices, 133
Tortuosity, 127–131
Toxicity, 3, 6–7, 24
  of methylene chloride and ethyl acetate, 36
  microsphere development and, 34–35
  of PEG-coated nanospheres, 174
  of skin following electroporation, 232–235
Toxins, 263–264
Transcellular pathway, microparticulate absorption and, 265–267
Transderm Nitro, 119
Transdermal administration, of insulin, 381–384; *see also* Iontophoresis
Transforming growth factor-β, 78, 95
Transmission electron microscopy, 178
Transport mechanisms of protein through hydrophilic polymer slab geometry, 159
TRH: *see* Thyroid releasing hormone
Triacetin, 54
Tributyl citrate, 310
Tricalcium phosphate, 4
Trichloroacetic acid, 106
Triethyl citrate, 54
Trilysyl insulin, 367
Tripartite Biocompatibility Testing Guidelines, 110
Tripeptides, 259
Triphasic release, 14*f*, 27, 52–53
Triptorelin, 7
Tris salt vesicles, 301
[D-Trp^{6},des-Gly^{10}]-LHRH diethylamide, 56

Trypsin, 63–64, 78, 95, 106–107
relative activity in ATRIGEL formulation, 108*f*
release kinetics in ATRIGEL formulation, 107*f*
Tucker, I., 79
Tumor growth factor-a-DCys-Pseudomonas exotoxin recombinant fusion protein, 141
Tumor necrosis factor-β, 95, 109
Tween, 34
2D-PAGE: *see* Two-dimensional polyacrylamide gel electrophoresis
Two-dimensional polyacrylamide gel electrophoresis, 183
2H3MB; *see* 2-Hydroxy-3-methoxybenzaldehyde
TX-4, 200
Tyle, P., 301

U.S. Food and Drug Administration, 7, 34, 53, 63, 121–122, 170
U-500 Insulin, 249
Ulashik, V.S., 338
Ultralente, 346, 357
Ultrasound
insulin administration and, 324–325, 383–384
modulated systems and, 366
Uncoated implants, somatotropins and, 303–305
Upjohn, 292, 294
Uptake kinetics, PEG-PLA nanospheres, 188*f*
Urea, 351
Urist, M., 75
UV analysis, 157, 160

Vaccine adjuvants, 208
Vaccines, 2, 11, 14, 24, 32
delivery of, 65–69
MN rgp120 controlled release, 26–30
oral, 270, 272
Vaginal administration, of insulin, 373–374
Vancomycin solid-state emulsion, oral administration, 207–208
Vasculature, percentage of insulin dose reaching, 329
Vasopressin, 216, 223, 225, 271
Verapamil, 330
Verrecchia, T., 183
Vial, 248
Virus neutralization titers for MN rgp120 controlled release vaccine in guinea pigs, 30*f*
Viswanathan, R., 292
Vitamin B_{12} uptake, 264
Volkheimer, G., 267

Wall, D.A., 380
Wallace, D., 72
Wang, P., 79
Wang, P.Y., 364
Water-in-oil emulsion method, 365
Water-in-oil-in-water emulsion, 12, 144, 199, 200*f*; *see also* Multiple emulsions
efficacy of, in rats, 302*f*
Wearley, L.L., 313
Weiner, A., 73
Welmed, 248
Wigley, F.M., 380
Wolany, G., 371
Woodley, J.F., 369
Wound healing, somatotropins and, 312
WOW: *see* Water-in-oil-in-water emulsion
Wyeth-Ayerst Laboratories, 119

XPS: *see* X-ray photoelectron spectroscopy
X-ray photoelectron spectroscopy, 181, 186

Yamahira, Y., 74, 80
Yamamoto, A., 374
Yamazaki, H., 78
YoKosuka, T., 371

Zein, 79
Zein microspheres, 271–272
Zeneca, 53
Zentner, G.M., 152
Zero-order release, 147, 303
Zero-order release kinetics, 145
Zeta potential, effect of absorption of PEG–PLA copolymers and poloxamine 908 on, 182
Zinc, 1
insulin and, 346–350
Zinc somatotropin, 80
ZnMBS: *see* Methionine N-terminated bovine somatotropin zinc salt
Zoladex, 7, 53
Zona pellucida protein antigen, effect of polymer type on release of, 96*f*